BIOSENSORS - MICRO AND NANOSCALE APPLICATIONS

Edited by **Toonika Rinken**

Biosensors - Micro and Nanoscale Applications
http://dx.doi.org/10.5772/59391
Edited by Toonika Rinken

Contributors

Tatiana Duque Martins, Antonio Carlos Chaves Ribeiro, Geovany De Souza, Henrique Camargo, Diogo Dias, Paulo Costa-Filho, Diericon Cordeiro, Flavio Colmati, Hun-Kuk Park, Gi-Ja Lee, Sung Wook Kang, Bochan Seo, Jae Ho Shin, Zihni Onur Onur Uygun, Hilmiye Deniz Deniz Ertuğrul, Nihal Ermiş, Erhan Canbay, Robert Valentin Sandulescu, Vernon Somerset, Luleka Luzi-Thafeni, Pere. Ll. Miribel-Català, Jaime Punter Villagrasa, Joan Cid, Jordi Colomer-Farrarons, Ivón Rodríguez-Villarreal, Nada F. Atta, Ahmed Galal, Ekram El-Ads, Andrey Bratov, Sergi Brosel-Oliu, Naroa Uria, Natalia Abramova, Jadwiga Sołoducho, Lei Zhang, Argoul, Toonika Rinken, Kairi Kivirand, Margarita Kagan, Rosa Fireman Dutra, Leonardo Moreira, Óscar Castillo-Fernandez, Francesc X Muñoz-Pascual

CBS, Edition **2017**

Published by InTech
Janeza Trdine 9, 51000 Rijeka, Croatia

Publishing Process Manager Ivona Lovric

Technical Editor InTech DTP team

Cover InTech Design team

Additional hard copies can be obtained from orders@intechopen.com

Biosensors - Micro and Nanoscale Applications, Edited by Toonika Rinken
p. cm.
ISBN 978-953-51-2173-2

INTECH
open science | open minds

Contents

Preface

Biosensors, combining the specificity of biological recognition and the sensitivity of physico-chemical detection, represent a flexible and rapid alternative to conventional analytical technologies. They have been applied to solve diverse analytical problems in medicine and biomedical research, drug discovery, the environment, food and process industries, security, and defence.

Fast technological progress has boosted the implementation of novel ideas and hi-tech solutions in all analytical equipment. Along with extending the selection of detectable compounds, the development of biosensors has been focused on making the systems more compact and easy to handle. Scaling down the dimensions, new qualitative features can be achieved and the analyses can be carried out with remarkably low detection limits. Micro and nanoscale biosensors show high potential to be used for carrying out automated online monitoring of minute concentrations of different chemical compounds, including various biomarkers and potentially hazardous compounds in a wide selection of matrixes. In addition, the footprint of analyses is small and enables sensing in living systems. Due to high specificity and sensitivity, biosensors can be used for the characterization of metabolic processes even in single cells.

The present book, including 16 chapters, introduces novel approaches and materials for the micro- and nanoscale biosensor construction and application for the detection of different biological and synthetic compounds in living organisms and natural matrixes in vitro. I would like to express my appreciation and gratitude to all the authors for their contribution and cooperation and wish them success in their forthcoming activities. Special thanks to the InTech team, particularly publishing process managers Ana Pantar and Ivona Lovric for their professional commitment and dedication.

Toonika Rinken, PhD
University of Tartu
Estonia

Chapter 1

New Materials for the Construction of Electrochemical Biosensors

Robert Săndulescu, Mihaela Tertiş,
Cecilia Cristea and Ede Bodoki

Additional information is available at the end of the chapter

http://dx.doi.org/10.5772/60510

Abstract

The development of electrochemical sensors has attracted great interest due to these sensors' high sensitivity and selectivity. Here, we present the general concept and the classification of biosensors, their advantages and drawbacks, the main strategies in electrochemical biosensor technology and the materials used in electrochemical sensors, such as electrodes and supporting substrates, materials for improved sensitivity and selectivity, materials for bioreceptor immobilization, and biological recognition elements. Various nanomaterials, such as carbon-based materials (carbon nanotubes, graphene, carbon nanoparticles), inorganic and organic nanoparticles (magnetic and metal nanoparticles, nanosized clays), conductive and insulating polymers (nanosized and nanostructured polymers, molecularly imprinted polymers), and hybrid materials, etc., have been successfully applied for the enhancement of the electroanalytical performance of biosensors and for the immobilization of biorecognition elements. Among these, due to their unique physiochemical features, carbon-based materials, such as carbon nanotubes and graphenes, have received special attention in recent years, and examples of surface functionalization using various types of nanoparticles are presented. The future trends in sensor research activities and areas of development that are expected to have an impact in biosensor performance, like immobilization techniques, nanotechnology, miniaturization and multisensor array determinations, are also examined.

Keywords: Carbon-based nanomaterials, metal nanoparticles, magnetic nanoparticles, nanostructure, molecularly imprinted polymers

1. Introduction

The development of electrochemical sensors has attracted a great deal of interest due to their high sensitivity and selectivity, and they are being increasingly used in many fields, such as analytical chemistry, industrial process monitoring and control, clinical diagnostics, environmental monitoring and security, and food safety. By adding an enzyme at the electrode surface, they can also gain high specificity: thus the biosensor research area was born.

Biosensors (analytical devices coupling a transducer with a biorecognition element) are attractive for pharmaceutical and biomedical analysis due to their sensitivity (ng/mL or less) and high selectivity, and sometimes their specificity, high benefit/cost ratio, simple use and rapidity of data collection [1].

The major advantages of biosensors over traditional analytical methods, which will certainly lead in the near future to their even more pronounced use in the biomedical field, are: the fact that analyte detection can very often be made without prior separation; the short response times that make possible the real-time monitoring of biological and manufacturing processes; their ease of use, allowing in-field or point-of-care measurements; the flexibility and simplicity of preparation; the possibility of mass production and low production costs; and the possibility of miniaturization and automatization. Miniaturization is of great importance because biological samples are available in small amounts, and tissue damage must be minimized in cases of in-vivo monitoring. Therefore, the use of biosensors as components of modern medical devices has improved their portability, functionality and reliability for point-of-care analysis and real-time diagnosis [2].

However, many amperometric biosensors described in the recent literature still display a few drawbacks compared to other analytical methods. The most difficult problems to overcome for biosensors with biomedical applications are: the reduced stability, the electrochemical interferences, and the lack of or low response reproducibility. Removed from their natural environment, most biocomponents tend to rapidly lose their activity and thus limit the lifetime of the sensor. Modifying electrode surfaces in a way that exclusively favours one single electrochemical process is a difficult task and sample matrices in the biomedical field are very complex. Last but not least, in the case of in-vivo measurements biocompatibility and biofouling are critical issues [3].

The selectivity of the biosensor for the target analyte is mainly determined by the biorecognition element, whilst the sensitivity of the biosensor is greatly influenced by the transducer.

New sensing elements like genetically transformed cells and genetically engineered receptor molecules can improve affinity and specificity; therefore, genetic engineering and mass production of the molecular recognition may ultimately dictate the success of biosensor technologies. Enzyme variants that are specific for individual analytes have already been obtained by genetic modification, and novel gene fusions will lead to more sensitive biosensors [4].

As for the electrochemical transducer, important advances have been recently made thanks to the introduction of new platforms for biosensor design, such as nanotechnological materials and nanostructured architectures (i.e., nanoparticles, carbon nanotubes and nanofibres, graphenes, nanostructured surfaces, etc.), which have improved the sensitivity of the assembly [5].

2. Strategies of electrochemical biosensor technology

The biosensor development and construction strategy includes five features: 1) the detected or measured parameter and the matrix, 2) the working principle of the transducer, 3) the chemical/biochemical model, 4) the field of application and 5) the technology and materials for sensor fabrication [6].

Materials generally used for electrochemical sensors are classified as: (1) materials for the electrode and supporting substrate, (2) materials for improving electroanalytical performances, (3) materials for the immobilization of biological recognition elements and (4) biological elements; the last two are applicable for electrochemical biosensors [5].

Materials used for the electrode and supporting substrate are usually conductive materials exhibiting low currents in an electrolyte solution, free of any electroactive species, over a relatively wide potential window. Among the most frequently used materials for the electrode and supporting substrate, the following may be mentioned: metals (mercury, platinum, gold, silver and stainless steel), metal oxides (indium tin oxide, ITO), carbon-based materials (glassy carbon, graphite, carbon black and carbon fibre), new hybrid materials, and organic electroconductive polymers or salts. Boron-doped diamond is a special case of material, because bare diamond is a non-conducting, allotropic form of carbon. In the last decade, in order to avoid mercury-based electrodes (which are highly toxic), different metallic films (Au, Hg, Bi, Ag, Pb, Sn, Sb, Co, Ga, Se) that are relatively simple to achieve and easy to reproduce were developed. Some strategies to develop two- and three-dimensional nanostructured electrodes with larger surfaces (mesoporous silicates, metal oxides, polymers or carbons) were also recently reported [7]. The major challenge in the field of electrochemical sensors, consisting in the improvement of their electroanalytical performances, mainly in terms of sensitivity and selectivity, was addressed by the development of nanotechnology (especially nanoparticles, carbon nanotubes and graphenes) and nanostructured architectures, and this will be detailed in section 3.

The stable immobilization of a bioelement on an electrode surface, with complete retention of its biological activity and good diffusion properties for substrates, is a crucial problem for the commercial development of biosensors. A successful matrix should stably immobilize or integrate biomolecules at the transducer's surface, it should efficiently maintain their functionality, while providing accessibility for the target analyte, and it should also ensure an intimate contact with the transducer surface [8]. The immobilization matrix may function purely as a support or may also be involved in the mechanism of signal transduction mediation. The selection of an appropriate immobilization method depends on the nature of the biological element, the type of transducer used, the physicochemical properties of the analyte and the biosensor's operating conditions [9]. The choice of materials for the immobilization of biological recognition elements strongly depends on the employed procedure. The procedures that may be employed are as follows: 1) Entrapment behind a permeable membrane as a thin film covering the electrochemical detector (dialysis membrane) [10]; 2) Entrapment within a polymeric matrix (polyacrylonitrile, agar gel, polyurethane or poly(vinyl-alcohol) membranes, sol gels or redox hydrogels with redox centres such as $[Os(bpy)_2Cl]^+$, conducting polymers [11]; 3) Entrapment within self-assembled monolayers (SAMs) [12] or bilayer lipid membranes (BLMs) [13]; 4) Covalent bonding on membranes or surfaces activated by means of bifunctional groups or spacers (glutaraldehyde, carbodiimide, SAMs or multilayers, avidin-biotin, silanization) [14]; 5) Bulk modification of the electrode material (graphite epoxy resin or carbon paste modified with polyoxometallates, clays and double layered hydroxides, zeolites, functionalized silica, sol-gel-derived inorganic and hybrid materials, nanomaterials, sparingly soluble or insoluble inorganic salts, molecular and macrocyclic ligands, complex compounds, organic polymers, surfactants and lipids) [7].

The **biorecognition elements** used for biosensor development are classified into biological and artificial (biomimetic) receptors. They are selected depending on the application and the desired performance criteria, based on the following interaction principles: enzyme/substrate, antigen/antibody, mRNA/DNA, PNA/DNA or RNA, microorganism/substrate or toxic molecules, cell receptor/hormone and (semi)synthetic receptor/target molecule [15]. Initially, three types of biorecognition elements were identified [16]: 1) Enzyme (mono or multienzyme), the most common and best developed recognition system, including antibodies, natural receptors and nucleic acids; 2) Whole cells (microorganisms, such as bacteria, fungi, eukaryotic cells or yeast), or cell organelles or particles (mitochondria, cell walls); 3) Tissue (plant or animal tissue slice). More recently, the range of biochemical receptors was extended with biomimetic recognition elements (aptamers, ribosymes and molecularly imprinted polymers) [17]. The biological recognition elements are based either on catalytic or bioaffinity features, whilst biomimetic ones are based on biocomplexing and bioaffinity.

The **biocatalytic recognition sensors** are based on a reaction catalysed by macromolecules, either in their original biological environment, in their isolated form, or as a product of semisynthesis. Thus, a continuous consumption of substrate(s) is achieved by the immobilized biocatalyst incorporated into the sensor, and transient or steady-state responses are monitored by the integrated detector. In the construction of biosensors, **enzymes** are the most commonly used biological elements, despite their rather high extraction, isolation and purification costs, as they rapidly and cleanly form selective bonds with the substrate. Among the enzymes commercially available the most commonly used are the oxidases, such as glucose oxidase (GOx) and horseradish peroxidase (HRP), beta-lactamase, urease, tyrosinase, and acetylcholine esterase or choline oxidase.

The **biocomplexing or bioaffinity sensors** are based on **antibodies, natural biological chemoreceptors, nucleic acids, aptamers or molecularly imprinted polymers (MIPs)**. These biosensors provide selective interactions of the analyte with a given ligand to form a thermodynamically stable complex. The analyte interacts with macromolecules or organized molecular assemblies that have either been isolated from their original biological environment or engineered. Thus, equilibrium is usually reached and there is no further consumption of the analyte by the immobilized biocomplexing agent. These equilibrium responses are either monitored by the integrated detector or by using a complementary biocatalytic reaction. The most developed examples of biosensors using biocomplexing receptors are based on immunochemical reactions, i.e., the binding of an antigen (Ag) to a specific antibody (Ab), and will be discussed in detail in the chapter concerning immunosensors.

The *biological chemoreceptors* [18] are used for drug and neurotransmitter detection, and they are based on an affinity reaction. Using *nucleic acids* as a biocomponent, one can detect the interaction between the synthetic drug and the genetic material and also the hybridization processes by working with single-strand DNA [19, 20].

The use of *microorganisms and of whole vegetal or animal tissues* is complex and very often their selectivity is limited [21].

3. Nanomaterials involved in biosensor construction

Advances in nanotechnology have led to the recent development of many nanomaterials, including carbon nanomaterials, magnetic and metallic nanoparticles [22]. Nanomaterials, generally defined as materials with feature sizes smaller than 100 nm, have a remarkable impact on various fields of application due to their properties (enhanced electrical conductivity, tensile strength and chemical reactivity), which are imparted by their increased surface area per unit weight. Nanomaterials have already been applied in electronics, foods, cosmetics, electronics, drug development and sensing devices. The use of nanomaterials, especially nanoparticles and nanostructured films, offers advantageous properties that can be exploited in order to maximize the interactions with specific bioelements, to maintain their activity, minimize structural changes, and enhance the catalytic step. In the biosensor field, the analytical exploitation of such protein-nanomaterial interactions are an emerging trend that spans many disciplines [23]. Nanomaterials based on metals, semiconductors and organic compounds contribute to the enhancement of optical, electrical, chemical and magnetic properties that are relevant in the case of sensing devices, and due to these facts they have been frequently used in this research field [24, 25].

Nanomaterials are mainly used *for electrode construction or modification* and *as biomolecule tracers*. Nanoparticles (NPs) are very stable (compared to enzyme labels), offering high sensitivity (thousands of atoms can be released from one nanoparticle) and a wide variety are available on the market. Nanoparticles are used nowadays as electrochemical labels or as vehicles containing several hundreds or thousands of electroactive labels, pushing detection limits down to several hundreds of biomolecules [26].

Various nanomaterials have been successfully applied for the immobilization of bioelements, such as metallic nanoparticles, carbon or metallic nanotubes, nanosilver-coated magnetic beads, magnetic particles, functionalized conductive polymers, etc.

3.1. Carbon-based materials

Different types of carbon nanostructures are becoming more and more popular due to their specific structures, properties and the possibility of being used for many applications. In fact, a wide variety of carbon-based materials are available, such as nanoparticles, nanodiamonds, nano-onions, peapods, nanofibers, nanorings, fullerenes and nanotubes, which have been extensively used in analytical applications. The basic structure of fullerenes and nanotubes consists of a layer of sp^2-bonded carbon atoms. This configuration, which resembles that of graphene, is responsible for their good electrical conductivity and their ability to form charge-transfer complexes when in contact with electron donor groups [27]. This configuration is responsible for the development of strong van der Waals forces that significantly hamper the solubility and dispersion of carbon-based nanoparticles. In order to avoid these problems, different pretreatment methods have been proposed [28, 29, 30], such as the addition of polar groups (oxygen, hydroxyl, phenyl and polyvinylpyrrolidone). The surface defects could also affect the stability, the mechanical and electrical properties of carbon nanostructures [31, 32, 33, 34].

The carbon nanomaterials (Figure 1) cover a broad range of structures, beginning with zero-dimensional structures (fullerenes, diamond clusters), continuing with one-dimensional (nanotubes), two-dimensional (graphene), and three-dimensional structures (nanocrystalline diamond, fullerite).

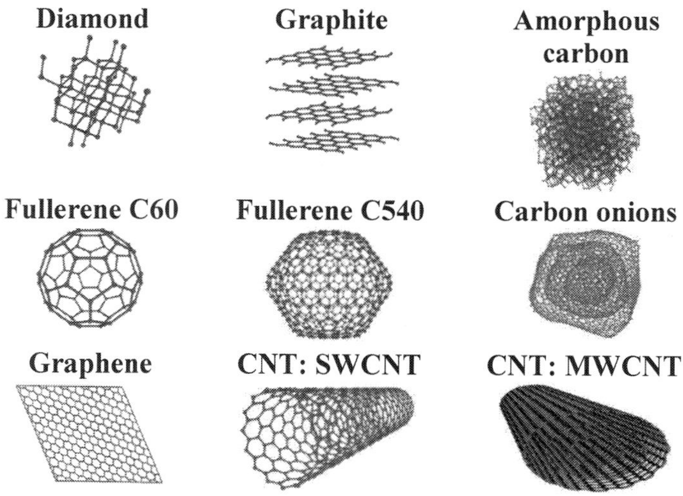

Figure 1. Main carbon entities at nanoscale level

Carbon nanomaterials, including fullerenes, graphene and carbon nanotubes, have many technological advantages such as facile modification by functional groups, high carrier capacity, incorporating both hydrophilic and hydrophobic substances, and high chemical stability.

3.1.1. Carbon nanotubes

Within the family of nanomaterials, carbon nanotubes (CNTs), described for the first time by Iijima [35], are arousing growing interest in relation to several modern devices, such as transducers, for the improvement of biosensing systems, mainly due to their unique structural, electronic and chemical properties, which include large surface area, high electrical conductivity, and excellent properties of electron transfer reactions. Furthermore, CNTs have a unique tubular structure, good biocompatibility and modifiable sidewall, making them ideal candidates for the construction of sensors with high performances [36].

Electrochemical nanobiosensors have been increasingly used in various types of nanomaterials, mainly CNTs, as electrochemical transducers, in order to improve them from the perspectives of automation, miniaturization and multiplexed analysis, but also to improve the analytical performances of such devices [37, 38]. For example, the detection of diseases at an early stage is one of the goals in developing and improving CNT biosensors, focusing on their sensitivity, fast response and small sample volume, in order to provide new strategies for the detection of specific biomarkers at low concentrations in complex sample media (e.g., serum) [39].

Many properties and the quality of CNTs are directly influenced by the way in which the graphene sheets are wrapped around [40], and they are affected by the operating conditions in their fabrication process. The production methods generally require transition metal nanoparticles (i.e., Fe, Co and Ni) to function as growth catalysts. When the catalysts are in contact with a gaseous carbon or a hydrocarbon source, the resulting carbon is deposited on the particle surface and the CNT grows rapidly at the surface of the catalyst. The diameter of the CNTs, their quantity and quality can be controlled by varying some reaction parameters (e.g., temperature, metal concentration, gas type and pressure). The diameter of the NPs defines the nature of the CNTs, whether single-walled (SWCNTs) with diameters in the range 0.4-3 nm, or a group of concentric CNTs sharing a common axis constituting double- and multiwalled (MWCNTs) with diameters in the range 1.4-100 nm [36].

In the case of CNT structures, the reactivity of atoms situated in the plane is different from those at the edges. For a number of biologically important compounds, the electrochemistry of atoms located at the edge-plane sites of the CNT is comparable to that of different planes of graphite [41], and the metal and non-metal impurities contained determine some of the electrochemical catalytic properties and limitations of CNTs [42, 43, 44]. It is also worth mentioning that the electrode surface modification with CNTs leads to a changed mass transport regime from semi-infinite diffusion to effective thin-layer diffusion in a porous layer, and the increased current may frequently be erroneously assigned to electrocatalysis [45, 46, 47].

Currently, SWCNTs and MWCNTs are synthesized by high-temperature processes (e.g., electric arc discharge and laser ablation, which are physical methods to convert carbon into

NTs) and chemical vapour deposition (CVD), which is performed at lower temperatures and atmospheric pressure.

Usually, the nanostructured materials obtained using CNTs were morphologically characterized using transmission electron microscopy (TEM) and scanning electron microscopy (SEM). Figure 2 presents the SEM images obtained for SWCNT [48] and MWCNT [49].

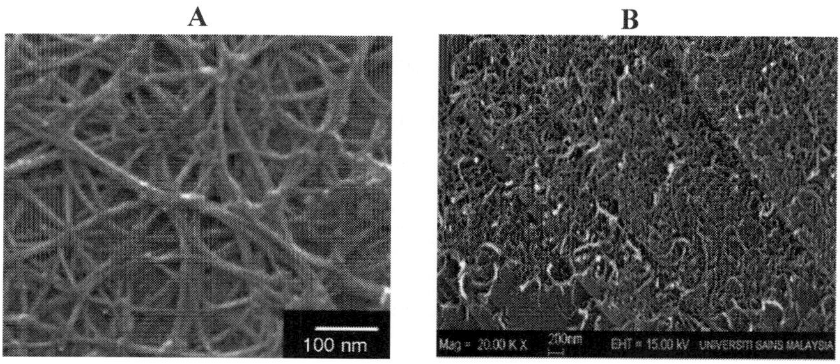

Figure 2. (A) SEM image of SWCNT composite film [48]; (B) FE-SEM images of MWCNT in high magnification [49]

Many papers published in recent years report CNT biosensors based on different oxidases applied for the detection of glucose [50, 51, 52, 53, 54], H_2O_2 [55, 56, 57], *L*-lactate [58, 59, 60], acetaminophen [61, 62, 63, 64, 65], etc. The construction and optimization of new biosensors with HRP immobilized onto the transducer surface, either screen-printed electrode (SPE) or glassy carbon electrode (GCE), by entrapment into polymer films (polyethylenimine (PEI) and polypyrrole (Ppy)) doped with SWCNT and MWCNT, was described [61, 66].

The obtained configurations were used to monitor the signal produced by the electrochemical reduction of the enzymatically generated electroactive *N*-acetylbenzoquinoneimine (the oxidized form of acetaminophen) in the presence of hydrogen peroxide using amperometry, cyclic voltammetry (CV) and differential pulse voltammetry (DPV) in synthetic and real samples, with the best limit of detection (LOD) obtained for acetaminophen of 1.36 μM [61].

3.1.2. Graphene

Graphene is another material that offers great prospects for the future of analytical chemistry, consisting of a one-atom-thick sheet of sp^2 hybridized carbon atom composed of six member rings that provide a surface area that is nearly twice as large as that of SWCNT [43]. This material presents high mechanical strength, high elasticity and high thermal conductivity [43], but also presents a major disadvantage: its poor dispersion in aqueous medium. Stable dispersions of graphene sheets have been achieved using amphiphilic polymers, alkylamines and hydrophilic carboxyl groups, among others, as dispersing agents [67]. The functionalization of graphene is considered an important route for improving its dispersion. Therefore, different authors have tried producing graphene oxide, a graphite derivative with hydroxyl, epoxy and carboxyl groups covalently attached to its layers, with a better dispersion in some solvents being obtained.

Graphene can be interlinked with CNT for the fabrication of high-performance transparent electrodes, and the resulting films present features comparable to indium tin oxide [68]. The electrochemical activity of crystalline graphene is different from that of reduced graphene oxide flakes [69]. A wide variety of methods are available for graphene fabrication, the first being published in 2004 [70], and consist in the mechanical exfoliation (repeated peeling) of small patches of highly ordered pyrolytic graphite. Many other fabrication methods have been developed, including unzipping MWCNT to form graphene ribbons [71, 72], substrate-independent methods using micromoulding inside a capillary [73], and spray deposition of graphene oxide-hydrazine dispersions [74].

SEM images of a graphene film deposited on GCE revealed the typical crumpled and wrinkled graphene sheet structure on the rough surface of the film (Figure 3 A). The TEM image presented in Figure 3B shows the wrinkled graphene sheet with no aggregation, indicating that the functionalized graphene sheets were well dispersed in ethanol, forming stable suspensions [75].

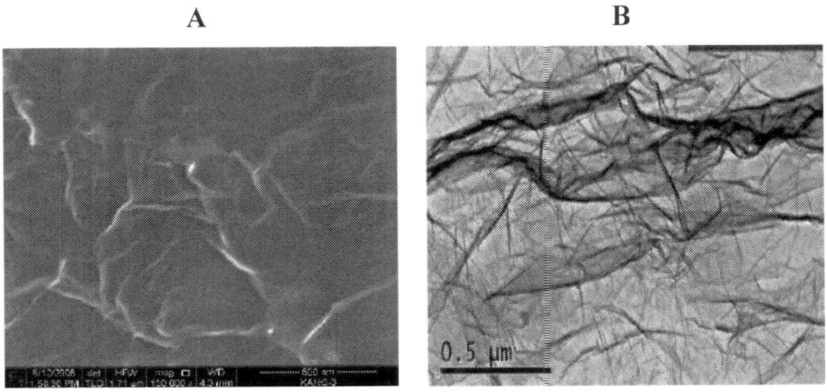

Figure 3. (A) SEM image of graphene-film-modified GCE. (B) TEM image of graphene in ethanol [75]

With regard to aptasensor development, graphene looks like it could have the necessary requirements to implement next-generation and high-performance aptasensors. Graphene naturally adsorbs the unfolded aptamer through the π–π stacking interaction between the purine and pyrimidine bases and the graphene plane, and hinders the adsorption of the folded aptamer [76, 77].

The currently developed strategies for graphene functionalization with aptamers include covalent and non-covalent approaches. The covalent approach often relies on the oxidation or the reduction of graphene oxide that presents carbonyl and carboxyl functional groups [78]. The density of carbonyl and carboxyl functional groups can be increased through chemical oxidation, which is widely used for the chemical modification of the inert graphitic structure [79]. Because this procedure affects the properties of graphene, the covalent modification of graphitic surface through a free radical addition reaction is used, without a preoxidation step and with less impact on the electronic properties [80]. Besides the immobilization methods mentioned above, amide [81] and phosphoramidate reactions [82] were employed towards the fabrication of graphene-based biosensors. Some graphene derivatives, such as graphene oxide, were also used as transducers in electrochemical biosensor development [83].

Graphene can also mediate electron transfer in view of the high-density edge-plane-like defects present within its nanostructure. In the absence of a target, aptamers immobilized on the gold electrode adsorb graphene nanosheets due to the strong π–π interaction that accelerates the electron transfer ratio between the electroactive species and the electrode surface. In the presence of a target, the binding reaction inhibits the adsorption of graphene and blocks the electron transfer. A label-free electrochemical aptasensor was constructed by taking advantage of the ultra-fast electron transfer ratio of graphene [84]. This detection strategy was further improved by nuclease for interferon gamma, yielding an LOD of 0.065 pM [85]. Other authors [86] employed graphene oxide nanoplatelets (GONPs) as electroactive labels for thrombin detection.

A simple label-free electrochemical immunosensor was constructed by modifying a graphite-based SPE with graphene oxide after functionalization with *N*-hydroxysuccinimide in the presence of 1-ethyl-3-(3-dimethyl aminopropyl) carbodiimide hydrochloride, Figure 4.

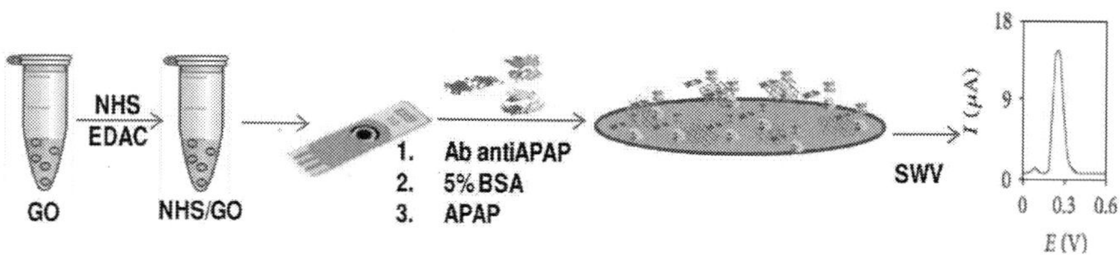

Figure 4. Steps involved in the antiacetaminophen Ab-based immunosensor development; GO=Graphene oxide; NHS=*N*-hydroxysuccinimide; EDAC=1-ethyl-3-(3-dimethyl aminopropyl) carbodiimide hydrochloride; Ab anti APAP=antibody antiacetaminophen; APAP=acetaminophen; SWV=square wave voltammetry [87]

The above-described immunosensor was applied with good results for the determination of acetaminophen in synthetic and real samples by using square wave voltammetry (SWV) with an LOD of 0.17 μM [87].

3.1.3. *Carbon nanoparticle-based electrochemical sensor*

A sensitive and selective electrochemical sensor fabricated via the drop-casting of a suspension of carbon nanoparticles (CNPs) onto a GCE was investigated in simultaneous determination of acetaminophen and tramadol in pharmaceuticals and human plasma. CV and DPV studies of acetaminophen and tramadol were carried out at the modified and bare GCE. The results of the electrochemical investigations showed that CNPs enhanced the electroactive surface area and caused a remarkable increase in the peak currents, due to the diffusion within the porous layer. Enhanced sensitivity and a considerable decrease in the anodic overpotential leading to negative shifts in peak potentials was obtained with the modified GCE. By using DPV, the sensor showed good sensitivity and selectivity. LODs for acetaminophen and tramadol were 0.05 and 1 μM, respectively [88].

3.2. Inorganic and organic nanoparticles

Nanoparticles of major environmental, pharmaceutical and biomedical importance include magnetic nanoparticles, quantum dots, metal nanoparticles, silica nanoparticles and polymeric types with intrinsic properties contributing to their use in specific applications. Typically, nanoparticles are classified into: *magnetic nanoparticles* (possessing a core of Fe_3O_4) frequently used in drug delivery and lately in immunosensing; *semiconductor/QDs* (based on CdS, CdSe) used in bio- and immunosensing; *metal nanoparticles* (Ag, Au, Pd) used in biosensing and drug delivery; *polymeric nanoparticles* (polystyrene) mainly used in drug and gene delivery; *hybrid nanoparticles* (carbon or Fe_3O_4 covered with different materials like metal oxides, polymers, amino acids, etc.) [89].

3.2.1. *Magnetic nanoparticles*

In recent years, considerable efforts were made to develop magnetic nanoparticles (MNPs), due to their inherent advantages (magnetism, nanosize) and low cost of production. MNPs with 10-20 nm diameters exhibit their best performance, due to supermagnetism, which makes them especially suitable for a fast response. Due to their inherent dimensions, they possess a large surface area and high mass transfer capacity [90].

MNPs can be integrated into the transducer materials and/or be dispersed in the sample followed by their attraction by an external magnetic field onto the active detection surface of the (bio)sensor. Biorecognition molecules like enzymes, antibodies or oligonucleotides immobilized onto magnetic particles can be easily trapped by magnets and retained close to or on an electrode surface [90, 91]. Since the properties of MNPs depend strongly on their dimensions, their synthesis and preparation have to be designed in order to obtain particles with adequate size-dependent physicochemical properties. MNPs possessing adequate physicochemistry and tailored surface properties have been synthesized under precise conditions and applied for sensors, biosensors and other detection systems [92, 93].

Magnetic particles consist of a magnetic core surrounded by a non-magnetic shell. Iron oxides, such as magnetite (Fe_3O_4) or maghemite (γ-Fe_2O_3), are often used as core materials instead of iron due to their higher stability [94]. For biological applications, magnetic particles generally consist of nanosized superparamagnetic iron oxide distributed through a matrix, mainly composed of silica [95] or an organic polymer [96]. Silica-based magnetic beads can be prepared in different shapes and sizes with different degrees of porosity, which are much more attractive for the immobilization of biomolecules due to their chemical and mechanical stability and their resistance to bacterial attack [97]. Many types of magnetic particles, functionalized with different groups like carboxyl, amino, hydroxyl or epoxy, are currently available for different applications.

The majority of MNP-based systems use inorganic nanocrystals as magnetic cores. The composition of these inorganic nanocrystal ranges from metals and alloys to metal oxides. Among the metal oxide compounds, superparamagnetic iron oxide nanoparticles, including magnetite (Fe_3O_4) and maghemite (γ-Fe_2O_3), are the most popular [98]. The surfaces of the particles can be modified with organic polymer or inorganic compounds (e.g., gold, silica, alumina), suitable for further functionalization by the attachment of various bioactive molecules [99]. The functionalization of MNPs with bioelements exhibits specificity, due to both the bioreceptor and the magnetism.

MNPs amplify the electrochemical signal, improving the sensitivity of electrochemical devices [90], through their contact with the electrode surface, transport of a redox-active species to the electrode surface, and formation of a thin film on the electrode surface. The detection for MNP-based electrochemical biosensors can be performed by potentiometry [100, 101], amperometry [102, 103], electrochemiluminescence [104, 105], electrochemical impedance spectroscopy [106, 107] and especially voltammetry [108, 109, 110, 111, 112, 113, 114, 115].

Protein G-MNP-based label-free electrochemical immunosensor was elaborated for the ultrasensitive and specific detection of acetaminophen using a carbon-based SPE as immobilization platform (Figure 5). An LOD of 1.76 μM was calculated based on the calibration data. The performances of the optimized immunoassay were tested using two pharmaceutical products containing acetaminophen with excellent recoveries [116].

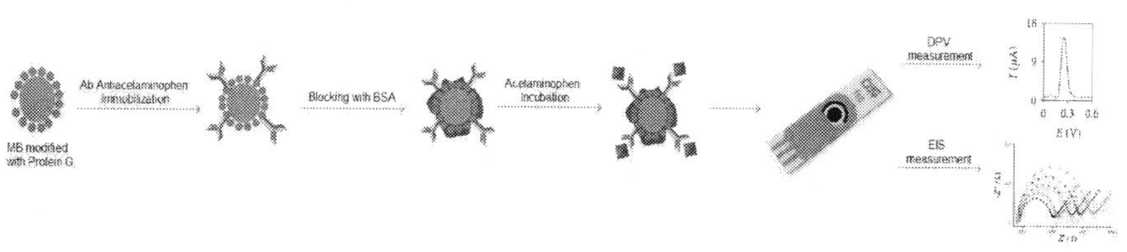

Figure 5. Protein G-MNP-based label-free electrochemical immunosensor development [116]

Protein G-functionalized magnetic beads and graphite-based SPEs were also used to develop an HRP-labelled sandwich type immunosensor for Mucin 1 (MUC1) electrochemical detection with LOD of 1.34 ppb [117].

Fe_3O_4 magnetic dipolar attraction and its large surface area/volume ratio may lead to aggregation in clusters when exposed to biological solutions. Functionalization can overcome this problem and also enhance biocompatibility. Many strategies were used for the functionalization of MNPs, such as core-shell Au-Fe_3O_4 [109], core-shell Au-Fe_3O_4@SiO_2 [102], core-shell Fe_3O_4@SiO_2 [112], Au-Fe_3O_4composite NPs [108], Fe_3O_4@SiO_2/MWCNTs [103], Fe_3O_4 anchored on reduced graphene oxide [113] and Fe_3O_4@ Au-MWCNT-chitosan [114]. Among all these, core-shell Fe_3O_4@SiO_2 contributes to the stabilization of MNPs in solution and enhances the binding of ligands at the surface of MNPs. Electrode surface modified with core-shell Fe_3O_4@SiO_2 shows good electrical conductivity and more electroactive interaction sites, providing enhanced mass transport and easier accessibility to the active sites, thus increasing the analytical signal and sensitivity [90]. A comparison between the analytical performances of some recently developed electrochemical sensors and biosensors based on MNPs is presented in Table 1.

Detection method	MNPs type	Analyte	LOD	Reference
Voltammetry	Au-Fe_3O_4	Organochloride pesticides	56 pg/mL	[108]
Voltammetry	Core-shell Au-Fe_3O_4	Carcinoembryonic Ag	10 ng/mL	[109]
Voltammetry	Fe_3O_4 Au NP	Clenbuterol	220 ng/mL	[110]
Voltammetry	Fe_3O_4 Au NP	H_2O_2	20 μM	[111]
Voltammetry	Core-shell Fe_3O_4@SiO_2	Metronidazole	18 nM	[112]
Voltammetry	Fe_3O_4-RGO	Cr(III)	-	[113]
Voltammetry	Fe_3O_4@Au-MWCNT-chitosan	Streptomycin	1.5 nM	[114]
Voltammetry	Core-shell Fe_3O_4@SiO_2/MWCNT	Uric acid	0.13 μM	[115]
Potentiometry	MaMB Dynabeads Protein G	Zearalenone	7 ng/mL	[100]
Potentiometry	Core-shell-Fe_3O_4	Glucose	0.5 μM	[101]
Amperommetry	Core-shell Au-Fe_3O_4@SiO_2	Glucose	0.01 mM	[102]
Amperommetry	Fe_3O_4@SiO_2-MWCNT	Glucose	800 nM	[118]
EIS	Fe_3O_4-COOH/MNP	OchratoxinA	10 pg/mL	[106]
EIS	Fe@Au NP-2-aminoethanethiol - graphene NP	DNA	2 fM	[107]
Detection method	MNPs type	Analyte	LOD	Reference

Table 1. Selected examples of sensors and biosensors based on magnetic nanoparticles

3.2.2. *Metal nanoparticles*

Metal nanoparticles and CNTs, have been employed to immobilize biomolecules for the construction of sol-gel biosensors, due to their large surface area, high catalytic activity and good biocompatibility. Moreover, the excellent conductivity of gold nanoparticles (AuNPs) provides the potential to construct amperometric biosensors based on redox enzymes (as sensing elements) and nanoparticle arrays (as conductive matrices where enzyme molecules are immobilized) [118, 119]. Metal nanoparticles, such as Pd and Pt, were deposited on

SWCNTs and MWCNTs in order to increase the electrocatalytic activity of CNTs. An improved conductivity and electrocatalytic behaviour was obtained by combining sol-gel materials with CNTs, the electron transfer between the redox enzyme and the electrode being facilitated [120].

Gold nanomaterials have unusual optical and electronic properties, high stability and biological compatibility. Aptamer-conjugated gold nanomaterials provide powerful platforms to facilitate targeted recognition, detection and therapy applications, due to their controllable morphology, size dispersion and easy surface functionalization [121].

Many applications of nanoparticles, such as metal NPs (gold, silver) or semiconductor dots in biosensing, have been recently reported [122, 123]. A DNA array detection system based on oligonucleotide targets labelled with AuNPs was developed [124], and AuNPs or silver-enhanced colloidal gold were used as labels for electrochemical detection of DNA hybridization [125, 126, 127]. These NP-based electrochemical biosensors employed the stripping voltammetric technique as the read-out principle due to the preconcentration step resulting in ultra-trace-level LODs.

AuNPs are irreversibly immobilized on the transducer, the electrode surface being difficult to regenerate. To overcome this problem, Chen *et al.* [128] have developed a novel electrochemical system for the sensitive detection of glucose (LOD of 1 mM) as well as single-nucleotide polymorphism, which has the AuNPs directly in the supporting electrolyte. This one-pot detection system can be operated and regenerated very easily, since all the components are integrated in the electrolyte of the gold colloid and the unmodified electrode can be reused.

The sensing mechanism of potentiometric biosensors using NPs is based on a sandwich assay (Figure 6), the target being captured by the immobilized primary receptor and then attached to the NP-labelled secondary receptor. Potentiometry is finally used to detect the dissolution products of NPs after a release step with H_2O_2 [123].

Miniaturized ion-selective electrodes (ISEs) designed to detect ions in microvolume samples, with a concentration of fM, were reported [129, 130], and an LOD of 12.5 pmol/50μL was obtained for IgG using an ISE based on silver-enhanced AuNPs [131]. Metal NPs can also be used as a catalyst for the reduction of metal ion, via the enzymatically generated reducing agent, and the biometallization processes can be monitored by ISEs in real time [132]. Thereby, an ISE based on AuNPs was developed and applied for highly sensitive detection of glucose [132, 133], while the NADH potentiometric detection was successfully obtained by monitoring the NADH-stimulated catalytic reduction of Cu^{2+} in the presence of AuNPs (which have a crucial role in improving the sensitivity) with a Cu^{2+}-ISE [134]. Wang *et al.* [135] reported a highly sensitive potentiometric biosensor for detecting DNA hybridization based on the depletion of Ag^{+} ions induced by the biocatalytic reaction of alkaline phosphatase, recorded in the presence of AuNPs. The sensor has been used to detect DNA and 16S rRNA of *Escherichia coli* with LOD of 50 fM, using only 4 μL of sample solution.

Two simple electrochemical aptasensors were developed for the detection of Mucin 1 (MUC1) tumour marker based on its specific recognition by thiolated aptamers immobilized on Au NP-modified graphite and gold SPEs. The quantitative detection of MUC1 protein was electrochemically achieved by electrochemical impedance spectroscopy (EIS) and DPV. The

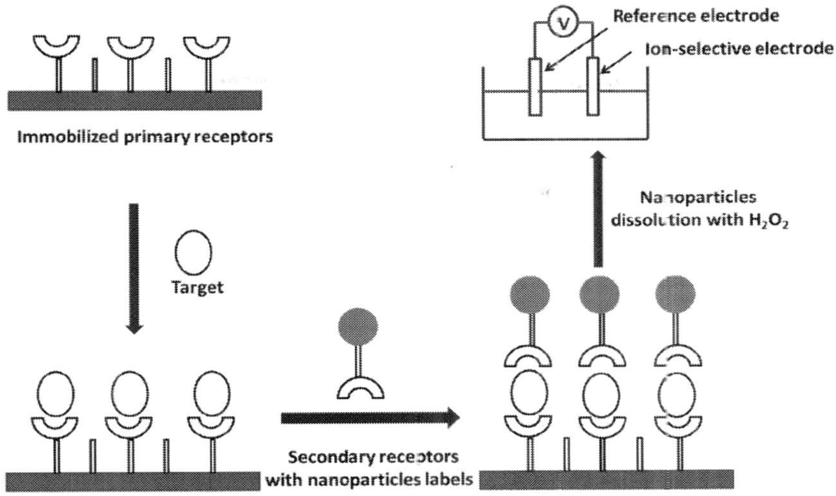

Figure 6. Potentiometric biosensors using nanoparticles as labels [123]

estimated detection limits for MUC1 were 3.6 ng/mL at AuNP modified graphite SPE by EIS and 0.95 ng/mL at AuNP modified gold SPE by DPV [136].

3.2.3. Nanosized clays

Nanosized clays can serve as matrices for electroactive species because they are usually able to incorporate ions by an ion exchange process. Moreover, adsorption of proteins on clay mineral surfaces plays a very important role in the development of biosensors. By modifying the electrode surface with clays, the heterogeneous electron transfer process between the protein and the electrode surface is facilitated.

The entrapment of biomolecules in the clay matrix represents an inexpensive, fast and easy method for the elaboration of enzyme electrodes. The procedure consisting in the adsorption of an enzyme/clay aqueous colloid mixture onto the electrode surface offered the possibility of entrapping a large and known amount of enzyme into the clay film, and could also be adapted for interdigitated microelectrodes [137].

Mesoporous silica nanoparticles can be easily synthesized by using surfactant templating methods [138]. Silica nanoparticles have become an important platform in biomedical applications and bioanalysis because of their high density, which facilitates separation through centrifugation during their synthesis, and their modification and detection steps [139]. The silica surface also provides a biocompatible and versatile substrate for biomolecule immobilization, required for further utilization in biosystems and for drug delivery [140]. Following typical DNA silica surface conjugate chemistry, the aptamer immobilization on silica nanoparticles has been well developed for various applications [141].

Biosensors based on clay materials can play an important role as efficient tools in the field of forensic analysis considering their well-known advantages of sensitivity, selectivity, simple functioning and affordability [142]. Thereby, a bienzyme biosensor based on acid phosphatase (AcP) and polyphenol oxidase (PPO) simultaneously entrapped in anionic clays was constructed and applied for the specific amperometric determination of As(V). The biosensor relied on the successive hydrolysis of phenyl phosphate to phenol catalysed by AcP and the oxidation of phenol to *o*-quinone catalysed by PPO. The electrochemical reduction of the formed *o*-quinone was accomplished at a detection potential of -0.2 V vs Ag/AgCl with a detection limit of 0.15 ng/mL [143]. DNA-functionalized, layer-by-layer assembled SWCNT hybrids with clay nanoparticles were proposed for the detection of As(III) achieving an LOD of 0.05 mg L^{-1} [144]. Another simple amperometric biosensor for cyanide involved the immobilization of PPO into ion exchanger clay based on Zn-Al double-layered hydroxides (LOD of 0.1 nM). The low LOD was a consequence of the anion accumulation in the ion exchanger clay which also contained the immobilized enzyme [145].

3.2.4. *Nanosized and nanostructured polymeric films*

One of the most important aspects of biosensor development is the bioelement immobilization platform, which needs to ensure high stability and maintain its activity as long as possible. Several types of **conductive** or **insulating polymers** were used to obtain a nano/micropatterned surface with applications in the biosensor field [146]. Electropolymerization in the presence of a template, removed after the polymer formation, is an efficient route for the synthesis of 2D and 3D conductive polymers (CPs). Among the conductive polymers, polypyrrole and polyaniline [147, 148] were extensively studied because of their electrochemical properties, aqueous compatibility and their ability to form nanostructures like nanowires, nanotubes or microcontainers [149].

Hydrogels are cross-linked hydrophilic polymer structures that can hold large amounts of water or biological fluids in their pores. As one of the newest classes of polymer-based systems, target-responsive hydrogels have found numerous biomedical and pharmaceutical applications [121, 150]. Polyacrylamide/acrylate is a copolymer of acrylamide and acrylic acid, and acrydite-modified oligonucleotides can form hydrogels by the introduction of proper complementary DNA as a cross-linker. He *et al.* [151] reported an aptamer-based reversible DNA induced hydrogel system for molecular recognition and separation.

Different types of hydrogel nanoparticles were used together with various biocomponents, such as enzymes, proteins and nucleic acids, during the development of electrochemical biosensing systems, in order to increase their sensing performances [152].

CPs and in particular functionalized CPs are useful for biosensor construction as a platform to immobilize the primary Ab on the electrode surface. Poly *o*-aminobenzoic acid (PABA) can be considered a carboxyl-functionalized aniline CP, capable of self-doping, and is also a soluble derivative of polyaniline. Its carboxyl group can be exploited as a linker, either for PABA immobilization onto the electrode, or for covalent bond formation with biomolecules such as proteins and antibodies [153].

An electrochemical aptamer-based biosensing assay for MUC1 protein detection by the modification of the graphite-based SPE's surface using a functionalized conductive polymer (PABA), and by using methylene blue (MB) as an electrochemical indicator was developed. The recognition of immunoreactions and aptamer binding events was assessed by monitoring the interfacial electron transfer resistance with EIS, while CV and DPV were employed for sensitive indirect quantification of MUC1 via MB with a detection limit of 0.62 ppb [154]. In order to establish the optimal configuration of HRP-based biosensors able to detect acetaminophen in various matrices, another conductive polymer, polypyrrole, was deposited on GCE and SPE by several methods. Chronoamperometric studies performed proved the interaction between the immobilized enzyme and the electroactive species of paracetamol, NAPQI, enzymatically generated in the presence of H_2O_2 [155].

The combination of nanosphere lithography, electrodeposition and click chemistry was applied for the immobilization of ferrocene at GCE. The nanostructuration of the surface was obtained by adsorption of 100 and 900 nm diameter polystyrene nanospheres followed either by electropolymerization of *N*-(10-azidodecyl) pyrrole or by electrografting of 4-azidobenzenediazonium. This approach offered potential applications such as engineering of porous materials and was used for HRP immobilization. The nanostructured biosensor was applied for acetaminophen detection in the presence of hydrogen peroxide, and showed better performances in terms of sensitivity and LOD in comparison with a non-structured electrode [156]. Another template based on nanostructured polymeric film was developed combining latex nanosphere lithography with electropolymerization of *N*-substituted pyrrole monomer in order to immobilize both tyrosinase and GOx and to obtain a biosensor with higher sensing performances [148].

3.2.5. Molecularly imprinted polymers

MIPs can be successfully used in applications relying on selective molecular binding as they are fully artificial macromolecular structures with biomimetic properties, imitating receptor-ligand, Ab-Ag, or enzyme-substrate biorecognition [157].

MIPs possess several advantages over their biological counterparts, including low cost, ease of preparation, long term stability and shelf life, inherent reusability, high mechanical strength, and remarkable chemical and thermal stability under various experimental conditions, allowing the very specific binding of the analyte in "harsh" chemical media [158, 159].

The rational design of MIPs is still considered a challenge due to the numerous experimental variables to be addressed, and it has been previously reviewed in detail by several groups of authors [157, 158, 160, 161] along with their most important analytical applications.

MIPs prepared by bulk polymerization have proven to be poorly compatible with transducers in terms of efficient immobilization strategies, mass transfer and rebinding kinetics, even if they are ground before use. Therefore, several aspects related to the in-situ polymerization, and more specifically to the electrosynthesis of MIPs and hybrid technologies involving MIPs in the fabrication of biosensors, as well as their most recent advances in addressing two important challenges for MIP-based electrochemical biosensors, the determination of biomacromolecules and enantiospecific analysis, will be very briefly presented.

Electroactive organic functional monomers can polymerize to form conducting, semiconducting or insulating polymers. Polymerization may be performed chemically (using an oxidizer as external reagent) or electrochemically (by electropolymerization). Electropolymerization is the more convenient approach, in terms of simplicity and time-effectiveness, for electrochemical biosensor fabrication as it leads to a tight binding of an MIP film onto the surface of the transducer. This procedure ensures both the synthesis and the surface immobilization of the biomimetic layer, along with the possibility of controlling polymer nucleation and growth, film thickness and morphology by the proper selection of the electropolymerization parameters. The reproducible preparation of thin MIP layers is of the utmost importance in sensing applications, because it reduces response time by significantly shortening diffusional path lengths. Solvent swelling and the inclusion of the supporting electrolyte's ions tunes the rigidity and porosity of the MIP film. Electropolymerization may be performed under galvanostatic, potentiometric or most frequently potentiodynamic conditions, in both non-aqueous and aqueous solutions. The first two techniques are able to generate well-ordered, charged polymeric layers doped with counter ions for charge neutralization, whereas the latter generates a more tangled, charged or neutral polymeric matrix due to the continuous influx and release of solvated ions during its growth upon film charging and discharging [157].

Using electroinsulating MIP films in the fabrication of electrochemical sensors carries a major disadvantage due to the high resistance to charge transfer and the lack of direct path for electron conduction from the recognition sites to the electrode transducer, suffering from an incomplete template removal upon their overoxidation. Using electronically conducting polymers for MIP synthesis avoids these undesirable effects.

Some of the most common electroinsulating layers employed in MIP technology are acrylic- and vinylic-based polymers, but the most frequently reported electrosynthesized insulating MIPs are polyphenylenediamine, polyphenol, polyaminophenol and polythiophenol. Due to their low conductivity, sensors based on such MIPs usually use optical, piezoelectric or impedimetric signal transduction. However, as long as they are electrodeposited in a thickness that does not insulate the underlying conductive electrode surface (<10 nm) [162], a successful voltamperometric signal transduction may also be performed [163].

In the case of electrosynthesis of conducting MIP layers for biosensing, electroactive functional monomers such as pyrrole, thiophene, aniline, 3-aminophenylboronic acid and porphyrin derivatives are most often reported. Analyte determination using such conducting layers may be performed by any available signal transduction procedure, such as optical, piezoelectric, voltammetric, potentiometric, chronoamperometric, impedimetric or conductometric procedures. Poly(pyrrole), poly(thiophene), poly(aniline) and their derivatives are amongst the most widely studied conducting polymers [157, 160], due to their high conductivities in their oxidized form and their ability to reversibly switch between conducting and insulating states by doping and undoping [160]. As mentioned earlier, in the non-covalent molecular imprinting, a wide variety of templates are amenable to be used with elevated imprinting efficiencies, ranging from small analytes to macromolecules (proteins, nucleotides). Additional promising approaches include implementing surface MIP for cells (yeast), bacteria, virus or blood [164, 165, 166]. In the case of using small organic molecules as templates, MIP synthesis is well established, however the imprinting of much larger molecules is still considered to be a challenge. Larger templates are less rigid, which does not ensure the formation of well-defined binding cavities during macromolecular imprinting, and, furthermore, the secondary and tertiary structures of biomacromolecules (i.e., proteins) may be also affected under the imprinting conditions. The large size hinders the protein both in reaching and leaving the artificial binding sites, and their molecular complexity increases the chances of non-specific binding with the MIP, leading to poor selectivity and cross-reactivity. More importantly, proteins are often incompatible with the organic solvents used in MIPs synthesis. The use of an aqueous solution for macromolecular imprinting greatly restricts the choice of reactive functional monomers and cross-linkers. Additionally, water could compete for and potentially disrupt any hydrogen bonds between the template and the functional monomers. The protein as template in macromolecular imprinting may be introduced by three different approaches: it may be dissolved into the prepolymerization mixture (bulk imprinting), grafted to the surface of the transducer followed by a partial inclusion into the bulk polymer with the resulting binding sites situated on the MIP surface (surface imprinting), or only a small epitope part of the protein (epitope imprinting) may be used as a template, resulting in an MIP that is able to recognize the whole target protein. Surface and epitope imprinting enhance the imprinting effect for biomolecules, leading to higher affinity and capacity for the template, as well as to reduced non-specific binding; they improve protein binding kinetics, overcoming difficulties to mass transfer and protein removal in imprinted matrices. While surface imprinting is an efficient technique, the conditions used to accomplish the imprinting process must be carefully considered. The conformational stability of proteins is sensitive to their surrounding conditions, including temperature, pH, ion concentration, surfactant and its concentration, the ratio of template to functional monomers, the types of functional monomers used, and initiator type [167]. MIPs generated by epitope imprinting manifest minimal non-specific binding and improved affinity. Furthermore, it is less costly working with short peptides as templates, and more importantly, because of the template's higher stability, organic solvents may also be used during the imprinting process. The detection of the bound non-electroactive target macromolecule is performed indirectly, in the presence or absence of a redox probe, by various signal transduction principles, such as voltammetry, piezomicrogravimetry, impedance spectroscopy, etc.

The high selectivity of such biomimetic polymers is also demonstrated by their efficient use for the molecular imprinting of optically active templates. This special application of MIP-based chiral electrochemical sensing has not been fully exploited, even though there is a great necessity in the pharmaceutical industry for developing fast and cost-effective methods of chiral analysis applicable from the early stages of drug development. The vast majority of MIP-based electrochemical sensors were developed for non-chiral analysis. Only around 20 articles report MIP-based sensors designed for the chiral analysis of different small-molecules, but not pharmaceutically active compounds, like amino acids [168, 169, 170] or monosaccharides [171]. Moreover, enantiospecific signal transduction was only reported in the case of amino acids.

Nevertheless, very recently an MIP-based electrochemical sensor has been reported for the simultaneous enantiospecific recognition of several β-blocker enantiomers [163].

In spite of the electroinsulating properties of the R(+)-atenolol imprinted acrylate-based MIP, voltamperometric signal transduction was successfully performed, due to the resulting very thin (less than 4 nm) and porous MIP layer. The sensor exhibited distinctive enantiospecific oxidation peaks towards the R-antipodes of four β-blocker representatives and additional oxidation peaks common to both enantiomers of each studied β-blocker, thus allowing the simultaneous analysis of all of their enantiomers in a single determination.

3.3. Hybrid materials

Metal NPs decorating CNTs or graphene sheets, carbon-coated magnetic NPs, nanoparticles included in carbon paste electrodes, MWCNT-alumina-coated silica, and insulating MIPs combined with various nanostructures are examples of some hybrid nanomaterials used for the enhancement of biosensor sensitivity.

3.3.1. CNT-based hybrid materials

Electrodes modified by hybrid materials, consisting of CNTs and metal nanoparticles, have been developed for use as fuel cell catalysts and biosensors [172]. For example, platinum nanoparticles (PtNPs) catalyse the electrochemical oxidation of H_2O_2, which is generated by the enzymatic reaction, while MWCNTs can be employed to modify the working electrode for the enlargement of the electroactive surface area [173]. A highly sensitive Pt-CNT-glucose biosensor was developed using the incorporation of GOx on a Pt-CNT electrode but with poor enzyme stability after storage [174]. Electrodes modified by CNTs/PtNPs hybrids protected by dendrimers were first developed for methanol electrochemical oxidation [175], and then adapted for the detection of inorganic phosphorous pesticide through the inhibition of the acetylcholinesterase [172]. These electrochemical biosensors are based on a hybrid material, consisting of MWCNTs, PtNPs and poly(amidoamine) dendrimer (DEN) as a binder for enzyme immobilization (Figure 7).

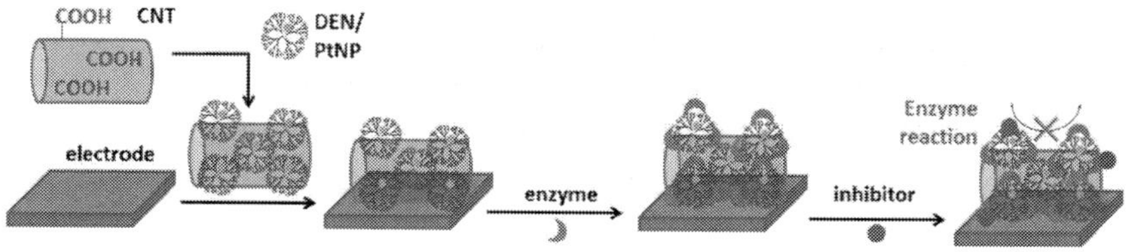

Figure 7. Biosensor preparation by loading MWCNT, PtNPs and enzyme through dendrimer binders [172]

MWCNTs increase the surface area and promote the electron transfer properties, PtNPs intensify the electron transferability in addition to catalytic performance, and dendrimers play a role in the uptake of the redox agent as well as the binder functionality.

Two new xanthine biosensors, based on GC modified with chitosan (CS), Co_3O_4NPs and MWCNTs were developed [176]. Two different enzyme systems were investigated involving monoenzyme xanthine oxidase (XO) immobilization and bienzyme xanthine oxidase (XO)/HRP co-immobilization. Xanthine amperometric determination was achieved by both electrooxidation and electroreduction of the enzymatically generated hydrogen peroxide, with LODs of 0.2 μM and 20 μM with anodic and cathodic processes, respectively. The proposed biosensors were applied for detecting xanthine in fish with average recoveries of 96.2±3.5 % with XO/Co_3O_4MWCNTs/CS/GCE and 95.4±2.1 % with XO/HRP/Co_3O_4MWCNTs/CS/GCE, respectively.

3.3.2. Graphene-based hybrid materials

Graphene can be decorated with AuNPs to enhance the signal transduction or to increase the grafting area for biocomponent immobilization. AuNPs are the most commonly used metal nanoparticles in graphene-based biosensing systems, mainly due to their ability to form strong covalent bonds with the thiol groups [26].

An electrochemical aptasensor using graphene-AuNPs composite obtained by the reduction of tetrachloroauric acid with sodium citrate in a graphene water suspension was reported [109]. Direct electrodeposition technique of AuNPs was also applied to graphene-based electrochemical transducers yielding a high density of nanoparticles that provide an ultra-large specific surface area available for aptamer immobilization [177].

A graphene-CNTs-incorporating zinc oxide NPs hybrid composite (GR-CNT-ZnO) was synthesized and applied to fabricate an enzyme-based glucose biosensor (Figure 8).

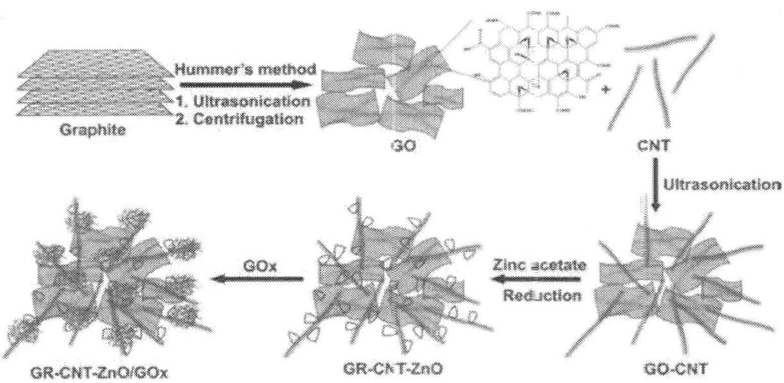

Figure 8. The preparation of GR-CNT-ZnO composite [178]

By using CV, authors have concluded that the modified film in the presence of oxygen has high electrocatalytic ability in glucose detection with a wide linear detection range and a LOD of 4.5 μM of glucose. The sensor was successfully applied for human serum samples, and the results were in good agreement with those determined using a standard photometric method [178].

3.3.3. Molecularly imprinted polymer-based hybrid materials

Hybrid materials combining insulating MIPs with electronically conducting polymers tend to correct the drawbacks of the former and feature networks of molecular wires connecting recognitions sites to the electrode surface [179]. Molecularly imprinted nanoparticles, nanospheres, nanoshells or nanofibres obtained by various synthesis procedures (precipitation or emulsion/suspension polymerization, mini/emulsion precipitation, etc.), along with various organic/inorganic hybrid materials [165, 180, 181], metal nanoparticles, such as Ag [182] and Au [183] or carbon nanomaterials [173], may be combined with electrosynthesized MIP films, leading to composite MIP/nanomaterial biosensing systems [160, 161] with larger specific surface area, higher sensibility and specificity, and higher biocompatibility for trace electrochemical analysis of small analytes and biomacromolecules.

4. Perspectives and future development

The selection of materials and fabrication techniques is crucial for adequate sensor function and the performance of a biosensor often ultimately depends upon these, rather than upon other factors. Consequently, future developments in biosensor design will inevitably focus upon the technology of new materials, especially the new copolymers that promise to solve the biocompatibility problem and offer the prospect of more widespread use of biosensors in clinical monitoring. The future trends in sensor development that are expected to have an impact on biosensor performances mainly concern immobilization techniques, nanotechnology, miniaturization and design of multisensor arrays.

Author details

Robert Săndulescu*, Mihaela Tertiş, Cecilia Cristea and Ede Bodoki

*Address all correspondence to: rsandulescu@umfcluj.ro

Analytical Chemistry Department, "Iuliu Haţieganu" University of Medicine and Pharmacy Cluj-Napoca, Romania

References

[1] Cristea C, Hârceagă V, Săndulescu R. Electrochemical sensor and biosensors. In: Moretto LM, Kalcher K, editors. Environmental Analysis by Electrochemical Sensors and Biosensors. Vol. 1 Fundamentals. 1st ed. New York: Springer; 2014. p. 155-165.

[2] Morrison DWG, Dokmeci MR, Demirci U, Khademhosseini A. Clinical applications of micro- and nanoscale biosensors. In: Gonsalves KE, Halberstadt CR, Laurencin CT,

Nair LS, editors. Biomedical Nanostructures. 1st ed. New York: John Wiley & Sons Inc.; 2008. p. 1-10.

[3] Castillo J, Gaspar S, Leth S, Niculescu M, Mortari A, Botidean I et al. Biosensors for life quality design, development and applications. Sensor. Actuat. B-Chem. 2004;102(2):179-194.

[4] Campas M, Prieto-Simon B, Marty JL. A review of the use of genetically engineered enzymes in electrochemical biosensors. Semin. Cell. Dev. Biol. 2009;20(1):3-9.

[5] Farré M, Kantiani L, Pérez S, Barceló D. Sensors and biosensors in support of EU directives. TrAC Trends Anal. Chem. 2009;28(2):170-185.

[6] Zhang S, Wright G, Yang Y. Materials and techniques for electrochemical biosensor design and construction. Biosens. Bioelectron. 2000;15(5-6):273-282.

[7] Walcarius A, Etienne M, Herzog G, Urbanova V, Vila N. Electrode materials (bulk materials and modification). In: Moretto LM, Kalcher K, editors. Environmental Analysis by Electrochemical Sensors and Biosensors. Vol. 1 Fundamentals. 1st ed. New York: Springer; 2014. p. 403-495.

[8] Xu Z, Chen X, Dong S. Electrochemical biosensors based on advanced bioimmobilization matrices. TrAC Trends Anal. Chem. 2006;25(9):899-908.

[9] Mello LD, Kubota LT. Review of the use of biosensors as analytical tools in the food and drink industries. Food Chem. 2002;77(2):237-256.

[10] Lojou E, Bianco P. Membrane electrodes for protein and enzyme electrochemistry. Electroanal. 2004;16(13-14):1113-1121.

[11] Teles FRR, Fonseca LP. Applications of polymers for biomolecule immobilization in electrochemical biosensors. Mat. Sci. Eng. C Biomin. 2008;28(8):1530-1543.

[12] Samanta D, Sarkar A. Immobilization of bio-macromolecules on self-assembled monolayers: Methods and sensor application. Chem. Soc. Rev. 2011;40(5):2567-2592.

[13] Tien HT. Bilayer lipid membrane-based electrochemical biosensors. J. Clin. Lab. Anal. 2005;2(4):256-264.

[14] Walcarius A. Electroanalytical applications of microporous zeolites and mesoporous(organo) silicas: Recent trends. Electroanal. 2008;20(7):711-738.

[15] Chambers JP, Arulanandam BP, Matta LL, Weis A, Valdes JJ. Biosensor recognition elements. Curr. Issues. Mol. Biol. 2008;10(1-2):1-12.

[16] Thevenot DR, Toth K, Durst RA, Wilson GS. Electrochemical biosensors: Recommended definitions and classification. Biosens. Bioelectron. 16;1-2(121-131)

[17] Bard AJ, Inzelt G, Scholz F, editors. Electrochemical Dictionary. 2nd ed. Berlin: Springer; 2012.

[18] Mascini M, Minunni M, Guilbault GG, Carter R. Methods in biotechnology. In: Rogers KR, Mulchandani A, editors. Affinity Biosensors: Techniques and Protocols. 1st ed. Totowa, NJ: Humana Press Inc.; 1998.

[19] Palecek E, Fojta M. DNA hybridization and damage. Anal. Chem. 2001;73(75A-83A).

[20] Rauf S, Gooding JJ, Akhtar K, Ghauri MA, Rahman M, Anwar MA, et al. Electrochemical approach of anticancer drugs-DNA interaction. J. Pharm. Biomed. Anal. 2005;37(2):205-217.

[21] Mulchandani A, Rogers K. Enzyme and microbial biosensors: Techniques and protocol. Totowa, NJ. 1st ed. Totowa, NJ: Humama Press; 1998.

[22] Yamashita T, Yamashita K, Naeshi H, Yoshikawa T, Yoshioka Y, Tsunoda S et al. Carbon nanomaterials: Efficacy and safety for nanomedicine. Materials. 2012;5:350-363.

[23] Bhakta SA, Evans E, Benavidez TE, Garcia CD. Protein adsorption onto nanomaterials for the development of biosensors and analytical devices: A review. Anal. Chim. Acta. Forthcoming. DOI: Doi:10.1016/j.aca.2014.10.031

[24] Chan CPY, Mak WC, Cheung KY, Sin KK, Yu CM, Rainer TH et al. Evidence-based point-of-care diagnostics: Current status and emerging technologies. Annu. Rev. Anal. Chem. 2013;6:191-211.

[25] Carregal-Romero S, Caballero-Díaz E, Beqa L, Abdelmonem AM, Ochs M, Hühn D et al. Multiplexed sensing and imaging with colloidal nano- and microparticles. Annu. Rev. Anal. Chem. 2013;6:53-81.

[26] Pumera M, Sanchez S, Ichinose I, Tang J. Electrochemical nanobiosensors. Sensor. Actuat. B-Chem. 2007;123(2):1195-1205.

[27] Barnes TM, van de Lagemaat J, Levi D, Rumbles G, Coutts TG, Weeks CL et al. Optical characterization of highly conductive single-wall carbon-nanotube transparent electrodes. Phys. Rev. B. 2007;75(23):23541001-22354110.

[28] Zhao J, Chen X, Xie JRH. Optical properties and photonic devices of doped carbon nanotubes. Anal. Chim. Acta. 2006;568:161-170.

[29] Wooten M, Gorski W. Facilitation of NADH electrooxidation at treated carbon nanotubes. Anal. Chem. 2010;82(4):1299-1304.

[30] Bose S, Khare RA, Moldenaers P. Assessing the strengths and weaknesses of various types of pretreatment of carbon nanotubes on the properties of polymer carbon nanotubes composites: A critical review. Polymer. 2010;51:975-993.

[31] Wang HZ, Huang ZP, Cai QJ, Kulkarni K, Chen CL, Carnahan D et al. Reversible transformation of hydrophobicity and hydrophilicity of aligned carbon nanotube arrays and buckypapers by dry processes. Carbon. 2010;48(3):868-875.

[32] Soetedjo H, Mora MF, Garcia CD. Optical properties of single-wall carbon nanotube films deposited on Si/SiO_2 wafers. Thin Solid Films. 2010;518(14):3954-3959.

[33] Qureshi A, Kang WP, Davidson JL, Gurbuz Y. Review on carbon-derived, solid-state, micro and nano sensors for electrochemical sensing applications. Diam. Relat. Mater. 2009;18(12):1401-1420.

[34] Masheter AT, Abiman P, Wildgoose GG, Wong E, Xiao L, Rees NV et al. Investigating the reactive sites and the anomalously large changes in surface pKa values of chemically modified carbon nanotubes of different morphologies. Mater. Chem. 2007;17(25):2616-2626.

[35] Iijima S. Helical microtubules of graphitic carbon. Nature. 1991;354:56-58.

[36] Hu C, Hu S. Carbon nanotube-based electrochemical sensors: Principles and applications in biomedical systems. Journal of Sensors. 2009;2009:187615. DOI: 10.1155/2009/187615.

[37] Sadik OA, Mwilu SK, Aluoch A. Smart electrochemical biosensors: From advanced materials to ultrasensitive devices. Electrochim. Acta 2010;55(14):4287-4296.

[38] Erdem A. Nanomaterials based sensor development towards electrochemical sensing of bioreactions. In: Vaseashta A, Braman E, Susmann P, editors. Technological innovation in sensing and detection of chemical, biological, radiological, nuclear threats and ecological terrorism. NATO Science for Peace and Security Series A: Chemistry and Biology. 1st ed. New York: Springer; 2012. p. 165-169.

[39] Giljohann DA, Mirkin CA. Drivers of biodiagnostic development. Nature. 2009;462:461-464.

[40] Belin T, Epron F. Characterization methods of carbon nanotubes: A review. Mat. Sci. Eng. B-Solid. 2005;119:105-118.

[41] Banks CE, Davies TJ, Wildgoose GG, Compton RG. Electrocatalysis at graphite and carbon nanotube modified electrodes: Edge-plane sites and tube ends are the reactive sites. Chem. Commun. 2005;7(21):829-841.

[42] Banks CE, Crossley A, Salter C, Wilkins SJ, Compton RG. Carbon nanotubes contain metal impurities which are responsible for the "electrocatalysis" seen at some nanotube-modified electrodes. Angew. Chem. Int. Ed. 2006;45(16):2533-2537.

[43] Pumera M, Ambrosi A, Bonanni A, Chng ELK, Poh HL. Graphene for electrochemical sensing and biosensing. TrAC Trends Anal. Chem. 2010;29(9):954-965.

[44] Ambrosi A, Pumera M. Nanographite impurities dominate electrochemistry of carbon nanotubes. Chem.-A Eur. J. 2010;16(36):10946-10949.

[45] Streeter I, Wildgoose GG, Shao LD, Compton RG. Cyclic voltammetry on electrode surfaces covered with porous layers: An analysis of electron transfer kinetics at single-walled carbon nanotube modified electrodes. Sensor. Actuat. B-Chem. 2008;133:462-466.

[46] Sims MJ, Rees NV, Dickinson EJF, Compton RG. Effects of thin-layer diffusion in the electrochemical detection of nicotine on basal plane pyrolytic graphite (BPPG) elec-

trodes modified with layers of multi-walled carbon nanotubes (MWCNT-BPPG). Sensor. Actuat. B-Chem. 2010;144(1):153-158.

[47] Henstridge MG, Dickinson EJF, Aslanoglu M, Batchelor-McAuley C, Compton RG. Voltammetric selectivity conferred by the modification of electrodes using conductive porous layers or films: The oxidation of dopamine on glassy carbon electrodes modified with multiwalled carbon nanotubes. Sensor. Actuat. B-Chem. 2010;145(1): 417-427.

[48] Li W, Jung H, Hoa ND, Kim D, Hong S-K, Kim H. Nanocomposite of cobalt oxide nanocrystals and single-walled carbon nanotubes for a gas sensor application. Sensor. Actuat. B-Chem. 2010;150(1):160-166.

[49] Ghadimi H, Tehrani RMA, Mohamed Ali AS, Mohamed N, Ab Ghani S. Sensitive voltammetric determination of paracetamol by poly (4-vinylpyridine)/multiwalled carbon nanotubes modified glassy carbon electrode. Anal. Chim. Acta. 2013;765:70-76.

[50] Gao Q, Guo Y, Liu J, Yuan X, Qi H, Zhang C. A biosensor prepared by co-entrapment of a glucose oxidase and a carbon nanotube within an electrochemically deposited redox polymer multilayer. Bioelectrochemistry. 2011;81(2):109-113.

[51] Salimi A, Compton RG, Hallaj R. Glucose biosensor prepared by glucose oxidase encapsulated sol–gel and carbon nanotube modified basal plane pyrolytic graphite electrode. Anal. Biochem. 2004;333(1):49-56.

[52] Sato N, Okuma H. Development of single-wall carbon nanotubes modified screen-printed electrode using a ferrocene-modified cationic surfactant for amperometric glucose biosensor applications. Sensor. Actuat. B-Chem. 2008;129(1):188-194.

[53] Rahman MM, Umar A, Sawada K. Development of amperometric glucose biosensor based on glucose oxidase co-immobilized with multi-walled carbon nanotubes at low potential. Sensor. Actuat. B-Chem. 2009;137(1):327-333.

[54] Deng C, Chen J, Chen X, Xiao C, Nie L, Yao S. Direct electrochemistry of glucose oxidase and biosensing for glucose based on boron-doped carbon nanotubes modified electrode. Biosens. Bioelectron. 2008;23(8):1272-1277.

[55] Li S, Zhu X, Zhang W, Xie G, Feng W. Hydrogen peroxide biosensor based on gold nanoparticles/thionine/gold nanoparticles/multi-walled carbon nanotubes – chitosan composite film-modified electrode. Appl. Surf. Sci. 2012;258(7):2802-2807.

[56] Liu Y, Yuan R, Chai Y, Tang D, Dai J, Zhong X. Direct electrochemistry of horseradish peroxidase immobilized on gold colloid/cysteine/Nafion-modified platinum disk electrode. Sensor. Actuat. B-Chem. 2006;115(1):109-115.

[57] Zhang Y, Yuan R, Chai Y, Xiang Y, Hong C, Ran X. An amperometric hydrogen peroxide biosensor based on the immobilization of HRP on multi-walled carbon nanotubes/electro-copolymerized nano-Pt-poly (neutral red) composite membrane. Biochem. Eng. J. 2010;51(3):102-109.

[58] Huang J, Li J, Yang Y, Wang X, Wu B, Anzai J-I et al. Development of an amperometric L-lactate biosensor based on L-lactate oxidase immobilized through silica sol-gel film on multi-walled carbon nanotubes/platinum nanoparticle modified glassy carbon electrode. Mater. Sci. Eng. C. 2008;28(7):1070-1075.

[59] Wang YT, Yu L, Wang J, Lou L, Du WJ, Zhu ZQ et al. A novel L-lactate sensor based on enzyme electrode modified with ZnO nanoparticles and multiwall carbon nanotubes. J. Electroanal. Chem. 2011;661(1):8-12.

[60] Goran JM, Lyon JL, Stevenson KJ. Amperometric detection of L-lactate using nitrogen-doped carbon nanotubes modified with lactate oxidase. Anal. Chem. 2011;83(21): 8123-8129.

[61] Tertiş M, Florea A, Săndulescu R, Cristea C. Carbon based electrodes modified with horseradish peroxidase immobilized in conducting polymers for acetaminophen analysis. Sensors Basel. 2013;13:4841-4854. DOI:10.3390/s130404841

[62] Baghayeri M, Namadchian M. Fabrication of a nanostructured luteolin biosensor for simultaneous determination of levodopa in the presence of acetaminophen and tyramine: Application to the analysis of some real samples. Electrochim. Acta. 2013;108(1):22-31.

[63] Amiri-Aref M, Raoof JB, Ojani R. A highly sensitive electrochemical sensor for simultaneous voltammetric determination of noradrenaline, acetaminophen, xanthine and caffeine based on a flavonoid nanostructured modified glassy carbon electrode. Sensor. Actuat. B-Chem. 2014;192:634-641.

[64] Babaei A, Afrasiabi M, Babazadeh M. A glassy carbon electrode modified with multi-walled carbon nanotube/chitosan composite as a new sensor for simultaneous determination of acetaminophen and mefenamic acid in pharmaceutical preparations and biological samples. Electroanal. 2010;22(15):1743-1749.

[65] Shahrokhian S, Asadian E. Simultaneous voltammetric determination of ascorbic acid, acetaminophen and isoniazid using thionine immobilized multi-walled carbon nanotube modified carbon paste electrode. Electrochim. Acta. 2010;55(3):666-672.

[66] Tertiş M, Florea A, Feier B, Marian IO, Silaghi-Dumitrescu L, Cristea A et al. Electrochemical impedance studies on single and multi-wall carbon nanotubes-polymer nanocomposites for biosensors development. J. Nanosci. Nanotechnolog. 2015;15(5): 3383-3393.

[67] Bustos-Ramírez K, Martínez-Hernández AL, Martínez-Barrera G, de Icaza M, Castaño VM, Velasco-Santos C. Covalently bonded chitosan on graphene oxide via redox reaction. Materials. 2013;6:911-926.

[68] Guoqing X, Hwang W, Kim N, Cho SM, Chae H. A graphene sheet exfoliated with microwave irradiation and interlinked by carbon nanotubes for high-performance transparent flexible electrodes. Nanotechnol. 2010;21:405201.

[69] Lim CX, Hoh HY, Ang PK, Loh KP. Direct voltammetric detection of DNA and pH sensing on epitaxial graphene: An insight into the role of oxygenated defects. Anal. Chem. 2010;82(17):7387-7393.

[70] Novoselov KS, Geim AK, Morozov SV, Jiang D, Zhang Y, Dubonos SV et al. Electric field effect in atomically thin carbon films. Science. 2004;306(5696):666-669.

[71] Bai J, Huang Y. Fabrication and electrical properties of graphene nanoribbons. Mat. Sci. Eng. R. 2010;70(3-6):341-353.

[72] Kim WS, Moon SY, Bang SY, Choi BG, Ham H, Sekino T et al. Fabrication of graphene layers from multiwalled carbon nanotubes using high DC pulse. Appl. Phys. Lett. 2009;95:083103-183103.

[73] He Q, Sudibya HG, Yin Z, Wu S, Li H, Boey F et al. Centimeter-long and large-scale micropatterns of reduced graphene oxide films: Fabrication and sensing applications. ACS Nano. 2010;4(6):3201-3208.

[74] Pham VH, Cuong TV, Hur SH, Shin EW, Kim JS, Chung JS et al. Fast and simple fabrication of a large transparent chemically-converted graphene film by spray-coating. Carbon. 2010;48:1945-1951.

[75] Kang X, Wang J, Wu H, Liu J, Aksay IA, Lin Y. A graphene-based electrochemical sensor for sensitive detection of paracetamol. Talanta. 2010;81:754-759.

[76] Liu YX, Dong XC, Chen P. Biological and chemical sensors based on graphene materials. Chem. Soc. Rev. 2012;41(6):2283-2307.

[77] Park JW, Lee SJ, Choi EJ, Kim J, Song JY, Gu MB. An ultra-sensitive detection of a whole virus using dual aptamers developed by immobilization-free screening. Biosens. Bioelectron. 2014;51:324-329.

[78] Biju V. Chemical modifications and bioconjugate reactions of nanomaterials for sensing, imaging, drug delivery and therapy. Chem. Soc. Rev. 2014;43:744-764.

[79] Maiti UN, Lee WJ, Lee JM, Oh Y, Kim JY, Kim E et al. 25th anniversary article: Chemically modified/doped carbon nanotubes & graphene for optimized nanostructures & nanodevices. Adv. Mater. 2014;26(1):40-67.

[80] Bekyarova E, Itkis ME, Ramesh P, Berger C, Sprinkle M, de Heer WA et al. Chemical modification of epitaxial graphene: Spontaneous grafting of aryl groups. J. Am. Chem. Soc. 2009;131(4):1336-1337.

[81] Liu S, Xing XR, Yu JH, Lian WJ, Li J, Cui M et al. A novel label-free electrochemical aptasensor based on graphene–polyaniline composite film for dopamine determination. Biosens. Bioelectron. 2012;36(1):186-191.

[82] Yan M, Sun GQ, Liu F, Lu JJ, Yu JH, Song XR. An aptasensor for sensitive detection of human breast cancer cells by using porous GO/Au composites and porous PtFe alloy as effective sensing platform and signal amplification labels. Anal. Chim. Acta. 2013;798:33-39.

[83] Zhao J, Chen GF, Zhu L, Li GX. Graphene quantum dots-based platform for the fabrication of electrochemical biosensors. Electrochem. Commun. 2011;13(1):31-33.

[84] Wang L, Xu M, Han L, Zhou M, Zhu CZ, Dong SJ. Graphene enhanced electron transfer at aptamer modified electrode and its application in biosensing. Anal. Chem. 2012;84(17):7301-7307.

[85] Yan GP, Wang YH, He XX, Wang KM, Liu JQ, Du YD. A highly sensitive label-free electrochemical aptasensor for interferon-gamma detection based on graphene controlled assembly and nuclease cleavage-assisted target recycling amplification. Biosens. Bioelectron. 2013;44:57-63.

[86] Loo AH, Bonanni A, Pumera M. Thrombin aptasensing with inherently electroactive graphene oxide nanoplatelets as labels. Nanoscale. 2013;5:4758-4762.

[87] Tertis M, Hosu O, Fritea L, Farcau C, Cernat A, Sandulescu R et al. A novel label-free immunosensor based on activated graphene oxide for acetaminophen detection. Electroanal. 2015; 27(3):638-647. DOI: 10.1002/elan.201400583

[88] Ghorbani-Bidkorbeh F, Shahrokhian S, Mohammadi A, Dinarvand R. Simultaneous voltammetric determination of tramadol and acetaminophen using carbon nanoparticles modified glassy carbon electrode. Electrochim. Acta. 2010;55(8):2752-2759.

[89] Weingart J, Vabbilisetty P, Sun XL. Membrane mimetic surface functionalization of nanoparticles: Methods and applications. Adv. Colloid. Interfac. Sci. 2013;197-198:68-84.

[90] Rocha-Santos TAP. Sensors and biosensors based on magnetic nanoparticles. TrAC Trends Anal. Chem. 2014;62:28-36.

[91] Rusling JF, Sotzing G, Papadimitrakopoulos F. Designing nanomaterials-enhanced electrochemical immunosensors for cancer biomarker proteins. Bioelectrochemistry. 2009;76(1-2):189-194.

[92] Aguilar-Arteaga K, Rodriguez JA, Barrado E. Magnetic solids in analytical chemistry: A review. Anal. Chim. Acta. 2010;674(2):157-165.

[93] Xu Y, Wang E. Electrochemical biosensors based on magnetic micro/nanoparticles. Electrochim. Acta. 2012;84:62-73.

[94] Nomura A, Shin S, Mehdi OO, Kauffmann JM. Preparation, characterization, and application of an enzyme-immobilized magnetic microreactor for flow injection Analysis. Anal. Chem. 2004;76(18):5498-5502.

[95] Zhang XF, Dong XL, Huang H, Lv B, Zhu KG, Lei JP et al. Synthesis, structure and magnetic properties of SiO_2-coated Fe nanocapsules. Mater. Sci. Eng. A. 2007;454-455:211-215.

[96] Gruttner C, Rudershausen S, Teller J.J. Improved properties of magnetic particles by combination of different polymer materials as particle matrix. J. Magn. Magn. Mater. 2001;225(1-2):1-7.

[97] Deere J, Magner E, Wall JG, Hodnett BK. Adsorption and activity of cytochrome c on mesoporous silicates. Chem. Commun. 2001;5:465-466.

[98] Yang L, Zhang X, Ye M, Jiang J, Yang R, Fu T. Aptamer-conjugated nanomaterials and their applications. Adv. Drug Deliv. Rev. 2011;63(14-15):1361-1370.

[99] Berry CC, Curtis ASG. Topical review: Functionalisation of magnetic nanoparticles for applications in biomedicine. J. Phys. D: Appl. Phys. 2003;36(13):R198–R206.

[100] Hervas M, Lopez MA, Escarpa A. Simplified calibration and analysis on screen-printed disposable platforms for electrochemical magnetic bead-based inmunosensing of zearalenone in baby food samples. Biosens. Bioelectron. 2010;25(7):1755-1760.

[101] Yang Z, Zhang C, Zhang J, Bai W. Potentiometric glucose biosensor based core-shell Fe_3O_4-enzyme-polypyrrole nanoparticles. Biosens. Bioelectron. 2014;51:268-273.

[102] Chen X, Zhu J, Chen Z, Xu C, Wang Y, Yao C. A novel bienzyme glucose biosensor based on three layers Au-Fe_3O_4@SiO_2 magnetic nanocomposite. Sensor. Actuat. B-Chem. 2011;159(1):220-228.

[103] Baby TT, Ramaprabhu R. SiO_2 coated Fe_3O_4 magnetic nanoparticle dispersed multi-walled carbon nanotubes based amperometric glucose biosensor. Talanta. 2010;80(5): 2016-2022.

[104] Zhou H, Gan N, Li T, Cao Y, Zeng S, Zheng L et al. The sandwich-type electroluminescence immunosensor for α-fetoprotein based on enrichment by Fe_3O_4-Au magnetic nano probes and signal amplification by CdS-Au composite nanoparticles labeled anti-AFP. Anal. Chim. Acta. 2012;746:107-113.

[105] Li J, Xu Q, Wei X, Hao Z. Electrogenerated chemiluminescence immunosensor for Bacillus thuringiensis Cry1Ac based on Fe_3O_4@Au nanoparticles. J. Agric. Food Chem. 2013;61(7):1435-1440.

[106] Zamfir L-G, Geana I, Bourigua S, Rotariu L, Bala C, Errachid A, et al. Highly sensitive label-free immunosensor for ochratoxin A based on functionalized magnetic nanoparticles and EIS/SPR detection. Sensor. Actuat. B-Chem. 2011;159(1):178-184.

[107] Yola ML, Eren T, Atar N. A novel and sensitive electrochemical DNA biosensor based on Fe@Au nanoparticles decorated graphene oxide. Electrochim. Acta. 2014;125:38-47.

[108] Gan N, Yang X, Xie D, Wu Y, Wen W. A disposable organophosphorus pesticides enzyme biosensor based on magnetic composite nano-particles modified screen printed carbon electrode. Sensors. 2010;10(1):625-638.

[109] Li J, Gao H, Chen Z, Wei X, Yang CF. An electrochemical immunosensor for carcinoembryonic antigen enhanced by self-assembled nanogold coatings on magnetic particles. Anal. Chim. Acta. 2010;665(1):98-104.

[110] Yang X, Wu F, Chen D-Z, Lin H-W. An electrochemical immunosensor for rapid determination of clenbuterol by using magnetic nanocomposites to modify screen

printed carbon electrode based on competitive immunoassay mode. Sensor. Actuat. B-Chem. 2014;192:529-535.

[111] Xin Y, Fu-Bing X, Hong-Wei L, Feng W, Di-Zhao C, Zhao-Yang W. A novel H_2O_2 biosensor based on Fe_3O_4-Au magnetic nanoparticles coated horseradish peroxidase and grapheme sheets-Nafion film modified screen-printed carbon electrode. Electrochim. Acta. 2013;109:750-755.

[112] Chen D, Deng J, Liang J, Xie J, Hue C, Huang K. A core-shell molecularly imprinted polymer grafted onto a magnetic glassy carbon electrode as a selective sensor for the determination of metronidazole. Sensor. Actuat. B-Chem. 2013;183:594-600.

[113] Prakash A, Chandra S, Bahadur D. Structural, magnetic, and textural properties of iron oxide-reduced graphene oxide hybrids and their use for the electrochemical detection of chromium. Carbon. 2012;50(11):4209-4212.

[114] Hu Y, Zang Z, Zhang H, Luo L, Yao S. Selective and sensitive molecularly imprinted sol-gel film-based electrochemical sensor combining mecaptoacetic acid modified PbS nanoparticles with Fe_3O_4@Au-multi-walled carbon nanotubes-chitosan. J. Solid State Electrochem. 2012;16(3):857–867.

[115] Arvand M, Hassannezhad M. Magnetic core-shell Fe_3O_4@SiO_2/MWCNT nanocomposite modified carbon paste electrode for amplified electrochemical sensing of uric acid. Mater. Sci. Eng. C. 2014;36:161-167.

[116] Hosu O, Tertis M, Cristea C, Sandulescu R. Protein G magnetic beads based immunosensor for sensitive detection of acetaminophen. Farmacia. 2015;63(1):143-146.

[117] Taleat Z, Cristea C, Marrazza G, Sandulescu R. Electrochemical sandwich immunoassay for the ultrasensitive detection of human MUC1 cancer biomarker. Int. J. Electrochem. 2013; Article ID 740265, http://dx.doi.org/10.1155/2013/740265

[118] Jia J, Wang B, Wu A, Cheng G, Li Z, Dong S. A method to construct a third-generation horseradish peroxidase biosensor: Self-assembling gold nanoparticles to three-dimensional sol-gel network. Anal. Chem. 2002;74:2217-2223.

[119] Di J, Shen C, Peng S, Tu Y, Li S. A one-step method to construct a third-generation biosensor based on horseradish peroxidase and gold nanoparticles embedded in silica sol–gel network on gold modified electrode. Anal. Chim. Acta. 2005;553:196-200.

[120] Xu Z, Chen X, Dong S. Electrochemical biosensors based on advanced bioimmobilization matrices. TrAC Trends Anal. Chem. 2006;25(9):899-908.

[121] Yang L, Zhang X, Ye M, Jiang L, Yang R, Fu T et al. Aptamer-conjugated nanomaterials and their applications. Adv. Drug Deliv. Rev. 2011;63(14-15):1361-1370.

[122] Merkoçi A. Nanoparticles-based strategies for DNA, protein and cell sensors. Biosens. Bioelectron. 2010;26(4):1164-1177.

[123] Yin T, Qin W. Applications of nanomaterials in potentiometric sensors. TrAC Trends Anal. Chem. 2013;51:79-86.

[124] Taton TA, Mirkin CA, Letsinger RL. Scanometric DNA array detection with nanoparticle probes. Science. 2000;289(5485):1757-1760.

[125] Wang J, Polsky R, Xu D. Silver-enhanced colloidal gold electrochemical stripping detection of DNA hybridization. Langmuir. 2001;17(19):5739-5741.

[126] Wang J, Xu D, Kawde A-N, Polsky R. Metal nanoparticle-based electrochemical stripping potentiometric detection of DNA hybridization. Anal. Chem. 2001;73(22): 5576-5581.

[127] Wang J, Liu G, Merkoçi A. Electrochemical coding technology for simultaneous detection of multiple DNA targets. J. Am. Chem. Soc. 2003;125(11):3214-3215.

[128] Chen G, Tong H, Gao T, Chen Y, Li G. Direct application of gold nanoparticles to one-pot electrochemical biosensors. Anal. Chim. Acta. 2014;849:1-6.

[129] Malon A, Vigassy T, Bakker E, Pretsch E. Potentiometry at trace levels in confined samples: Ion-selective electrodes with subfemtomole detection limits. J. Am. Chem. Soc. 2006;128(25):8154-8155.

[130] Rubinova N, Chumbimuni-Torres K, Bakker E. Solid-contact potentiometric polymer membrane microelectrodes for the detection of silver ions at the femtomole level. Sensor. Actuat. B-Chem. 2007;121(1):135-141.

[131] Chumbimuni-Torres K, Dai Z, Rubinova N, Xiang Y, Pretsch E, Wang J et al. Potentiometric biosensing of proteins with ultrasensitive ion-selective microelectrodes and nanoparticle labels. J. Am. Chem. Soc. 2006;128(42):13676-13677.

[132] Willner I, Baron R, Willner B. Growing metal nanoparticles by enzymes. Adv. Mater. 2006;18:1109-1120.

[133] Ngeontae W, Janrungroatsakul W, Maneewattanapinyo P, Ekgasit S, Aeungmaitrepirom W, Tuntulani TW. Novel potentiometric approach in glucose biosensor using silver nanoparticles as redox marker. Sensor. Actuat. B-Chem. 2009;137(1):320-326.

[134] Chumbimuni-Torres KY, Wang J. Nanoparticle-induced potentiometric biosensing of NADH at copper ion-selective electrodes. Analyst. 2009;134(8):1614-1617.

[135] Wu J, Chumbimuni-Torres KY, Galik M, Thammakhet C, Haake DA, Wang J. Potentiometric detection of DNA hybridization using enzyme-induced metallization and a silver ion selective electrode. Anal. Chem. 2009;81(24):10007-10012.

[136] Florea A, Taleat Z, Cristea C, Mazloum-Ardakani M, Săndulescu R. Label free MUC1 aptasensors based on electrodeposition of gold nanoparticles on screen printed electrodes. Electrochem. Comm. 2013;33:127-130.

[137] Mousty C. Sensors and biosensors based on clay-modified electrodes-new trends. Appl. Clay Sci. 2004;27:159-177.

[138] Direnzo F, Cambon H, Dutartre R. A 28-year-old synthesis of micelle-templated mesoporous silica. Micropor. Mater. 1997;10:283-286.

[139] Qhobosheane M, Santra S, Zhang P, Tan WH. Biochemically functionalized silica nanoparticles. Analyst. 2001;126:1274-1278.

[140] Yan L, Chen X. Nanomaterials for drug delivery. In: Tjong SC, editor. Nanocrystalline Materials. 2nd ed. London: Elsevier; 2014. p. 221-268.

[141] Cheng SH, Lee CH, Yang CS, Tseng FG, Mou CY, Lo LW. Mesoporous silica nanoparticles functionalized with an oxygen-sensing probe for cell photodynamic therapy: Potential cancer theranostics. J. Mater. Chem. 2009;19(9):1252-1257.

[142] Yáñez-Sedeño P, Agüí L, Villalonga R, Pingarrón JM. Biosensors in forensic analysis. A review. Anal. Chim. Acta. 2014;823:1-19.

[143] Cosnier S, Mousty C, Cui XQ, Yang X, Dong S. Specific determination of As(V) by an acid phosphatase polyphenol oxidase biosensor. Anal Chem. 2006;78(14):4985-4989.

[144] Liu YX, Wei WZ. Layer-by-layer assembled DNA functionalized single-walled carbon nanotube hybrids for arsenic(III) detection. Electrochem. Comm. 2008;10(6): 872-875.

[145] Shan D, Mousty C, Cosnier S. Subnanomolar cyanide detection at polyphenol oxidase/clay biosensors. Anal. Chem. 2004;76(1):178-183.

[146] Santos L, Martin P, Ghilane L, Lacaze P-C, Randriamahazaka H, Abrantes LM et al. Electrosynthesis of well-organized nanoporous poly(3,4-ethylenedioxythiophene) by nanosphere lithography. Electrochem. Commun. 2010;12(7):872-875.

[147] Cernat A, Griveau S, Richard C, Bedioui, Săndulescu R. Horseradish peroxidase nanopatterned electrodes by click chemistry: Application to the electrochemical detection of paracetamol. Electroanal. 2013;25(6):1369-1372.

[148] Cernat A, Le Goff A, Holzinger M, Sandulescu R, Cosnier S. Micro- to nanostructured poly(pyrrole-nitrilotriacetic acid) films via nanosphere templates: Applications to 3D enzyme attachment by affinity interactions. Anal. Bioanal Chem. 2014;406(4): 1141-1147.

[149] Stetter JR, Penrose WR, Yao S. Sensors, chemical sensors, electrochemical sensors, and ECSJ. Electrochem. Soc. 2003;150(2):S11-S16.

[150] Ravi Kumar MNV, Kumar N, Kashyap N. Hydrogels for pharmaceutical and biomedical applications. Crit. Rev. Ther. Drug Carrier Syst. 2005;22(2):107-150.

[151] He X, Wei B, Mi Y. Aptamer based reversible DNA induced hydrogel system for molecular recognition and separation. Chem. Commun. 2010;46(34):6308-6310. DOI: DOI:10.1039/c0cc01392g.

[152] Hasanzadeh M, Shadjou N, de la Guardia M. Electrochemical biosensing using hydrogel nanoparticles. TrAC Trends Anal. Chem. 2014;52:11-19.

[153] Sriwichai S, Baba A, Phanichphant S, Shinbo K, Kato K, Kaneko F. Electrochemically controlled surface plasmon resonance immunosensor for the detection of human im-

munoglobulin G on poly(3-aminobenzoic acid) ultrathin films. Sensor. Actuator. B-Chem. 2010;147(1):322-329.

[154] Taleat Z, Cristea C, Marrazza G, Mazloum-Ardakani M, Săndulescu R. Electrochemical immunoassay based on aptamer–protein interaction and functionalized polymer for cancer biomarker detection. J. Electroanal. Chem. 2014;717-718:119-124.

[155] Cernat A, Tertis M, Griveau S, Bedioui F, Săndulescu R. New modified electrodes with HRP immobilized in polymeric films for paracetamol analysis. Farmacia. 2012;1:1-12.

[156] Cernat A, Griveau S, Martin P. Lacroix JC, Farcau C, Săndulescu R et al. Electrografted nanostructured platforms for click chemistry. Electrochem. Comm. 2012;23:141-144.

[157] Sharma PS, Pietrzyk-Le A, D'Souza F, Kutner W. Electrochemically synthesized polymers in molecular imprinting for chemical sensing. Anal. Bioanal. Chem. 2012;402:3177-3204.

[158] Yan H, Row KH. Characteristic and synthetic approach of molecularly imprinted polymer. Int. J. Mol. Sci. 2006;7:155-178.

[159] Moreira FTC, Dutra RAF, Noronha JPC, Sales MGF. Electrochemical biosensor based on biomimetic material for myoglobin detection. Electrochim. Acta. 2013;107:481-487.

[160] Malitesta C, Mazzotta E, Picca RA, Poma A, Chianella I, Piletsky SA. MIP sensors – the electrochemical approach. Anal. Bioanal. Chem. 2012;402:1827-1846.

[161] Irshad M, Iqbal N, Mujahid A, Afzal A, Hussain T, Sharif A et al. Molecularly imprinted nanomaterials for sensor applications. Nanomaterials. 2013;3:615-637.

[162] Herlem G, Gharbi T. Electrodeposition of insulating thin film polymers from aliphatic monomers as transducers for biosensor applications. In: Serra PA, editor. Biosensors – Emerging Materials and Applications. 1st edition. InTech; 2011. DOI: 10.5772/17584

[163] Iacob BC, Bodoki E, Florea A, Bodoki AE, Oprean R. Simultaneous enantiospecific recognition of several β-blocker enantiomers using molecularly imprinted polymer-based electrochemical sensor. Anal. Chem. Forthcoming. DOI: 10.1021/ac504036m

[164] Ge Y, Turner APF. Too large to fit? Recent developments in macromolecular imprinting. Trends Biotechnol. 2008;26:218-224.

[165] Takeuchi T, Hishiya T. Molecular imprinting of proteins emerging as a tool for protein recognition. Org. Biomol. Chem. 2008;6:2459-2467.

[166] Hussain M, Wackerlig J, Lieberzeit PA. Biomimetic strategies for sensing biological species. Biosensors. 2013;3:89-107.

[167] Ding XC, Heiden PA. Recent developments in molecularly imprinted nanoparticles by surface imprinting techniques. Macromol. Mat. Engineer. 2014;299:268-289.

[168] Chen Y, Chen L, Bi R, Xu L, Liu Y. A potentiometric chiral sensor for L-phenylalanine based on crosslinked polymethylacrylic acid-polycarbazole hybrid molecularly imprinted polymer. Anal. Chim. Acta. 2012;754:83-90.

[169] Prasad BB, Pandey I, Srivastava A, Kumar D, Tiwari MP. Multiwalled carbon nanotubes-based pencil graphite electrode modified with an electrosynthesized molecularly imprinted nanofilm for electrochemical sensing of methionine enantiomers. Sensor. Actuat. B-Chem. 2013;176:863-874.

[170] Prasad BB, Pandey I. Metal incorporated molecularly imprinted polymer-based electrochemical sensor for enantio-selective analysis of pyroglutamic acid isomers. Sensor. Actuat. B-Chem. 2013;186:407-416.

[171] Granot E, Tel-Vered R, Lioubashevski O, Willner I. Stereoselective and enantioselective electrochemical sensing of monosaccharides using imprinted boronic acid-functionalized polyphenol films. Adv. Funct. Mater. 2008;18:478-484.

[172] Siriviriyanun A, Imae T, Nagatani N. Electrochemical biosensors for biocontaminant detection consisting of carbon nanotubes, platinum nanoparticles, dendrimers, and enzymes. Anal. Biochem. 2013;443(2):169-171.

[173] Ahmed MU, Hossain MM, Tamiya E. Electrochemical biosensors for medical and food applications. Electroanal. 2008;20:616-626.

[174] Liu G, Riechers SL, Mellen MC, Lin Y. Sensitive electrochemical detection of enzymatically generated thiocholine at carbon nanotube modified glassy carbon electrode. Electrochem. Commun. 2005;7(11):1163-1169.

[175] Siriviriyanun A, Imae T. Advantages of electrodes with dendrimer-protected platinum nanoparticles and carbon nanotubes for electrochemical methanol oxidation. Phys. Chem. Chem. Phys. 2013;15:4921-4929.

[176] Dalkiran B, Kacar C, Erden PE, Kilic E. Amperometric xanthine biosensors based on chitosan-Co_3O_4-multiwall carbon nanotube modified glassy carbon electrode. Sensor. Actuat. B-Chem. 2014;200:83-91.

[177] Jiang BY, Wang M, Chen Y, Xie JQ, Xiang Y. Highly sensitive electrochemical detection of cocaine on graphene/AuNP modified electrode via catalytic redox-recycling amplification. Biosens. Bioelectron. 2012;32(1):305-308.

[178] Hwa K-Y, Subramani B. Synthesis of zinc oxide nanoparticles on graphene-carbon nanotube hybrid for glucose biosensor applications. Biosens. Bioelectron. 2014;62:127-133.

[179] Lakshmi D, Bossi A, Whitcombe MJ, Chianella I, Fowler SA, Subrahmanyam S et al. Electrochemical sensor for catechol and dopamine based on a catalytic molecularly imprinted polymer-conducting polymer hybrid recognition element. Anal. Chem. 2009;81:3576-3584.

[180] Lei R, Guo C, Xiong H, Dong C, Zhang X, Wang S. A novel electrochemical sensor for β_2-agonists with high sensitivity and selectivity based on surface molecularly imprinted sol-gel doped with antimony-doped tin oxide. Electroanal. 2014;26:1004-1012.

[181] Fatoni A, Numnuam A, Kanatharana P, Limbut W, Thavarungkul P. A novel molecularly imprinted chitosan–acrylamide, graphene, ferrocene composite cryogel biosensor used to detect microalbumin. Analyst. 2014;139:6160-6167.

[182] Du D, Chen S, Cai J, Tao Y, Tu H, Zhang A. Recognition of dimethoate carried by bilayer electrodeposition of silver nanoparticles and imprinted poly-o-phenylenediamine. Electrochim. Acta. 2008;53:6589-6595.

[183] Kan X, Liu T, Zhou H, Li C, Fang B. Molecular imprinting polymer electrosensor based on gold nanoparticles for theophylline recognition and determination. Microchim. Acta. 2010;171:423-429.

Chapter 2

Graphene — A Platform for Sensor and Biosensor Applications

Nada F. Atta, Ahmed Galal and Ekram H. El-Ads

Additional information is available at the end of the chapter

http://dx.doi.org/10.5772/60676

Abstract

Graphene, mother of all carbon materials, has opened up new era of exploration due to its unique properties. Graphene, one-atom thick, exhibits a unique chemical structure and outstanding electronic, optical, thermal, and mechanical properties that made it compelling for various engineering applications. Graphene and graphene-based materials are promising candidates for fabricating state-of-the-art nano-scale sensors and biosensors. They featured with good conductivity and large specific surface area thereby; graphene-based sensors/biosensors performed well with good accuracy, rapidness, high sensitivity and selectivity, low detection limits, and long-term stability. They are ideally used as gas sensors, electrochemical sensors for heavy metal ions, immunosensors and dihydronicotinamide dinucleotide NADH, DNA, catecholamine neurotransmitters, paracetamol, glucose, H_2O_2, hemoglobin, and myoglobin biosensors. This chapter reviews the applications of graphene in nano-technology since it came to the field particularly in sensing and biosensing applications. It updates the reader with the scientific progress of the current use of graphene as sensors and biosensors. There is still much room for the scientific research and application development of graphene-based theory, materials, and devices. Despite the vast amount of research already conducted on graphene for various applications, the field is still growing and many questions remain to be answered.

Keywords: Graphene, Graphene-based materials, Nano-composites, Sensors, Bio-sensors

1. Introduction

Carbon is an abundant element playing a major role in the chemistry of life [1]. Some of the advantageous properties of carbon-based electrodes include wide potential windows, fairly inert electrochemistry, and good electrocatalytic activity for many redox reactions [2]. Various carbon allotropes including diamond (the carbon atoms are bonded together in a tetrahedral lattice arrangement, 3D), graphite (the carbon atoms are bonded together in sheets of a hexagonal lattice, 3D), graphene (2D), carbon nanotubes (single-walled carbon nanotubes (SWCNT), multi-walled carbon nanotubes (MWCNT), 1D), and fullerenes (the carbon atoms are bonded together in spherical, tubular, or ellipsoidal formations, 0D) have very different physical and electrochemical properties from each other (see Table 1 and Figure 1) [2-4]. The advents of sp^2 and/or sp^3 hybridized structures, such as fullerenes, carbon nanotubes, and conducting diamond, provide a route for surface modification and are very beneficial to electrochemical research, especially electrocatalysis [2]. The former 3D allotropes have been known and widely used for centuries whereas fullerenes and nanotubes have been only discovered and studied in the last two decades. With the exception of diamond, it is possible to think of fullerenes, nanotubes, and graphite as different structures built from the same hexagonal array of sp^2 carbon atoms, namely, graphene [3].

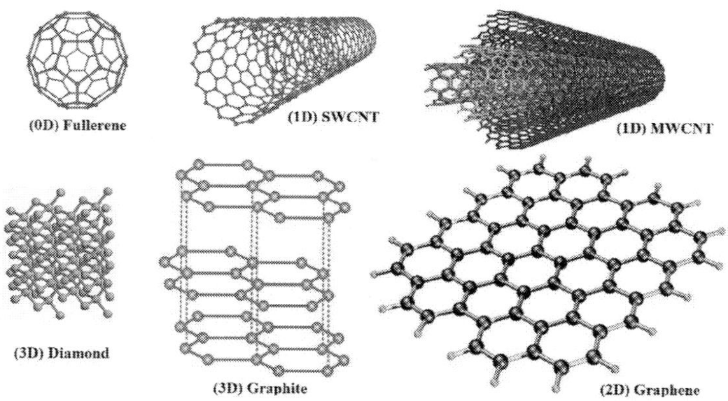

Figure 1. Different allotropes of carbon.

Graphene consists of a single sheet of carbon atoms that bind with each other in a honeycomb lattice. Graphene sheets are one-atom thick and 2D layers of sp^2-bonded carbon [4]. It is the basic building block for graphitic materials of all other dimensionalities. Graphene can be wrapped up into spherical shape to form 0D fullerenes (buckyballs), rolled into a cylindrical shape to form 1D nanotubes or stacked in layers to form 3D graphite as illustrated in Figure 2 [1, 3, 5-8]. The electronic structure of graphene sheets, individual layers of graphite, was first discussed by Wallace in 1947 [2]. Boehm and co-workers separated thin lamellae of carbon by heating and chemical reduction of graphite oxide in 1962 [6]. Until 2004, single layers of graphite were believed to be thermodynamically unstable under ambient conditions [2, 6]. Since the historical observation of single layer graphene in 2004 by two Nobel Laureates in physics, Prof. Andre Geim and Prof. Konstantin Novoselov, this atomically thin carbon film has received ever increasing attention and become the hottest topics and a rapidly "rising star" on the horizon of materials science and condensed-matter physics attracting enormous interests [8-10]. This revolutionary discovery has added a new dimension of research in the fields of physics, chemistry, biotechnology, and materials science [6]. Technologists and materials scientists have rapidly grabbed some of the assets of graphene and are already exploring the ways of incorporating graphene into applied devices and materials [3].

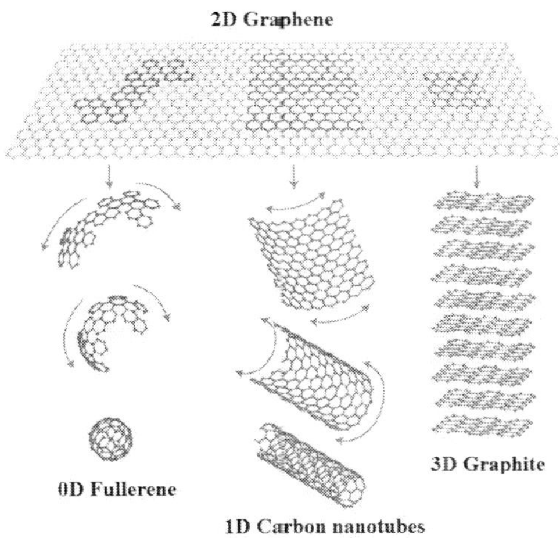

Figure 2. Basis of all graphitic forms. Graphene is a 2D building material for carbon materials of all other dimensionalities. It can be wrapped up into 0D buckyballs, rolled into 1D nanotubes or stacked into 3D graphite [6].

Dimension	0D	1D	2D	3D
Isomer	Fullerene	Nanotube	Graphene	Diamond
Hybridization	sp^2	sp^2	sp^2	sp^3
Density (g/cm³)	1.72	1.2-2.0	2.26	3.515
Bond length (Å)	1.40 (C=C) 1.46 (C-C)	1.44 (C=C)	1.42 (C=C)	1.54 (C-C)
Electronic properties	Semiconductor E_g = 1.9 eV	Meta/semiconductor E_g =~ 0.3-1.1 eV	Zero-gap semiconductor	Insulator

Table 1. Important parameters of carbon materials of different dimensionalities [6].

2. Graphene: Promising material for different applications

Graphene, a monolayer of sp^2-bonded carbon atoms arranged in a honeycomb lattice, has attracted tremendous attention from both the theoretical and experimental scientific communities in recent years because of its unique nanostructure and extraordinary properties [11, 12]. It has become a novel and very promising material for nanoelectronics, nanocomposites, optoelectronic devices, electrochemical supercapacitor devices, fabricated field-effect transistors, drug delivery systems, solar cells, memory devices, and constructed ultrasensitive chemical sensors such as pH sensors, gas sensors, biosensors, etc. [6, 11, 12]. The rapid adoption of graphene as a material of interest lies in its actual availability by a range of techniques and methods and may also be principally because monolayer and few-layer graphene possess a diverse set of unusual properties. These properties happen to be matching the short-comings of other materials, such as carbon nanotubes, graphite, or indium tin oxide (ITO) that have been studied and used for some time [3]. In the following points, we will give an exhaustive account of the unique physical and chemical properties of graphene.

- Graphene implies large theoretical specific surface area (~ 2630 m^2 g^{-1}, for single-layer graphene), exhibiting high electrocatalytic activities and ultrahigh loading capacity for biomolecules and drugs [6, 8, 13].
- Graphene is a unique bendable material with excellent mechanical flexibility and high crystallographic quality. It is strictly 2D and stable under ambient conditions [6, 11, 14].
- Graphene exhibits high optical transparency [6, 11] with an absorption of ~ 2.3% toward visible light [15]. Indeed, in the visible range, thin graphene films have a transparency that decreases linearly with the film thickness. For 2-nm thick films, the transmittance is higher than 95% and remains above 70% for 10-nm thick films. This optical characteristic combined with the excellent conductivity of graphene-based materials holds promises as a replacement to the cost-raising standard ITO [3].
- The breaking strength of graphene is 200 times higher than steel making it the strongest material ever tested. Graphene exhibits high values of Young's modulus (~1.1 Tpa) [6, 11].
- It is one of the thinnest known materials. The carbon atoms are densely packed in a honeycomb crystal lattice with a bond length of 0.141 nm. Different research groups have measured the thickness of graphene from 0.35 nm to 1.00 nm. Novoselov et al. have determined platelet thicknesses of 1.00–1.60 nm [6].
- It is well established that the superior properties of graphene are associated with its single layer [6]. Figure 3 (a) shows the honeycomb lattice of graphene that is formed by a triangular lattice with a basis of 2 atoms per unit cell. Each atom has four valence electrons: one *s* and three *p* orbitals. The *s* and two *p* orbitals hybridize to form strong covalent bonds in the plane. The out of plane *p* orbital contributes to the conductivity [1, 3]. The hexagonal structure of graphene possesses an alternate double bond arrangement that makes it perfectly conjugated in sp^2 hybridization. In this case, its p_x and p_y orbitals contain one electron each and the remaining p_z has only one electron. This p_z orbital overlaps with the

p_z orbital of a neighbor carbon atom to form a π-bond while the remaining orbitals form σ-bonds with other neighboring carbons. The π-electrons in graphene are delocalized and are largely responsible for its conduction properties while π-orbitals are the most important for determining the solid state properties of graphene [4]. Thus, graphene has extraordinary electronic transport properties and high electrocatalytic activities owing to its subtle electronic characteristics and attractive π–π interaction [12].

- It exhibits excellent room temperature electrical conductivity (~ 7200 $S.m^{-1}$) [6, 11] with amazing intrinsic mobilities of ~ 200,000 $cm^2V^{-1}s^{-1}$. The room temperature thermal conductivity values have been determined to be ~ $(4.84 \pm 0.44)\times10^3$ to $(5.30 \pm 0.48)\times10^3$ WmK^{-1} for single-layer graphene [5, 6, 16].
- One of the factors that made graphene so attractive in the research field is its low energy dynamics of electrons. The 2D crystal of sp^2 hybridized carbon is a zero band gap semiconductor in which the π and π* bands touch in a single point at the Fermi energy at the corner of the Brillouin zone and close to this so-called Dirac point, the bands display a linear dispersion (see Figure 3(b)) [1]. This topology of the bands gives rise to exotic electronic transport properties – the charge carriers behave like relativistic particles – which manifest themselves in unusual phenomena such as room-temperature anomalous quantum Hall effect and electrons behaving like massless (zero effective mass) Dirac fermions [2, 10, 16]. The ballistic charge carrier transport at 300 K and at high charge carrier concentrations makes graphene also interesting for applications in electronic devices [14].
- Graphene exhibits strong adsorptive capability [12] since every atom in a graphene sheet is a surface atom thus molecular interaction and electron transport through graphene can be highly sensitive to adsorbed molecules [17]. The unique electrochemical responses of graphene to target molecules are from the planar geometric structure and special electronic character of graphene [10].
- Easy synthesis, low cost, and non-toxicity of graphene make this material a promising candidate for many technological applications [5, 16].
- Graphene is an ideal material for various applications because of its very large 2D electrical conductivity, large surface area, and low cost. In comparison with carbon nanotubes (CNTs), two advantages of graphene are obvious as follows:

  1. Graphene does not contain metallic impurities as CNTs do. In many cases, such impurities dominate the electrochemistry of CNTs and lead to misleading conclusions.
  2. The production of graphene can be achieved using graphite that is cheap and accessible [18].

3. Methods of graphene preparation

The rapid adoption of graphene as a material of interest lies in its actual availability by a range of techniques and methods [3, 5]. In 2004, Geim and coworkers first reported graphene sheets

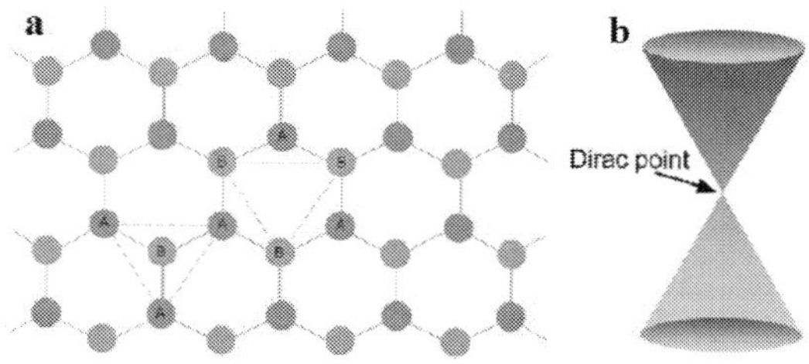

Figure 3. (a) Honeycomb lattice of graphene showing two triangular sublattices A and B. (b) Linear energy bands in the honeycomb lattice close to the Dirac point [1].

prepared by mechanical exfoliation (repeated peeling) of highly oriented pyrolytic graphite. This method, which is called the "scotch-tape" method, is still widely used in many laboratories to obtain pristine perfect structured graphene layer(s) for basic scientific research and for making proof-of-concept devices [18, 19]. However, this method is not suitable for the large-scale production of graphene necessary to fulfill the requirements in different areas. A large amount of recent effort has been devoted to develop methods to synthesize graphene at large scale for practical electronic applications. These methods include mechanical exfoliation and cleavage of natural graphite, chemical vapor deposition on metals, electric arc discharge, epitaxial growth on electrically insulating surfaces, unzipping of carbon nanotubes, solution-based chemical approaches (chemical or thermal reduction of graphite oxide), and thermal decomposition of SiC wafer under ultrahigh vacuum conditions [2, 5, 6, 19].

3.1. Chemical vapor deposition

Chemical vapor deposition (CVD) is a simple, scalable, and cost-efficient method to prepare single and few-layer of graphene on various substrates. It opens a new route to large-area production of high-quality graphene films for practical applications [20-23]. In general, CVD techniques involve the decomposition of fluid (gas and liquid sprays) at high temperature to form either thin films on substrates or powders through filters. There are many forms of CVD like hot wire CVD, thermal CVD [24], plasma-enhanced CVD [25], radio-frequency CVD, and ultrasonic spray pyrolysis. Evaporated Ni film on SiO_2/Si wafers or copper foils are ideal substrates for graphene synthesis. Synthesis of graphene on Ni supported on Si/SiO_2 wafers facilitated the breakthrough approach for large-scale graphene. This is attributed to the excellent geometrical fit of the ordered graphene/graphite phase of carbon to the crystalline metal surface that is provided by Ni films. Another reason is the convenient interactions that favor bond formation between carbon atoms at specific conditions. Carbon atoms dissolve into the Ni crystalline surface and they arrange epitaxially on the Ni (111) surface to form graphene at certain temperatures (Figure 4). CVD synthesis of graphene can be carried out at ambient pressure or vacuum by systematically varying parameters such as gas composition and flow rate, temperature, and deposition time [4, 19]. The carbon precursor in CVD is methane [26-29]. By using diluted methane, single and few-layers of graphene (less than 5 layers) were grown, while using concentrated methane results in multilayer growth [4]. Other carbon precursors can be used in CVD such as ethanol [30], isopropanol [27], ethylene, acetylene, and others [4].

http://dx.doi.org/10.5772/60676

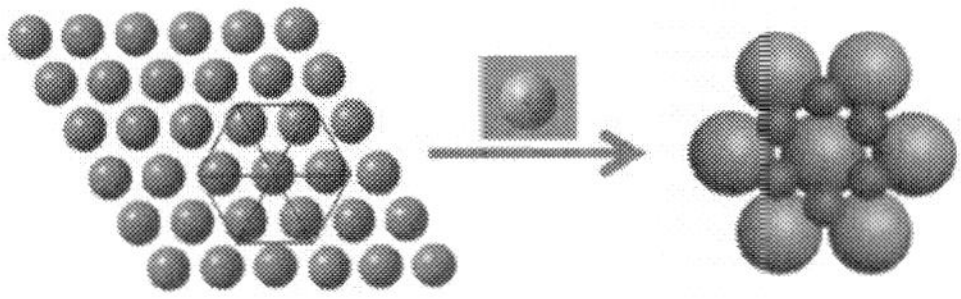

Figure 4. Schematic representation of the atomic arrangement of the hexagonal lattice of the (111) face of nickel (blue spheres) and how carbon atoms (gray spheres) would arrange on the Ni (111) surface to form graphene [4].

3.2. Unzipping of CNTs into monolayer graphene

The method of unzipping single-walled carbon nanotubes to form monolayer graphene has been reported recently accomplishing by harsh acids and the right thermodynamic conditions (Figure 5) [31]. Recently, Dai and coworker have taken a unique approach to fabricate graphene nanoribbons (GNRs) with well-controlled dimensions. Based on the concept that CNTs are considered to be GNRs rolled up into seamless tubes and the fact that the size of CNTs has been well controlled, Dai's group developed a method to produce GNRs through controlled unzipping of CNTs using the Ar plasma etching method. Shimizu and coworkers produced GNRs through the oxidization and longitudinal unzipping of MWCNTs in concentrated sulfuric acid followed by treatment with $KMnO_4$ [32]. Another approach involved the preparation of longitudinal unzipped MWCNTs by controlled oxidation and intercalation through reaction with concentrated H_2SO_4 and HNO_3 at room temperature in order to control the reaction rate. The unzipped MWCNTs are thermally treated up to 1000ºC and 2500ºC in an inert atmosphere to exfoliate and reduce the edge defects. The sonication process is responsible for chemo-mechanical breaking and the separation of multi-layer nanoribbons [33]. Furthermore, graphene nanoribbons were synthesized by the oxidative unzipping of SWCNTs using a mixture of concentrated HNO_3/H_2SO_4 in a 1:3 volume ratio followed by 8 hours of ultrasonic treatment performed at 50 W at 45ºC [34]. In addition, catalytic unzipping of single-, double-, and multi-walled carbon nanotubes (SWCNTs, DWCNTs, and MWCNTs) in the presence of Pd nanoparticles and an oxygen-containing liquid medium yielding few layer graphene sheets was performed under microwave irradiation [35].

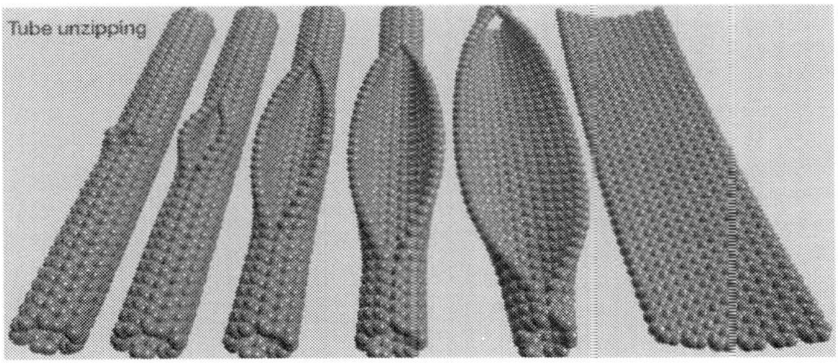

Figure 5. Modelling illustration of the SWCNT unzipped into monolayer graphene [31].

3.3. Arc discharge

Arc discharge is a versatile and low-cost method for producing various kinds of carbon nanomaterials. Compared to other methods, the graphene sheets synthesized by arc discharge method have merits of good dispersibility in organic solvent, few defects, and high thermal stability [36]. In the arc discharge technique, graphite rods (purity: 99.99 %) are used as electrodes for high voltage arcing. At extremely high voltages between the electrodes that are separated by very small distances, very high electric fields can be produced leading to instantaneous sparks like in a welding process. The fall-out during the discharge process is the end product that contains the carbon nano-structures [31]. To prepare pure few-layered graphene, a direct current arc discharge of graphite evaporation was carried out in a water-cooled stainless steel chamber filled with different buffer gases after evacuating the chamber by a mechanical pump [31, 36]. When the arc discharge process ended, the generated soot was collected under ambient conditions. The soot deposited on the inner and top wall of the reaction chamber was only collected in order to exclude relatively heavy products (amorphous carbon nanoparticles or multi-layered graphene sheets) dropped to the bottom of the chamber during the arc discharge process [36]. Different discharge atmospheres may induce different chemical bonds that cause the prepared graphene sheets to have different properties. Direct current hydrogen arc discharge evaporation of pure graphite electrodes in various gas mixtures can be used to mass-produce high-crystalline few-layer graphene sheets [40]. Various gas mixtures were reported such as H_2, He, mixture of H_2 and He, mixture of NH_3 and He, and mixture of CO_2 and He to synthesize few-layered graphene sheets with high electrical conductivity [36-38]. On the other hand, a novel method for the large-scale production of graphene flakes in an arc discharge enhanced with a specially shaped magnetic field and a custom-designed catalyst was reported [39].

3.4. Reduction of graphene oxide into graphene

Graphene oxide (GO) was synthesized by the strong oxidation of flake graphite using acid via Hummers' method [40, 41] or modified Hummers' method [42-44]. GO, an oxidized form of graphene, is decorated by hydroxyl and epoxy functional groups on the hexagonal network of carbon atoms with carbonyl and carboxyl groups at the edges. GO is highly hydrophilic and forms stable aqueous colloids due to the large number of oxygen-containing functional groups as well as the repulsive electrostatic interactions at the edges of the platelets [41, 45]. It was reported that GO can be reduced via thermal, chemical, electrochemical, and laser-scribing methods.

3.4.1. Thermal reduction

The thermal method is believed to be a green method in which no hazardous reductants are used. This process requires heating up to 1050°C in an oven under argon gas, up to 800°C under hydrogen gas, or up to 700°C in a quartz tube. The thermal reduction of GO was accompanied by the elimination of epoxy and carboxyl groups in the forms of O_2, CO, CO_2, and H_2O as predicted from thermogravimetry data [46-48]. Recently, graphene sheets were prepared by a solvothermal reduction of GO suspension in water, octadecylamine, butanol, or N,N-dimethylacetamide [46]. Wei Huang synthesized high-performance graphene nanosheets by thermal reduction of GO under ethanol atmosphere at 900°C. This method can effectively remove the oxygen-containing functional groups and restore its graphic structure compared to the ones obtained using hydrazine or hydrogen. It can provide a simple, low-cost, and high yield approach for the preparation of high performance graphene nanosheets [49]. Furthermore, graphene sheets with different reduction levels were produced through thermal reduction of GO in the temperature range of 200–900°C under N_2 atmosphere for supercapacitor applications [50, 51]. Thermal reduction can produce few layer graphene with less agglomeration, higher specific surface area, and higher electrical conductivity compared to other reduction routes [51].

3.4.2. Chemical reduction

GO formation was followed by subsequent chemical reduction with a reducing agent such as hydroquinone, sodium borohydride, hydrazine, and hydrazine with NH_3 [42, 47, 52]. Hydrazine hydrate was proved to be the best reducing agent for the chemical reduction of GO into graphene and very thin graphene-like sheets were obtained. Moreover, the hydrazine treatment resulted in the formation of unsaturated and conjugated carbon atoms, which in turn imparts electrical conductivity of the graphene sheets [41]. These reducing agents, particularly hydrazine, are highly toxic, explosive, and their use should be with extreme care and minimized [42, 52]. The chemical reduction may also leave some residual epoxide groups on the reduced GO sheets leading to some loss in electron mobility [52]. On the basis of these observations, a new approach was employed to develop a "green" synthesis method for the production of graphene by using environmentally friendly reducing agents such as vitamin C [44], reducing, sugars and l-glutathione under mild conditions [42].

On the other hand, the chemical reduction methods of exfoliated GO with reducing agents such as hydrazine hydrate provide a promising approach for the efficient large-scale production of chemically converted graphene (CCG) sheets. However, in most cases heating to nearly 100ºC over several hours is required. A novel, facile, convenient, and scalable method for the synthesis of CCG sheets as well as metallic and bimetallic nanoparticles supported on the CCG sheets using a simple household microwave oven was utilized [53-56]. It was reported that graphene is a good microwave-absorbing material and can reach a high temperature in minutes [57]. Microwave irradiation (MWI) has been demonstrated for the synthesis of a variety of nanomaterials with controlled size and shape without the need for high temperature or high pressure. The main advantages of MWI over other conventional heating methods are (i) rapid reaction velocity, (ii) clean and energy efficient, and (iii) uniform heating of the reaction mixture [47, 58-60]. Due to the difference in the solvent and reactant dielectric constants, selective dielectric heating can provide significant enhancement in the transfer of energy directly to the reactants causing an instantaneous internal temperature rise. Dried GO was sonicated in deionized water until homogenous dispersion was obtained. After that it was placed inside a conventional microwave after adding 100 μl of a reducing agent [hydrazine hydrate, ethylenediamine, or ammonium hydroxide] in 30 s cycles for a total reaction time of 60 s. A black color was obtained indicating the complete chemical reduction to graphene. The graphene sheets were separated by using centrifuge operated at 5000 rpm for 15 min and dried overnight under vacuum [53]. Microwave-assisted methods allowed the rapid chemical reduction of GO using a variety of reducing agents in either aqueous or organic media. By using this method, many types of metallic and bimetallic nanoparticles can be dispersed on the graphene sheets via simultaneous reduction of GO and a variety of metal salts to create novel nanocatalysts supported on the large surface area of the thermally stable 2D graphene [61]. Moreover, this method provides a low-cost approach to the facile production of nano-particles/graphene composites on a large scale for applications [57, 61, 62].

3.4.3. Electrochemical reduction

As pointed out by Kaner and co-workers, both chemical and thermal reduction techniques have some obvious drawbacks, namely, the highly toxic reductant hydrazine, lack control of the film thickness, and the incompatibility of the thermal reduction process under some conditions. Thus, the electrochemical method was adopted as an effective and controllable alternative technique for the modification of electronic states [63]. It has drawn great attention due to its fast and green nature [17, 52]. This is done by adjusting the external power source to change the Fermi energy level of the electrode surface that reduces GO in the presence of direct current bias [63]. Typically, the electrochemical synthesis of graphene was carried out via two steps, namely, GO being first assembled on the electrodes by solution deposition methods then being subjected to electrochemical reduction by scanning the potential (cyclic voltammetry) [45, 47, 52, 64, 65] or by applying constant potential (bulk electrolysis) [66, 67]. Another approach was achieved via a one-step technique in which graphene films can be prepared on electrodes directly from GO dispersions by simultaneous electrodeposition and electroreduction [17]. The reduction of GO can be confirmed from the color of GO electrodes that changes from yellow (before reduction) to black (after reduction) [64], involving the reduction of the high number of functional groups present in the GO sheets such as OH, COOH, and epoxides [52].

The properties of electrochemically converted graphene (ECG) are quite different from other nanostructured carbons (CNTs) and even different from CCG. The morphology of CCG and ECG on glassy carbon sheet was characterized by high field emission-scanning electron microscopy (FE-SEM) (Figure 6 (A, B)), respectively. The micrographs show that CCG sheets are rippled and crumpled with a dimension of several nm to few μm while ECG sheets are more homogeneous and more compact on the surface than CCG [66]. ECG exhibits greatly enhanced activity for the electrocatalytic reduction of O_2 and H_2O_2 and much higher electro-chemical capacitance for potential application in ultracapacitors [64]. Moreover, the ECG coating is very stable as a result of its poor insolubility in common solvents [17].

3.4.4. Laser-scribing technology

Compared to chemical and thermal methods, pulsed laser irradiation reduction is simple, rapid (shorten the reaction time from several hours to few minutes), energy efficient, and free from poisonous material and high temperature. Also, the pulsed laser reduction of GO in solution makes it particularly attractive for producing graphene composites by pre-mixing

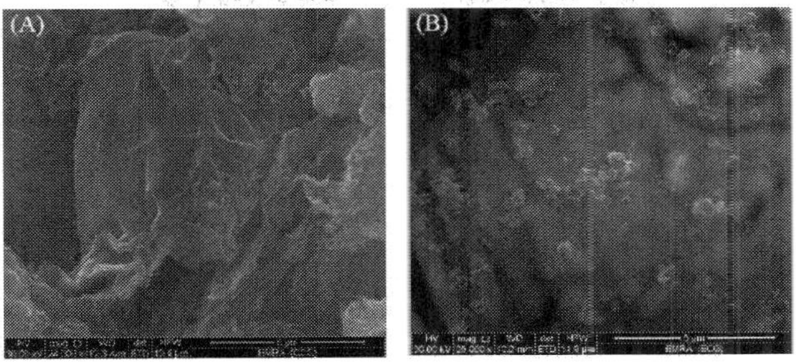

Figure 6. SEM images using FE-SEM for a) CCG, b) ECG [66].

with other materials [68]. Moreover, Feng-Shou Xiao fabricated graphene microcircuits by direct reduction and patterning of GO films using femtosecond laser representing well-conductivity for electrical applications [69]. Recently, Kaner and co-workers prepared graphene using standard laser-scribing technology more typically used to etch labels and images onto DVDs. Initially, a thin film of GO dispersed in water was drop-cast onto a flexible substrate (a DVD covered with a layer of plastic). Irradiation of the film with an infrared laser inside an inexpensive commercially available light scribe CD/DVD optical drive reduces the GO to laser-scribed graphene (LSG), as indicated by the change in film color from golden brown to black. Analysis of cross sections of the film with scanning electron microscopy showed that the initially stacked GO sheets were converted into well-exfoliated LSG sheets without any sticking together through laser irradiation [70].

4. Functionalization of graphene

GO contains oxygen functional groups such as epoxides, OH, and COOH that make it hydrophilic and well dispersed in water. However, GO is incompatible with most organic polymers [52]. It is well established that the superior properties of graphene are associated with its single layer. However, the fabrication of single-layer graphene is difficult at ambient temperature. Graphene sheets with a high specific surface area tend to form irreversible agglomerates or even restack to form graphite through π–π stacking and van der Waals interactions if the sheets are not well separated from each other. Aggregation can be reduced by the attachment of other small molecules or polymers to the graphene sheets. The presence of hydrophilic or hydrophobic groups prevents aggregation of graphene sheets by strong polar–polar interactions or by their bulky size. The attachment of functional groups to graphene also aids in dispersion in a hydrophilic or hydrophobic media as well as in the organic polymer [6, 52]. Furthermore, the oxidized rings of functionalized and defective graphene sheets contain abundant C–O–C (epoxide) and C–OH groups while the sheets are terminated with C–OH and –COOH groups. Defects of graphene may change its electronic and chemical properties. The functionalized and defective graphene sheets are more hydrophilic and can be easily dispersed in solvents with long-term stability [12]. Moreover, they are more easily produced in mass quantities as compared with the carbon nanotubes [12, 16, 52]. They may be used to prepare some novel graphene-based nanocomposite films that could facilitate the further manipulation and processing of these materials for developing novel electronic devices such as chemical sensors, biosensors [12], cellular imaging, and drug delivery [52]. On the other hand, various polymer and nanoparticle (metal, metal oxide, semiconductor) composites have been developed based on the unique properties of graphene. The bulk production of GO and graphene has given opportunities to explore this flat structure of carbon with polymer and nanoparticles in composites [32]. Pristine graphene is a hydrophobic material and has no appreciable solubility in most solvents. Nevertheless, the processing of graphene composites concerns itself foremost with the solubilization of graphene. To improve the solubility of graphene, different functional groups have been attached to the carbon backbone by chemical modification, covalent or non-covalent functionalization [15].

- Covalent modification: The structural alteration can take place at the end of the sheets and/or on the surface. Surface functionalization is associated with rehybridization of one or more sp^2 carbon atoms of the carbon network into the sp^3 configuration accompanied by simultaneous loss of electronic conjugation. Covalent functional groups can attach to the surface through the reaction between the functional groups (-OH, -COOH, -CO) present on the GO and reduced graphene oxide (rGO) surfaces and edges [15, 32]. The covalent modification of graphene can be achieved in four different ways: nucleophilic substitution, electrophilic addition, condensation, and addition [6].
- Non-covalent modification: It is based on the van der Waals force or the π–π interaction between reduced graphene oxide (rGO) and stabilizers that not only gives less negative impact on the structure of graphene and its derivatives, but also provides the feasibility to tune their solubility and electronic properties [15, 32].

5. Applications of graphene to sensors and biosensors

Graphene has recently attracted tremendous interest because of its unique thermal, mechanical, and electrical properties. One of the promising applications of graphene is electrochemical sensing. Since every atom in a graphene sheet is a surface atom, molecular interaction and thus electron transport through graphene can be highly sensitive to adsorbed molecules [17, 71]. Graphene is a strictly 2D material and as such enables devices based on graphene to have an identical performance on a large scale. In addition, with the 2D structure, the monolayer graphene has its whole volume exposed to the environment that can maximize the sensing effect. The principle of graphene devices is based on changes in device conductance due to chemical or biological species adsorbed on the surface of graphene acting as electron donors or acceptors [19]. On the other hand, graphene-based nanomaterials have recently shown fascinating applications in electrochemical sensors and biosensors. Owing to the extraordinary electronic transport property and high electrocatalytic activity of graphene, the electrochemical reactions of analyte are greatly promoted on graphene film resulting in enhanced voltammetric response. Moreover, the electrochemical properties of graphene can be effectively modified by integration with other functional nanomaterials such as catalyst nanoparticles to produce versatile electrochemical sensing performance. Comparing with CNTs, graphene has shown the advantages of high conductivity, ease of production and function, good biocompatibility, and abundance of inexpensive source material [71, 72]. Thus, graphene was a good candidate of advanced electrode materials and could be combined with other functional materials to fabricate the sensing interface for electroanalysis [3, 10].

5.1. Gas sensors

Graphene has potential use in gas sensors owing to its 2D structure with extremely high surface area. Gas sensing by graphene generally involves the adsorption and desorption of gaseous molecules (which act as electron donors or acceptors) on the graphene surface leading to change in the conductance of graphene. The high sensitivity of graphene toward different gaseous molecules has led to its use as gas sensor for hydrogen, H_2O, CO, NH_3, NO_2, Cl_2, ethanol, I_2, and O_2 [73]. Graphene/ZnO hybrid [74] and reduced graphene oxide (rGO) [75] were utilized as highly efficient gas sensors. Qingkai Yu et al. synthesized Pd-decorated graphene films to be used as hydrogen sensor. This gas sensor was fabricated on graphene film with a 1 nm Pd film deposited for hydrogen detection in the concentration range of 25–10,000 ppm. H_2 sensor based on Pd-decorated graphene films showed high sensitivity, fast response, and good recovery and can be used with multiple cycles [19]. Besides, Gupta et al. prepared a nanocomposite film based on poly(methyl methacrylate) (PMMA), rGO, and PMMA/rGO to be used as NH_3 gas sensor. The sensor performance in terms of selectivity and sensitivity was much better in the case of PMMA/rGO nanocomposite compared to the individual layers of rGO or PMMA [76]. On the other hand, an acetylene gas sensor was fabricated by synthesizing Ag-loaded ZnO-rGO hybrid via a facile chemical route (Figure 7). A well-structured crystalline nature and mixed phases of Ag, graphene (Gr), and ZnO was obtained. The morphological characterization revealed that particle-like nanostructures of ZnO and Ag were well distributed and closely distributed onto the surface of thin-layer RGO sheets. A low detection limit of 1 ppm, fast response of 25 s, recovery of 80 s, and good repeatability were obtained [77].

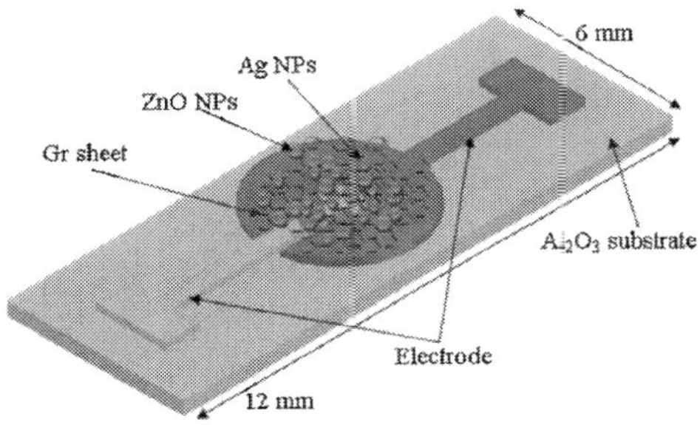

Figure 7. Schematic diagram of the fabricated sensor device [77].

5.2. Graphene-based electrochemical sensors for heavy metal ions

Metal ions such as lead, cadmium, silver, mercury, and arsenic have severe environmental and medical effects so that they require careful monitoring. Therefore, the development of a sensitive and selective detection method would benefit both the environmental and food chemists [5, 73]. A novel stannum film/poly(p-aminobenzene sulfonic acid, p-ABSA)/graphene composite modified glassy carbon electrode (GCE) was used for sensitive determination of trace Cd^{2+} ions by square wave anodic stripping voltammetry. Graphene layer presents enhanced electron transfer and enlarged specific surface area. The resulting matrix offers a good stripping performance for the analysis of Cd^{2+} in the linear range from 1.0 to 70.0 μg L^{-1} with a detection limit of 0.05 μg L^{-1} [78]. Furthermore, Erkang Wang used graphene nanosheets dispersed in Nafion (Nafion-G) solution in combination with in situ plated bismuth film electrode for fabricating the enhanced electrochemical sensing platform for Pb^{2+} and Cd^{2+} by differential pulse anodic stripping voltammetry. The prepared Nafion-G composite film exhibited improved sensitivity for the metal ion detections and alleviated the interferences as a result of the synergistic effect of Nafion-G. The stripping current signal is greatly enhanced and well distinguished on graphene electrodes with a wide linear range and 0.02 μg L^{-1} detection limit for both ions (Figure 8) [5, 79]. Moreover, various modified surfaces were constructed for the sensitive and simultaneous determination of Pb^{2+} and Cd^{2+} such as graphene/polyaniline/polystyrene nanoporous fiber [80] and graphene/ionic liquid/bismuth film composite modified screen-printed electrodes [81], functionalized graphene (GNS^{PF6}) with potassium hexafluorophosphate [82], and a hybrid nanocomposite of one-dimensional MWCNTs and graphene sheets [83]. Large surface area, good ionic and electronic conductivity, wide electrochemical window, as well as high stability, sensitivity, low detection limit, and applicability in real samples are the obvious characteristics of the proposed sensors. On the other hand, functionalized graphene sheets were fabricated and used for simultaneous removal of high concentration of inorganic species of arsenic (both trivalent and pentavalent) and sodium from aqueous solution using a supercapacitor-based water filter. Also, they were utilized for the desalination of sea water. High adsorption capacity for both inorganic species of arsenic and sodium along with the desalination ability of a graphene based supercapacitor provides a solution for a commercially feasible water filter. Additionally, the cost effective production of graphene sheets and better performance compared to other adsorbents like CNTs provides a platform for the development of commercially feasible supercapacitor-based water filter [84]. Furthermore, rGO–lead dioxide composite was presented as an excellent material for the detection of a low level of arsenic with a low detection limit of 10 nM. This study opens up the possibility of using the composite for simultaneous detection of arsenic and lead [85]. Furthermore, Zhuangjun Fan synthesized a type of graphene nanosheet/δ-MnO_2 (GNS/MnO_2) composite by a microwave-assisted method to be used as an adsorbent for the removal of nickel ions from waste water [86].

5.3. Biosensor for some biological compounds

5.3.1. Paracetamol

Paracetamol or acetaminophen (ACOP) is a long-established and one of the most extensively employed "over the counter" drugs in the world. It is non-carcinogenic and an effective

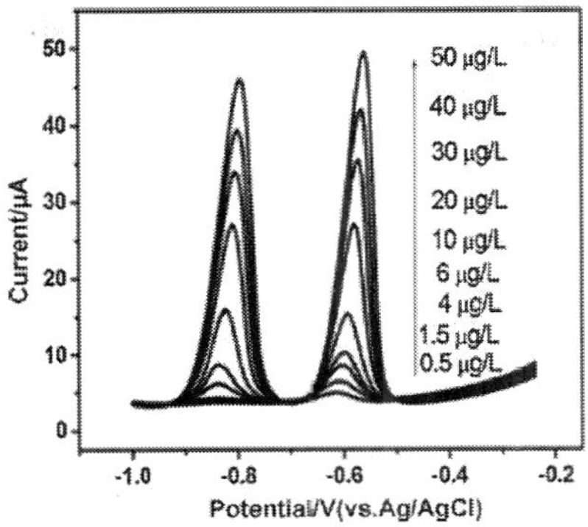

Figure 8. Striping voltammograms for the different concentrations of Cd^{2+} and Pb^{2+} on an in situ plated Nafion-G-BFE (bismuth film electrode) in solution containing 0.4 mg L^{-1} Bi^{3+}79]].

Electrode	Linear range	Detection limit	Reference
Nafion/TiO_2–graphene nanocomposite	1-100 μM	210 nM	[7]
Electrochemically reduced and deposited graphene onto GCE	5.0 nM-800 μM	2.13 nM	[91]
Electrochemically reduced graphene loaded nickel oxides (Ni_2O_3–NiO) nanoparticles coated onto GCE	0.04-100 μM	0.02 μM	[92]
Poly(4-aminobenzoic acid)/ electrochemically reduced graphene oxide composite film modified GCE	0.1–65 μM	0.01 μM	[93]

Table 2. Examples of modified electrodes for ACOP detection.

substitute for aspirin for patients with sensitivity to aspirin. It is used to reduce fever, cough, and cold and reduce mild to moderate pain. It is also useful in osteoarthritis therapy, protects hardening of arteries, relieves asthma patients, and protects against ovarian cancer. So it is very important to establish a simple, fast, sensitive, and accurate detection method for ACOP [87, 88]. Yuehe Lin et al. presented an electrochemical sensor for ACOP based on the electrocatalytic activity of functionalized graphene. Graphene-modified GCE obviously promotes the sensitivity of the determination of paracetamol with a low detection limit of 32 nM and a satisfied recovery from 96.4% to 103.3% [12]. Moreover, SWCNTs [89] or MWCNTs [90] graphene nanosheet nanocomposite modified GCE (SWCNTs-GNS/GCE or MWCNTs-GNS/ GCE) was utilized as a novel, simple, and highly selective electrochemical sensor for tyrosine and paracetamol. The proposed sensor exhibited perfect characteristics like high effective surface area, high porosity, more reactive sites, excellent electrochemical catalytic activity, and applicability in human blood serum and pharmaceutical samples. Very low detection limits of 0.19 μM and 0.10 μM were obtained for tyrosine and paracetamol, respectively [90]. Table 2 contains examples of various modified electrodes for ACOP detection [7, 91-93].

5.3.2. Catecholamine neurotransmitters

Catecholamines are a class of neurotransmitters and their detection in the human body has been of great interest to neuroscientists. They include dopamine (DA), epinephrine (EP), and norepinephrine (NE) playing important roles in various biological, pharmacological, and physical processes. They are widely distributed in the mammalian central nervous system for message transfer [94, 95].

EP, a hormone secreted by the medulla of adrenal glands, is an important catecholamine inhibitory neurotransmitter. The presence of EP in the body affects the regulation of blood pressure and the heart rate, lipolysis, immune system, and glycogen metabolism. Low levels of EP have been found in patients with Parkinson's disease. Therefore, the quantitative determination of EP in different human fluids such as plasma and urine is important for developing nerve physiology, pharmacological research, and life science [96]. Xiaoli Zhang fabricated graphene/Au nanocomposites modified GCE (GR/Au/GCE) for the detection of EP with high sensitivity. The oxidation of EP at this modified electrode occurred at less positive potentials than on bare GCE. In addition, the oxidation peaks of EP and ascorbic acid (AA) were separated from each other by approximately 180 mV. At the same time, this electrode also showed favorable electrocatalytic activity toward some other small biomolecules (such as DA, b-nicotinamide adenine dinucleotide, and uric acid) suggesting the potential applications of GR/Au nanocomposites for constructing biosensors [97]. Moreover, GR modified Pd was prepared to be used as a voltammetric sensor for NE, an important neurotransmitter. GR as well as Pd nanoparticles presented perfect characteristics for the proposed sensor-like excellent electrocatalytic activity for the oxidation of NE in terms of remarkable enhancement in the peak current and lowering of peak potential. Also, low detection and quantification limits were obtained that were 67.44 nM and 224.8 nM, respectively. The application of the proposed sensor in pharmaceutical dosage forms and human urine samples in the presence of high concentration of uric acid was achieved with excellent recovery results [98].

On the other hand, DA, ascorbic acid (AA), and uric acid (UA) are compounds of great biomedical interests playing determining roles in human metabolism. DA is one of the crucial catecholamine neurotransmitters playing an important role in the function of central nervous, renal, hormonal, and cardiovascular systems. Thus, a loss of DA-containing neurons may lead to neurological disorders such as Parkinsonism and schizophrenia [94, 95, 99]. AA (vitamin C), an antioxidant, is a vital vitamin in human diet existing in both animal and plant kingdoms. It has been used for the prevention and treatment of the common cold, mental illness, infertility, and cancer and in some clinical manifestations of human immunodeficiency virus infections. UA is the primary end product of purine metabolism. The extreme abnormalities of UA levels in the body are symptoms of several diseases including gout, hyperuricemia, and Lesch–Nyhan. Therefore, simultaneous detection of DA, AA, and UA is a challenge of critical importance not only in the field of biomedical chemistry and neurochemistry but also for diagnostic and pathological research [94, 95]. Feng Li et al. prepared a graphene doped carbon paste electrode (CPE) by the addition of graphene into the carbon paste mixture for the successful determination of AA. This sensor showed an excellent electrocatalytic oxidation activity toward AA with a lower overvoltage, pronounced current response, and good sensitivity. The proposed sensor exhibited a rapid response to AA within 5 s and a good linear calibration plot from 0.1 to 106 μM with low detection limit of 70 nM [8]. On the other hand, Jinghong Li et al. synthesized graphene chemically by the Hummers and Offeman method and the graphene-modified electrode was applied for the selective determination of DA in the linear range of 5–200 μM in a large excess of AA. Selective detection was realized in completely eliminating AA based on the unique planar structure of graphene. The interesting performance of graphene to DA and AA may be correlated with the following reasons:

1. The high quality of the sp^2 conjugated bond in the carbon lattice, graphene is highly conductive and shows metallic conductance even in the limit of zero carrier density. The unusual density of electronic states of graphene affects its electrochemical reactivity.

2. Molecular structures of DA and AA are distinct from each other. The π–π interaction between the phenyl structure of DA and 2D planar hexagonal carbon structure of graphene makes the electron transfer feasible. AA oxidation is inactive most likely because of its weak π–π interaction with graphene [10].

Several sensors were utilized for the simultaneous determination of DA and AA such as graphene modified electrode [100], graphene sheets/Congo red-molecular imprinted polymers [101], and Pt nanoparticles/polyelectrolyte-functionalized ionic liquid/graphene sheets [102]. On the other hand, TiO_2–graphene nanocomposite modified GCE exhibited remarkable electron transfer kinetics, electrocatalytic activity, and selective determination of DA in the presence of UA and AA. Differential pulse voltammetry (DPV) peaks of DA, AA, and UA are well resolved with peak potentials at 108 mV, 272 mV, and 403 mV for AA, DA, and UA, respectively. The separation of DPV peak potentials for AA–DA and DA–UA are calculated to be 164 mV and 131 mV, respectively, which are large enough to determine DA selectively [103]. Several electrochemical sensors based on graphene-modified surfaces were employed for the simultaneous determination of DA, AA, and UA like polystyrene-grafted-graphene hybrid [104] and nafion covered core–shell structured Fe_3O_4@graphene nanospheres (GNs) [105] modified GCE, graphene-AuAg (Au/Gr-AuAg) composite modified gold electrode [106], Au nanoplates and rGO modified GCE (Au/rGO/GCE) [107], three-dimensional reduced graphene oxide (3D-rGO) material [108], water-soluble sulfonated graphene [109], ferulic acid functionalized electrochemically reduced graphene [110], graphene/nickel hydroxide composite [111], and well-defined flower-like graphene-nanosheet clusters (f-FGO) modified GCE [112]. In addition, Pd–Pt bimetallic nanoparticles anchored on functionalized rGO nanomaterials [113], electrochemically reduced graphene oxide (ERGO) [114], nitrogen-doped graphene (NG) [115], and tryptophan-functionalized graphene nanocomposite (Trp-GR) [116] were fabricated for the simultaneous detection of DA, AA, and UA and multi-walled carbon nanotubes (MWNTs) bridged mesocellular graphene foam (MGF) nanocomposite (MWNTs/MGF) modified GCE for the simultaneous determination of DA, AA, UA, and tryptophan

[117]. As well, graphene (Gr)/copper(II) phthalocyanine-tetrasulfonic acid tetrasodium salt/ polyaniline nanocomposites [118] and NiO nanoparticles-graphene composite film were employed as electrochemical sensors for the selective determination of AA in the presence of DA and UA [118] and in the presence of folic acid [119], respectively.

On the other hand, serotonin (5-hydroxytryptamine, 5-HT) is an important neurotransmitter and it is important to sensitively determine its concentration. A double-layered membrane of rGO/polyaniline nanocomposites and molecularly imprinted polymers embedded with Au nanoparticles was employed as an electrochemical sensor for 5-HT. The prepared sensor displayed obvious selectivity to 5-HT and a low detection limit of 11.7 nM in the linear range of 0.2–10.0 μM [120]. Moreover, graphene nanosheets [121] and graphene grafted poly(lactic acid) with Pd nanoparticles [122] were applied as 5-HT sensor showing high sensitivity, good selectivity, low detection limit, fast response time, and stability. In addition, electrochemically reduced GO-P (graphene oxide and 5,15-pentafluorophenyl-10,20-p-aminophenyl porphyrin) was employed as an electrochemical sensor for the simultaneous and selective determination of DA and 5-HT in the presence of AA. The detection limits of DA and 5-HT were 35 nM and 4.9 nM, respectively [123].

Levodopa (LD), an important neurotransmitter, was determined at modified CPE modified with graphene (GR), 1-(6,7-dihydroxy-2,4-dimethylbenzofuran-3-yl) ethanone (DE), and ionic liquid (IL) (GR–DE–IL/CPE). GR–DE–IL/CPE showed enhanced electrocatalytic activity toward LD with a lower oxidation potential and good electrochemical performance with higher conductivity and lower electron transfer resistance. Very low detection limit of 5.0 nM in the linear range of 0.015 to 1000 μM was obtained [124]. An assembly of rGO, Au nanoparticles, and 2-(3,4-dihydroxy phenyl) benzothiazole [125] and graphene nanosheets modified GCE [126] and reduced graphene nanoribbons modified carbon screen-printed electrode [127] were employed for the simultaneous determination of LD, UA and folic acid, LD and carbidopa and AA, LD and UA, respectively.

5.3.3. Hydroquinone and catechol

Graphene has the great potential for distinguishing a diverse range of aromatic isomers. Chengbin Liu et al. had utilized graphene modified electrode (EG/GCE) for the simultaneous detection of hydroquinone HQ (1,4-dihydroxybenzene) and catechol CC (1,2-dihydroxybenzene). It showed enhanced electron transfer properties, good potential separation of oxidation peaks between HQ and CC (about 110 mV), wide linear concentration ranges, low detection limits, excellent reproducibility and stability, and high resolution capacity to the HQ and CC isomers [17]. On the other hand, a highly stable (pyridine-NG) was used as an electrochemical sensor for simultaneous determination of HQ and CC. Excellent electrocatalysis of pyridine-NG for simultaneous determination of HQ and CC was achieved due to the π–π interactions between the benzene ring of CC and graphene layer and the hydrogen bonds formed between hydroxyl in HQ molecule and pyridinic nitrogen atoms within graphene layers, especially the less density distribution of π electron cloud in pyridinic-NG in acidic condition [128]. Several novel sensors like poly(3,4-ethylenedioxythiophene)/nitrogen-doped graphene composite [129], MWCNTs-poly(diallyldimethylammonium chloride)-graphene [130], and laser reduced graphene [131] modified GCE were fabricated for the simultaneous detection of CC and HQ with low detection limit, high sensitivity, excellent potential peak separation and anti-interference ability. Table 3 contains examples of modified electrodes for HQ and CC detection [132-135].

Electrode	Linear range of HQ	Detection limit of HQ	Linear range of CC	Detection limit of CC	Reference
Graphene and TiO_2 nanocomposite	0.5–100 μM	0.082 μM	0.5–100 μM	0.087 μM	[132]
Electrochemically reduced graphene oxide- MWCNTs and terthiophene	---	35 nM	---	4.9 nM	[133]
1D MWCNTs and 2D graphene modified GCE using 1-butyl-3-methyl-imidazolium hexafluoro-phosphate ionic liquid	---	0.1 μM	---	0.06 μM	[134]
Au nanoparticle and graphene composite film modified carbon ionic liquid electrode	0.06-800 μM	0.018 μM	---	---	[135]

Table 3. Examples of modified electrodes for HQ and CC detection.

5.4. Drug biosensor

Morphine, noscapine, and heroin are three major alkaloids in heroin samples. Morphine and noscapine that are directly extracted from poppy are the most abundant constituents in opium. The widespread use of illicit drugs has led to an increased effort toward developing and improving methods for their determination in biological samples, which is still a very challenging task. Graphene nanosheets (GNSs) modified GCE was employed for the simultaneous determination of morphine, noscapine, and heroin exhibiting excellent electrocatalytic activity at reduced overpotentials in wide pH range. Fast response time, high sensitivity and stability, low cost, low detection limit, and ease of preparation method without using any specific electron-transfer mediator or specific reagent are the advantages of the proposed sensor [136]. In addition, electrochemically reduced MWNTs-doped GO (ER-MWNTs-doped-GO) composite film modified GCE was employed as morphine sensor combining the excellent conductivity of MWNTs and ER-GO with the film forming properties of GO [137]. Also, rGO–Pd (rGO–Pd) hybrid had higher current response toward morphine oxidation compared to the unmodified rGO with a low detection limit of 12.95 nM [138].

On the other hand, graphene and Nafion film modified GCE was fabricated for the detection of codeine displaying an excellent analytical performance and enhanced applicability for codeine detection in urine samples and cough syrup. The high electrocatalytic activity of the proposed sensor toward codeine was attributed to the outstanding electric conductivity of graphene and the high codeine loading capacity on the electrode surface [139]. Furthermore, a nanocomposite of Gr and $CoFe_2O_4$ nanoparticles modified CPE was proved to be an ultra-sensitive electrochemical sensor for codeine and ACOP with low detection limits of 0.011 and 0.025 μM, respectively. The proposed method was free from interference effects of glucose, ascorbic acid, caffeine, naproxen, alanine, phenylalanine, glycine, and others [140]. On the other hand, rapid and simultaneous determination of tramadol and ACOP was achieved based on $NiFe_2O_4$/graphene nanoparticles modified CPE with a low detection limit of 0.0036 and 0.0030 μM, respectively [141]. Table 4 contained a summary of some modified electrodes for drug sensing.

Electrode	Drug	Linear range	Detection limit	Reference
Graphene nanosheets modified GCE	morphine, noscapine and heroin	up to 65, 40 and 100 μM, respectively	0.4, 0.2 and 0.5 μM, respectively	[136]
electrochemically reduced MWNTs-doped GO modified GCE	morphine	---	---	[137]
rGO–Pd modified GCE	morphine	0.34-12 μM	12.95 nM	[138]
graphene and Nafion film modified GCE	codeine	5×10^{-8} to 3×10^{-5} M	15 nM	[139]
a nanocomposite of Gr and $CoFe_2O_4$ nanoparticles modified CPE	codeine and ACOP	0.03 to 12.0 μM for both	0.011 μM and 0.025 μM, respecively	[140]
$NiFe_2O_4$/graphene nanoparticles modified CPE	tramadol and ACOP	0.01–9 μM	0.0036 and 0.0030 μM, respectively	[141]

Table 4. Examples of modified electrodes for drug detection.

5.5. NADH biosensor

B-nicotinamide adenine dinucleotide (NAD) and its reduced form (NADH) are a cofactor of many dehydrogenases, which have received considerable interest in developing amperometric biosensors, biofuel cells, and bioelectronic devices [2, 5, 18, 73]. The oxidation of NADH serves as the anodic signal and regenerates the NAD cofactor, which is of great significance in biosensing important substrates such as lactate, alcohol, or glucose. Problems inherent to such anodic detection are the large overvoltage for NADH oxidation and surface fouling associated with the accumulation of reaction products. Graphene shows promise in addressing these problems [73, 142]. Jinghong Li studied the oxidation of NADH at GCE and reduced graphene sheet films (rGSF)/GC. The oxidation of NADH occurs at bare GC electrode at 0.75 V vs. Ag/AgCl, which decreased to 0.42 V at rGSF/GC [2]. Table 5 contained examples of several sensors for the sensitive and selective determination of NADH with low detection limit.

Electrode	Linear range	Detection limit	Reference
methylene green functionalized graphene modified GCE	---	---	[5]
ionic liquid functionalized graphene modified GCE	---	---	[73]
chemically reduced graphene oxides modified GCE	---	---	[18]
graphene assembled electrode modified GCE	---	---	[142]
electroreduced graphene oxide and polythionine modified GCE	0.01–3.9 mM	0.1 μM	[144]
nitrogen-doped graphene modified GCE	---	---	[145]
graphene–Au nanorods hybrid nanosheets modified GCE	5 to 377 μM	1.5 μM	[146]
Au nanoparticles /reduced graphene oxide nanocomposites modified GCE	50 nM to 500 μM	1.13 nM	[147]
graphene paste electrode modified GCE	---	---	[148]
GCE modified with deoxyribonucleic acid, graphene methylene blue	10 μM to 1.50 mM	1.0 μM	[143]

Table 5. Examples of modified electrodes for NADH detection.

5.6. DNA biosensors

Electrochemical DNA sensors offer high sensitivity, good selectivity, and low cost for the detection of selected DNA sequences or mutated genes associated with human disease and promise to provide a simple, accurate, and inexpensive platform of patient diagnosis. Electrochemical DNA sensors also allow device miniaturization for samples with a very small volume. Among all kinds of electrochemical DNA sensors, the one based on the direct oxidation of DNA is the simplest [5, 13]. Zhou et al. reported an electrochemical DNA sensor based on CR-GO. The current signals of the four free bases of DNA (i.e., guanine (G), adenine (A), thymine (T) and cytosine (C)) on the CR-GO/GC electrode are all separated efficiently indicating that CR-GO/GC can simultaneously detect four free bases but neither graphite nor

GCE can. This is attributed to the antifouling properties and the high electron transfer kinetics for bases oxidation on CR-GO/GC electrode resulting from a high density of edge-plane-like defective sites and oxygen containing functional groups on CR-GO. These functional groups are beneficial for accelerating electron transfer between the electrode and species in solution providing many active sites. CR-GO/GC is also able to efficiently separate all four DNA bases in both single-stranded DNA (ssDNA) and double-stranded DNA (ds-DNA), which are more difficult to oxidize than free bases at physiological pH without the need of a prehydrolysis step. This is attributed to the unique physicochemical properties of CR-GO (the single sheet nature, high conductivity, large surface area, antifouling properties, high electron transfer kinetics, etc.) [5, 73]. Furthermore, a zinc sulfur-coated poly (3,4-ethylenedioxythiophene)-reduced graphene oxide hybrid film was developed for the simultaneous determination of three deoxyribonucleic acid (DNA) bases: G, A, and T. This electrochemical sensor displayed a good photoelectronic effect and the peak currents of G, A, and T increased obviously. The proposed sensor was successfully applied to the analysis of G, A, and T contents in real-life samples such as herring sperm DNA samples, milk powder, and urine sample from human beings with satisfactory results [149]. As well, an electrochemical DNA biosensor for the detection of ssDNA sequence related to transgenic maize MON810 was fabricated based on electrochemical reduced graphene (ERG) modified carbon ionic liquid electrode (CILE) and methylene blue (MB). The presence of ERG increased the adsorption amounts of probe ssDNA sequence on the electrode resulting in the enhancement of the reduction peak current of MB that used as the hybridization indicator. Under optimal conditions, the linear range of ssDNA sequences was 1.0×10^{-11} to 1.0×10^{-6} M with the detection limit of 4.52×10^{-12} mol/L [150]. In addition, the simultaneous determination of adenine and guanine was achieved at the ionic liquid-functionalized graphene-modified carbon paste electrode (IL-graphene/CPE) with very low detection limits of 65 nM for guanine and 32 nM for adenine [13]. Moreover, TiO_2-graphene nanocomposite modified GCE exhibited high electrocatalytic activity, wide linear range, and low detection limit for the simultaneous determination of adenine and guanine. The excellent performance of this electrochemical sensor was attributed to the high adsorptivity and conductivity of TiO_2-graphene nanocomposite providing an efficient microenvironment for the electrochemical reaction of these purine bases [151]. On the other hand, a graphene-based electrochemical DNA sensor was developed for the detection of low concentrations of breast cancer-related BRCA1 gene. The sensor was stable, reproducible, and sensitive and it could detect up to 1 femtomolar BRCA1 gene (5.896 femtogram/ml) [152]. Moreover, a graphene sheets/polyaniline/gold nanoparticle based DNA sensor was employed for the electrochemical determination of BCR/ABL fusion gene in chronic myelogenous leukemia. Under optimum conditions, the DNA sensor exhibited a detection limit as low as 2.11 pM (S/N=3) with an excellent differentiation ability. It has been applied for assay of BCR/ABL fusion gene from real samples with satisfactory results [153]. Also, a riboflavin electrochemical sensor based on homoadenine single-stranded DNA/molybdenum disulfide–graphene nanocomposite modified gold electrode was fabricated. This sensor possessed a wide linear range from 0.025 to 2.25 μM with a low detection limit of 20 nM [154]. A brief summary of DNA biosensors is given in Table 6.

Electrode	DNA sequences and mutated genes	Linear range	Detection limit	Reference
zinc sulfur-coated poly (3,4-ethylenedioxythiophene)-reduced graphene oxide hybrids film	guanine, adenine and thymine	---	---	[149]
Chemically reduced graphene oxide modified GCE	guanine, adenine, thymine and cytosine	---	---	[5, 73]
electrochemical reduced graphene modified carbon ionic liquid electrode and methylene blue	ssDNA sequence related to transgenic maize MON810	1.0×10^{-11} to 1.0×10^{-6} M	4.52×10^{-12} mol/L	[150]
ionic liquid-functionalized graphene modified carbon paste electrode	adenine and guanine	---	65 nM and 32 nM for guanine and adenine, respectively	[13]
TiO_2-graphene nanocomposite modified GCE	adenine and guanine	0.5–200 μM	0.10 and 0.15 μM for adenine and guanine, respevtively	[151]
graphene-based electrochemical DNA sensor	breast cancer-related BRCA1 gene	---	---	[152]
graphene sheets/polyaniline/gold nanoparticles based DNA sensor	BCR/ABL fusion gene in chronic myelogenous leukemia	---	2.11 pM	[153]
homoadenine single-stranded DNA/molybdenum disulfide–graphene nanocomposite modified gold electrode	riboflavin	0.025 to 2.25 μM	20 nM	[154]

Table 6. Examples of modified electrodes for DNA detection.

5.7. Graphene-based immunosensor

Increasing attention has been focused on the development of immunoassay because it has become a major analytical tool in clinical diagnosis. In immunoassay, the determination of cancer markers associated with certain tumors in patients plays an important role in diagnosing cancer diseases. Carcinoembryonic antigen (CEA) is one of the most extensively used tumor markers. An elevated CEA level in serum may be an early indication of lung cancer, ovarian carcinoma, colon cancer, breast cancer, and cystadenocarcinoma. Hence, Sheng Yu used gold nanoparticles (AuNPs), reduced graphene oxide (R-GO) and poly(L-Arginine) composite material modified CILE as the platform for the construction of a new electrochemical CEA immunosensor. The poly(L-Arginine)/R-GO composite film was used to modify CILE to fabricate Arg/R-GO/CILE through electropolymerization of L-Arginine on R-GO/CILE. AuNPs were adsorbed on the modified electrode to immobilize the CEA antibody and to construct the immunosensor. By combining the specific properties such as the biocompatibility and big surface area of AuNPs, the excellent electron transfer ability of R-GO and the high conductivity of CILE, the synergistic effects of composite increased the amounts of CEA antibody adsorbed on the electrode surface resulting in the greatly increase of the electrochemical responses. The proposed immunosensor showed good reproducibility, selectivity, low detection limit, and acceptable stability [155]. As well, ultrathin Au-Pt nanowire-decorated thionine/reduced graphene oxide (AuPtNWs/THI/rGO) [156] and Au nanoparticles–graphene–chitosan nanocomposite cryogel electrode [157] were developed for CEA detection showing high sensitivity, reproducibility, stability, and applicability in real samples (Table 7). Moreover, Jiadong Huang developed an ultrasensitive electrochemical immunosensor based on nanogold particles (nano-Au), Prussian Blue (PB), polyaniline/poly(acrylic acid) (PANI (PAA)), and Au-hybrid graphene nanocomposite (AuGN) for the detection of salbutamol (SAL). SAL is the most widely used β_2-adrenergic receptor agonist that induces bronchodilation making it highly useful for curing bronchial asthma, chronic obstructive pulmonary disease, and other allergic diseases associated with respiratory pathway. Nano-Au, PB, and PANI (PAA)-incorporated film was used to enhance the electroactivity, stability, and catalytic activity for hydrogen reduction of the electrode. AuGN was used to immobilize chitosan, nano-Au, and horseradish peroxidase–anti-SAL antibody (HRP–AAb). The resulting nanostructure (AuGN–HRP–AAb) was used as the label for the immunosensor. This is attributed to the high surface-to-volume ratio of graphene that allows the immobilization of a high level of chitosan, nano-Au, and HRP–AAb and its good electrical conductivity that can improve the electron transfer among HRP, H_2O_2, and electrode. The proposed immunosensor showed low detection limit, excellent selectivity, good reproducibility, long-term stability, and high accuracy [158]. Furthermore, the synergistic effect between graphene sheet (GS) and cobalt hexacyanoferrate nanoparticle (CoNP) was investigated showing that the electroactivity of CoNP was greatly enhanced in the presence of GS due to the great electron-transfer ability of GS. A label-free electrochemical immunosensor for the sensitive detection of prostate specific antigen (PSA), it was widely used in prostate cancer diagnosis and screening, was fabricated. Molecule 1-pyrenebutanoic acid, succinimidyl ester (PBSE) was adsorbed onto GS and the colloidal solution containing GS–CoNP–PBSE was added onto the GCE surface to form a stable thin film with high electroactivity. After anti-PSA antibody was conjugated onto PBSE, the modified electrode could be used as an amperometric immunosensor for the detection of PSA. This immunosensor has the advantages of high sensitivity with a low detection limit of 0.01 ng/mL, good selectivity, and stability and could become a promising technique for cancer marker detection [159]. In addition, highly conductive graphene (GR)-gold (Au) composite modified electrode [160] and amino-functionalized graphene sheet–ferrocenecarboxaldehyde composite materials (NH_2-GS@FCA) and silver hybridized mesoporous silica nanoparticles (Ag@NH_2-MCM48) [161] were utilized as PSA immunosensor displaying an enhanced electron transfer and high sensitivity toward PSA (Table 7). On the other hand, ultrasensitive electrochemical immunosensor for carbohydrate antigen 19-9 was developed using Au nanoparticles functionalized porous graphene (Au-PGO) as sensing platform and Au@Pd core shell bimetallic functionalized graphene nanocomposites (Au@Pd-Gra) as signal enhancers [162].

Moreover, an electrochemical immunosensor for the sensitive detection of carbohydrate antigen 15-3 (CA15-3) was fabricated based on ionic liquid functionalized graphene and Cd^{2+}-functionalized nanoporous TiO_2. The good performance of the immunosensor was attributed to (i) high surface-to-volume ratio of graphene, (ii) excellent biocompatibility and electron transfer rate originating from ionic liquid functionalized graphene, and (iii) the highly specific surface area of nanoporous TiO_2 facilitating the adsorption of a high amount of Cd^{2+} for signal amplification [163].

Electrode	Immunosensor	Linear range	Detection limit	Reference
gold nanoparticles, reduced graphene oxide and poly(L-Arginine) composite material modified CILE	Carcinoembryonic antigen (CEA)	0.5 to 200 ng mL^{-1}	0.03 ng mL^{-1}	[155]
ultrathin Au-Pt nanowire-decorated thionine/reduced graphene oxide	CEA	50 fg/mL to 100 ng/mL	6 fg/mL	[156]
Au nanoparticles–graphene–chitosan nanocomposite cryogel electrode	CEA	1.0×10^{-6} to 1.0 ng/mL	2.0×10^{-7} ng/mL	[157]
nanogold particles, Prussian Blue, polyaniline/poly(acrylic acid) and Au-hybrid graphene nanocomposite	salbutamol	0.08 ng/mL and 1000 ng/mL	0.04 ng/mL	[158]
graphene sheet and cobalt hexacyanoferrate nanoparticle	prostate specific antigen (PSA)	---	0.01 ng/mL	[159]
graphene-gold composite modified electrode	PSA	0–10 ng/mL	0.59 ng/mL	[160]
amino-functionalized graphene sheet–ferrocenecarboxaldehyde composite materials and silver hybridized mesoporous silica nanoparticles	PSA	0.01–10.0 ng/mL	2 pg/mL	[161]
Au nanoparticles functionalized porous graphene and Au@Pd core shell bimetallic functionalized graphene nanocomposites	carbohydrate antigen 19-9	0.015 to 150 U/mL	0.006 U/mL	[162]
ionic liquid functionalized graphene and Cd^{2+}-functionalized nanoporous TiO_2	carbohydrate antigen 15-3	0.02–60 U/mL	0.008 U/mL	[163]

Table 7. Examples of modified electrodes used as immunosensors.

5.8. Hemoglobin and myoglobin biosensors

Hemoglobin (Hb) is the most important part of blood responsible for transporting O_2 throughout the circulatory system. Change of Hb concentration in blood can cause several diseases and even death. Therefore, accurate determination of Hb content in blood is medically very essential [73]. Chunming Wang and Jun-Jie Zhu synthesized functionalized graphene nanosheets (PDDA-G) with poly(diallyldimethylammonium chloride) (PDDA) and used to combine with room temperature ionic liquid (RTIL). The resulting RTIL/PDDA-G composite displayed an enhanced capability for the immobilization of Hb to realize its direct electrochemistry. The proposed biosensor showed a fast direct electron transfer of Hb. Moreover, the RTIL/PDDA-G based biosensor exhibited excellent electrocatalytic activity for the detection of nitrite with wide linear range and low detection limit. The present strategy definitely paves a way for the functionalization of graphene sheets with good biocompatibility and solubility, thus provides a novel and promising platform for the study of the biological application of graphene [164]. Moreover, Ke-Jing Huang fabricated a promising biosensor, hemoglobin (Hb)/chitosan (Chit)–ionic liquid (IL)–ferrocene (Fc)/graphene (Gr)/GCE, that exploited the synergistic beneficial characteristics of Fc, Gr, and IL for Hb detection. The proposed biosensor showed a strong electrocatalytic activity toward the reduction of H_2O_2, which could be attributed to the favored orientation of Hb in the well-confined surface as well as the high electrical conductivity of the resulting Chit–IL–Fc/Gr inorganic hybrid composite. The developed biosensor exhibited a fast amperometric response, a good linear response toward H_2O_2 over a wide range of concentration, and a low detection limit of 3.8 μM. High sensitivity and stability, technically simple, and the possibility of preparation at short period of time are of great advantage of the developed biosensors [165]. Moreover, GR and Mg_2Al layered double hydroxide composite was prepared and used for the immobilization of Hb on a CILE to obtain an electrochemical biosensor exhibiting electrocatalytic reduction to trichloroacetic acid [166]. As well, a biocompatible biosensor based on immobilized Hb on carboxymethyl cellulose functionalized reduced graphene oxide (rGO) nanocomposite was developed. The proposed Hb biosensor exhibited excellent bioelectrocatalytic activity toward the reduction of NO and H_2O_2. This biosensor opened up a promising platform for protein immobilization and biosensor preparation [167]. Also, a mediator-free biosensor was fabricated by the immobilization of Hb on TiO_2NPs–rGO with a double-layered structure. TiO_2NPs–rGO nanocomposite was an excellent matrix for immobilization and biocompatibility for redox protein presenting good protein bioactivity and stability. The direct electron transfer of Hb is facilitated due to the special double-layered structure of TiO_2NPs–rGO. The prepared biosensor showed good performance for H_2O_2 detection with a low detection limit of 10 nM in the linear range of 0.1–140 μM [168]. On the other hand, a novel biocompatible sensing strategy based on a graphene (GR), ionic liquid (IL) 1-ethyl-3-methylimidazolium tetrafluoroborate ($EMIMBF_4$) and chitosan (CTS) composite film for the immobilization of myoglobin (Mb) was adopted. The CTS–Mb–GR–IL modified electrode exhibited an excellent enzyme-like catalytic activity for the reduction of trichloroacetic acid [169]. Another nanocomposite based on Pt nanoparticle decorated GR nanosheet was developed for the direct electrochemistry and electrocatalysis of Mb. GR–Pt nanocomposite was integrated with Mb on the surface of CILE. Nafion/Mb–GR–Pt/CILE electrochemical biosensor had potential applications for the electrocatalytic reduction of trichloroacetic acid in the linear range of 0.9 to 9.0 mM with the detection limit of 0.32 mM [170]. Moreover, 1-hexylpyridinium hexafluorophosphate based CILE was used as a substrate for the in situ electrodeposition of GR and nickel oxide (NiO) to get NiO/GR/CILE. Nafion/Mb/NiO/GR/CILE electrochemical sensor was constructed by the immobilization of Mb on the surface of NiO/GR/CILE with a Nafion film. The presence of the NiO/GR nanocomposite on the electrode promoted the direct electron transfer of Mb. A third-generation electrochemical Mb biosensor based on NiO/GR/CILE was constructed for the electrocatalytic reduction of trichloroacetic acid and H_2O_2 [171]. Table 8 contained a brief summary of some hemoglobin and myoglobin biosensors.

Electrode	Hb and Mb biosensor	Linear range	Detection limit	Reference
poly(diallyldimethylammonium chloride) functionalized graphene nanosheets combined with room temperature ionic liquid and modified with Hb	nitrite	0.2 to 32.6 μM	0.04 μM	[164]
Hb/chitosan–ionic liquid–ferrocene/graphene/GCE	H_2O_2	50 μM to 1200 μM	3.8 μM	[165]
Hb/ GR and Mg_2Al layered double hydroxide composite	trichloroacetic acid	1.6 to 25.0 mM	0.534 mM	[166]
immobilized Hb on carboxymethyl cellulose functionalized reduced graphene oxide (rGO) nanocomposite	NO and H_2O_2	0.864 μM-19.8 μM and 0.083–13.94 μM for NO and H_2O_2, respectively	0.37 μM and 0.08 μM for NO and H_2O_2, respectively	[167]
Hb/ TiO_2NPs–rGO	H_2O_2	0.1–140 μM	10 nM	[168]
myoglobin (Mb)/graphene , 1-ethyl-3-methyl-imidazolium tetrafluoroborate and chitosan composite film	trichloroacetic acid	---	---	[169]
Mb/ Pt nanoparticle decorated GR nanosheet	trichloroacetic acid	0.9 to 9.0 mM	0.32 mM	[170]
Nafion/Mb/NiO/GR/1-hexylpyridinium hexafluorophosphate CILE	trichloroacetic acid and H_2O_2	---	---	[171]

Table 8. Examples of modified electrodes used as Hemoglobin and Myoglobin Biosensors.

5.9. Glucose and H_2O_2 biosensors

Glucose biosensors are becoming increasingly important due to their applications in biological and chemical analyses, clinical detection, and environmental monitoring. The immobilization of enzymes on electrodes is generally the first step in fabrication attracting significant efforts because enzymes are highly selective and quickly responsive to specific substrates. Glucose oxidase (GOD) is employed as the enzyme in most of the glucose biosensors. It catalyzes the electron transfer from glucose to oxygen accompanied by the production of gluconolactone and hydrogen peroxide. The quantification of glucose can be achieved via electrochemical detection of the enzymatically liberated H_2O_2 [172]. Hua Zhang proposed a direct electrochemical method to reduce single-layer GO adsorbed on the 3-aminopropyltriethoxysilane (APTES) modified GCE. The reduced GO was modified with GOD and the obtained GCE-APTES-rGO-GOD was successfully used to detect glucose opening up a potential application in biosensing [52]. In addition, a bionanocomposite film consisting of GOD/Pt/functional graphene sheets/chitosan (GOD/Pt/FGS/chitosan) was employed for glucose sensing. With the electrocatalytic synergy of FGS and Pt nanoparticles to H_2O_2, a sensitive biosensor with a detection limit of 0.6 μM for glucose was achieved displaying good reproducibility, long-term stability, and anti-interfering ability from AA and UA. The large surface area and good electrical conductivity of graphene suggests that graphene is a potential candidate as a sensor material [173]. Also, AuNPs–graphene nanocomposite was prepared and its electrocatalytic activity toward glucose oxidation was investigated. The results indicated that AuNPs–graphene nanocomposite had a higher catalytic activity than AuNPs or graphene alone indicating the synergistic effect of graphene and AuNPs [174]. Furthermore, Li Niu developed a novel glucose biosensor based on graphene/AuNPs/chitosan at gold electrode exhibiting good electrocatalytic activity toward H_2O_2 and O_2. With GOD as a model, the graphene/AuNPs/GOD/chitosan composite-modified electrode was constructed through a simple casting method. The resulting biosensor exhibited good amperometric response to glucose with linear range, good reproducibility, and low detection limit of 180 μM [175]. Also, amperometric glucose biosensors were fabricated by using Pt–Au and Au nanoparticle spacers decorated functionalized graphene (f-G) nanosheets. The immobilization of GOD over nafion-solubilized metal nanoparticles dispersed graphene f-G-(Pt–Au) and f-G-(Au) electrodes has been achieved by physical adsorption. The resultant bioelectrode retains its biocatalytic activity offering fast and sensitive glucose quantification. The fabricated f-G-(Au) based glucose biosensor exhibited best sensing performance with a linear response up to 30 mM with an excellent detection limit of 1 μM [172]. Moreover, a dual biosensor platform based on hemin-functionalized graphene nanosheets (H-GNs) was presented. The nanomaterial combines the features of both graphene (high conductivity and surface area) and hemin (excellent catalysis and intrinsic peroxidase-like activity). Based on these features, the simple, economical, and highly sensitive amperometric and colorimetric biosensors for H_2O_2 and glucose have been developed. The detection limits for H_2O_2 and glucose reached 0.2 μM and 0.3 μM by the amperometric method, respectively. By the colorimetric methods, the detection limits for H_2O_2 and glucose were as low as 20 nM and 30 nM, respectively. The rapid, simple, and sensitive sensing platform showed great promising applications in the pharmaceutical, clinical, and industrial detection of H_2O_2 and glucose [176]. Also, GCE modified with electrochemically reduced graphene oxide/sodium dodecyl sulfate (GCE/ERGO/SDS) was further modified with GOD to be used as an enzymatic glucose biosensor with a low detection limit of 40.8 μM [177]. On the other hand, graphene functionalized with poly(diallyldimethylammonium chloride) (PDDA) can be used as non-enzymatic H_2O_2 biosensor showing enhanced electrocatalytic activity toward H_2O_2 reduction. As well, this modified surface can be used as an enzymatic glucose sensor by the immobilization of the negatively charged GOD onto the positively charged PDDA-G matrix through electrostatic interaction. GOD/PDDA-G bionanocomposite can detect glucose with a low detection limit of 8 μM in the linear range of 0.02 to 1.8 mM [178]. In addition, a graphene and cobalt oxide nanoparticle composite was employed as non-enzymatic H_2O_2 and enzymatic glucose biosensor [179]. Several modified surfaces were utilized as non-enzymatic H_2O_2 biosensors, which were summarized in Table 9. On the other hand, a stable and sensitive non-enzymatic glucose biosensor based on CuO and rGO nanocomposite was developed. The proposed biosensor presented a high sensitivity of 2221 μA/mM.cm^2 and a wide linear range from 0.4 μM to 12 mM toward glucose with good selectivity and stability. The TEM of rGO sheets showed a typical wrinkled and crumpled shape with thin layers. Upon CuO modification, there is a uniform distribution of rod-like nanoparticles with an average length of 20–30 nm on the surface of rGO indicating the successful synthesis of CuO-rGO nanocomposite [186]. Also, Ni(OH)$_2$/electroreduced graphene oxide-MWCNTs film was employed as non-enzymatic H_2O_2 and glucose biosensors [187].

Electrode	H_2O_2 and glucose biosensor	Linear range	Detection limit	Reference
GOD modified reduced GO adsorbed on the 3-aminopropyltriethoxysilane (APTES) modified GCE	enzymatic glucose	---	---	[52]
GOD/Pt/functional graphene sheets/chitosan	enzymatic glucose	---	0.6 μM	[173]
AuNPs–graphene nanocomposite	enzymatic glucose	---	---	[174]
graphene/AuNPs/GOD/chitosan composite-modified electrode	enzymatic glucose	2 to 10 mM	180 μM	[175]
GOD modified Pt–Au and Au nanoparticle spacers decorated functionalized graphene (f-G) nanosheets	enzymatic glucose	up to 30 mM	1 μM	[172]
hemin-functionalized graphene nanosheets (H-GNs)	H_2O_2 and glucose	---	0.2 μM for H_2O_2 and 0.3 μM for glucose	[176]
GCE modified with electrochemically reduced graphene oxide/sodium dodecyl sulfate and GOD	enzymatic glucose	---	40.8 μM	[177]
graphene functionalized with poly(diallyldimethylammonium chloride)	Non-enzymatic H_2O_2 and enzymatic glucose	0.02 to 1.8 mM	8 μM	[178]

Electrode	H_2O_2 and glucose biosensor	Linear range	Detection limit	Reference
graphene and cobalt oxide nanoparticle composite	non-enzymatic H_2O_2 and enzymatic glucose	0.5 mM to 16.5 mM for glucose. 0.2 to 211.5 μM for H_2O_2	0.06 μM for H_2O_2	[179]
prussian blue nanocubes on nitrobenzene-functionalized reduced graphene oxide	non-enzymatic H_2O_2	1.2 μM to 15.25 mM	0.4 μM	[180]
layer-by-layer deposited low dimensional building blocks of graphene-prussian blue onto graphite screen-printed electrodes	non-enzymatic H_2O_2	---	---	[181]
Ag nanoparticles/graphene nanosheet	non-enzymatic H_2O_2	0.1 to 100 mM	0.5 μM	[61]
prussian blue-graphene modified GCE	non-enzymatic H_2O_2	10–1440 μM	3 μM	[182]
graphene/Nafion/azure I/Au nanoparticles composites modified GCE	non-enzymatic H_2O_2	30 μM to 5 mM	10 μM	[183]
rGO and Fe_3O_4 nanocomposites	non-enzymatic H_2O_2	0.1 mM to 6 mM	3.2 μM	[184]
cationic polyelectrolyte-functionalized ionic liquid decorated graphene sheets nanocomposite	non-enzymatic H_2O_2	---	---	[185]
CuO and rGO nanocomposite	non-enzymatic glucose	0.4 μM to 12 mM	---	[186]
$Ni(OH)_2$/electroreduced graphene oxide-MWCNTs film	non-enzymatic H_2O_2 and non-enzymatic glucose	10-1500 μM and 10 μM-9050 μM for glucose and H_2O_2, respectively	2.7 μM and 4.0 μM for glucose and H_2O_2, respectively	[187]

Table 9. Examples of modified electrodes used as Glucose and H_2O_2 Biosensors.

6. Conclusions

Graphene exhibits a unique chemical structure and outstanding properties making it very attractive for various engineering applications. Graphene and graphene-based materials have a profound impact on electroanalysis, electrocatalysis, sensors, and biosensors. In sensing applications, graphene-based materials featured with good conductivity and large specific surface area have demonstrated accurate, rapid, selective, sensitive, and even single-molecular sensing abilities. Graphene-based materials are ideally used as gas sensors, electrochemical sensors for heavy metal ions, immunosensors and NADH, DNA, catecholamine neurotransmitters, paracetamol, glucose, H_2O_2, hemoglobin, and myoglobin biosensors. Future investigations on graphene-based sensing platforms, combined with versatile sensing strategies, are expected to continuingly lower the detection threshold.

7. Permissions

Figure number (in chapter)	Figure Source (reference number)	Permission number
2	[6]	3556440341708
3	[1]	3556440556169
4	[4]	Open access
5	[31]	Open access
6	[66]	Open access
7	[77]	3555441025155
8	[79]	3555441165370
Table 1	[6]	3555460926032

Author details

Nada F. Atta*, Ahmed Galal and Ekram H. El-Ads

*Address all correspondence to: Nada_fah1@yahoo.com

Departement of Chemistry, Faculty of Science, Cairo University, 12613, Giza, Egypt

References

[1] Deshpande A, LeRoy B. Scanning probe microscopy of graphene. Physica E 2012;44:743–759. DOI:10.1016/j.physe.2011.11.024

[2] Tang L, Wang Y, Li Y, Feng H, Lu J, Li J. Preparation, structure, and electrochemical properties of reduced graphene sheet films. Advanced Functional Materials 2009;19:2782–2789. DOI:10.1002/adfm.200900377

[3] Soldano C, Mahmood A, Dujardin E. Production, properties and potential of graphene. Carbon 2010;48:2127–2150. DOI:10.1016/j.carbon.2010.01.058

[4] Arco L, Zhang Y, Zhou C. Large Scale Graphene by Chemical Vapor Deposition: Synthesis, Characterization and Applications. In: Gong J, editor. Graphene–Synthesis, Characterization, Properties and Applications. Croatia: InTech; 2011. pp. 161–184. DOI:10.5772/1742.ch10

[5] Shao Y, Wang J, Wu H, Liu J, Aksay I, Lin Y. Graphene based electrochemical sensors and biosensors: A review. Electroanalysis 2010;22:1027–1036. DOI:10.1002/elan.200900571

[6] Kuila T, Bose S, Mishra A, Khanra P, Kim N, Lee J. Chemical functionalization of graphene and its applications. Progress in Materials Science 2012;57:1061–1105. DOI:10.1016/j.pmatsci.2012.03.002

[7] Fan Y, Liu J, Lu H, Zhang Q. Electrochemical behavior and voltammetric determination of paracetamol on Nafion/TiO_2–graphene modified glassy carbon electrode. Colloids and Surfaces B: Biointerfaces 2011;85:289–292. DOI:10.1016/j.colsurfb.2011.02.041

[8] Li F, Li J, Feng Y, Yang L, Du Z. Electrochemical behavior of graphene doped carbon paste electrode and its application for sensitive determination of ascorbic acid. Sensors and Actuators B 2011;157:110–114. DOI:10.1016/j.snb.2011.03.033

[9] Odahara G, Ishikawa T, Fukase K, Otani S, Oshima C, Suzuki M, Yasue T, Koshikawa T. Self-Standing Graphene Sheets Prepared with Chemical Vapor Deposition and Chemical Etching. In: Gong J, editor. Graphene–Synthesis, Characterization, Properties and Applications. Croatia: InTech; 2011. pp. 3–20. DOI:10.5772/1742.ch1

[10] Wang Y, Li Y, Tang L, Lu J, Li J. Application of graphene-modified electrode for selective detection of dopamine. Electrochemistry Communications 2009;11:889–892. DOI:10.1016/j.elecom.2009.02.013

[11] Craciun M, Russo S, Yamamoto M, Tarucha S. Tuneable electronic properties in graphene. Nano Today 2011;6:42–60. DOI:10.1016/j.nantod.2010.12.001

[12] Kang X, Wang J, Wu H, Liu J, Aksay I, Lin Y. A graphene-based electrochemical sensor for sensitive detection of paracetamol. Talanta 2010;81:754–759. DOI:10.1016/j.talanta.2010.01.009

[13] Du M, Yang T, Ma S, Zhao C, Kui J. Ionic liquid-functionalized graphene as modifier for electrochemical and electrocatalytic improvement: Comparison of different carbon electrodes. Analytica Chimica Acta 2011;690:169–174. DOI:10.1016/j.aca.2011.01.051

[14] Wintterlin J, Bocquet M. Graphene on metal surfaces. Surface Science 2009;603:1841–1852. DOI:10.1016/j.susc.2008.08.037

[15] Huang X, Yin Z, Wu S, Qi X, He Q, Zhang Q, Yan Q, Boey F, Zhang H. Graphene-based materials: Synthesis, characterization, properties, and applications. Small 2011;7:1876–1902. DOI:10.1002/smll.201002009

[16] Wang X, Chen S. Graphene-Based Nanocomposites. In: Mikhailov S, editor. Physics and Applications of Graphene – Experiments. Croatia: InTech; 2011. pp. 135–168. DOI:10.5772/590.ch8

[17] Chen L, Tang Y, Wang Ke, Liu C, Luo S. Direct electrodeposition of reduced graphene oxide on glassy carbon electrode and its electrochemical application. Electrochemistry Communications 2011;13:133–137. DOI:10.1016/j.elecom.2010.11.033

[18] Pumera M, Ambrosi A, Bonanni A, Chng E, Poh H. Graphene for electrochemical sensing and biosensing. Trends in Analytical Chemistry 2010;29:954–965. DOI: 10.1016/j.trac.2010.05.011

[19] Wu W, Liu Z, Jauregui L, Yu Q, Pillai R, Cao H, Bao J, Chen Y, Pei S. Wafer-scale synthesis of graphene by chemical vapor deposition and its application in hydrogen sensing. Sensors and Actuators B 2010;150:296–300. DOI:10.1016/j.snb.2010.06.070

[20] Kalita G, Qi L, Namba Y, Wakita K, Umeno M. Femtosecond laser induced micropatterning of graphene film. Materials Letters 2011;65:1569–1572. DOI:10.1016/j.matlet.2011.02.057

[21] Feng T, Xie D, Tian H, Peng P, Zhang D, Fu D, Ren T, Li X, Zhu H, Jing Y. Multilayer graphene treated by O_2 plasma for transparent conductive electrode applications. Materials Letters 2012;73:187–189. DOI:10.1016/j.matlet.2011.12.121

[22] Choi Y, Kang S, Kim H, Choi W, Na S. Multilayer graphene films as transparent electrodes for organic photovoltaic devices. Solar Energy Materials & Solar Cells 2012;96:281–285. DOI:10.1016/j.solmat.2011.09.031

[23] Somani P, Somani S, Umeno M. Planer nano-graphenes from camphor by CVD. Chemical Physics Letters 2006;430:56–59. DOI:10.1016/j.cplett.2006.06.081

[24] An H, Lee W, Jung J. Graphene synthesis on Fe foil using thermal CVD. Current Applied Physics 2011;11:S81–S85. DOI:10.1016/j.cap.2011.03.077

[25] Terasawa T, Saiki K. Growth of graphene on Cu by plasma enhanced chemical vapor deposition. Carbon 2012;50:869–874. DOI:10.1016/j.carbon.2011.09.047

[26] Juang Z, Wu C, Lu A, Su C, Leou K, Chen F, Tsai C, Graphene synthesis by chemical vapor deposition and transfer by a roll-to-roll process. Carbon 2010;48:3169–3174. DOI:10.1016/j.carbon.2010.05.001

[27] Guermoune A, Chari T, Popescu F, Sabri S, Guillemette J, Skulason H, Szkopek T, Siaj M. Chemical vapor deposition synthesis of graphene on copper with methanol, ethanol, and propanol precursors. Carbon 2011;49:4204–4210. DOI:10.1016/j.carbon.2011.05.054

[28] Gautam M, Jayatissa A. Gas sensing properties of graphene synthesized by chemical vapor deposition. Materials Science and Engineering C 2011;31:1405–1411. DOI: 10.1016/j.msec.2011.05.008

[29] Liu W, Chung C, Miao C, Wang Y, Li B, Ruan L, Patel K, Park Y, Woo J, Xie Y. Chemical vapor deposition of large area few layer graphene on Si catalyzed with nickel films. Thin Solid Films 2010;518:S128–S132. DOI:10.1016/j.tsf.2009.10.070

[30] Negishi R, Hirano H, Ohno Y, Maehashi K, Matsumoto K, Kobayashi Y. Layer-by-layer growth of graphene layers on graphene substrates by chemical vapor deposition. Thin Solid Films 2011;519:6447–6452. DOI:10.1016/j.tsf.2011.04.229

[31] Mwakikunga B, Hillie K. Graphene Synthesis, Catalysis with Transition Metals and Their Interactions by Laser Photolysis. In: Gong J, editor. Graphene–Synthesis, Characterization, Properties and Applications. Croatia: InTech; 2011. pp. 59–78. DOI: 10.5772/1742.ch5

[32] Singh V, Joung D, Zhai L, Das S, Khondaker S, Seal S. Graphene based materials: Past, present and future. Progress in Materials Science 2011;56:1178–1271. DOI: 10.1016/j.pmatsci.2011.03.003

[33] Dhakate S, Chauhan N, Sharma S, Mathur R. The production of multi-layer graphene nanoribbons from thermally reduced unzipped multi-walled carbon nanotubes. Carbon 2011;49:4170–4178. DOI:10.1016/j.carbon.2011.05.060

[34] Cataldo F, Compagnini G, Patané G, Ursini O, Angelini G, Ribic P, Margaritondo G, Cricenti A, Palleschi G, Valentini F. Graphene nanoribbons produced by the oxidative unzipping of single-wall carbon nanotubes. Carbon 2010;48:2596–2602. DOI: 10.1016/j.carbon.2010.03.063

[35] Janowska I, Ersen O, Jacob T, Vennégues P, Bégin D, Ledoux M, Pham-Huu C. Catalytic unzipping of carbon nanotubes to few-layer graphene sheets under microwaves irradiation. Applied Catalysis A: General 2009;371:22–30. DOI:10.1016/j.apcata.2009.09.013

[36] Shen B, Ding J, Yan X, Feng W, Li J, Xue Q. Influence of different buffer gases on synthesis of few-layered graphene by arc discharge method. Applied Surface Science 2012;258:4523–4531. DOI:10.1016/j.apsusc.2012.01.019

[37] Chen Y, Zhao H, Sheng L, Yu L, An K, Xu J, Ando Y, Zhao X. Mass-production of highly-crystalline few-layer graphene sheets by arc discharge in various H_2-inert gas mixtures. Chemical Physics Letters 2012;538:72–76. DOI:10.1016/j.cplett.2012.04.020

[38] Li N, Wang Z, Zhao K, Shi Z, Gu Z, Xu S. Large scale synthesis of N-doped multi-layered grapheme sheets by simple arc-discharge method. Carbon 2010;48:255–259. DOI:10.1016/j.carbon.2009.09.013

[39] Levchenko I, Volotskova O, Shashurin A, Raitses Y, Ostrikov K, Keidar M. The large-scale production of graphene flakes using magnetically-enhanced arc discharge between carbon electrodes. Carbon 2010;48:4556–4577. DOI:10.1016/j.carbon.2010.07.055

[40] Hummers W, Offeman R. Preparation of graphitic oxide. Journal of the American Chemical Society 1958;80(6):1339–1339. DOI:10.1021/ja01539a017

[41] Stankovich S, Dikin D, Piner R, Kohlhaas K, Kleinhammes A, Jia Y, Wu Y, Nguyen S, Ruoff R. Synthesis of graphene-based nanosheets via chemical reduction of exfoliated graphite oxide. Carbon 2007;45:1558–1565. DOI:10.1016/j.carbon.2007.02.034

[42] Phama T, Kimb J, Kima J, Jeong Y. One-step reduction of graphene oxide with l-glutathione. Colloids and Surfaces A: Physicochemical and Engineering Aspects 2011;384:543–548. DOI:10.1016/j.colsurfa.2011.05.019

[43] Marcano D, Kosynkin D, Berlin J, Sinitskii A, Sun Z, Slesarev A, Alemany L, Lu W, Tour J. Improved synthesis of graphene oxide. AcsNano 2010;4:4806–4814. DOI: 10.1021/nn1006368

[44] Kai-xuan S, Yu-xi X, Chun L, Gao-quan S. High-performance self-assembled graphene hydrogels prepared by chemical reduction of graphene oxide. New Carbon Materials 2011;26(1):9–15. DOI:10.1016/S1872-5805(11)60062-0

[45] Ramesha G, Sampath S. Electrochemical reduction of oriented graphene oxide films: An in situ Raman spectroelectrochemical study. The Journal of Physical Chemistry C 2009;113:7985–7989. DOI:10.1021/jp811377n

[46] Chen W, Yan L. Preparation of graphene by a low-temperature thermal reduction at atmosphere pressure. Nanoscale 2010;2:559–563. DOI:10.1039/B9NR00191C

[47] Pei S, Cheng H. The reduction of graphene oxide. Carbon 2012;50:3210–3228. DOI: 10.1016/j.carbon.2011.11.010

[48] Cuong T, Pham V, Tran Q, Chung J, Shin E, Kim J, Kim E. Optoelectronic properties of graphene thin films prepared by thermal reduction of graphene oxide. Materials Letters 2010;64:765–767. DOI:10.1016/j.matlet.2010.01.009

[49] Wei A, Wang J, Long Q, Liu X, Li X, Dong X, Huang W. Synthesis of high-performance graphene nanosheets by thermal reduction of graphene oxide. Materials Research Bulletin 2011;46:2131–2134. DOI:10.1016/j.materresbull.2011.06.036

[50] Le L, Ervin M, Qiu H, Fuchs B, Lee W. Graphene supercapacitor electrodes fabricated by inkjet printing and thermal reduction of graphene oxide. Electrochemistry Communications 2011;13:355–358. DOI:10.1016/j.elecom.2011.01.023

[51] Zhao B, Liu P, Jiang Y, Pan D, Tao H, Song J, Fang Tao, Xu W. Supercapacitor performances of thermally reduced graphene oxide. Journal of Power Sources 2012;198:423–427. DOI:10.1016/j.jpowsour.2011.09.074

[52] Wang Z, Zhou X, Zhang J, Boey F, Zhang H. Direct electrochemical reduction of single-layer graphene oxide and subsequent functionalization with glucose oxidase. The Journal of Physical Chemistry C 2009;113:14071–14075. DOI:10.1021/jp906348x

[53] Hassan H, Abdelsayed V, Khder A, AbouZeid K, Terner J, El-Shall M, Al-Resayes S, El-Azhary A. Microwave synthesis of graphene sheets supporting metal nanocrystals in aqueous and organic media. Journal of Materials Chemistry 2009;19:3832–3837. DOI:10.1039/b906253j

[54] Wang S, Jiang S, Wang X. Microwave-assisted one-pot synthesis of metal/metal oxide nanoparticles on graphene and their electrochemical applications. Electrochimica Acta 2011;56:3338–3344. DOI:10.1016/j.electacta.2011.01.016

[55] Hu H, Zhao Z, Zhou Q, Gogotsi Y, Qiu J. The role of microwave absorption on formation of graphene from graphite oxide. Carbon 2012;50:3267–3273. DOI:10.1016/j.carbon.2011.12.005

[56] Liao C, Liao C, Tso C, Shy H. Microwave-polyol synthesis and electrocatalytic performance of Pt/graphene nanocomposites. Materials Chemistry and Physics 2011;130:270–274. DOI:10.1016/j.matchemphys.2011.06.038

[57] Xin Y, Liu J, JiX, Liu W, Liu F, Yin Y, Gu J, Zou Z. Preparation and electrochemical characterization of nitrogen doped graphene by microwave as supporting materials for fuel cell catalysts. Electrochimica Acta 2012;60:354–358. DOI:10.1016/j.electacta.2011.11.062

[58] Galal A, Atta N, Ali S. Optimization of the synthesis conditions for $LaNiO_3$ catalyst by microwave assisted citrate method for hydrogen production. Applied Catalysis A: General 2011;409–410:202–208. DOI:10.1016/j.apcata.2011.10.005

[59] Galal A, Atta N, Ali S. Investigation of the catalytic activity of $LaBO_3$ (B = Ni, Co, Fe or Mn) prepared by the microwave-assisted method for hydrogen evolution in acidic medium. Electrochimica Acta 2011;56:5722–5730. DOI:10.1016/j.electacta.2011.04.045

[60] Galal A, Darwish S, Atta N, Ali S, Abd El Fatah A. Synthesis, structure and catalytic activity of nano-structured Sr–Ru–O type perovskite for hydrogen production. Applied Catalysis A: General 2010;378:151–159. DOI:10.1016/j.apcata.2010.02.015

[61] Liu S, Tian J, Wang L, Sun X. Microwave-assisted rapid synthesis of Ag nanoparticles/graphene nanosheet composites and their application for hydrogen peroxide detection. Jornal of Nanoparticle Research 2011;13:4539–4548. DOI:10.1007/s11051-011-0410-3

[62] Siamaki A, Khder A, Abdelsayed V, El-Shall M, Gupton B. Microwave-assisted synthesis of palladium nanoparticles supported on graphene: A highly active and recyclable catalyst for carbon–carbon cross-coupling reactions. Journal of Catalysis 2011;279:1–11. DOI:10.1016/j.jcat.2010.12.003

[63] Liu S, Wang J, Zeng J, Ou J, Li Z, Liu X, Yang S. "Green" electrochemical synthesis of Pt/graphene sheet nanocomposite film and its electrocatalytic property. Journal of Power Sources 2010;195:4628–4633. DOI:10.1016/j.jpowsour.2010.02.024

[64] Shao Y, Wang J, Engelhard M, Wang C. Lin Y. Facile and controllable electrochemical reduction of graphene oxide and its applications. Journal of Materials Chemistry 2010;20:743–748. DOI:10.1039/B917975E

[65] Harima Y, Setodoi S, Imae I, Komaguchi K, Ooyama Y, Ohshita J, Mizota H, Yano J. Electrochemical reduction of graphene oxide in organic solvents. Electrochimica Acta 2011;56:5363–5368. DOI:10.1016/j.electacta.2011.03.117

[66] Galal A, Atta N, Hassan H. Graphene supported-Pt-M (M = Ru or Pd) for electrocatalytic methanol oxidation. International Journal of Electrochemical Science 2012;7:768–784

[67] Peng X, Liu X, Diamond D, Lau K. Synthesis of electrochemically-reduced graphene oxide film with controllable size and thickness and its use in supercapacitor. Carbon 2011;49:3488–3496. DOI:10.1016/j.carbon.2011.04.047

[68] Huang L, Liu Y, Ji L, Xie Y, Wang T, Shi W. Pulsed laser assisted reduction of graphene oxide. Carbon 2011;49:2431–2436. DOI:10.1016/j.carbon.2011.01.067

[69] Zhang Y, Guo L, Wei S, He Y, Xia H, Chen Q, Sun H, Xiao F. Direct imprinting of microcircuits on grapheme oxides film by femtosecond laser reduction. Nano Today 2010;5:15–20. DOI:10.1016/j.nantod.2009.12.009

[70] El-Kady M, Strong V, Dubin S, Kaner R, Laser scribing of high-performance and flexible graphene-based electrochemical capacitors. Science 2012;335:1326–1330. DOI: 10.1126/science.1216744

[71] Artiles M, Rout C, Fisher T. Graphene-based hybrid materials and devices for biosensing. Advanced Drug Delivery Reviews 2011;63:1352–1360. DOI:10.1016/j.addr.2011.07.005

[72] Fan Y, Liu J, Yang C, Yu M, Liu P. Graphene–polyaniline composite film modified electrode for voltammetric determination of 4-aminophenol. Sensors and Actuators B 2011;157:669–674. DOI:10.1016/j.snb.2011.05.053

[73] Kuila T, Bose S, Khanra P, Mishra A, Kim N, Lee J. Recent advances in graphene-based biosensors. Biosensors and Bioelectronics 2011;26:4637–4648. DOI:10.1016/j.bios.2011.05.039

[74] Khadem S, Abdi Y, Darbari S, Ostovari F. Investigating the effect of gas absorption on the electromechanical and electrochemical behavior of graphene/ZnO structure, suitable for highly selective and sensitive gas sensors. Current Applied Physics 2014;14(11):1498–1503. DOI:10.1016/j.cap.2014.07.020

[75] Zhou Y, Jiang Y, Xie T, Tai H, Xie G. A novel sensing mechanism for resistive gas sensors based on layered reduced graphene oxide thin films at room temperature. Sensors and Actuators B 2014;203:135–142. DOI:10.1016/j.snb.2014.06.105

[76] Mishra S, Tripathi S, Choudhary V, Gupta B. SPR based fibre optic ammonia gas sensor utilizing nanocomposite film of PMMA/reduced graphene oxide prepared by in situ polymerization. Sensors and Actuators B 2014;199:190–200. DOI:10.1016/j.snb.2014.03.109

[77] Uddin A, Phan D, Chung G. Low temperature acetylene gas sensor based on Ag nanoparticles-loaded ZnO-reduced graphene oxide hybrid. Sensors and Actuators B 2015;207:362–369. DOI:10.1016/j.snb.2014.10.091

[78] Wang Z, Wang H, Zhang Z, Yang X, Liu G. Sensitive electrochemical determination of trace cadmium on astannum film/poly(p-aminobenzene sulfonic acid)/electrochemically reduced graphene composite modified electrode. Electrochimica Acta 2014;120:140–146. DOI:10.1016/j.electacta.2013.12.068

[79] Li J, Guo S, Zhai Y, Wang E. High-sensitivity determination of lead and cadmium based on the nafion-graphene composite film. Analytica Chimica Acta 2009;649:196–201. DOI:10.1016/j.aca.2009.07.030

[80] Promphet N, Rattanarat P, Chailapakul O, Rangkupan R, Rodthongkum N. An electrochemical sensor based on graphene/polyaniline/polystyrene nanoporous fibers modified electrode for simultaneous determination of lead and cadmium. Sensors and Actuators B 2015;207:526–534. DOI:10.1016/j.snb.2014.10.126

[81] Wang Z, Wang H, Zhang Z, Liu G. Electrochemical determination of lead and cadmium in rice by a disposable bismuth/electrochemically reduced graphene/ionic liquid composite modified screen-printed electrode. Sensors and Actuators B 2014;199:7–14. DOI:10.1016/j.snb.2014.03.092

[82] Deng X, Lu L, Li H, Luo F. The adsorption properties of Pb(II) and Cd(II) on functionalized grapheme prepared by electrolysis method. Journal of Hazardous Materials 2010;183:923–930. DOI:10.1016/j.jhazmat.2010.07.117

[83] Huang H, Chen T, Liu X, Ma H. Ultrasensitive and simultaneous detection of heavy metal ions based on three-dimensional graphene-carbon nanotubes hybrid electrode materials. Analytica Chimica Acta 2014;852:45–54. DOI:10.1016/j.aca.2014.09.010

[84] Mishra A, Ramaprabhu S. Functionalized graphene sheets for arsenic removal and desalination of sea water. Desalination 2011;282:39–45. DOI:10.1016/j.desal.2011.01.038

[85] Ramesha G, Sampath S. In-situ formation of graphene-lead oxide composite and its use in trace arsenic detection. Sensors and Actuators B 2011;160:306–311. DOI:10.1016/j.snb.2011.07.053

[86] Ren Y, Yan N, Wen Q, Fan Z, Wei T, Zhang M, Ma J. Graphene/δ-MnO_2 composite as adsorbent for the removal of nickel ions from waste water. Chemical Engineering Journal 2010;175:1–7. DOI:10.1016/j.cej.2010.08.010

[87] Atta N, Galal A, Abu-Attia F, Azab S. Simultaneous determination of paracetamol and neurotransmitters in biological fluids using a carbon paste sensor modified with gold nanoparticles. Journal of Materials Chemistry 2011;21:13015–13024. DOI: 10.1039/c1jm11795e

[88] Atta N, El-Kady M, Galal A. Palladium nanoclusters-coated polyfuran as a novel sensor for catecholamine neurotransmitters and paracetamol. Sensors and Actuators B 2009;141:566–574. DOI:10.1016/j.snb.2009.07.002

[89] Chen X, Zhu J, Xi Q, Yang W. A high performance electrochemical sensor for acetaminophen based on single-walled carbon nanotube–graphene nanosheet hybrid films. Sensors and Actuators B: Chemical 2012;161(1):648–654. DOI:10.1016/j.snb.2011.10.085

[90] Arvand M, Gholizadeh T. Simultaneous voltammetric determination of tyrosine and paracetamol using a carbon nanotube-graphene nanosheet nanocomposite modified electrode in human blood serum and pharmaceuticals. Colloids and Surfaces B: Biointerfaces 2013;103:84–93. DOI:10.1016/j.colsurfb.2012.10.024

[91] Adhikari B, Govindhan M, Chen A. Sensitive detection of acetaminophen with graphene-based electrochemical sensor. Electrochimica Acta. DOI:10.1016/j.electacta.2014.10.028

[92] Liu G, Chen H, Lin G, Ye P, Wang X, Jiao Y, Guo X, Wen Y, Yang H. One-step electrodeposition of graphene loaded nickel oxides nanoparticles for acetaminophen detection. Biosensors and Bioelectronics 2014;56:26–32. DOI:10.1016/j.bios.2014.01.005

[93] Zhu W, Huang H, Gao X, Ma H. Electrochemical behavior and voltammetric determination of acetaminophen based on glassy carbon electrodes modified with poly(4-aminobenzoic acid)/electrochemically reduced graphene oxide composite films. Materials Science and Engineering: C 2014;45:21–28. DOI:10.1016/j.msec.2014.08.067

[94] Atta N, El-Kady M, Galal A. Simultaneous determination of catecholamines, uric acid and ascorbic acid at physiological levels using poly(N-methylpyrrole)/Pd-nanoclusters sensor. Analytical Biochemistry 2010;400:78–88. DOI:10.1016/j.ab.2010.01.001

[95] Atta N, El-Kady M. Poly(3-methylthiophene)/palladium sub-micro-modified sensor electrode. Part II: Voltammetric and EIS studies, and analysis of catecholamine neurotransmitters, ascorbic acid and acetaminophen. Talanta 2009;79:639–647. DOI: 10.1016/j.talanta.2009.04.040

[96] Atta N, Galal A, El-Ads E. A novel sensor of cysteine self-assembled monolayers over gold nanoparticles for the selective determination of epinephrine in presence of sodium dodecyl sulfate. Analyst 2012;137:2658–2668. DOI:10.1039/c2an16210e

[97] Cui F, Zhang X. Electrochemical sensor for epinephrine based on a glassy carbon electrode modified with graphene/gold nanocomposites. Journal of Electroanalytical Chemistry 2012;669:35–41. DOI:10.1016/j.jelechem.2012.01.021

[98] Rosy, Yadav S, Agrawal B, Oyama M, Goyal R. Graphene modified palladium sensor for electrochemical analysis of norepinephrine in pharmaceuticals and biological fluids. Electrochimica Acta 2014;125:622–629. DOI:10.1016/j.electacta.2014.01.160

[99] Atta N, El-Kady M. Novel poly(3-methylthiophene)/Pd, Pt nanoparticle sensor: Synthesis, characterization and its application to the simultaneous analysis of dopamine and ascorbic acid in biological fluids. Sensors and Actuators B 2010;145:299–310. DOI:10.1016/j.snb.2009.12.014

[100] Kim Y, Bong S, Kang Y, Yang Y, Mahajan R, Kim J, Kim H. Electrochemical detection of dopamine in the presence of ascorbic acid using graphene modified electrodes. Biosensors and Bioelectronics 2010;25:2366–2369. DOI:10.1016/j.bios.2010.02.031

[101] Mao Y, Bao Y, Gan S, Li F, Niu L. Electrochemical sensor for dopamine based on a novel graphene-molecular imprinted polymers composite recognition element. Biosensors and Bioelectronics 2011;28:291–297. DOI:10.1016/j.bios.2011.07.034

[102] Li F, Jia Chai, Yang H, Han D, Niu Li. Synthesis of Pt/ionic liquid/graphene nanocomposite and its simultaneous determination of ascorbic acid and dopamine. Talanta 2010;81:1063–1068. DOI:10.1016/j.talanta.2010.01.061

[103] Fan Y, Lu H, Liu J, Yang C, Jing Q, Zhang Y, Yang X, Huang K. Hydrothermal preparation and electrochemical sensing properties of TiO_2–graphene nanocomposite. Colloids and Surfaces B: Biointerfaces 2011;83:78–82. DOI:10.1016/j.colsurfb.2010.10.048

[104] Liu W, Xiao J, Wang C, Yin H, Xie H, Cheng R. Synthesis of polystyrene-grafted-graphene hybrid and its application in electrochemical sensor of dopamine. Materials Letters 2013;100:70–73. DOI:10.1016/j.matlet.2013.02.075

[105] Zhang W, Zheng J, Shi J, Lin Z, Huang Q, Zhang H, Wei C, Chen J, Hu S, Hao A. Nafion covered core–shell structured Fe_3O_4@graphene nanospheres modified electrode for highly selective detection of dopamine. Analytica Chimica Acta 2015;853:285–280. DOI:10.1016/j.aca.2014.10.032

[106] Pruneanu S, Biris A, Pogacean F, Socaci C, Coros M, Rosu M, Watanabe F, Biris A. The influence of uric and ascorbic acid on the electrochemical detection of dopamine using graphene-modified electrodes. Electrochimica Acta 2015;154:197–204. DOI: 10.1016/j.electacta.2014.12.046

[107] Wang C, Du J, Wang H, Zou C, Jiang F, Yang P, Du Y. A facile electrochemical sensor based on reduced graphene oxide and Au nanoplates modified glassy carbon electrode for simultaneous detection of ascorbic acid, dopamine and uric acid. Sensors and Actuators B: Chemical 2014;204:302–309. DOI:10.1016/j.snb.2014.07.077

[108] Yu B, Kuang D, Liu S, Liu C, Zhang T. Template-assisted self-assembly method to prepare three-dimensional reduced graphene oxide for dopamine sensing. Sensors and Actuators B: Chemical 2014;205:120–126. DOI:10.1016/j.snb.2014.08.038

[109] Li S, He J, Zhang M, Zhang R, Lv X, Li S, Pang H. Electrochemical detection of dopamine using water-soluble sulfonated graphene. Electrochimica Acta 2013;102:58–65. DOI:10.1016/j.electacta.2013.03.176

[110] Han H, Seol H, Kang D, Ahmed M, You J, Jeon S. Electrochemical oxidation and determination of dopamine in the presence of AA using ferulic acid functionalized electrochemically reduced graphene. Sensors and Actuators B: Chemical 2014;204:289–296. DOI:10.1016/j.snb.2014.07.075

[111] Nancy T, Kumary V. Synergistic electrocatalytic effect of graphene/nickel hydroxide composite for the simultaneous electrochemical determination of ascorbic acid, dopamine and uric acid. Electrochimica Acta 2014;133:233–240. DOI:10.1016/j.electacta.2014.04.027

[112] Wang H, Ren F, Yue R, Wang C, Zhai C, Du Y. Macroporous flower-like graphene-nanosheet clusters used for electrochemical determination of dopamine. Colloids and Surfaces A: Physicochemical and Engineering Aspects 2014;448:181–185. DOI:10.1016/j.colsurfa.2014.02.028

[113] Yan J, Liu S, Zhang Z, He G, Zhou P, Liang H, Tian L, Zhou X, Jiang H. Simultaneous electrochemical detection of ascorbic acid, dopamine and uric acid based on graphene anchored with Pd–Pt nanoparticles. Colloids and Surfaces B: Biointerfaces 2013;111:392–397. DOI:10.1016/j.colsurfb.2013.06.030

[114] Yang L, Liu D, Huang J, You T. Simultaneous determination of dopamine, ascorbic acid and uric acid at electrochemically reduced graphene oxide modified electrode. Sensors and Actuators B: Chemical 2014;193:166–172. DOI:10.1016/j.snb.2013.11.104

[115] Li S, Yang S, Wang Y, Lien C, Tien H, Hsiao S, Liao W, Tsai H, Chang C, Ma C, Hu C. Controllable synthesis of nitrogen-doped graphene and its effect on the simultaneous electrochemical determination of ascorbic acid, dopamine, and uric acid. Carbon 2013;59:418–429. DOI:10.1016/j.carbon.2013.03.035

[116] Lian Q, He Z, He Q, Luo A, Yan K, Zhang D, Lu X, Zhou X. Simultaneous determination of ascorbic acid, dopamine and uric acid based on tryptophan functionalized graphene. Analytica Chimica Acta 2014;823:32–39. DOI:10.1016/j.aca.2014.03.032

[117] Li H, Wang Y, Ye D, Luo J, Su B, Zhang S, Kong J. An electrochemical sensor for simultaneous determination of ascorbic acid, dopamine, uric acid and tryptophan based on MWNTs bridged mesocellular graphene foam nanocomposite. Talanta 2014;127:255–261. DOI:10.1016/j.talanta.2014.03.034

[118] Pakapongpan S, Mensing J, Phokharatkul D, Lomas T, Tuantranont A. Highly selective electrochemical sensor for ascorbic acid based on anovel hybrid graphene-cop-

per phthalocyanine-polyanilinenanocomposites. Electrochimica Acta 2014;133:294–301. DOI:10.1016/j.electacta.2014.03.167

[119] Liu B, Luo L, Ding Y, Si X, Wei Y, Ouyang X, Xu D. Differential pulse voltammetric determination of ascorbic acid in the presence of folic acid at electro-deposited NiO/graphene composite film modified electrode. Electrochimica Acta 2014;142:336–342. DOI:10.1016/j.electacta.2014.07.126

[120] Xue C, Wang X, Zhu W, Han Q, Zhu C, Hong J, Zhou X, Jiang H. Electrochemical serotonin sensing interface based on double-layered membrane of reduced graphene oxide/polyaniline nanocomposites and molecularly imprinted polymers embedded with gold nanoparticles. Sensors and Actuators B: Chemical 2014;196:57–63. DOI: 10.1016/j.snb.2014.01.100

[121] Kim S, Kim D, Jeon S. Electrochemical determination of serotonin on glassy carbon electrode modified with various graphene nanomaterials. Sensors and Actuators B: Chemical 2012;174:285–291. DOI:10.1016/j.snb.2012.08.034

[122] Han H, You J, Jeong H, Jeon S. Synthesis of graphene oxide grafted poly(lactic acid) with palladium nanoparticles and its application to serotonin sensing. Applied Surface Science 2013;284:438–445. DOI:10.1016/j.apsusc.2013.07.116

[123] Han H, Lee H, You J, Jeong H, Jeon S. Electrochemical biosensor for simultaneous determination of dopamine and serotonin based on electrochemically reduced GO-porphyrin. Sensors and Actuators B: Chemical 2014;190:886–895. DOI:10.1016/j.snb.2013.09.022

[124] Mazloum-Ardakani M, Khoshroo A, Hosseinzadeh L. Application of graphene to modified ionic liquid graphite composite and its enhanced electrochemical catalysis properties for levodopa oxidation. Sensors and Actuators B: Chemical 2014;204:282–288. DOI:10.1016/j.snb.2014.07.069

[125] Benvidi A, Dehghani-Firouzabadi A, Mazloum-Ardakani M, Mirjalili B, Zare R. Electrochemical deposition of gold nanoparticles on reduced graphene oxide modified glassy carbon electrode for simultaneous determination of levodopa, uric acid and folic acid. Journal of Electroanalytical Chemistry 2015;736:22–29. DOI:10.1016/j.jelechem.2014.10.020

[126] Wang Q, Das M, Li M, Boukherroub R, Szunerits S. Voltammetric detection of l-dopa and carbidopa on graphene modified glassy carbon interfaces. Bioelectrochemistry 2013;93:15–22. DOI:10.1016/j.bioelechem.2012.03.004

[127] Martín A, Hernández-Ferrer J, Martínez M, Escarpa A. Graphene nanoribbon-based electrochemical sensors on screen-printed platforms. Electrochimica Acta 2014; in press. DOI:10.1016/j.electacta.2014.11.090

[128] Guo H, Peng S, Xu J, Zhao Y, Kang X. Highly stable pyridinic nitrogen doped graphene modified electrode in simultaneous determination of hydroquinone and cate-

chol. Sensors and Actuators B: Chemical 2014;193:623–629. DOI:10.1016/j.snb.2013.12.018

[129] Si W, Lei W, Han Z, Hao Q, Zhang Y, Xia M. Selective sensing of catechol and hydroquinone based on poly(3,4-ethylenedioxythiophene)/nitrogen-doped graphene composites. Sensors and Actuators B: Chemical 2014;199:154–160. DOI:10.1016/j.snb.2014.03.096

[130] Song D, Xia J, Zhang F, Bi S, Xiang W, Wang Z, Xia L, Xia Y, Li Y, Xia L. Multiwall carbon nanotubes-poly(diallyldimethylammonium chloride)-graphene hybrid composite film for simultaneous determination of catechol and hydroquinone. Sensors and Actuators B: Chemical 2015;206:111–118. DOI:10.1016/j.snb.2014.08.084

[131] Lai T, Cai W, Dai W, Ye J. Easy processing laser reduced graphene: A green and fast sensing platform for hydroquinone and catechol simultaneous determination. Electrochimica Acta 2014;138:48–55. DOI:10.1016/j.electacta.2014.06.070

[132] Zhang Y, Xiao S, Xie J, Yang Z, Pang P, Gao Y. Simultaneous electrochemical determination of catechol andhydroquinone based on graphene–TiO_2 nanocomposite modified glassy carbon electrode. Sensors and Actuators B 2014;204:102–108. DOI:10.1016/j.snb.2014.07.078

[133] Han H, You J, Seol H, Jeong H, Jeon S. Electrochemical sensor for hydroquinone and catechol based on electrochemically reduced GO–terthiophene–CNT. Sensors and Actuators B: Chemical 2014;194:460–469. DOI:10.1016/j.snb.2014.01.006

[134] Wang X, Wu M, Li H, Wang Q, He P, Fang Y. Simultaneous electrochemical determination of hydroquinone and catechol based on three-dimensional graphene/MWCNTs/BMIMPF6 nanocomposite modified electrode. Sensors and Actuators B: Chemical;192:452–458. DOI:10.1016/j.snb.2013.11.020

[135] Hu S, Wang Y, Wang X, Xu L, Xiang J, Sun W. Electrochemical detection of hydroquinone with a gold nanoparticle and graphene modified carbon ionic liquid electrode. Sensors and Actuators B 2012;168:27–33. DOI:10.1016/j.snb.2011.12.108

[136] Navaee A, Salimi A, Teymourian H. Graphene nanosheets modified glassy carbon electrode for simultaneous detection of heroine, morphine and noscapine. Biosensors and Bioelectronics 2012;31:205–211. DOI:10.1016/j.bios.2011.10.018

[137] Li Y, Zou L, Li Y, Li K, Ye B. A new voltammetric sensor for morphine detection based on electrochemically reduced MWNTs-doped graphene oxide composite film. Sensors and Actuators B: Chemical 2014;201:511–519. DOI:10.1016/j.snb.2014.05.034

[138] Atta N, Hassan H, Galal A. Rapid and simple electrochemical detection of morphine on graphene–palladium-hybrid-modified glassy carbon electrode. Analytical and Bioanalytical Chemistry 2014;406:6933–6942. DOI:10.1007/s00216-014-7999-x

[139] Li Y, Li K, Song G, Liu J, Zhang K, Ye B. Electrochemical behavior of codeine and its sensitive determination on graphene-based modified electrode. Sensors and Actuators B: Chemical 2013;182:401–407. DOI:10.1016/j.snb.2013.03.023

[140] Afkhami A, Khoshsafar H, Bagheri H, Madrakian T. Facile simultaneous electrochemical determination of codeine and acetaminophen in pharmaceutical samples and biological fluids by graphene–$CoFe_2O_4$ nancomposite modified carbon paste electrode. Sensors and Actuators B: Chemical 2014;203:909–918. DOI:10.1016/j.snb.2014.07.031

[141] Afkhami A, Khoshsafar H, Bagheri H, Madrakian T. Preparation of $NiFe_2O_4$/graphene nanocomposite and its application as a modifier for the fabrication of an electrochemical sensor for the simultaneous determination of tramadol and acetaminophen. Analytica Chimica Acta 2014;831:50–59. DOI:10.1016/j.aca.2014.04.061

[142] Guo K, Qian K, Zhang S, Kong J, Yu C, Liu B. Bio-electrocatalysis of NADH and ethanol based on graphene sheets modified Electrodes. Talanta 2011;85:1174–1179. DOI:10.1016/j.talanta.2011.05.038

[143] Ferreira G, Oliveira F, Leite F, Maroneze C, Kubota L, Damos F, Luz R. DNA and graphene as a new efficient platform for entrapment of methylene blue (MB): Studies of the electrocatalytic oxidation of β-nicotinamide adenine dinucleotide. Electrochimica Acta 2013;111:543–551. DOI:10.1016/j.electacta.2013.08.037

[144] Li Z, Huang Y, Chen L, Qin X, Huang Z, Zhou Y, Meng Y, Li J, Huang S, Liu Y, Wang W, Xie Q, Yao S. Amperometric biosensor for NADH and ethanol based on electroreduced graphene oxide–polythionine nanocomposite film. Sensors and Actuators B: Chemical 2013;181:280–287. DOI:10.1016/j.snb.2013.01.072

[145] Gai P, Zhao C, Wang Y, Abdel-Halim E, Zhanga J, Zhua J. NADH dehydrogenase-like behavior of nitrogen-doped graphene and its application in NAD+-dependent dehydrogenase biosensing. Biosensors and Bioelectronics 2014;62:170–176. DOI: 10.1016/j.bios.2014.06.043

[146] Li L, Lu H, Deng L. A sensitive NADH and ethanol biosensor based on graphene–Au nanorods nanocomposites. Talanta 2013;113:1–6. DOI:10.1016/j.talanta.2013.03.074

[147] Govindhan M, Amiri M, Chen A. Au nanoparticle/graphene nanocomposite as a platform for the sensitive detection of NADH in human urine. Biosensors and Bioelectronics 2015;66:474–480. DOI:10.1016/j.bios.2014.12.012

[148] Gasnier A, Pedano M, Rubianes M, Rivas G. Graphene paste electrode: Electrochemical behavior and analytical applications for the quantification of NADH. Sensors and Actuators B: Chemical 2013;176:921–926. DOI:10.1016/j.snb.2012.09.092

[149] Ye X, Du Y, Duan K, Lu D, Wang C, Shi X. Fabrication of nano-ZnS coated PEDOT-reduced graphene oxidehybrids modified glassy carbon-rotating disk electrode and

its application for simultaneous determination of adenine, guanine, and thymine. Sensors and Actuators B 2014;203:271–281. DOI:10.1016/j.snb.2014.06.135

[150] Sun W, Lu Y, Wu Y, Zhang Y, Wang P, Chen Y, Li G. Electrochemical sensor for transgenic maize MON810 sequence with electrostatic adsorption DNA on electrochemical reduced graphene modified electrode. Sensors and Actuators B 2014;202:160–166. DOI:10.1016/j.snb.2014.05.072

[151] Fan Y, Huang K, Niu D, Yang C, Jing Q. TiO_2-graphene nanocomposite for electrochemical sensing of adenine and Guanine. Electrochimica Acta 2011;56:4685–4690. DOI:10.1016/j.electacta.2011.02.114

[152] Abdul Rasheed P, Sandhyarani N. Graphene-DNA electrochemical sensor for the sensitive detection of BRCA1 gene. Sensors and Actuators B: Chemical 2014;204:777–782. DOI:10.1016/j.snb.2014.08.043

[153] Wang L, Hua E, Liang M, Ma C, Liu Z, Sheng S, Liu M, Xie G, Feng W. Graphene sheets, polyaniline and AuNPs based DNA sensor for electrochemical determination of BCR/ABL fusion gene with functional hairpin probe. Biosensors and Bioelectronics 2014;51:201–207. DOI:10.1016/j.bios.2013.07.049

[154] Wang Y, Zhuang Q, Ni Y. Fabrication of riboflavin electrochemical sensor based on homoadenine single-stranded DNA/molybdenum disulfide–graphene nanocomposite modified gold electrode. Journal of Electroanalytical Chemistry 2015;736:47–54. DOI:10.1016/j.jelechem.2014.10.028

[155] Yu S, Cao X, Yu M. Electrochemical immunoassay based on gold nanoparticles and reduced graphene oxide functionalized carbon ionic liquid electrode. Microchemical Journal 2012;103:125–130. DOI:10.1016/j.microc.2012.02.002

[156] Lu W, Ge J, Tao L, Cao X, Dong J, Qian W. Large-scale synthesis of ultrathin Au-Pt nanowires assembled on thionine/graphene with high conductivity and sensitivity for electrochemical immunosensor. Electrochimica Acta 2014;130:335–343. DOI: 10.1016/j.electacta.2014.03.065

[157] Samanman S, Numnuam A, Limbut W, Kanatharana P, Thavarungkul P. Highly-sensitive label-free electrochemical carcinoembryonic antigen immunosensor based on a novel Au nanoparticles–graphene–chitosan nanocomposite cryogel electrode. Analytica Chimica Acta 2015;853:521–532. DOI:10.1016/j.aca.2014.10.006

[158] Huang J, Lin Q, Zhang X, He X, Xing X, Lian W, Zuo M, Zhang Q. Electrochemical immunosensor based on polyaniline/poly (acrylic acid) and Au-hybrid graphene nanocomposite for sensitivity enhanced detection of salbutamol. Food Research International 2011;44:92–97. DOI:10.1016/j.foodres.2010.11.006

[159] Li T, Yang M, Li H. Label-free electrochemical detection of cancer marker based on graphene–cobalt hexacyanoferrate nanocomposite. Journal of Electroanalytical Chemistry 2011;655:50–55. DOI:10.1016/j.jelechem.2011.02.009

[160] Jang H, Kim S, Chang H, Choi J. 3D label-free prostate specific antigen (PSA) immunosensor based on graphene–gold composites. Biosensors and Bioelectronics 2015;63:546–551. DOI:10.1016/j.bios.2014.08.008

[161] Li Y, Han J, Chen R, Ren X, Wei Q. Label electrochemical immunosensor for prostate-specific antigen based on graphene and silver hybridized mesoporous silica. Analytical Biochemistry 2015;469:76–82. DOI:10.1016/j.ab.2014.09.022

[162] Yang F, Yang Z, Zhuo Y, Chai Y, Yuan R. Ultrasensitive electrochemical immunosensor for carbohydrate antigen 19-9 using Au/porous graphene nanocomposites as platform and Au@Pd core/shell bimetallic functionalized graphene nanocomposites as signal enhancers. Biosensors and Bioelectronics 2015;66:356–362. DOI:10.1016/j.bios.2014.10.066

[163] Zhao L, Wei Q, Wu H, Dou J, Li H. Ionic liquid functionalized graphene based immunosensor for sensitive detection of carbohydrate antigen 15-3 integrated with Cd^{2+}-functionalized nanoporous TiO_2 as labels. Biosensors and Bioelectronics 2014;59:75–80. DOI:10.1016/j.bios.2014.03.006

[164] Liu K, Zhang J, Yang G, Wang C, Zhu J. Direct electrochemistry and electrocatalysis of hemoglobin based on poly(diallyldimethylammonium chloride) functionalized graphene sheets/room temperature ionic liquid composite film. Electrochemistry Communications 2010;12:402–405. DOI:10.1016/j.elecom.2010.01.004

[165] Huang K, Miao Y, Wang L, Gan T, Yu M, Wang L. Direct electrochemistry of hemoglobin based on chitosan–ionic liquid–ferrocene/graphene composite film. Process Biochemistry 2012;47:1171–1177. DOI:10.1016/j.procbio.2012.04.014

[166] Sun W, Guo Y, Lu Y, Hu A, Shi F, Li T, Sun Zhenfan. Electrochemical biosensor based on graphene, Mg_2Al layered double hydroxide and hemoglobin composite. Electrochimica Acta 2013;91:130–136. DOI:10.1016/j.electacta.2012.12.088

[167] Cheng Y, Feng B, Yang X, Yang P, Ding Y, Chen Y, Fei J. Electrochemical biosensing platform based on carboxymethyl cellulose functionalized reduced graphene oxide and hemoglobin hybrid nanocomposite film. Sensors and Actuators B: Chemical 2013;182:288–293

[168] Liu H, Duan C, Su X, Dong X, Shen W, Zhu Z. Titania nanoparticles modified reduced graphene oxide nanocomposite with a double-layered structure encapsulating hemoglobin for a mediator-free biosensor. Ceramics International 2014;40(7):9867–9874. DOI:10.1016/j.snb.2013.03.007

[169] Ruan C, Li T, Niu Q, Lu M, Lou J, Gao W, Sun W. Electrochemical myoglobin biosensor based on graphene–ionic liquid–chitosan bionanocomposites: Direct electrochemistry and electrocatalysis. Electrochimica Acta 2012;64:183–189. DOI:10.1016/j.electacta.2012.01.005

[170] Suna W, Li L, Lei B, Li T, Ju X, Wang X, Li G, Sun Z. Fabrication of graphene–platinum nanocomposite for the direct electrochemistry and electrocatalysis of myoglo-

bin. Materials Science and Engineering: C 2013;33(4):1907–1913. DOI:10.1016/j.msec. 2012.12.077

[171] Suna W, Gong S, Deng Y, Li T, Cheng Y, Wang W, Wang L. Electrodeposited nickel oxide and graphene modified carbon ionic liquid electrode for electrochemical myglobin biosensor. Thin Solid Films 2014;562:653–658. DOI:10.1016/j tsf.2014.05.002

[172] Baby T, Aravind S, Arockiadoss T, Rakhi R, Ramaprabhu S. Metal decorated graphene nanosheets as immobilization matrix for amperometric glucose biosensor. Sensors and Actuators B 2010;145:71–77. DOI:10.1016/j.snb.2009.11.022

[173] Wu H, Wang J, Kang X, Wang C, Wang D, Liu J, Aksay I, Lin Y. Glucose biosensor based on immobilization of glucose oxidase in platinum nanoparticles/graphene/ chitosan nanocomposite film. Talanta 2009;80:403–406. DOI:10.1016/j.talanta. 2009.06.054

[174] Hu Y, Jin J, Wu P, Zhang H, Cai C. Graphene–gold nanostructure composites fabricated by electrodeposition and their electrocatalytic activity toward the oxygen reduction and glucose oxidation. Electrochimica Acta 2010;56:491–500. DOI:10.1016/ j.electacta.2010.09.021

[175] Shan C, Yang H, Han D, Zhang Q, Ivaska A, Niu L. Graphene/AuNPs/chitosan nanocomposites film for glucose biosensing. Biosensors and Bioelectronics 2010;25:1070–1074. DOI:10.1016/j.bios.2009.09.024

[176] Guo Y, Li J, Dong S. Hemin functionalized graphene nanosheets-based dual biosensor platforms for hydrogen peroxide and glucose. Sensors and Actuators B 2011;160:295–300. DOI:10.1016/j.snb.2011.07.050

[177] Shamsipur M, Tabrizi M. Achieving direct electrochemistry of glucose oxidase by one step electrochemical reduction of graphene oxide and its use in glucose sensing. Materials Science and Engineering C 2014;45:103–108. DOI:10.1016/j.msec.2014.09.002

[178] Jia L, Liu J, Wang H. Preparation of poly(diallyldimethylammoniumchloride)-functionalized graphene and its applications for H_2O_2 and glucose sensing. Electrochimica Acta 2013;111:411–418. DOI:10.1016/j.electacta.2013.07.218

[179] Karuppiah C, Palanisamy S, Chen S, Veeramani V, Periakaruppan P. A novel enzymatic glucose biosensor and sensitive non-enzymatichydrogen peroxide sensor based on graphene and cobalt oxidenanoparticles composite modified glassy carbon electrode. Sensors and Actuators B 2014;196:450–456. DOI:10.1016/j.snb.2014.02.034

[180] Wang L, Ye Y, Lu X, Wu Y, Sun L, Tan H, Xu F, Song Y. Prussian blue nanocubes on nitrobenzene-functionalized reducedgraphene oxide and its application for H_2O_2 biosensing. Electrochimica Acta 2013;114:223–232. DOI:10.1016/j.electacta.2013.10.073

[181] Michopoulos A, Kouloumpis A, Gournis D, Prodromidis M. Performance of layer-by-layer deposited low dimensional building blocks of graphene-prussian blue onto

graphite screen-printed electrodes as sensors for hydrogen peroxide. Electrochimica Acta 2014. DOI:10.1016/j.electacta.2014.09.031

[182] Jiang Y, Zhang X, Shan C, Hua S, Zhang Q, Bai X, Dan L, Niu L. Functionalization of graphene with electrodeposited Prussian blue towards amperometric sensing application. Talanta 2011;85:76–81. DOI:10.1016/j.talanta.2011.03.028

[183] Zhang Y, Liu Y, He J, Pang P, Gao Y, Hu Q. Electrochemical behavior of graphene/Nafion/Azure I/Au nanoparticles composites modified glass carbon electrode and its application as nonenzymatic hydrogen peroxide sensor. Electrochimica Acta 2013;90:550–555. DOI:10.1016/j.electacta.2012.12.068

[184] Ye Y, Kong T, Yu X, Wu Y, Zhang K, Wang X. Enhanced nonenzymatic hydrogen peroxide sensing with reduced graphene oxide/ferroferric oxide nanocomposites. Talanta 2012;89:417–421. DOI:10.1016/j.talanta.2011.12.054

[185] Mao Y, Bao Y, Wang W, Li Z, Li F, Niu L. Layer-by-layer assembled multilayer of graphene/Prussian blue toward simultaneous electrochemical and SPR detection of H_2O_2. Talanta 2010;85:2106–2112. DOI:10.1016/j.talanta.2011.07.056

[186] Wang X, Liu E, Zhang X. Non-enzymatic glucose biosensor based on copper oxide-reducedgraphene oxide nanocomposites synthesized from water-isopropanolsolution. Electrochimica Acta 2014;130:253–260. DOI:10.1016/j.electacta.2014.03.030

[187] Gao W, Tjiu W, Wei J, Liu T. Highly sensitive nonenzymatic glucose and H_2O_2 sensor based on $Ni(OH)_2$/electroreduced graphene oxide Multiwalled-carbon nanotube film modified glassy carbon electrode. Talanta 2014;120:484–490. DOI:10.1016/j.talanta.2013.12.012

Chapter 3

Molecularly Imprinted Sensors — New Sensing Technologies

Zihni Onur Uygun, Hilmiye Deniz Ertuğrul Uygun,
Nihal Ermiş and Erhan Canbay

Additional information is available at the end of the chapter

http://dx.doi.org/10.5772/60781

Abstract

In this chapter we discus molecular imprinting technology (MIT), molecular imprinted polymers (MIPs), and their compatibility on a proper transducer to construct a sensing system. Molecularly imprinted sensors (MISens), in other words, artificial receptor-based sensors synthesized in the presence of the target molecule, are capable of sensing target molecules by using their specific cavities and are compatible with the target molecule. This MIP technology is a viable alternative of artificial receptor technology, and the sensor technology is capable of detecting any kind of molecule without pre-analytic preparations. In this chapter, you can find examples, sensor construction techniques and fundamentals of MIP and sensor combinations to look forward in your studies. For sensor technology, we explained and discussed the new sensing technologies of MIP-based electrochemical, optical (especially surface plasmon resonance, SPR), and piezoelectric techniques. Therefore, this chapter presents a short guideline of MISens.

Keywords: molecular imprinting, sensors, artificial receptors, impedimetric sensors, capacitive sensors, potentiometric sensors, amperometric sensors, fluorimetric sensors, SPR sensors, QCM sensors, piezoelectric sensors

1. Introduction

Nowadays, technology develops exponentially and the rate of article publishing and patent applications are immensely high reflecting the growth of technological improvements and discoveries. The purpose of all developmental statuses is to simplify human life. Therefore, every system is being designed in an "all in one fashion" for devices and equipment. You can imagine, for instance, that if we compare the current mobile phone systems with 10 years earlier, we will see that there is a big jump. This technological development has a wide area of application ranging from health to food applications, from environment to space. The best providence of this technology is its ability to provide people a simpler life, which comes within the simplified new treatment systems for medicine, or with the control of the environmental balance more easily by simple devices. These simple devices are being developed especially for personal use such as chemical sensors and biosensors. These devices are famous for their features such as that they are easy to use, cost-effective with a high sensitivity and selectivity. Therefore, sensor technology is widely used in different platforms ranging from health technologies to environmental analysis methods [1-6]. Generally, sensors consist of three parts: transducers, recognition elements, and an analytical device. Transducers are part of sensors which convert energy from one form to another. For example, piezoelectric transducers convert electrical charges produced by piezoelectric solid materials into energy. Recognition elements (enzyme, DNA, antibodies, etc.) and a proper analytical device to show noticeable signals formed by transducers are the elements of chemical and biochemical sensors [7-11]. Chemical sensors use chemically formed materials (nanomaterials, MIPs, etc.) as recognition elements, whereas biochemical sensors use biomolecules (enzymes, antibodies, DNAs, receptors, proteins, etc.) as recognition elements. If we compare these sensor types, although biosensors are more selective then chemical sensors, chemical sensors have advantages such as the capacity to resist harsh conditions such as strong pH, extreme ionic strength, and a wide variety of organic solvents. In this chapter, we especially point out these differences together with the concepts of construction of molecularly imprinted sensors. Just like biosensors, the chemical sensors are divided into two main functional combinations: affinity-based sensors and catalytic sensors. Catalytic sensors are modified with different molecules that show catalytic properties [12]. Affinitybased sensors, which are MIP-based sensors, have a specific recognition patter mechanism of recognition for the target analyte which mimics to recognize target analyte [13].

2. Molecular imprinting technology

In brief, molecular imprinting is defined as the formation of artificial receptors for a specific target molecule on a polymer or on self-assembled materials. Natural receptors are widely used for sensor technology to target the analyte, leading to electrochemical, optical, and mass or magnetic changes on transducers [14-18]. MIPs are obtained by polymerization of a monomer and a cross-linker, which are located around the target molecule (Fig. 1). This assembly of a monomer around a target molecule is encouraged by covalent and non-covalent interactions. It is easier to remove the target molecule from a non-covalently formed MIP-target molecule complex than removing from a covalently formed MIP-target molecule complex. Just as we described, MIPs are synthetic polymers, which can only be used as plastic antibodies for now. It means that, currently, MIPs on sensors are only used as affinity sensors and not as a catalytic biochemical enzyme mimicking sensors.

http://dx.doi.org/10.5772/60781

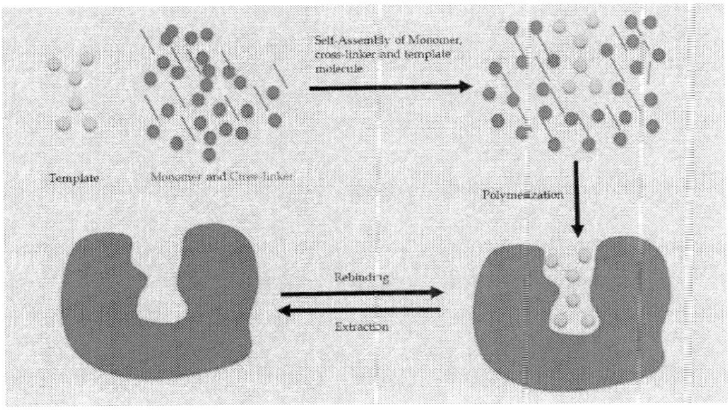

Figure 1. A schematic of MIP preparation steps.

Molecular imprinting technology, in general, is developed as bulk polymerization (3-D) and surface imprinting polymerization (2-D). Bulk polymerization is prepared as bulk materials, so they have further preparation steps before use to recognize desired materials, for example, grinding of bulky MIPs, which cause disruptive heterogeneous binding sites leading to poor site accessibility. Moreover, there are embedded target molecules inside the bulk polymers [19]. Surface imprinting is more advantageous due to controlled surface imprinting and its convenience for sensor technology. For the sensor and the analytical device preparation by using MIPs, there are two methods to design MIPs; one is an in situ technique that imprinted polymer is prepared on the transducer, whereas the other is an ex situ technique that imprinted polymer is prepared separately from transducers, where MIPs are immobilized on transducers after this preparation to construct MISens

3. Sensors

Chemical sensors are a major class of sensors, which have many applications, such as environmental and food analysis, process control, and medical diagnosis. A chemical sensor is a device that transforms chemical information, ranging from the concentration of a specific sample component to the analysis of the total composition of a sample, into an analytically useful signal [20] and [21]. Like many fields in science, chemical sensors have benefited from the growing power of computers, integrated electronics, new materials, novel designs, and processing tools. Breakthroughs over the last decade have pushed chemical sensors into new markets, as well as to new applications within existing markets [22].

When operated, a chemical sensor performs two functions: recognition and transduction. First, the analyte interacts in a more or less selective way with the recognition (or sensing) element, which shows affinity for the analyte. The sensing element may be composed of distinct molecular units called recognition receptors. Alternatively, the recognition element can be a material that includes certain recognition sites in its composition. Beyond this, the recognition element can be formed of a material with no distinct recognition sites, but capable of interacting with the analyte. In a chemical sensor, the recognition and transduction functions are integrated within the same device. An analytical device with no recognition function is not a chemical sensor but a concentration transducer [20, 23]. The signal from a sensor is typically electronic in nature, being a current, voltage, or impedance/conductance change caused by the change in analyte composition or quality. While chemical sensors contain a physical transducer and a chemically sensitive layer or recognition layer, the micro-instrument or spectrometer sends out an energy signal, which can be thermal, electrical, or optical, and reads the change in this same property caused by the intervening chemical and this is close to molecular spectroscopy [24].

Biosensors have specific recognition elements of the proper chemical substances, which is performed as an analytical devise. The biological material that serves as recognition element is used in combination with a transducer. The transducer transforms the concentration of substrate or product to electrical signal that is amplified and further processed. The biosensors may utilize enzymes, antibodies, nucleic acids, organelles, plant and animal tissue, whole organism, or organs. Biosensors that contain biological catalysts (enzymes) are called catalytic biosensors. These types of biosensors are the most abundant, and they have their largest application area in medicine, ecology, and environmental monitoring [25-27].

Molecularly imprinted polymers (MIPs) are synthetic materials used as recognition elements in the design of sensors due to their higher thermal stability than biological receptor, reusability, and selectivity compared to biological receptors. These polymeric materials bind to the target molecules causing variations in physical parameters, such as mass, absorbance, or refractive index depending upon the shape, charge, and functionality of the target molecule leading to Ref. [28]. The design of these synthetic materials, which are able to mimic the recognition processes found in nature, has become an important and active area of research making molecular imprinting one of the strategies followed to create materials with recognition ability comparable to the natural systems in recent years.

4. Molecularly imprinted sensors

A combination of molecularly imprinted polymers and transducers form a synergistic device. Just as we mentioned before, MIPs have the ability to resist pH, organic environment, and ionic strength. Therefore, their usage in sensor technology is very beneficial. Because of this, studies including molecular imprinting are increasing year by year, which can be clearly seen in Fig. 2. Moreover, this technology is quite suitable and advantageous for non-electroactive molecule detection. Non-electroactive species are molecules that cannot be transformed by electrochemical reactions such as pesticides, drugs, etc. Therefore, they can be measured by affinity techniques, or catalytic secondary molecule usage. Secondary molecule usage, however, has disadvantages such as secondary molecule and target molecule interaction, solvent problems, where template and secondary molecule may not be solved in the same solvent or harsh conditions can affect the reaction of target molecule, hence the measurement. Therefore, affinity measurement is very beneficial for these kinds of molecules. Affinity measurement is used to detect molecules depending on the affinity between target molecule and the molecule it shows affinity. In biosensor technology DNA, antibody, protein, and receptor-based systems are designed which could be collectively called affinity-based systems. However, these bio-compounds are expensive, hard to immobilize onto transducers, and challenging to study on their optimum conditions. Then, an idea came up to the scientists to avoid these disabilities for use of MIPs on transducer surfaces. MIP-based sensors have been constructed since then as electrochemically, optically, and piezoelectrically.

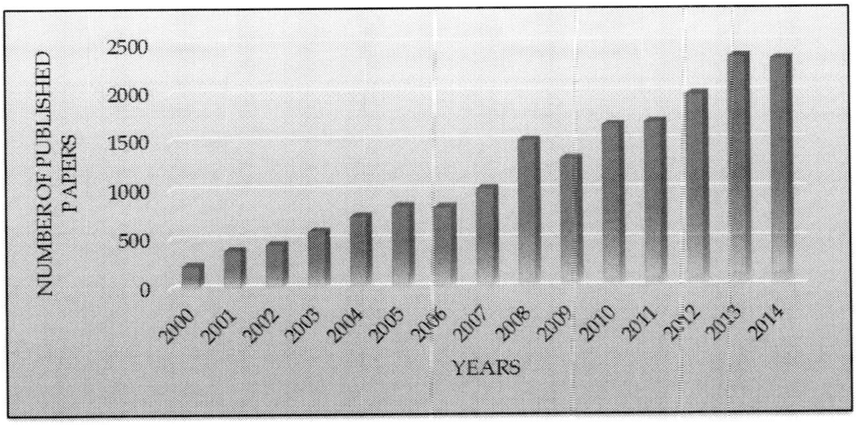

Figure 2. The number of papers referring to biosensor based on MIPs in the last 15 years (searching was performed using "molecularly imprinted sensors" as search key terms on Google Scholar [29]).

4.1. Electrochemical MISens

The fundamentals of electrochemistry are to study the interaction between matter and electricity (Fig. 3). This interaction gives information and provides quantitative measurement of the analyte. Electrochemical techniques of MISens mostly measure surface properties of the transducers, binding kinetics and polymer rearrangements. In this section we gave examples of electrochemical MISens.

Silva and co-workers designed a novel electrochemical sensor for the determination of trimethoprim by electropolymerization of pyrrole (PY) and molecularly imprinted polymer (MIP) which was synthesized onto a glassy carbon electrode (GCE) in aqueous solution using cyclic voltammetry. In their study, they used graphene (GNPs) in order to enhance the sensitivity of the sensor by an increase in the electrochemical conductivity. The performance of the imprinted and non-imprinted (NIP) films was investigated by electrochemical impedance spectroscopy (EIS) and the cyclic voltammetry (CV) of a ferric solution. The sensor they

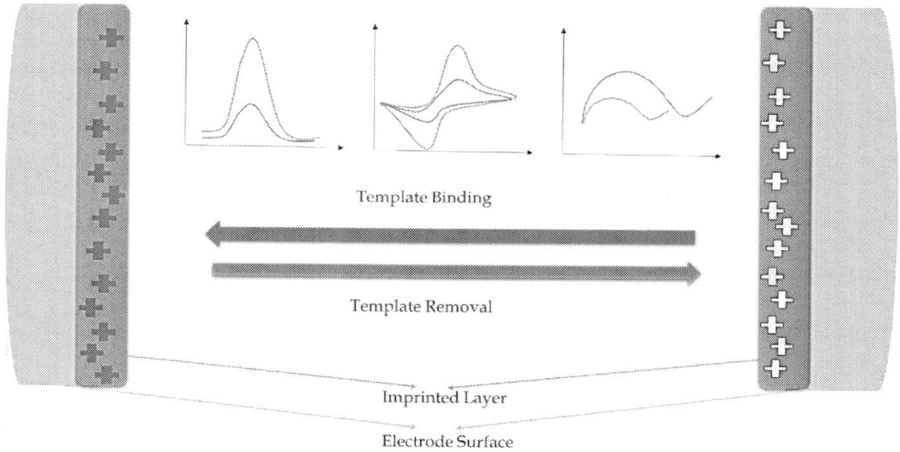

Figure 3. A representative molecularly imprinted electrochemical sensor system.

developed presented a linear range between peak current intensity and logarithm of TMP (trimethoprim) concentration with a range from 10^{-6} to 10^{-4} M. The results were accurate (with recoveries higher than 94%), precise (with standard deviations less than 5%), and the detection limit was 1.3×10^{-7} M [30].

Xue et al. reported an electrochemical sensor for the amperometric detection of dopamine that was carried out via gold nanoparticles doped MIP. In this work, dopamine (DA) was used as the template molecule, functionalized AuNPs (F-AuNPs) as functional monomers and p-aminobenzenethiol (p-ATP) as the cross-linker. They synthesized MIP following these steps: An electrolyte solution containing 1 mmol L^{-1} DA, 10 mmol L^{-1} F-AuNPs, 7 mmol L^{-1} p-ATP, and 0.1 mol L^{-1} ABS (acetate buffer solution) (pH 5.0) was kept in the dark under a nitrogenous atmosphere at room temperature for 6 h to complete the pre-assembly between DA and F-AuNPs through the hydrogen-bond interaction. The AuNPs-modified electrode was immersed into the electrolyte solution and the AuNPs@MIES (gold nanoparticle and MIPs modified sensor) was prepared by the electropolymerization at a constant potential of 1.0 V for 400 s. After that, the electrode was immersed in 0.5 mol L^{-1} H_2SO_4 and treated with a constant potential of −0.5 V for 400 s to remove the templates and dried under nitrogen flow. The developed sensor effectively minimized the interferences caused by ascorbic acid (AA) and uric acid (UA). Also according to linear range (0.02 μmol L^{-1} to 0.54 μmol L^{-1}) and detection limit (with the detection limit of 7.8 nmol L^{-1}) of reported dopamine sensor, it can be said that the developed sensor exhibited high sensitivity and high selectivity [31].

Yu et al. designed a molecularly imprinted electrochemical sensor based on nickel nanoparticle-modified electrodes for phenobarbital determination. Reported electrochemical sensor was developed by thermal polymerization with the use of methacrylic acid (MAA) as the functional monomer, 2,2-azobisisobutyronitrile (AIBN) and ethylene glycol maleic rosinate (EGMRA) acrylate as the crosslinking agent, phenobarbitals (PBs) as the template molecule, and dimethyl sulfoxide (DMSO) as an organic solvent. In the sensor fabrication process, 0.0464 g PB and 0.0688 g MAA were mixed in 3 mL DMSO and sonicated for 10 min. After 5 h, 1.0244 g EGMRA and 0.0074 g AIBN were added into the mixture and sonicated for 30 min to obtain PB-imprinted polymer solutions. After that, 10 μL of 2.0 mg mL^{-1} Ni nanoparticle solution dropped on the GCE surface and then the sensor was dried at room temperature. Approximately 5 μL of the prepared PB-imprinted polymer solution was then coated on the Ni nanoparticle-modified GCE and vacuum dried at 75°C for 6 h. Following the thermal polymerization, the imprinted sensor was washed with (acetic acid) HAc/methanol (volume ratio, 3:7) for 7 min to remove the template molecules. The electrochemical properties of the modified molecularly imprinted and non-imprinted polymer sensors were investigated by cyclic voltammetry, differential pulse voltammetry, electrochemical impedance spectroscopy, and chronoamperometry. Under optimized conditions, the currents were found to be proportional to the PB concentrations within a range of 1.4×10^{-7} mol L^{-1} to 1.3×10^{-4} mol L^{-1} ($r^2 = 0.9976$), with a detection limit of 8.2×10^{-9} mol L^{-1}. The developed sensor was used to determine PB in actual fish samples [32].

Anirudhan and co-workers reported molecularly imprinted polymer-based potentiometric sensor from the surface modified multiwalled carbon nanotube (MWCNT) for the determination of an organochlorine pesticide that is lindane (γ-hexachlorocyclohexane). A MWCNT modified imprinted electrochemical sensor was developed by the following these steps: MWCNT-CH=CH2 was added to the solvent mixture of 60 mL of acetonitrile and 10 mL of toluene in a 500 mL round-bottom flask. After that the mixture was purged with N_2 gas under a constant magnetic stirring. A mixture of γ-HCCH (γ-hexachlorocyclohexane) and MAA was prepared and dissolved in 35 mL of N,N-dimethylformamide. It was stirred for 30 min to get a compound of template molecule and functional monomer. To that mixture, the cross linker ethylene glycol dimethacrylate (EGDMA) and initiator AIBN were also added; the reaction was allowed to proceed for 16 h at 70°C. Ethanol was used to remove template molecules. A MWCNT was grafted using glycidyl methacrylate (GMA). The reaction of MWCNT with GMA produces MWCNT-g-GMA and the epoxide ring present in the GMA upon reaction with allylamine produces the vinylated MWCNT (MWCNT-CH=CH2). MWCNT-based imprinted polymer (MWCNT-MIP) was synthesized by means of methacrylic acid (MAA) as the monomer, EGDMA as the cross linker, α,α′ azobisisobutyronitrile (AIBN) as the initiator, and γ-HCCH, an organochlorine pesticide molecule, as the template. The properties of the modified molecularly imprinted and non-imprinted polymer sensors were investigated by linear sweep voltagrams, FTIR, XRD, Raman spectra, and TEM analyses. This developed sensor presented a linear range of 10^{-10}–10^{-3} M and the detection limit of 10^{-10} M [33].

Patra and co-workers developed a molecular imprinting-based sensor for medullary thyroid carcinoma marker. The fabrication of the sensor was made by the following steps. Accordingly, bipyridyl (0.2 mmol) and $CuCl_2$ (0.1 mmol) were dissolved in 2 mL DMSO (dimethyl sulfoxide) to obtain a solution of Cu(II)-complex. Subsequently, this complex was mixed with a ZnO nanostructure modified monomer (10 mg, 1.0 mL DMSO), template (calcitonin, 2.0 mg, 1.0 mL DMSO), and EGDMA (ethylene glycol dimethacrylate) (1 mmol, 180mL) in the presence of ascorbic acid (0.1 mmol) as the reducing agent. A sharp colour change from light blue to green indicated the in situ reduction of Cu(II)-complex to Cu(I)-complex that catalysed the chain propagation in the presence of the ethyl-2-bromo isobutyrate (2 mmol, 300mL) as initiator. The whole mixture was purged with N2 gas for 10 min. A drop of this mixture (5.0μL) was spread over the protruding tip of the functionalized PGE and kept in a pre-heated oven for half an hour at 45°C, resulting in calcitonin adduct polymer modified electrochemical sensor. The morphologies and properties of the developed sensor were characterized by scanning electron microscopy, cyclic voltammetry, difference pulse voltammetry, and chronocoulometry. Linear responses of the imprinted sensor to calcitonin were observed for concentrations ranging from 9.99 ng L^{-1} to 7.919 mg L^{-1} and the detection limit was as low as 3.09± 0.01 ng L^{-1}. The reported imprinted electrochemical sensor was used to determine the concentration of calcitonin in the human blood serum samples [34].

Karimian and co-workers reported an on/off-switchable molecularly imprinted polymer (MIP) affinity sensor for folic acid using copolymerization of poly(N-isopropylacrylamide) (PNI-PAAm) with a cross-linker (N,N'-methylenebisacrylamide) (MBA) and additional monomer (o-phenylenediamine (o-PD)), in the presence of folic acid as template. Polymerization was carried out following these steps: The folic acid molecularly imprinted film was prepared by the electrochemical polymerization of PNIPAAm and o-PD on the surface of gold electrode, using cyclic voltammetry in the potential range between 0 and 1.1 V (versus Ag/AgCl), for 20 cycles at a scanning rate of 50 mV s^{-1}. The polymerization mixture consisted of an aqueous solution containing 10 mM o-PD, 2.5 mM PNIPAAm, 2.5 mM MBA, and 0.2 mM folic acid. For the preparation of the polymers, the components were dissolved in acetate buffer (0.5 M, pH 5.8). For the washing procedure, the polymer film was rinsed in methanol-acetic acid (9:1, v/v) solution for 20 min at 50°C, followed by subsequent washing with methanol to remove the template entrapped in the polymeric matrix. The electrochemical behaviour of the thin film (MIP) was characterized using differential pulse voltammetry and cyclic voltammetry. Reported sensor response shows a limit of detection of 0.9 μM with linear range from 1.0 μM to 200 μM [35].

Wang et al. have developed an electrochemical sensor for the determination of aflatoxin B_1 based on MWCNT-supported Au/Pt bimetallic nanoparticles. This study involves a molecularly imprinted sensor technology, which was a modification of glassy carbon electrode (GCE) by o-phenylenediamine (OPD), electrochemically. Carbon nanotubes were used as support material and supported by Au/Pt bimetallic nanoparticles. Moreover, this layer formation was monitored by cyclic voltammetry (CV). Amine groups on OPD were the donor of hydrogen to form hydrogen bonds between AFB1's oxygen. After MWCNT coating, Au/PtNPs were deposited onto MWCNTs-GCE. DP and CV measurements were carried out by using $Fe(CN)_6$ redox solution. Template molecule was removed by using HCl solution pH=2 for 9 min. A linear relationship between the sensor response signal and the logarithm of AFB1 concentrations ranging from 1×10^{-10} to 1×10^{-5} mol L^{-1} was obtained with a detection limit of 30 pikomol L^{-1}. It was applied to detect AFB1 in hogwash oil successfully [1]. As you can understand, the main objective of this study is based on examination of the surface characteristics of the modified electrode. Ferricyanide oxidation/reduction peaks altered, when selective cavities of OPD/MWCNT-Au/Pt layer bind the AFB1 [36].

Uygun and Dilgin developed a novel impedimetric sensor based on molecularly imprinted polypyrrole (PPy) modified pencil graphite electrode (PGE) for trace level determination of chlorpyrifos (CPF), which is a pesticide. In this study, they used PGE as transducer, and PGE was modified by pyrrole electrochemically formed polymers on electrode by cyclic voltammetry, and CPF was used in polymerization process simultaneously. CPF was used as template and removed after polymerization by using pH=2 HCl solution to remove H bonds between PPy and CPF. The whole surface polymerization steps and measurement steps were examined by electrochemical impedance spectroscopy, which is an electrochemical electron resistance-based surface characterization technique, by using ferri/ferrocyanide redox probes. Under experimental conditions, the proposed impedimetric sensor has a linear response range from 20 to 300 μg L^{-1} CPF with a detection limit of 4.5 μg L^{-1} (based on 3sb). Furthermore, the fabricated sensor was successfully applied to determine CPF in CPF-added artificial corn leaves, tap water, and soil samples. Two types of organophosphates and two metabolite of CPF that chlorpyrifos oxon (CPFO) and 3,5,6-trichloro-2-pyridinol (TPD) and 2,4-dichlorophenoxyacetic acid (2,4D) which is a common systemic pesticide/herbicide were selected for the control experiments [13].

Zhong et al. have developed a pyrrole–phenyl boronic acid: a novel monomer for dopamine (DA) recognition and detection based on imprinted electrochemical sensor. They used a new monomer for MIP by synthesizing pyrrole-phenyl boronic acid. In this study GCE used as a transducer. Dopamine was used as template and polymerization was performed by CV. DA was extracted by H_2SO_4 and applied electrical force 0–0.15 V to remove DA from imprinted polymer cavities. Differential pulse voltammetry (DPV) was used as measurement method, a linear ranging from $5.0x10^{-8}$ to $1.0x10^{-5}$ mol L^{-1} for the detection of DA was obtained with a detection limit of 3.3×10^{-8} mol L^{-1} (S/N=3). For the recovery tests, the samples were spiked with 4.0×10^{-6} mol L^{-1}, 6.0×10^{-6} mol L^{-1}, and 8.0×10^{-6} mol L^{-1} DA varied from 91.5% to 105.2% [37].

For another study of electrochemical sensor, Wang et al. developed a sensor technology that is an ultrasensitive molecularly imprinted electrochemical sensor based on magnetic graphene oxide/β-cyclodextrin (CD)/Au nanoparticle composites for chrysoidine, which is an azoic dye. As you can read, a magnetic graphene oxide (MGO), cyclodextrin, which has hydrophobic and hydrophilic residues, and gold nanoparticles as electrical conductive material were used. GCE was used as the transducer for measurements. MGO/CD@AuNPs modified GCE was put in a solution, which contains pyrrole and chrysoidine together to form an imprinted material by employing CV method. After polymerization, the template molecule was removed from the surface by soaking modified GCE in ethanol. The surface of both non-imprinted and imprinted sensor system was characterized by SEM (scanning electron microscopy), EIS, and CV measurements. The measurement system was based on the differential pulse voltammetry (DPV) to quantify chrysoidine. The calibration curve data was between 5.0×10^{-8} and 5.0×10^{-6}mol L^{-1}. The detection limit was estimated to be 1.7×10^{-8} mol L^{-1} at a signal-to-noise ratio of 3σ (where σ is the standard deviation of the blank, n = 6) [38].

Yola et al. reported a study where a molecularly imprinted electrochemical biosensor based on Fe@Au nanoparticles involved in 2-amino ethanethiol (2-AET) functionalized multiwalled carbon nanotubes was developed for the sensitive determination of cefixime (CEF) in human plasma. In this study, they modified a GCE by p-nitro phenyl diazonium tetra fluoro borate (p-NPDEFB) salt in MeCN with TBATFB (Tetrabutylammonium tetrafluoroborate) using CV, reduced the formed nitro groups by applying negative voltage, activated MWCNT tubes that were attached onto the modified electrode surface, and 2-AET and Fe@Au layers were formed by self-assembling, respectively. After electrode surface modification, the modified electrode was soaked in a solution, which contains pyrrole and CEF, to form CEF imprinted layers by using CV. NaCl solution was used as a desorption agent of CEF. For the measurement, square wave voltammograms were used as a function of concentration. Limit of detection(LOD) was calculated as 2.2×10^{-11}M and the calibration curve was created from 0.1 nM to 10nM [39]. Oxygen groups on the CEF and N groups on the PPy were the fundamentals of attraction of specified cavities.

4.2. Surface plasmon resonance MISens

Surface plasmons are formed by an electromagnetic wave, which propagate along the surface of a thin metal layer. According to Abbas et al., surface plasmon resonance (SPR) is a collective oscillation of conduction electrons, which present at the interface of metal-dielectric media. SPR have three features, important in terms of any new sensor: firstly the enhancement of the electric field, secondly the propagation length, and the lastly the penetration depth [40].

The SPR phenomenon was recognized in 1960s after Otto and Kretschmann had invented surface plasma with invisible light. The SPR sensor technique has been used in very different areas for immunosensors, the determination of interaction between immunoglobulin G (IgG) protein and antigen, monitoring of the interactions between drugs and biological molecules, and so on [41]. There are plenty of planar configurations of SPR biosensors. Among these, in general, Otto configuration is used. Generally, SPR sensor is formed of six parts, including a light source, a detector, a transduction surface, a prism, biomolecule, and a flow system.

A typical SPR system (immunoassay technique is described), as mentioned above and can be seen in Fig. 4., uses microfluids to pass controlled amounts of analyte across the sensor surface to which the antibody is immobilized. With reflecting a beam of polarized light to the back surface of the metal film, the analysis is made through a prism. After the beam of light hits the noble metal surface, not all the light is reflected. Some of the energy of photons is absorbed by the metal and causes electron oscillations at the interface of two materials. When molecules are bound to the sensor surface, the refractive index (RI) changes. RI affects also reflected light intensity, angle, and wavelength. It is measured as resonance units (RU). In general, 1 RU is equal to 1 pg mm-2 of analyte concentration [41].

As can be seen from Fig. 4, there are two mediums and an interface. One medium is optically denser. When light passing from the optically denser medium is exposed to the light-thinning medium, at the interface of two mediums, total reflection will occur, if an appropriate range of incident angles are inherent in the medium, change of resonance amplitude occurs, with a penetration depth. For example, if antibody was hold on sensor chip and was let to interact with antigen solution, the refractive index (RI) of the metal film surface would change. The change of SPR resonance angle would change with a change of refractive index. The change

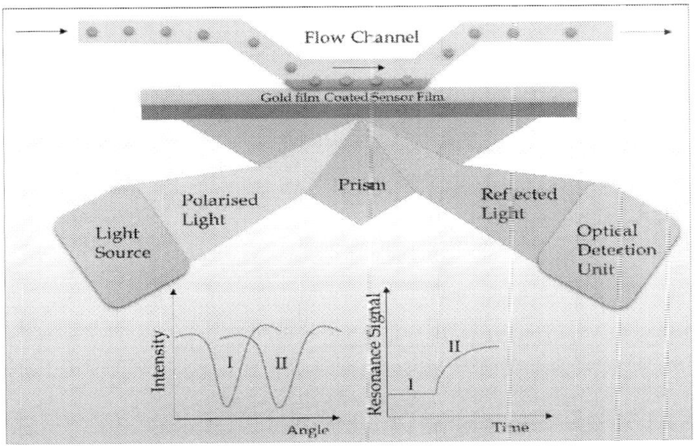

Figure 4. Schematic representation of SPR.

of refractive index will be proportional to mass change, due to the absorption of antigens to antibodies. This means that the mass change of biological macromolecules causes refractive index change with SPR resonance angle change [42].

Optical sensor research has very advantageous features that it allows label-free analysis, it is simple to construct, and has the ease of use – inexpensive and highly sensitive [41].

Carlucci et al. made a study to determine Vitamin D (25OHD) with a novel optical and electrochemical-based biosensor. For SPR measurements, first, gold SPR disks were cleaned with fresh piranha solution (3:1 H_2SO_4 98% : H_2O_2 30%). Then, a self-assembled monolayer with 11-mercaptoundecanoic acid (MUA) was formed on gold surface. The carboxyl functions on the SAM layer were activated with a mixture of N-(3-dimethylaminopropyl)-N′-ethylcarbodiimide (EDC) and N-hydroxysuccinimide (NHS). After removing the mixture, making later steps, a LOD of 2 μg mL^{-1} but when vitamin D was modified with gold nanoparticles (AuNPs) a lower LOD of 1 μg mL^{-1} was reached. With an electrochemical biosensor, which was based on the reaction of vitamin D with 4-ferrocenylmethyl-1,2,4-triazoline-3,5-dione (FMTAD), vitamin D was determined with a LOD of 10 ng mL^{-1} [44].

In a study of Choi et al., Zearalenone, found in a number of cereal crops, was determined via surface plasmon resonance sensor. Zearalenone is a mycoestrogen, which acts like an endocrine disruptor. For determination, pyrrole was electropolymerized in the presence of Zearalenone. Electropolymerization was made with a three-electrode electrochemical system. Au film was used as the working electrode, Ag/AgCl as reference, and Pt grid as counter. After that, PPy-coated Au chips were mounted on the SPR cell and change of incident angle of laser was measured. According to the results, the sensor exhibited a linear response in the range of 0.3–3000 ng mL^{-1} with a LOD of 0.3 ng g^{-1}. For selectivity test, structural analogues of Zearalenone were used (α-Zearalenone, Zearalenone, β-Zearalenone). Among these compounds, the sensor showed the highest selectivity to Zearalenone, due to the strong binding capacity [45].

Yao et al. made a SPR sensor to determine pesticide, which has high toxicity and binds irreversibly to acetylcholinesterase (AChE). Because of this, it causes serious harm in the respiratory tract, human nervous system, and cardiovascular system. To detect and enhance detection sensitivity, magnetic molecularly imprinted nanoparticles (NPs) were used. Magnetic NPs were prepared through the self-polymerization of dopamine on the Fe_3O_4 NP surface in the presence of template, chlorpyrifos (CPF). Using these NPs, pesticide was detected in a range from 0.001 to 10 μM with a detection limit of 0.76 nM [46].

Like other examples, the SPR technique is also used for the detection of another organic molecule, domoic acid (DA), which is a neurotoxic amino acid. This toxin accumulates in mussels (such as Mytilus edulis), crabs or anchovies and when DA-contaminated shellfish is taken,) called an intoxication syndrome known as amnesic shellfish poisoning (ASP) can occur. The typical symptoms of ASP are vomiting, cramps, and diarrhoea and neurological symptoms including severe headache, seizures, and either temporary or permanent memory loss. Lotierzo et al. prepared a MIP film by direct photo-grafting onto a gold chip. Firstly, the gold surface was functionalized with a self-assembled monolayer of 2-mercaptoethylamine and subsequent carbodiimide. This provided covalent attachment of the photo initiator 4,4′-azobis(cyanovaleric acid). After proper steps, DA containing polymerization solution was deposited on the gold surface, and for polymerization the chip was irradiated with ultraviolet light. Non-printed control chips were prepared with the same procedure but without the template, DA. According to the results, DA was detected in a range of 2-3300 μg L^{-1}. After a number of tests, molecularly imprinted DA sensor protected its stability until 30 (±5) analysis [47].

Molecular imprinting based surface plasmon resonance technology can be applied also for enzyme detection. Matsunaga et al. prepared molecularly imprinted polymers for lysozyme with acrylic acid (AAc) as the functional monomer, and N,N′-methylenebisacrylamide (MBAA) as the cross-linker. For preparation of SPR sensor, firstly an Au-coated SPR sensor chip was immersed in N,N′-bis(acryloyl)cystamine to bear vinyl groups on Au surface. A polymerization mixture (including template, lysozyme) was applied on Au and polymerized by radical polymerization. For polymerization, the vinyl group grafted SPR sensor chip was poured on glass, on which the polymerization mixture had been poured. After that another glass plate and a weight were placed on the sensor. A non-imprinted polymer thin film was prepared with the same procedure but without adding template. After proper steps, SPR measurements were made. To examine the effect of salt concentration on the rebinding of lysozyme, imprinted sensors were prepared in various concentrations of NaCl (0, 20, 40 mM). It was seen that the bound amounts of proteins were decreased with the increasing concentrations of NaCl in the rebinding buffer. From this result, it can be thought that electrostatic interactions took place between proteins and acrylic acid residues. With examples of different proteins, it was seen that binding changed upon isoelectric points of amino acid residues. For example at pH 7.4, lysozyme (pI:11), Cytochrome C (pI:10), and RNAse (ribonucleotidase) (pI: 9,5) were positively charged and strongly bound to the films via acrylic acid residues. At the same pH value, myoglobin (pI:7) and lactalbumin (pI:4,5) were negatively charged and that is why they showed almost no binding because of this electrostatic repulsion. This study is also an example for selective protein sensors with SPR sensing technique [48].

Enterotoxins can be detected with SPR technique just in the same way as with quartz crystal microbalance (QCM) sensors. Homola et al. developed a new SPR sensor for *Staphylococcal enterotoxin B* (SEB), which is a soluble protein, secreted by *Staphylococcus aureus*. According to the results SEB could be detected at low concentrations, such as 5.0 μg L^{-1}, in pure samples, directly. But, by using a sandwich assay, this limit has been decreased to 0.5 μg L^{-1} in both pure samples and in milk [49].

Food allergens can be detected via SPR technology. Yman et al. detected peanut allergen protein with optical sensor with both direct and sandwich immunoassays. By these methods they detected milk, hazelnut, sesam, egg, and peanut proteins in food samples. They used polyclonal antibodies to detect these allergens. According to the results, allergens were detected in the range of 1.0–12.5 μg g^{-1} in food samples [50].

As it can be seen from these examples, SPR sensing technology can be used in a variety of areas changing from protein detection to environmental pollutants with a low detection range, faster attainment of results, and selectivity.

4.3. Quartz Crystal Microbalance (QCM) MISens

The QCM consists of a thin piezoelectric plate, which has acoustic resonances in the MHz range. When the crystal comes into contact with the sample, the resonance properties change. QCM technology was first recognized by Sauerbrey in 1959. He indicated usefulness of the method for measuring the characteristic frequency of an oscillator circuit. The frequency changes were determined by using a piezoelectric crystal and as can be seen in Fig. 5, the oscillating frequency of the crystal decreases with the adsorption of foreign substances on the surface.

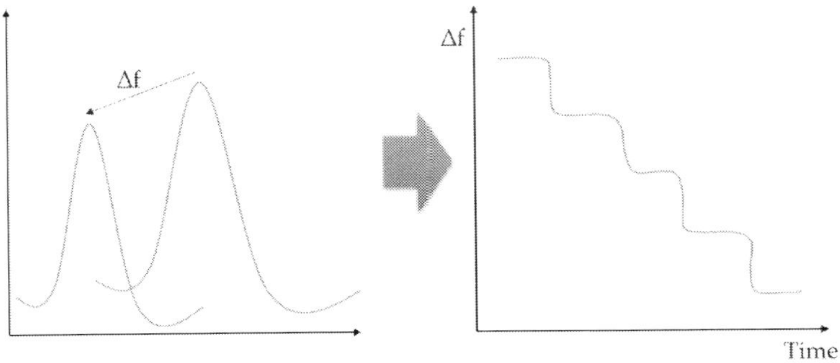

Figure 5. Frequency change on QCM electrode, while interacting with sample.

Because of the sensitive nature of quartz crystal, this method was described as a very precise method. The results of this work are embodied in the equation. According to the Sauerbrey equation, the mass change per unit area at the QCM electrode surface and frequency changes are proportional.

The observed change in oscillation frequency of the crystal:

$$\Delta f = -Cf \times \Delta m \tag{1}$$

where

Δf = the observed frequency change (Hz)

Δm= the change in mass per unit area (g/cm^2)

Cf = the sensitivity factor for the crystal (56.6 Hz μg^{-1} cm^2 for a 5 MHz AT-cut quartz crystal at room temperature)

As mentioned above, the Sauerbrey equation relies on a linear sensitivity factor, Cf, which is a fundamental property of the QCM crystal. The method was also utilized for the direct weighed of a mass [51]. QCM, which is used for the biosensor experiment, is consisted of piezoelectric crystal, oscillator, and frequency counter(Fig 6.).

The piezoelectric quartz crystal is driven by a low-frequency transistor oscillator. The frequency of the vibrating crystal is monitored by the frequency counter. The crystal, which is mounted on its holder, is connected to the oscillator circuit. The frequency counter is connected to the oscillatory device. By frequency counter, frequency changes are recorded after each step in coating or in interaction with the sample.

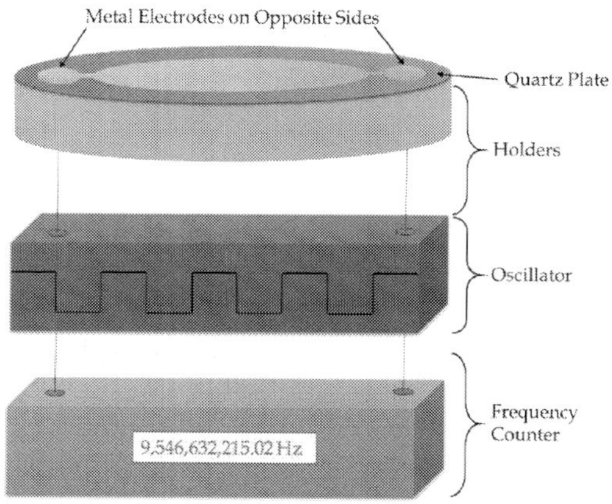

Figure 6. The schematic diagram of experimental piezoelectric sensor.

The advantages of the technique are surface specificity, monolayer sensitivity, and high acoustic contrast for dilute adsorbents [52]. Besides QCM does not require any labelling, has low barriers of entry, ease-of-use, low cost, and speed to result [54].

The theoretical detection limit of oscillating quartz crystals is about 10–12 g, which means a detection in pictogram range. With this low detection limit, the QCM can be used in trace analysis, immunosensors, DNA biosensors, and drug analysis. Piezoelectric crystals can also be used in microbalances for thin film technology [53].

Because of expense, sensitivity, and short shelf life of biological materials, molecularly imprinted polymers (MIPs) are used on sensor surface. In addition in QCM analysis, MIPs are used to achieve a specific binding site and a high affinity. MIPs are advantageous because of their features, like high similarity to natural receptors, physicochemical, mechanic, thermal stability, simple preparation, and easy adaptation of application [53].

The crystal frequency changes when the interaction occurs between imprinted polymer and template solution. It can be seen that before and after the interaction crystal frequency decreases due to the uptake of template by the imprinted polymer. Liao et al. made a study for stereospecific L-histidine sensor with imprinted polyacrylamide membranes. According to the study when the crystals were interacted with L-histidine, the net frequency shifts of the crystal modified with L-histidine is found much more than the shifts, which belongs to D-histidine. It can be concluded that the L-histidine imprinted membrane showed better selectivity to L-histidine. Besides specificity, selectivity is also an important feature of imprinted polymers. In this study, L-tyrosine and L-arginine were tested with the L-histidine imprinted membrane and DL-phenylalanine was tested with D-histidine imprinted membrane. Under the same reaction conditions (time, concentration, etc.), the imprinted membrane showed much more affinity to the same molecule, which was used as template because of specific cavities, formed in the polymer. From these results, it can be concluded that this imprinted piezoelectric sensor can be used for the chiral separation of histidine [55].

In another study, Liu et al. reported a novel method for the separation of D- and L-tryptophan using molecularly imprinted quartz crystal microbalance (QCM) sensor. They fabricated the sensor by using molecularly imprinted polymers, which was prepared by using acrylamide (AM) as monomer and 1,1,1-trimethylolpropane trimethacrylate (TRIM) as cross-linker in different molar ratios. With the fabricated optimum imprinted polymer, the binding of template L-tryptophan was about four-fold to three-fold larger than that obtained with the D-tryptophan enantiomer. It was calculated that the enantiomeric selectivity coefficient of the fabricated molecularly imprinted sensor was 6.4. Moreover, it was observed that the binding of L- and D-tryptophan enantiomers on the non-imprinted polymer (NIP) was almost the same. This indicated the sensitivity and enantioselectivity of molecularly imprinted polymer [52].

It is also possible to make mass determinations in protein mixtures via MIP-QCM sensor. Lin et al. prepared an albumin imprinted copolymer of 3-dimethylaminopropyl methacrylamide (DMAPMA) and different acrylate series cross-linking agents. Gold surface was used and four kinds of Au-coated crystals were prepared. One of them was bare and the other three were prepared with different functional groups bonded to the surface of the sensor. As functional groups, they used –NH2, –OH, –COOH on Au surface. According to the results, the greatest adsorption capacity belonged to Au-OH electrode and to a lesser extent, to bare Au electrode. However, according to the time effectiveness to obtain stability and according to the adsorbed albumin amount, Au-NH_2 and Au-OH were the optimal electrodes. Between NIP and MIP electrodes prepared, the molecularly imprinted electrode showed more efficient albumin determination than non-imprinted electrode with the time taken to receive a steady state frequency and adsorbed amount of albumin. In addition to this, the prepared albumin-imprinted QCM sensor showed largest adsorption of albumin among similar molecules like cytochrome c, lysozyme, and myoglobin, whose molecular sizes were far smaller than albumin. In the range between 60 and 150 ppm, albumin was obtained. With results of this study, it can be indicated that the presence of albumin-specific cavities in the prepared electrode gave a greater adsorption and a smaller diffusion resistance, which makes response time shorter [53].

Sun et al. used piezoelectric quartz crystal for sensing taste-causing molecules by using molecular imprinting technology. A PQC (piezoelectric quartz crystal) sensor array, which is MIP coated, is developed to quickly and more sensitively detect taste-causing compounds in beverage. They studied quinine, which is a bitter-taste causing compound, and usually flavoured with saccharine to reduce its unpleasant bitter taste. Because of this they used quinine and saccharine as template molecules. Methacrylic acid (MAA) is used as a monomer. The MIP coated PQC sensor array was studied under flow injection analysis and results were compared with the results of volunteer human taste panellists. With the satisfactory repeatability, and with a high sensitivity to detect the change in bitter taste in tonic water with much less suppressing effect in the presence of saccharine, the developed sensor was very comprehensive. According to results, the quinine-MIP modified PQC sensor displayed a linear working range for quinine from 10 mg L^{-1} to 1080 mg L^{-1} and for saccharine from 51 mg L^{-1} to 3420 mg L^{-1}. The calculated limit of detection is 2.04 mg/L for quinine and 32.8 mg L^{-1} for saccharine [56].

MIP-PQC is used not only for taste application but also to distinguish different taste causing compounds [57] and to detect organic pollutants with taste implication such as organic/ inorganic acids and amines in drinking water [58].

Due to the use of QCM in the gas phase, it has application as odour sensors. Ji et al. used 2-methylisoborneol (MIB) and geosmin (GEO) as off-flavour compounds which cause odour problems in drinking water. They are produced by some microorganisms. These odour chemicals were analysed with GC-MS or Enzyme-Linked Immunosorbent Assay(ELISA) and detection limits were ca. 1 ng/L and 1 μg/L, respectively. But these methods need a high budget in terms of chemicals and equipment. But with the piezoelectric sensor, the analysis could be done at a lower cost and more sensitively. In their study, they made pre-treatment with nylon layers to QCM electrode. After that, they used MIB and GEO as template molecules and imprinted polymers are prepared with methacrylic acid as functional monomer, ethylene glycol dimethacrylate as cross-linker, and 2,2′-azobis (2,4-dimethyl)valeronitrile as initiator. They were all dissolved in hexane, used as porogen, under nitrogenous atmosphere. Five microliters of this solution were pipetted onto prepared QCM and polymerized at 40°C for 48

h. Non-imprinted polymers were synthesized under the same conditions except the use of the template. The prepared QCMs were interacted with template molecules in a thermostat chamber in a stream of nitrogen flowing. After interaction, imprinted sensors showed an average frequency change of 2864 Hz ± 6.26 % (n=3) and NIP-sensors 3014 Hz ± 5.14%. It indicates a similar amount of substrate immobilization on sensors (ca. 3 μg). It was observed that the frequency change after MIP application to the nylon sublayer was about 50% higher than after application of the MIP to a bare QCM. To analyse selectivity except MIB and GEO, some other odorants like terpinol, β-ionone, and citronellol were also interacted with MIB-imprinted sensor and it was found that the highest frequency change was observed at MIB-sensor after interaction with MIB. In spite of their previous sensor with an LOD of 200 ppb, these synthesized sensors could detect above 10 ppb. This means approximately 20-fold more sensitive detection capacity [59].

QCM sensor could be used for influenza detection. Either influenza virus can be detected or influenza virus binding capabilities can be analysed. Diltemiz et al. have developed a sensor for recognition of the hemagglutinin (HA) protein, which occurs by influenza virus with infection and causes hemagglutination. For this, they used 4-aminophenyl boronic acid (4-APBA) as a new ligand for binding of sialic acid (SA), which has a valuable role in the binding of HA through boronic acid sugar interaction. QCM sensor surface was modified with thiol groups and then 4-aminophenyl boronic acid and sialic acid were immobilized on sensor surfaces, respectively. To do these, first QCM electrodes were cleaned with alkaline piranha solution (1:1:5 deionized water : H_2O_2: NH_3 v/v). After cleaning, electrode surfaces were modified with 11-mercaptoundecanoic (MUA) acid. By using MUA, carboxyl fictionalization were achieved. After that, QCM electrodes were modified with imide groups by using N-(3-dimethylaminopropyl)-N'-ethylcarbodiimide and N-hydroxysuccinimide for immobilization of 4-APBA and SA. After interaction with the samples, the binding capacity and limit of detection of QCM sensors were found to be 4.7×10^{-2} μM and 0.26 μM ml^{-1}, respectively [60].

As mentioned above, QCM finds lots of application areas in terms of low cost, speed to result, and low detection limit. Because of these advantages, studies made with QCM sensor are increasing day by day.

5. Comparison of MI-sensors

In this section we compared MISens by showing Table 1 to describe polymer type, measurement type, LOD, and detection range.

Sensor Type	Modification	Target Molecule	Detection Range	LOD	Reference
Impedimetric	PGE/PPy	CPF	20 to 300 μg L^{-1}	4.5 ug L^{-1}	13
Impedimetric/ voltammetric	GCE/PPy	Trimethoprim	10^{-6}–10^{-4} M	0.13 μM	30

Sensor Type	Modification	Target Molecule	Detection Range	LOD	Reference
Amperometric	AuE/ AuNPs@MIES	Dopamine		7.8 nM	31
CV/DPV/ Amperometric	GCE/Ni/MIP	Phenobarbital	0.14uM–1.3mM	8.2 nM	32
Potentiometric detection	CuE/MWCNT-MIP	γ-hexachlorocyclohexane	10^{-9}–10^{-3} M	10^{-10} M	33
CV/DPV/ chronocoulometry	PGE/MAA	Calcitonin	9.99ngL^{-1}–7.919mgL^{-1}	3.09 ng L^{-1}	34
DPV/CV	AuE/ PNIPAAm/o-PD	Folic Acid	1–200 μM	0.9 μM	35
DPV/CV	GCE/ MWCNT/Au-Pt/oPD	Aflatoxin B1	1×10–10 – 1×10–5 mol L^{-1}	30 pmol L^{-1}	36
DPV	GCE/ AuNPs/oCD/MGO	Chrysodine	5.0×10^{-8} – 5.0×10^{-6} mol L^{-1}	1.7×10^{-8} mol L^{-1}	38
SWV	GCE/MWCNT/p-NPDEFB/2-AET-Fe@Au	Cefexime	0.1 nM–10nM	2.2×10^{-11}M	39

Table 1. Modification, measurement type, LOD, and detection range

Sensor Type	Modification	Target Molecule	Detection Range	LOD	Reference
Optical	GOx/PtOEP	Glucose	2–120 mg dL^{-1}	1.5±0.2 mg dL^{-1}	44
Optical	Au/MUA	Vitamin D	0.05–1.0 μg mL^{-1}	0.045 μg mL^{-1}	45
Optical	Au/Pyrrole	Zearalenone	0.3–3000 ng mL^{-1}	0.3 ng g^{-1}	46
Optical	Au/MUA	Chlorpyrifos (CPF)	0.001–10 μM	0.76 nM	47
Piezoelectric	Au electrode/Acrylamide	L-Tryptophan	1–4 mM	8.8 μM	53
Piezoelectric	Au electrode/Methacrylic Acid	Quinine Saccharine	10–1080 mg L^{-1} 51–3420 mg L^{-1}	2,04 mg L^{-1} 32.8 mg L^{-1}	56

PGE: Pencil Graphite Electrode, PPy: Polypyrrole, CPF: Chlorpyrifos, GCE: Glassy Carbon Electrode, AuE: Gold Electrode, MIES: MIP modified electrode, Ni: Nickel, MWCNT: Multi-Walled Carbon Nanotube, MAA: Methacrylic Acid, o-PD: o-Phenylenediamine, o-CD: o-Cyclodextrin, MGO: Magnetic graphene oxide, AET: Aminoethenethiol, MUA: Mercaptoundecanoicacid

Table 2. Modification type, measurement type, LOD, and detection range

6. Conclusion

As a result of these examples and studies, molecularly imprinted sensor systems have been developing, and they will continue to be developed. Just as we mentioned above, biological receptors are restricted to detect analyte by environmental parameters. Therefore, the combination of molecular imprinting technology and the chemical sensor technologies useful to be employed as bio-mimicking measurement system, and these combinations are easy to construct as well as they have a low cost causing them to become more prominent to focus on.

Acknowledgements

We, the authors of this chapter, would like to thank PhD. Elif KARADADAŞ, who studies in Ege University Medical Biochemistry Department. Thanks to her efforts, comments, and revisions in the publishing process.

Author details

Zihni Onur Uygun[1*], Hilmiye Deniz Ertuğrul Uygun[2], Nihal Ermiş[3] and Erhan Canbay[1]

*Address all correspondence to: onur_uygun@hotmail.com

1 Ege University, Faculty of Medicine, Medical Biochemistry Department, Bornova, İzmir, Turkey

2 Dokuz Eylül University, Faculty of Science, Chemistry Department, Buca, İzmir, Turkey", Turkey

3 Ondokuz Mayıs University, Faculty of Science and Arts, Chemistry Department, Kurupelit, İzmir, Tu rkey

References

[1] Rodriguez-Mozaz S, Lopez de Alda MJ, Marco MP, Barcelo D. Biosensors for environmental monitoring: a global perspective. Talanta. 2005; 65: 291–297.

[2] Kirsch J, Siltanen C, Zhou Q, Revzin A, Simonian A. Biosensor technology: recent advances in threat agent detection and medicine. Chem Soc Rev. 2014; 42(22): 8733-68.

[3] Liu SQ, Zheng ZZ, Li XY. Advances in pesticide biosensors: current status, challenges, and future perspectives. Anal Bioanal Chem. 2013; 405(1): 63-90.

[4] Yadav R, Dwivedi S, Kumar S, Chaudhury A. Trends and perspectives of biosensors for food and environmental virology. Food Environ Virol. 2010; 2(2): 53-63.

[5] Van Dorst B, Mehta J, Bekaert K, Rouah-Martin E, De Coen W, Dubruel P, et al. Recent advances in recognition elements of food and environmental biosensors: A review. Biosens Bioelectron. 2010 Dec 15; 26(4): 1178-94.

[6] Thakur MS, Ragavan KV. Biosensors in food processing. J Food Sci Tech Mys. 2013; 50(4): 625-41.

[7] Park JW, Lee C, Jang J. High-performance field-effect transistor-type glucose biosensor based on nanohybrids of carboxylated polypyrrole nanotube wrapped graphene sheet transducer. Sensor Actuat B-Chem. 2015 Mar 1; 208: 532-7.

[8] Banerjee S, Sarkar P, Turner APF. Amperometric biosensor based on Prussian Blue nanoparticle-modified screen-printed electrode for estimation of glucose-6-phosphate. Anal Biochem. 2013 Aug 15; 439(2): 194-200.

[9] Krieg AK, Gauglitz G. An optical sensor for the detection of human pancreatic lipase. Sensor Actuat B-Chem. 2014 Nov; 203: 663-9.

[10] Ali MA, Singh N, Srivastava S, Agrawal VV, John R, Onoda M, et al. Chitosan-modified carbon nanotubes-based platform for low-density lipoprotein detection. Appl Biochem Biotech. 2014 Oct; 174(3): 926-35.

[11] Massad-Ivanir N, Segal E. Porous silicon for bacteria detection. Woodh Publ Ser Biom. 2014; (68): 286-303.

[12] Li RY, Zhang JJ, Wang ZP, Li ZJ, Liu JK, Gu ZG, et al. Novel graphene-gold nanohybrid with excellent electrocatalytic performance for the electrochemical detection of glucose. Sensor Actuat B-Chem. 2015 Mar 1; 208: 421-8.

[13] Uygun ZO, Dilgin Y. A novel impedimetric sensor based on molecularly imprinted polypyrrole modified pencil graphite electrode for trace level determination of chlorpyrifos. Sensor Actuat B-Chem. 2013 Nov; 188: 78-84.

[14] Taghdisi SM, Danesh NM, Lavaee P, Emrani AS, Ramezani M, Abnous K. Aptamer biosensor for selective and rapid determination of insulin. Anal Lett. 2015 Mar 4; 48(4): 672-81.

[15] Zhu GC, Zhang CY. Functional nucleic acid-based sensors for heavy metal ion assays. Analyst. 2014 Dec 21; 139(24): 6326-42.

[16] De Avila BEF, Escamilla-Gomez V, Campuzano S, Pedrero M, Salvador JP, Marco MP, et al. Ultrasensitive amperometric magnetoimmunosensor for human C-reactive protein quantification in serum. Sensor Actuat B-Chem. 2013 Nov; 188: 212-20.

[17] Date Y, Aota A, Sasaki K, Namiki Y, Matsumoto N, Watanabe Y, et al. Label-free impedimetric immunoassay for trace levels of polychlorinated biphenyls in insulating oil. Anal Chem. 2014 Mar 18; 86(6): 2989-96.

[18] Baykov IK, Matveev AL, Stronin OV, Ryzhikov AB, Matveev LE, Kasakin MF, et al. A protective chimeric antibody to tick-borne encephalitis virus. Vaccine. 2014 Jun 17; 32(29): 3589-94.

[19] Mosbach K. The promise of molecular imprinting. Sci Am. 2006 Oct; 295(4): 86-91.

[20] Hulanicki A, Geab S, Ingman F. Chemical sensors: definitions and classification. Pure Appl. Chem. 1991; (63): 1247–50.

[21] Cammann K, Ross B, Katerkamp A, Reinbold J, Grundig B, Renneberg R. Chemical and Biochemical Sensors, Ullmann's Encyclopedia of Industrial Chemistry Wiley-VCH Verlag GmbH & Co. KGaA, Weinheim, 2012, 109–221.

[22] Praveen K.S., Eric L.B., Rangachary M., Fernando H.G. Chemical sensors for environmental monitoring and homeland security. The Electrochemical Society. 2010; 35-40.

[23] Florinel-Gabriel B. Chemical Sensors and Biosensors; Fundamantels and Applications, Wiley, 2012.

[24] Joseph RS, William RP, Sheng Y. Sensors, Chemical sensors, electrochemical sensors, and ECS. Journal of the Electrochemical Society. 2003; (150): 11-6.

[25] Canbay E, Habip A, Kara G, Eren Z, Akyılmaz E. A microbial biosensor based on Lactobacillus delbruecki sp. bacterial cells for simultaneous determination of lactic and pyruvic acid. Food Chemistry. 2015; (169): 197-202.

[26] Wilson JS, Sensor Technology Handbook, Elsevier, Amsterdam/ Boston, 2005.

[27] Blum LJ, Coulet PR (Eds.). Biosensor Principles and Applications, Marcel Dekker, New York, 1991.

[28] Avila M, Zougagh M, Ríos Á. Molecularly Imprinted Polymers for Selective Piezoelectric Sensing of Small Molecules. Trends Anal. Chem. 2008; (27): 54–65.

[29] www.scholar.google.com

[30] Silva H, Júlia JGP, Subramanian MCSM, Matos VCD. A MIP-graphene-modified glassy carbon electrode for the determination of trimethoprim. Biosensors and Bioelectronics. 2014; (52): 56–61.

[31] Xue C, Han Q, Wang Y, Wu J, Wen T, Wang R, Hong J, Zhou X, Jiang H. Amperometric detection of dopamine in human serum by electrochemical sensor based on gold nanoparticles doped molecularly imprinted polymers. Biosensors and Bioelectronics. 2013; (49): 199–203.

[32] Yu HC, Huang XY, Lei FH, Tan XC, Wei YC, Li H. Molecularly imprinted electrochemical sensor based on nickel nanoparticle-modified electrodes for phenobarbital determination. Electrochimica Acta. 2014, (141), 45–50.

[33] Anirudhan TS, Alexander S. Design and fabrication of molecularly imprinted polymer-based potentiometric sensor from the surface modified multiwalled carbon

nanotube for the determination of lindane (γ-hexachlorocyclohexane), an organochlorine pesticide. Biosensors and Bioelectronics. 2015; (64): 586–93.

[34] Patra S, Roy E, Madhuri R, Sharma PK. Imprinted ZnO nanostructure-based electrochemical sensing of calcitonin: A clinical marker for medullary thyroid carcinoma. Analytica Chimica Acta. 2015; (853): 271–284.

[35] Karimian N, Zavar MHA, Chamsaz M, Turner APF, Tiwari A. On/off-switchable electrochemical folic acid sensor based on molecularly imprinted polymer electrode. Electrochemistry Communications. 2013; (36): 92–5.

[36] Wang ZH, Li JS, Xu LJ, Feng YJ, Lu XQ. Electrochemical sensor for determination of aflatoxin B-1 based on multiwalled carbon nanotubes-supported Au/Pt bimetallic nanoparticles. J Solid State Electr. 2014 Sep; 18(9):2487-96.

[37] Zhong M, Teng Y, Pang SF, Yan LQ, Kan XW. Pyrrole-phenylboronic acid: A novel monomer for dopamine recognition and detection based on imprinted electrochemical sensor. Biosens Bioelectron. 2015 Feb 15; 64: 212-8.

[38] Wang XJ, Li XJ, Luo CN, Sun M, Li LL, Duan HM. Ultrasensitive molecularly imprinted electrochemical sensor based on magnetism graphene oxide/beta-cyclodextrin/Au nanoparticles composites for chrysoidine analysis. Electrochim Acta. 2014 Jun 1; 130: 519-25.

[39] Yola ML, Eren T, Atar N. Molecularly imprinted electrochemical biosensor based on Fe@Au nanoparticles involved in 2-aminoethanethiol functionalized multi-walled carbon nanotubes for sensitive determination of cefexime in human plasma. Biosens Bioelectron. 2014 Oct 15; 60: 277-85.

[40] Abbas A, Linman MJ, Cheng QA. New trends in instrumental design for surface plasmon resonance-based biosensors. Biosens Bioelectron. 2011 Jan 15; 26(5): 1815-24.

[41] Holford TRJ, Davis F, Higson SPJ. Recent trends in antibody based sensors. Biosens Bioelectron. 2012 Apr 15; 34(1): 12-24.

[42] Xu X, Ye ZZ, Wu JA, Ying YB. Application and research development of surface plasmon resonance-based immunosensors for protein detection. Chinese J Anal Chem. 2010 Jul; 38(7): 1052-9.

[43] Stein EW, Grant PS, Zhu HG, McShane MJ. Microscale enzymatic optical biosensors using mass transport limiting nanofilms. 1. Fabrication and characterization using glucose as a model analyte. Anal Chem. 2007 Feb 15; 79(4): 1339-48.

[44] Carlucci L, Favero G, Tortolini C, Di Fusco M, Romagnoli E, Minisola S, et al. Several approaches for vitamin D determination by surface plasmon resonance and electrochemical affinity biosensors. Biosens Bioelectron. 2013 Feb 15; 40(1): 350-5.

[45] Choi SW, Chang HJ, Lee N, Kim JH, Chun HS. Detection of mycoestrogen zearalenone by a molecularly imprinted polypyrrole-based surface plasmon resonance (SPR) sensor. J Agr Food Chem. 2009 Feb 25; 57(4): 1113-8.

[46] Yao GH, Liang RP, Huang CF, Wang Y, Qiu JD. Surface plasmon resonance sensor based on magnetic molecularly imprinted polymers amplification for pesticide recognition. Anal Chem. 2013 Dec 17; 85(24): 11944-51.

[47] Lotierzo M, Henry OYF, Piletsky S, Tothill I, Cullen D, Kania M, et al. Surface plasmon resonance sensor for domoic acid based on grafted imprinted polymer. Biosens Bioelectron. 2004 Sep 15; 20(2): 145-52.

[48] Matsunaga T, Hishiya T, Takeuchi T. Surface plasmon resonance sensor for lysozyme based on molecularly imprinted thin films. Anal Chim Acta. 2007 May 15; 591(1): 63-7.

[49] Homola J, Dostalek J, Chen SF, Rasooly A, Jiang SY, Yee SS. Spectral surface plasmon resonance biosensor for detection of staphylococcal enterotoxin B in milk. Int J Food Microbiol. 2002 May 5; 75(1-2): 61-9.

[50] Yman IM, Eriksson A, Johansson MA, Hellenas KE. Food allergen detection with biosensor immunoassays. J Aoac Int. 2006 May-Jun; 89(3): 856-61.

[51] Sauerbrey G. Verwendung von Schwingquarzen zur Wägung dünner Schichten und zur Mikrowagung. Zeitschrift für Physik 1959; 155(2): 206-222.

[52] Liu F, Liu X, Ng SC, Chan HSO. Enantioselective molecular imprinting polymer coated QCM for the recognition of L-tryptophan. Sensor Actuat B-Chem. 2006 Jan 17; 113(1): 234-40.

[53] Lin TY, Hu CH, Chou TC. Determination of albumin concentration by MIP-QCM sensor. Biosens Bioelectron. 2004 Jul 30; 20(1): 75-81.

[54] Uludag Y, Piletsky SA, Turner APF, Cooper MA. Piezoelectric sensors based on molecular imprinted polymers for detection of low molecular mass analytes. Febs J. 2007 Nov; 274(21): 5471-80.

[55] Liao HP, Zhang ZH, Nie LH, Yao SZ. Electrosynthesis of imprinted polyacrylamide membranes for the stereospecific L-histidine sensor and its characterization by AC impedance spectroscopy and piezoelectric quartz crystal technique. J Biochem Bioph Meth. 2004 Apr 30; 59(1): 75-87.

[56] Sun H, Mo ZH, Choy JTS, Zhu DR, Fung YS. Piezoelectric quartz crystal sensor for sensing taste-causing compounds in food. Sensor Actuat B-Chem. 2008 Apr 14; 131(1): 148-58.

[57] Leonte II, Sehra G, Cole M, Hesketh P, Gardner JW. Taste sensors utilizing high-frequency SH-SAW devices. Sensor Actuat B-Chem. 2006 Oct 25; 118(1-2): 349-55.

[58] Rosler S, Lucklum R, Borngraber R, Hartmann J, Hauptmann P. Sensor system for the detection of organic pollutants in water by thickness shear mode resonators. Sensor Actuat B-Chem. 1998 May 30; 48(1-3): 415-24.

[59] Ji HS, McNiven S, Lee KH, Saito T, Ikebukuro K, Karube I. Increasing the sensitivity of piezoelectric odour sensors based on molecularly imprinted polymers. Biosens Bioelectron. 2000 Oct; 15(7-8): 403-9.

[60] Diltemiz SE, Ersoz A, Hur D, Kecili R, Say R. 4-Aminophenyl boronic acid modified gold platforms for influenza diagnosis. Mat Sci Eng C-Mater. 2013 Mar 1; 33(2): 824-30.

Chapter 4

Silicon-based Integrated Microarray Biochips for Biosensing and Biodetection Applications

Lei Zhang, Cheng Zhu, Jinwen Geng, Xizeng Shi, Yunhua Gao, Zhijie Chang and He Qian

Additional information is available at the end of the chapter

http://dx.doi.org/10.5772/60441

Abstract

The silicon-based integrated microarray biochip (IMB) is an inter-disciplinary research direction of microelectronics and biological science. It has caught the attention of both industry and academia, in applications such as deoxyribonucleic acid (DNA) and immunological detection, medical inspection and point-of-care (PoC) diagnosis, as well as food safety and environmental surveillance. Future biodetection strategies demand biochips with high sensitivity, miniaturization, integration, parallel, multi-target and even intelligence capabilities. In this chapter, a comprehensive investigation of current research on state-of-the-art silicon-based integrated microarray biochips is presented. These include the electrochemical biochip, magnetic tunnelling junction (MTJ) based biochip, giant magnetoresistance (GMR) biochip and integrated oscillator-based biochip. The principles, methodologies and challenges of the aforementioned biochips will also be discussed and compared from all aspects, e.g., sensitivity, fabrication complexity and cost, compatibility with silicon-based complementary metal-oxide-semiconductor (CMOS) technology, multi-target detection capabilities, signal processing and system integrations, etc. In this way, we discuss future silicon-based fully integrated biochips, which could be used for portable medical detection and low cost PoC diagnosis applications.

Keywords: Integrated microarray biochip, silicon, electrochemical, GMR, oscillator

INTECH
open science | open minds

1. Introduction

Silicon-based integrated microarray biochips (IMB) are an inter-disciplinary research direction of microelectronics and biological science. They have caught the attention of both industry and academia, in applications such as deoxyribonucleic acid (DNA) and immunological detection, medical inspection and point-of-care (PoC) diagnosis, as well as food safety and environmental surveillance. Future biodetection strategies demand biochips with high sensitivity, miniaturization, integration, parallel, multi-target and even intelligence capabilities [1, 2]. Recent studies focus on the high sensitivity and multi-target microarray biochips for accurate and early disease diagnosis. They also discuss treatment that is based on silicon compatible technologies due to its low cost, large-scale and readily integration with signal extraction and processing front-end and back-end circuits monolithically for a lab-on-a-chip solution [3-16]. Such studies oppose conventional assays that are currently used in hospitals and laboratories, e.g., enzyme-linked immunosorbent assay (ELISA) [17] and fluorescent detections [18], etc.

Biosensor microarray assays are one type of popular method to perform comparable biological detections. These are typically carried out for the analysis of large-scale gene and protein expression changes in a biological sample. Although there are many different forms of microarray biochips, they can be generally classified into two categories, based on the sensing methodologies: labelled and label-free. Labelled techniques tag an analyte of interest with an externally observable label such as a magnetic tag [9-16] or fluorophore [18]. On the other hand, label-free techniques detect an intrinsic property of the biomolecule such as the electrochemical charge/current [3-8], mass, thermal reactivity, etc. One typical example of label-free detection is electrochemical microarray, which senses the current induced in the binding reaction during detection. Meanwhile, magnetically labelled assays with ultra-high sensitivity, such as GMR and MTJ microarrays, are labelled techniques.

Recent studies of IMB focus on high sensitivity and multi-target microarray biochips, based on silicon compatible technologies. For example, CMOS, due to its low cost, large-scale and readily integration with signal extraction and processing front-end and back-end circuits monolithically for a lab-on-a-chip solution. State-of-the-art silicon-based IMB assays include the electrochemical microarray biochip [3-8], GMR and MTJ microarray biochips [9-14] and the fully CMOS compatible oscillator microarray biochip [15-17].

This chapter is organized as follows. Section 2 presents the electrochemical microarray biochip and illustrates examples of state-of-the-art electrochemical microarray biochips. Furthermore, design issues and considerations are elaborately analysed and discussed. Section 3 discusses magnetically labelled detection assay that are based on GMR and MTJ sensors and biochips. Methodology, trade-offs and calibration strategies are also presented, as well as a system prototype with performance evaluations and comparisons. Section 4 discusses biochip assay based on on-chip oscillator microarray that is fully compatible with modern aggressively scaled CMOS technology, with a proposed in-situ calibration strategy. Finally, the chapter concludes in Section 5, where future directions for IMB research are envisioned and reviewed.

2. Electrochemical microarray biochips

An electrochemical microarray biochip is one type of sensor used for DNA and protein detection. It has caught the attention of both academia and industry due to its applications in DNA and immunological analysis, analyte detection, drug delivery, trends of miniaturization and large-scale integration with CMOS technology for a system-on-a-chip (SoC) solution. In this strategy, the target detection and hybridization action are converted into electric current signals on the microarray electrodes on the chip surface. This is readily extracted by the high sensitivity analogue circuit front-end of the CMOS biochip. The integrated circuit (IC) interfacing, which is composed of an analogue circuit front-end for signal extraction, multiplexor (mux) and demultiplexor (demux) circuits for electrode enabling and decoding, as well as a potentiostat for electrode voltage stabilization, governs the overall performance of the microarray biochip.

2.1. Design challenges and considerations

Along with the development of micro- and nano-fabrication technologies, such as lithograph, nano-imprint, electron beam, etc., the feature size of recent electrode microarray tends to shrink towards a nanometre scale, together with the scaling down of CMOS technology. However, electrodes of nanoscale seriously suffer from background noise and random currents in a biochemical sample solution medium. This degrades the sensitivity and makes the design of IC interfacing difficult. Previous research suggests that the minimal sensitivity of signal extraction front-end IC should be at least at the order of pico-ampere (pA, 10^{-12} A) or lower. Moreover, for an effective current monitoring and response, the bandwidth must be several MHz or higher - an even more difficult task. Prior applications of interfacing IC, based on potentiostat [19], trans-impedance amplifier [20] and current amplifier [21, 22] have been reported.

Large-scale integration also presents challenges regarding the signal processing back-end. For example, there are at least 20 known amino acids, while there are over 500 types of haemoglobin in human blood. A large amount of data has to be extracted and analysed by the back-end circuitry during a single detection phase. This heavily burdens the digital signal processor (DSP) or field-programmable gate array (FPGA). Recent research on neuromorphic circuits and systems, based on memristive devices and circuit, may provide a novel pathway for the large-scale, parallel signal processing and mode recognition in the near future.

The scaling down of electrode microarray is also a challenging task. Investigators have proposed a post-CMOS fabrication methodology [8]. However, when shrinking the device sizes, this methodology suffers from keyhole issues, resulting in the deterioration of the surface roughness and eventually, the sensitivity. Therefore, a co-design and fabrication regarding the microarray floor planning, lithography and etching are currently under investigation.

2.2. Current research

In previous studies, the electrochemical detection and sensing of DNA has been reported [3]. Here, the viability of the methodology has been verified. The first fully CMOS integrated DNA microarray biochip was published by Infineon. In this case, a 128-spot electrode structure with in-pixel analogue-to-digital conversion was realized, achieving 100 dB of a dynamic range for direct current (DC) detection [4]. The chip was improved to a sub-nanoampere level [5] by introducing differential architecture. However, limited by the in-pixel integrator, the scale of single electrode could not be reduced beyond 100 μm. This prevented the downscaling of the electrode feature size and the upscaling of the microarray's integration level. A fluorescent-based biosensor microarray was introduced [6], which utilized the in-pixel photo-diode. This was followed by a preamplifier and analogue-to-digital converter (ADC), which were used to detect the fluorescence on the target. A 10^{-12}A level current sensitivity and a detection limit of 0.5 nM were achieved. Although a microarray is extremely sensitive in DC detection, it can hardly respond to transient signals with a bandwidth of 1 kHz and above. This is mandatory in broadband sensing (10 kHz and above) and low noise alternating current (AC) detections.

In this section, we present a fully integrated 64-pixel CMOS biosensor microarray, which is based on memory addressing architecture. Thanks to the pixel-decoding scheme, the embedded potentiostat operational amplifier (opamp) and current amplifiers are shared by all of the pixels in the microarray. This helps to reduce the pixel complexity and provides a pathway to the miniaturization of electrodes towards a nanoscale. An electrode feature size of 600 nm has been successfully demonstrated. To the best of our knowledge, this is the smallest size among published works. Moreover, the embedded current amplifier is realized with pico-ampere sensitivity in order to reconcile with the current level of the nanoscale electrodes. This is over 15 kHz bandwidth, which supports both the high-speed sensing and low noise AC detections [7, 21].

The system architecture of the proposed biosensor microarray is shown in Figure 1. It is designed with a 64-pixel of 8 × 8 working electrode microarray, together with interdigital references and counter electrodes. A two-stage address decoding approach (pre-decoding and pixel-decoding) is introduced to selectively activate the microarray's working electrodes. An on-chip potentiostat opamp P is implemented to stably control the electrode potential and provide electrochemical current to the electrolyte from a negative feedback mechanism during biosensing detections. Meanwhile, an ultra-low-current amplifier (ULCA) is also integrated in order to sense the detection current and introduce amplification, before being sent out to off-chip devices. The proposed ULCA achieved a gain of 20 dB, a bandwidth of 15.5 kHz and a current sensitivity of less than 38 pA [21]. The potentiostat serving for stabilizations of electrolyte potential and accommodation of electrochemical current is realized by an opamp with a rail-to-rail input common-mode range and Class-AB output stage. The proposed biosensor microarray is realized in standard 0.18 μm CMOS technology. Figure 2(a) shows the die micrograph of the chip, which occupies 2 mm by 2 mm of the chip area. The top electrodes of the microarray are fabricated following the procedures in previous studies [8]. Limited by the etching resolution, the minimal electrode feature size demonstrated was 600 nm, which is the smallest among published works. A biosensor detection system, based on magnetic nanoparticle (MNP), biotinlated polystyrene and streptavidin, is utilized to demonstrate the functionalities of the proposed microarray. The fluorescent results are shown in Figure 2(b) and zoomed in by Figure 2(c) and (d). The normalized fluorescent intensities are measured as 127 and 136, while the noise background is 12.3. The translating to signal-to-noise ratio (SNR) is 10.1 dB and 10.4 dB, respectively. The measured currents start from 1~2 μA and gradually reduce in the first 20 seconds, due to the electric shielding effect of the targets. They eventually saturate at about 20 nA and 7.3 nA, respectively.

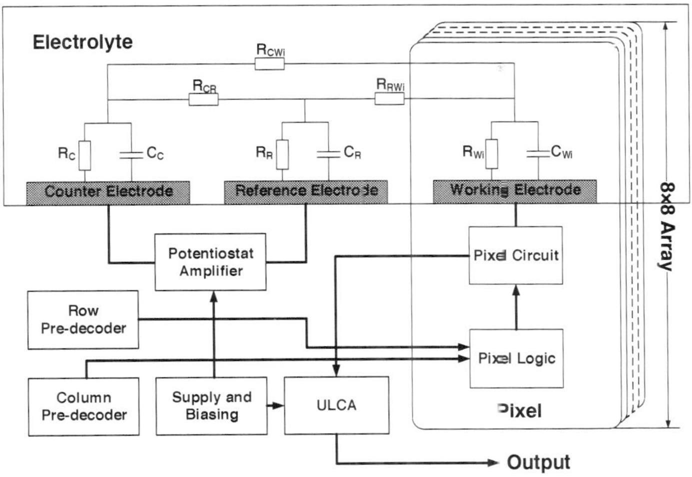

Figure 1. (a): Circuit architecture of proposed biosensor microarray.

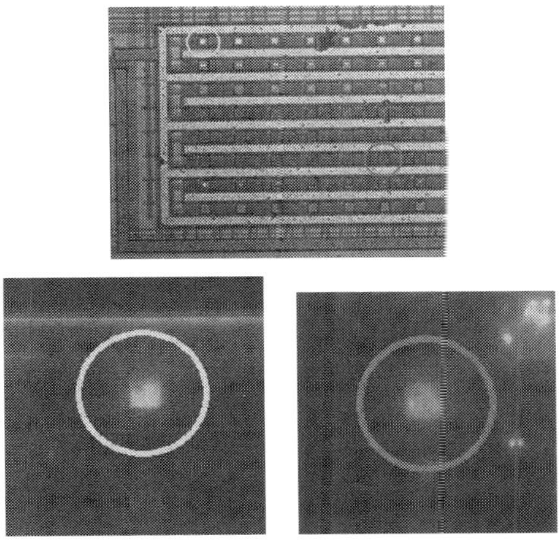

Figure 2. (a): Addressed working electrodes on the microarray, (b) fluorescent detection results on the addressed electrodes, (c) zoom in on the 8 μm square electrode and (d) zoom in on the 600 nm spiral electrode.

2.3. Discussions

Electrochemical microarray biochips are one type of widely interested assay that are capable of mid-sensitivity multi-target detection. They have important applications in multi-target DNA and immunological analysis, as well as in analyte detection and drug delivery. Along with the trends of miniaturization and large-scale integration with CMOS technology for a SoC solution, various issues need to be solved in future research. Such issues include background noise and random currents in a biochemical sample solution medium during the shrinking of electrode feature size, post-CMOS fabrication technologies, parallel back-end signal processing and mode recognition. The current sensitivity of recent reported applications falls in the nano-ampere range, which is limited by background noise and random currents in the biochemical medium and buffer. This prevents its application for higher sensitivity sensing and detection applications.

3. GMR and MTJ microarray biochips

In recent years, rapid biological detection based on GMR and MTJ sensors and biochips is a hot issue. This is due to its interesting features such as being fast, cheap, user-friendly and acceptable high sensitivity [9-14]. As there is less magnetic material in body fluid, the magnetic signal detected by GMR and MTJ biosensors suffers minimally from the interferences of body fluid and therefore, exhibits a high sensitivity.

3.1. Overview

The GMR and MTJ biosensor detection system generally consists of several steps. These include the biochemical reaction on the surface of the sensor, analyte tagging by super-paramagnetic nano-particles, receiving signals from the top of the sensor, small signal extraction circuit front-end, digital domain data processing and correction back-end. These are shown in Figure 3 and Figure 4.

Figure 3 and Figure 4 show the detailed detection scheme and signal processing steps. Firstly, probe molecules that target multiple specific analytes are immobilized on the biosensor chip surface through biochemical bonding and/or reactions. During the detection phase, the analyte molecules are introduced, which are specifically captured by probes that are readily immobilized on the sensor surface. In the labelling phase, a labelling molecule tagging with MNP is subsequently bonded with analyte molecules, which are captured by probes in the detection phase. The analytes, acting as a bridge between the probe molecule and the MNP label, are therefore fixed on the surface of the biosensor chip. Finally, MNPs and analytes that are not fixed on the top of the sensor surface are washed away. The number/density of MNPs immobilized on the biosensor surface is proportional to the analyte concentration to be detected. The probe and analyte in the aforementioned detection assay can either be DNA or antigen/antibody protein, depending on the applications.

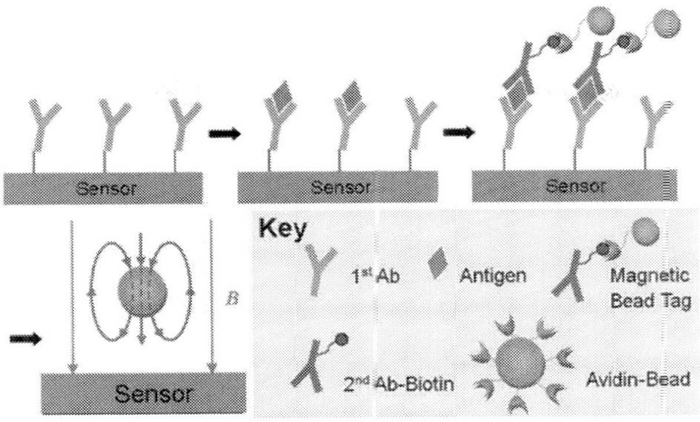

Figure 3. Biochemical detections paradigms based on MNP-labelled targets.

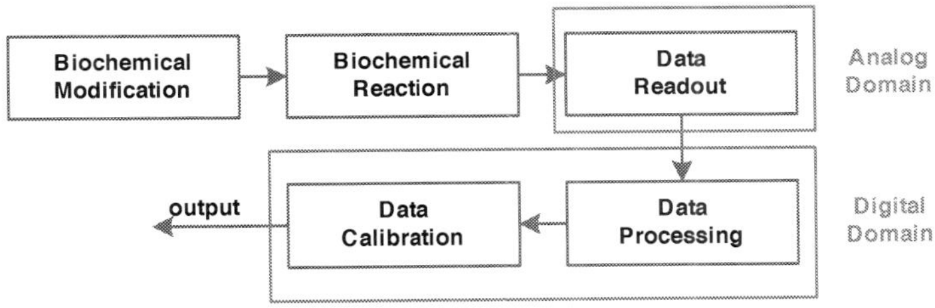

Figure 4. Signal flow and processing steps of GMR and MTJ biosensor detection system.

3.2. MTJ microarray biochips

Recent progress on the MTJ biosensor multi-target microarray biochip has been reported [9]. In such studies, a fully automated in vitro diagnostic (IVD) system for diagnosing acute myocardial infarction was developed. This uses high sensitivity MTJ array as sensors and 300 nm nano-magnetic particles as tags. An array of 12 × 10^6 MTJ devices are integrated with a three metal layer CMOS circuit. Consequently, 48 different types of bio targets can be analysed simultaneously. A micro-fluidic cartridge, which contains all of the reagents necessary for completing the assaying process, is also assembled. Electrical, mechanical and micro-fluidic pumping devices are also integrated. The system yields results of the three major acute myocardial infarction (AMI) biomarkers (cTnI, MYO, CK-MB) in 15 minutes. An extremely high sensitivity of 0.1 ng/ml is realized and the dynamic range of detention is over three orders. However, system solutions that are based on MTJ microarray biochips are pricy in terms of the MTJ sensor itself and the circuit board/microfluidic channel integration. Meanwhile, over 100 wiring connections have to be realized and therefore lowering the yield. In the present form, this can hardly be commercialized.

3.3. GMR microarray biochips

GMR biosensors and biochips are proposed and utilized in the above detection assay for their acceptable high sensitivity responses to MNP labels, as well as their lower fabrication cost, compared with MTJ microarray biochips. This makes them potentially suitable for mass production and commercialization. Recent progress has been reported by researchers in Stanford [10]. In their study, 256 separate GMR sensors were fabricated and a CMOS signal extraction front-end with 16 channel ΣΔ modulators was designed and realized in 180 nm CMOS process. This has a sensitivity of around 10 picomolar (pM), roughly equivalent to about 50 ppm of GMR resistance change. However, a number of issues remain unsolved. Firstly, temperature calibrations must be performed during the detection phase in order to cancel out the local temperature variation effects around the GMR sensor detection region. The calibration is made with one single fixed temperature coefficient over a wide range of GMR sensors. This incurs inaccurate results. Secondly, the biochemical buffer solution on the detection region is manually controlled on a disposable test stick. This can hardly be extended for multi-target applications with 10 or more samples and thus, is impractical with regard to efficiently controlling the biochemical reactions on multiple targets.

In this section, a complete prototype of GMR microarray biochips and system is presented. The GMR biochip, signal extraction IC, back-end signal processing, microfluidic channels, novel calibration strategies and mechanical parts with microfluidic channels are all integrated. The system performance shows a sub-50 ppm sensitivity, corresponding to roughly two 500 nm MNPs within a 1000 μm^2 GMR sensor detection area on average. This advances the present application [11].

The physical structures of the GMR biosensor are illustrated in Figure 5.

The GMR biosensor is a multilayer structure device. It has two major functionality layers: a pinned layer and a free layer. The pinned layer is a layer with fixed magnetic moment direction. Meanwhile, the free layer features a changeable magnetic moment direction, which can be modulated by an externally applied magnetic field. Both of the layers are ferromagnetic and are separated by a non-ferromagnetic layer in between them.

A GMR sensor is generally designed with two alternative structures, as shown in Figure 5. In the left subfigure, the electrons flow vertically through both the pinned and free layer (Current Perpendicular to Plane, CPP). The electrons experience scattering in ferromagnetic layers, depending on the magnetic moment of the layer and the electron spin. When the pinned and free layers exhibit parallel magnetic moment direction, the electrons experience the lowest scattering through both layers, corresponding to the lowest resistance state. If the magnetic moment directions of the pinned and free layer are antiparallel, electrons experience large scattering though both layers. This results in the highest resistance state. In the right subfigure, the electrons flow horizontally (Current In Plane, CIP). The physics of CIP is similar to the CPP structure. Furthermore, the sensor exhibits different resistances when the magnetic moment direction of the free layer is modulated by the external magnetic field. As the length of the GMR sensor is far larger than the thickness, the CPP structure normally has a higher magneto-

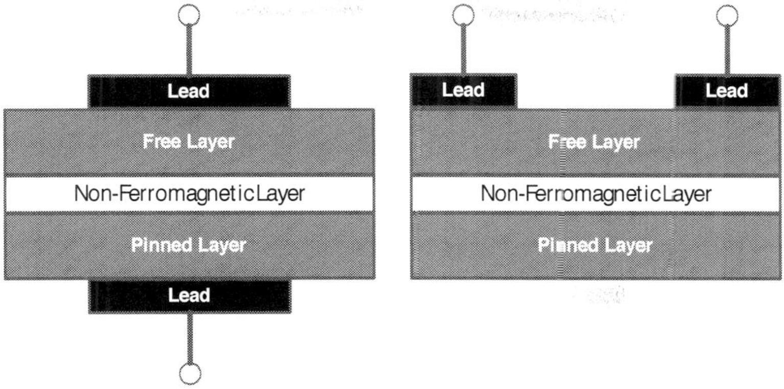

Figure 5. GMR biosensor structures.

resistance (MR) than its CIP counterpart. However, it is hard to fabricate and not easily integrated with the planar process. Modern GMR biosensors are generally fabricated with a CIP structure.

Super-paramagnetism MNP is utilized. This can be magnetized by the applied magnetic field and produce a small magnetic field around the MNPs, modulating the resistance of the sensor.

Firstly, a DC biasing magnetic field is applied to the sensor. This magnetic field magnetizes the MNPs on top of the sensor and forces them to produce small magnetic fields in the same direction, with the applied magnetic field on top of the sensor. An extra AC magnetic field is further applied in order to modulate the magnetic field of the MNP and consequently, induce the resistance change of the sensor itself at the same frequency.

To readout the change of the resistance of GMR sensor, a voltage is applied to the sensor. The output current flowing through the sensor is measured and readout. Considering that the flicker noise is large at a low frequency and that the resistance change caused by the MNPs is only several tens ppm of the initial resistance (chosen as a few kΩ) of the sensor, an AC voltage running at several kHz, together with an AC magnetic field at a few hundred Hz, is utilized to read out the signal. Therefore, the output current through the sensor can be calculated by the following formula:

$$I_{GMR} = \frac{Vcos(2\pi f_c t)}{R_0 + \Delta Rcos(2\pi f_s t)} = \frac{V}{R_0}\cos(2\pi f_c t) - \frac{V\Delta R}{2R_0^2}(\cos(2\pi(f_c + f_s)t) + \cos(2\pi(f_c - f_s)t) \tag{1}$$

where V is the amplitude of the applied voltage, R_0 is the initial resistance of the sensor, ΔR is the resistance change of the sensor due to the applied magnetic field, MNPs, f_c is the frequency of the external AC voltage and f_s is the frequency of the AC magnetic field.

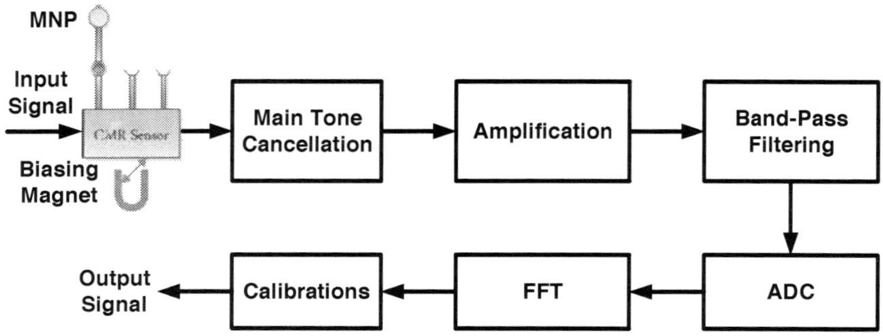

Figure 6. Signal processing flow of the proposed GMR sensing system, including the main tone cancellation, amplification, filtering, ADC, FFT and output correction.

It can be seen from Figure 6 that the sensor acts as a mixer and produces three tones at f_c, f_c+f_s and f_c-f_s. As ΔR is much smaller than R_0, the main tone amplitude at f_c is higher than the side tones amplitudes at f_c+f_s and f_c-f_s, which are signals of interest, since the effects of MNPs only relate to ΔR.

The MR of a GMR sensor is generally less than 15 %. The operation point is chosen at the middle point with maximal slope and the ΔR is generally about 1~3 % of R_0. Meanwhile, the resistance change due to MNPs can be as small as 0.1 % of ΔR. The signal is very small compared to the centre tone, requiring an ADC with a dynamic range of over 100 dB - an extremely difficult number to achieve. To overcome this issue, cancellation of the main tone is necessary. This can be implemented by applying an AC voltage with the opposite phase to a reference resistor of the same resistance as the GMR sensor. The centre tone can be reduced by at least 20 dB, considering a 10 % deviation between the resistance of the GMR sensor and the cancellation resistor.

After cancellation, a filter is introduced to suppress the noise out of the band. It is followed by an amplifier before ADC in order to transform the signal from an analogue to digital domain. In the digital domain, Fast Fourier Transform (FFT) is used to calculate the frequency spectrum and eventually extract the signal from the side tones.

The signal can be extracted by the following formulas:

$$\mathrm{MR} = \frac{4ST}{CT - 2ST} \tag{2}$$

$$\mathrm{Signal} = \frac{MR(t) - MR(0)}{MR_{max}} \tag{3}$$

where CT is the centre tone or main tone of the GMR signal before cancellation, ST is the side tone. The centre tone amplitude before cancellation can be easily recovered based on the gain, cancelling resistance and source amplitude:

$$CT_{before-cancellation} = \frac{VR_s}{R_c} \pm CT_{after-cancellation} \tag{4}$$

with the sign decided by the resistances of the GMR sensor and the cancellation resistor.

As the sensor resistances are sensitive to the environment, output correction and calibrations of temperature variations are always necessary in order to ensure the veracity of the detection. Therefore, the temperature-calibrated scheme is proposed in this section.

Firstly, considering the temperature variations of the initial resistance of the GMR sensor and the resistance change caused by the magnetic field and MNPs, the GMR current can be written as,

$$I_{GMR} = \frac{V\cos\left(2\pi f_c t\right)}{R_0(1+\alpha\Delta T)+\Delta R(1+\beta\Delta T)\cos\left(2\pi f_s t\right)} \tag{5}$$

where α is the temperature coefficient of the initial resistance of GMR sensor and β is the coefficient factor of the resistance change due to MNPs. It follows that

$$I_{GMR} \approx \frac{V}{R_0}\cos\left(2\pi f_c t\right)\left(\frac{1}{1+\alpha\Delta T}\right) - \frac{V\Delta R}{2R_0^2}\left(\cos\left(2\pi(f_c+f_s)t\right)+\cos\left(2\pi(f_c-f_s)t\right)\right)\left(\frac{1+\beta\Delta T}{\left(1+\alpha\Delta T\right)^2}\right) \tag{6}$$

Then, the CT and ST can be calculated by:

$$\begin{aligned} \mathrm{CT}(\mathrm{t}) &= \frac{V}{R_0}\left(\frac{1}{1+\alpha\Delta T}\right) \approx \frac{V}{R_0}\left(1-\alpha\Delta T\right) \quad (\beta\Delta T \ll 1) \\ \mathrm{ST}(\mathrm{t}) &= \frac{V\Delta R}{2R_0^2}\left(\frac{1+\beta\Delta T}{\left(1+\alpha\Delta T\right)^2}\right) \approx \frac{V\Delta R}{2R_0^2}\left(1+\beta\Delta T\right)\left(1-2\alpha\Delta T\right) \approx \frac{V\Delta R}{2R_0^2}\left(1+\left(\beta-2\alpha\right)\Delta T\right) \\ &(\alpha\Delta T \ll 1, \beta\Delta T \ll 1) \end{aligned} \tag{7}$$

In normalizing the CT and ST to them at initial time 0, we have,

$$\begin{aligned} \frac{CT(t)}{CT(0)} &= \left(1-\alpha\Delta T\right) \\ \frac{ST(t)}{ST(0)} &= 1+\left(\beta-2\alpha\right)\Delta T \end{aligned} \tag{8}$$

We can define κ as the following relationship between CT and ST,

$$\kappa = \frac{\left(\frac{ST(t)}{ST(0)} - 1\right)}{\left(\frac{CT(t)}{CT(0)} - 1\right)} = \frac{\beta - 2\alpha}{-\alpha} \tag{9}$$

One can obtain κ by linear fitting and utilizing the centre tone to calibrate the side tone,

$$\begin{aligned} \mathrm{CF} &= \frac{1}{1+\kappa\left(\frac{CT(t)}{CT(0)} - 1\right)} = \frac{1}{1+(\beta - 2\alpha)\Delta T} \\ \mathrm{ST} &= \mathrm{ST}(\mathrm{t}) \times \mathrm{CF} \approx \frac{V\Delta R}{2R_0^2}\left(1+(\beta - 2\alpha)\Delta T\right) \times \frac{1}{1+(\beta - 2\alpha)\Delta T} = \frac{V\Delta R}{2R_0^2} \end{aligned} \tag{10}$$

From the above equation, the temperature variation can be easily calibrated out during the detection. The major improvement is that the data obtained at the beginning of the test are utilized to calculate the temperature coefficient κ for every single sensor instead of using a standard number. Consequently, this produces a better correction and more accurately reduces the temperature effect over the whole chip.

A complete circuit and system solution prototype, including the GMR biosensor design and fabrication, small signal extraction front-end, data conversion and signal processing back-end for the GMR biosensing and detection assay, has been proposed and realized by the experiment.

The proposed GMR biosensor is a NiFe/CoFe/Cu/CoFe/PtMn multilayer structure, as shown in Figure 7(a). A 40 nm thick SiO_2 layer is fabricated on top of the multilayer sensor in order to keep it from oxidization and prevent electrical shorting. Furthermore, a thinner passivation layer achieves closer proximity between the magnetic particle and sensor surface for higher sensitivity, resulting in an easier electric broken down.

A biosensor chip containing 12 separate GMR sensors was fabricated, as shown in Figure 7(b). Taking into consideration the trade-off between the statistical fluctuation of magnetic particle distributions and uneven magnetic field effect during detection, the physical size of each sensor is 120 μm × 120 μm.

In the system prototype, the signal extraction front-end with main tone cancellation, amplification and filtering functions has been realized by CMOS ICs. The sampling frequency of the ADC is 1 MHz and each packet of data contains a signal of six seconds. First, the data is divided into 60 pieces and each piece is sent to a Hanning Window for data smoothing before FFT. An average of over 10 groups of data is used to reduce the noise. An adaptive filter algorithm is used to further reduce the noise and a 50 % noise reduction can be achieved. To further improve the sensitivity of the sensing system, the proposed temperature calibration scheme is also introduced and realized.

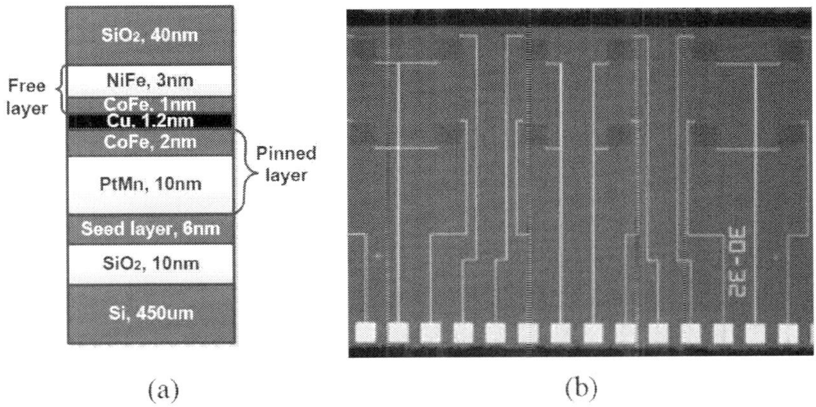

Figure 7. (a): Multilayer structure of the proposed GMR sensor and (b) the GMR biosensor chip containing 12 sensors.

Figure 8(a) shows the α prototype detector utilizing the GMR biosensor detection scheme that is proposed in this section. The structure of the detector is shown in Figure 8(b), including the front-end circuit board, signal processing DSP, PC interface and mechanical parts of the disposable microfluidic box.

Figure 8. (a): α prototype detector system, utilizing the proposed GMR biosensor detection scheme and microarray biochip and (b) structure of the α prototype detector.

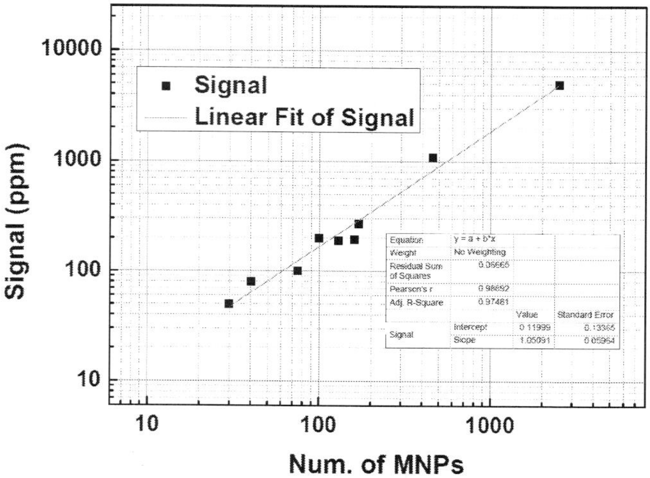

Figure 9. Magnetic signal versus the number of magnetic particles. Each magnetic particle produces a signal of about 2 ppm.

In the experiment, magnetic particles with diameters of 500 nm are utilized to produce the target magnetic signal. The current leakage between the two connection wires can introduce a signal as large as several hundred ppm. As a result, the data has to be collected when the surface is dry. A microscope is used to observe and count the magnetic particles on top of the sensor. The relationship between the number of magnetic particles and the output signal is shown in Figure 9. It can be seen that a sensitivity of sub-50 ppm has been achieved in the proposed system. In this case, about 30 magnetic particles on a 120 μm × 120 μm sensor surface are observed. Therefore, on average, each magnetic particle roughly produces a signal of about 2 ppm. Table 1 shows comparisons with other GMR biosensing systems. This work achieves superior performance versus prior arts. In summary, biological detections based on the proposed GMR biochips and system have been verified by experiments and a sub-50 ppm sensitivity has been reached, advancing the present work.

	Sensor Size (μm × μm)	MNP Diameter (nm)	Min. Detectable Num. N	N per 1000 μm2 Area
[10]	100 × 100	50	2000	200
[12]	120 × 67.5	30	N/A	N/A
[13]	160 × 58.5	1000	N/A	N/A
[14]	1500	300	3 (theoretical)	2
This Work	120 × 120	500	30	2.08

Table 1. Comparisons with state-of-the-art GMR biosensing systems.

3.4. Discussions

Magneto-sensitive detection assays are proved with extremely high sensitivities and provide full support to multi-target (~100 targets) biochemical analysis applications, e.g., DNA and immunological detections that are widely used in laboratories and hospitals. Biosensing systems and chips based on MTJ and GMR microarrays are possible candidates. Previous research suggests that a MTJ microarray biochip has a pricy sensor with unsatisfied yield. Thus, in the present form, it can hardly be commercialized. On the other hand, a GMR microarray biochip features a sub-50 ppm high sensitivity response to MNP labels and a lower fabrication cost compared to a MTJ microarray biochip. This makes it potentially suitable for mass production and commercialization in the near future. However, both biochips require an extra fabrication flow beyond the CMOS process, preventing the monolithic integration with acquisition IC front-end, signal processing back-end and DSP for a SoC solution to further improve system performance and reduce costs.

4. Oscillator-based microarray biochips

In the area of medical diagnosis, high accuracy, fast speed and low cost biosensor systems are always in demand. However, conventional biosensor systems, such as fluorescent labelled biosensor and MTJ and GMR biosensors, require complicated extra-process steps or fabrications. These are not fully compatible with CMOS technology and, consequently, the cost and design complexity can hardly be reduced. Moreover, both MTJ and GMR sensors require external biasing magnetic field, which complicates implementing the system.

4.1. Overview

An oscillator-based microarray biochip, which relies on the frequency-shift of LC oscillators, is one type of scheme that is potentially capable of solving these problems [15-17]. It uses an electromagnetic mechanism to detect the changes of output frequency, which directly corresponds to the number of targets tagged by MNPs. It potentially features high sensitivity, low cost, multi-target and easy integration, compared to GMR and MTJ biosensing strategies.

An LC oscillator is a basic component for local oscillation (LO) generation in radio-frequency (RF) integrated circuits [23]. An inductor and a fixed capacitor oscillate at a free running frequency f_0, driven by a negative resistance from a cross-coupled transistor pair. We suppose that some probe molecules are deposited on the inductor surface. During the detection phase, some target molecules with MNPs labels are also deposited on the inductor and bonded with the probes. Therefore, the MNPs change the magnetic field around the detection region defined by the inductor, effectively changing the inductance L_0 to $L_0+\Delta L$. This eventually shifts the oscillation frequency f_0, as shown in Eq. (11), where M is the target concentration and α is a factor.

$$f = \frac{1}{2\pi\sqrt{(L_0 + \Delta L)C}} \approx f_0\left(1 - \frac{\Delta L}{2L_0}\right) = f_0\left(1 - \frac{\alpha M}{2L_0}\right) \tag{11}$$

The frequency shift Δf is proportional to the inductance change ΔL and therefore, reflects the target molecule concentration in a sample solution.

The principle of the oscillator-based high sensitivity multi-target biochip detection is illustrated in Figure 10. Firstly, the CMOS biochip surface with multiple detection regions is treated by various probe molecules, which are a target of interest for probe immobilization, as shown in Figure 10(a) and (b). During the detection phase, the sample solutions with target molecules labelled with MNPs are brought to the detection region by either droplet or microfluidic channel for chemical bonding and hybridization. The chip is then rinsed by DI water for surface cleaning in order to remove non-specific bonding or contaminations. Eventually, the electric signal induced by the tag MNPs on the target of interest is detected by the buried CMOS signal extraction front-end circuit and conveyed off-chip (or to the on-chip ASIC) for signal processing by the back-end circuits. Since the chemical bonding and hybridization between the probe and target are specific, multiple target detections from one sample droplet can, therefore, be realized in parallel on the single biochip. This increases the detection efficiency and reduces the cost. Depending on the non-specific bonding rate between the different probes and targets, the number of targets supported by a single detection has to be traded with the fake positive rate and detection sensitivity.

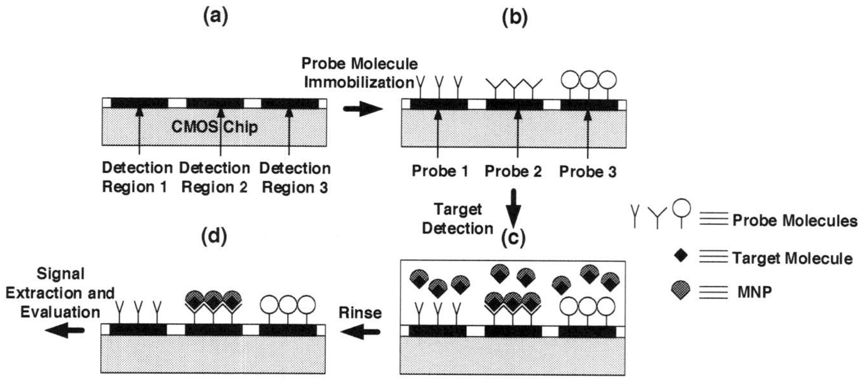

Figure 10. Oscillator-based high sensitivity multi-target biochip detection sketch.

4.2. Design challenges and considerations

4.2.1. Sensor design

4.2.1.1. Running frequency

The free running frequency f_0 of the oscillator has to be optimized, based on the considerations of phase noise, inductor detection region geometry and the response frequency of MNPs. Most MNPs with a diameter of 500 nm or below exhibit a response frequency of lower than 1.5 GHz [24], defining the upper boundary of f_0. Although a low f_0 would present a lower phase noise, other factors include increasing the number of turns of the inductor itself, shrinking the effective detection region size and degrading the uniformity of magnetic field on the surface. Among the above factors, experiments suggest an f_0 of around 1 GHz - 1.5 GHz, with a proper trade-off.

4.2.1.2. Q of the inductor

The quality factor (Q) of the inductor also has to be optimized, considering the sensor dimension and phase noise of the LO oscillator, which directly correlate to the sensitivity of the biochip. Sensor dimensions are normally established from the chip size, multi-target capabilities and microfluidic channel structures, etc. For a given sensor size, the Q of the inductor can be improved by optimizing the number of metal layers, turns, trace width and spacing of the traces. For example, in a standard 180 nm CMOS process, the Q of the inductor can be maximized as 5.2 by paralleling the top three to four layers of metals with a reasonable inductance of 1.7 nH in a chip area of 130 μm x 130 μm. This is a fairly desirable value in the design, as shown in Figure 11.

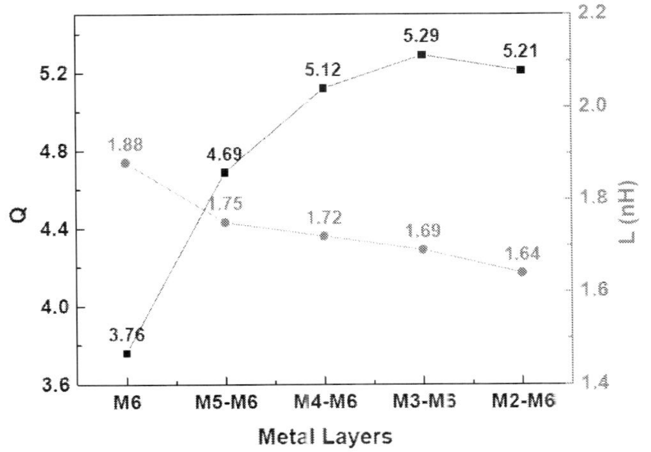

Figure 11. Q and inductance of an optimized inductor versus metal layers.

4.2.1.3. *Uniformity of the magnetic field*

Uniformity of the surface magnetic field becomes important when the sensitivity is pushed aggressively to an extremely low level or even for single molecule/MNP detection applications. In this case, it has to be co-optimized with Q and the inductance of the inductor itself. Bowl-shape [16] and multi-layer structured [25] round shape inductor sensors are investigated to improve the field uniformity. However, the Q of the inductor of interest is substantially degraded. A non-uniform inductor trace width strategy has been investigated to improve the field uniformity. For example, as the trace width of the inner winding has been adjusted to 9 μm, as opposed to a width of 8 μm, for the rest of the outer windings in a three turn inductor, the field uniformity can be decently improved, as shown in Figure 12. It can be seen that the magnetic field intensity varies by roughly 20 % over the inductor surface before optimization. With the proposed technique, it is reduced to about 5 %.

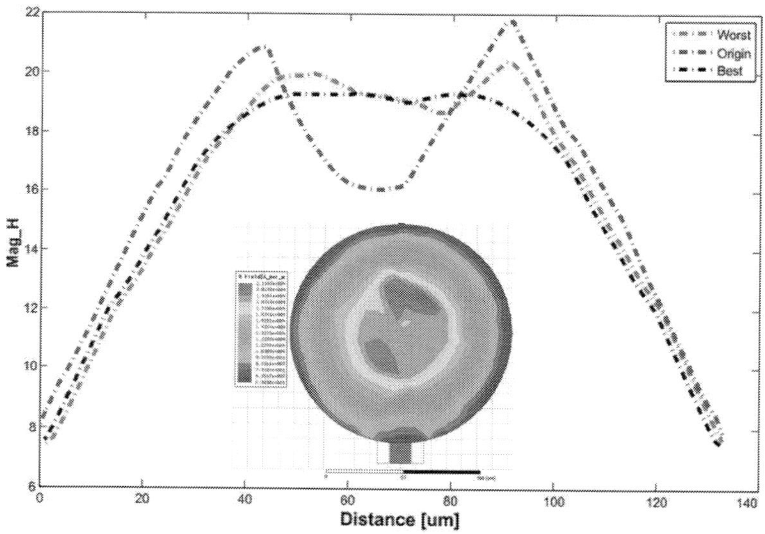

Figure 12. Simulated magnetic field uniformity on the optimized inductor surface.

4.2.2. *Biochip design*

The block diagram of a CMOS high sensitivity oscillator-based biochip is shown in Figure 13. On different detection regions in a microarray, the output oscillation signal from the LC oscillators is firstly processed by a multiplexer for selection. It is then down-converted to an intermediate frequency (IF) and subsequently extracted by a low noise band-pass filter (BPF) for out-of-band noise and spur suppression. The frequency synthesizer and circuitry supply can be realized either on-chip or off-chip, depending on the applications. The IF signal is further

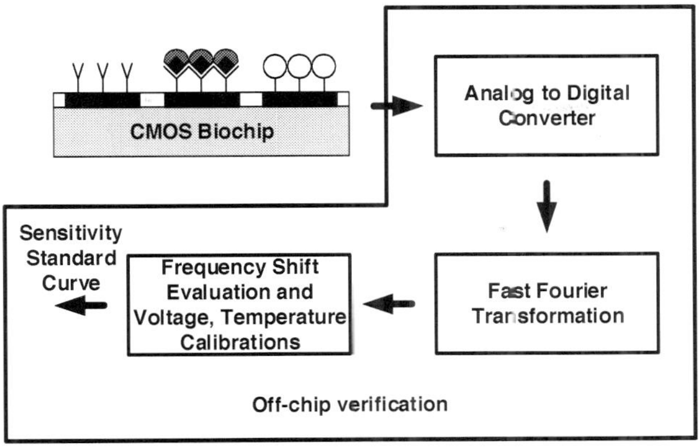

Figure 14. Block diagram of a CMOS high sensitivity oscillator-based biochip front-end and signal processing back-end.

digitized by an ADC off-chip, as shown in Figure 14. FFT is applied on the digitized IF signal to calculate the frequency shift induced by the attached MNPs and eventually, obtain the standard curve for biodetection and evaluation purposes. The ADC, FFT and following calculation can be verified through off-chip and commercial chips/FPGA/Matlab. It is then considered to be realized on-chip for a system-on-a-chip (SoC) solution.

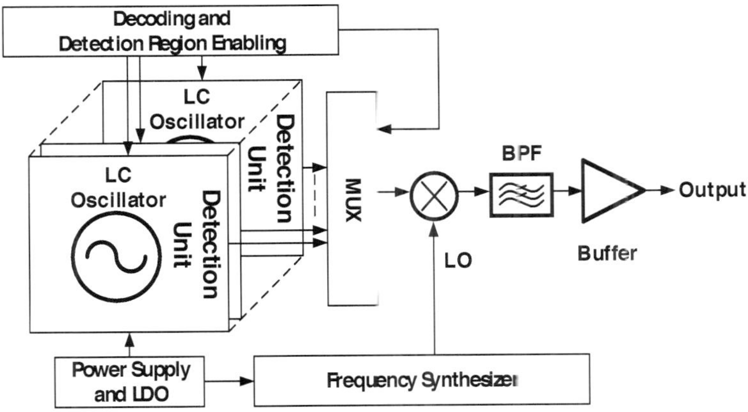

Figure 13. Circuit schematic of the oscillator-based biochip.

The challenging parts from the front-end biochip and back-end circuits are from the following aspects.

4.2.2.1. Architecture of the signal down-conversion chain

The down-conversion chain of the proposed oscillator-based biochip is similar to that of the conventional RF front-end circuits for wireless communication, including direct conversion and heterodyne architectures. Since the output from a detection unit is a large signal with high output swing and considering the injection pulling effects, the LO signal from the frequency synthesizer should be at a different frequency from the unit. On the other hand, since the frequency of interest is around 1 GHz - 1.5 GHz, a double conversion is less efficient, owing to the power hungry dividers, buffers, etc. Therefore, a low IF architecture is preferred, considering the trade-off among architectures, complexity and power.

4.2.2.2. Optimization of LO frequency

Being different from conventional receivers, there are no imaging effects in the proposed biochip. Unfortunately, the selection of LO should be traded with the dynamic range and sensitivity. A high LO would induce a reduction in the dynamic range and sensitivity, while an over-low LO frequency would increase the IF frequency, burdening the design of BPF and increasing power. The LO frequency has to be optimized based on the considerations of sensitivity, dynamic range and power.

As shown in Figure 14, FFT is applied on the digitized IF signal from the ADC in order to calculate the frequency shift induced by the attached MNPs and eventually, obtain the standard curve for biodetection and evaluation purposes.

In this part, we need to apply a Hanning window on the signal to suppress the spectrum leakage of FFT, before translating it into frequency domain. This improves the spectrum uniformity and sensitivity.

4.2.3. Calibrations

Frequency drifting is a major issue for oscillator-based microarray biochips. Conventional assays introduce an additional LC tank for reference in order to cancel out the common-mode noises including drifts and PVT variations, however, increasing the chip area due to the large footprint of the on-chip inductor [24]. In this section, an "in-situ switched capacitor (ISSC) based calibration" scheme is presented by introducing only one switched capacitor in the LC tank for common-mode noise rejection, as shown in Figure 15. An enabling switch is used for the state switching and the measured frequencies between the adjacent time cycles are subtracted for the cancellation of common-mode voltage and temperature variations. The obtained frequency offsets in the initial state f_{offset} and after introducing the MNPs f'_{offset} follow that:

$$f'_{offset} = f_{offset}\left(1 - \frac{\Delta L}{2L_0}\right) \qquad (12)$$

Then, we have:

$$\frac{\Delta L}{2L_0} = 1 - \frac{f'_{offset}}{f_{offset}} = \frac{\Delta f}{f_0} = \frac{\alpha M}{2L_0} \tag{13}$$

Since all of the frequency drifting relating to the voltage and temperature variations are common-mode and cancelled out, the output signal, which is proportional to the MNP density within the inductor detection region, can therefore be obtained from the above relation.

The switching time has to be optimized, considering the typical voltage and temperature variation rates for the best cancellation. In this prototype, an empirical value of 0.1 s is found. In comparison to conventional dual-site cancellation scheme, this method is advantageous in terms of both circuit complexity and cost areas, supporting more detection regions on a single chip. It is capable of swiping out both drifts and PVT variation and does not require complicated external processing. Therefore, it improves the sensitivity and reliability of the multi-target detection, as well as simplifying the implementation.

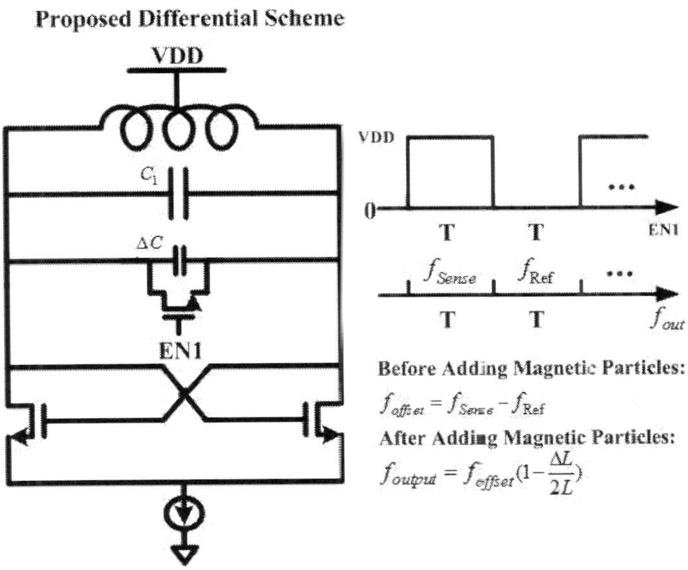

Figure 15. ISSC differential reference scheme for the voltage and temperature calibrations.

4.2.4. Experimental results and discussions

An oscillator-based microarray biochip prototype has been designed and fabricated in the standard 180 nm CMOS process. The die micrograph is shown in Figure 16(a). The chip is designed with two inductor detection regions, which are 130 μm in diameter. The measured results show that the oscillator works properly at a centre frequency of 1.45 GHz, with a phase noise of -120 dBc/Hz at 1 MHz offset and an output swing of larger than 1.25 V, while consuming 55 mW of power.

In the experiment, a standard 500 nm MNP sample solution has been utilized for verification and evaluation of the oscillator-based microarray biochip. The microscope picture of the distributed MNPs on the inductor detection region of the biochip is shown in Figure 16(b). The sample solution has been diluted by 50, 500, 5000 and 50000 times, respectively. The measured output frequency offset versus the dilution ratio is shown in Figure 17.

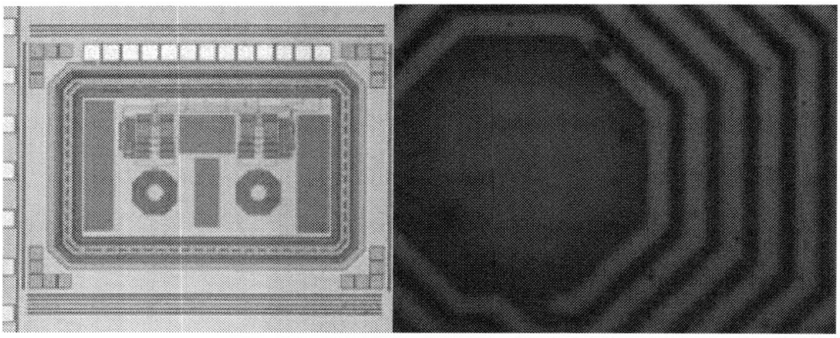

Figure 16. (a): Die micrograph of the oscillator-based microarray prototype and (b) the distributed MNPs on the inductor detection region of the oscillator-based microarray biochip.

It is found that the oscillator-based biochip works properly and a frequency shifting sensitivity of about 17 kHz has been achieved, roughly corresponding to about 10 ppm. The proposed ISSC calibration scheme is further applied and characterized through experiment. The output frequency shifts corresponding to different numbers of MNPs on the detection region are measured as shown in Figure 18, in which the exact quantity of MNPs is obtained from scanning electron microscope (SEM) images. The effective noise floor (defined as the standard deviation of measured frequency shift) of the sensor after ISSC calibration is 896 Hz, corresponding to about 0.62 ppm, and the extrapolated frequency shift for one single 500 nm MNP is 138.5 Hz which is related to the size and material of the MNPs, running frequency, and inductor structure. The maximal frequency shift is measured as 9.2 MHz, implying a dynamic range of at least 80.2 dB, while the area of the sensor detection region is only 60000 μm^2, well below other calibration schemes. The comparisons with prior arts are shown in Table 2.

Compared with the electrochemical, GMR and MTJ biosensing strategies, the oscillator-based microarray biochip potentially features high sensitivity, good compatibility with CMOS technology, low cost, and multi-target capabilities.

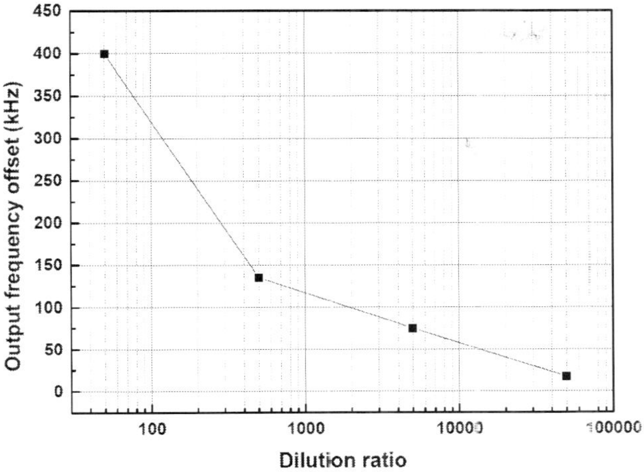

Figure 17. The measured output frequency offset versus the dilution ratio.

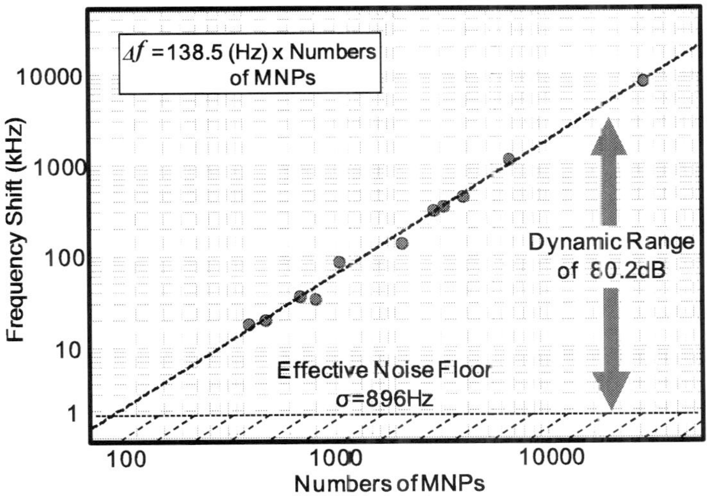

Figure 18. Frequency shifts corresponding to different numbers of MNPs.

	ISSC scheme This work	Differential scheme [17]	CDC scheme [17]
Operation freq. f_0	1.44 GHz	1.13 GHz	1.13 GHz
Phase noise at 1MHz	-123.4 dBc/Hz	-120.3 dBc/Hz	-120.3 dBc/Hz
Dynamic range	80.2 dB	85.4 dB	85.4 dB
Effective noise floor/ operation freq.	0.62 ppm	1.07 ppm	0.35 ppm
On-chip area per sensing region	60000 μm^2	>72000 μm^2	72000 μm^2
Technology	0.18-μm CMOS	45-nm SOI CMOS	45-nm SOI CMOS

Table 2. Comparisons with state-of-the-art oscillator-based biochips with calibrations.

5. Conclusions and future directions of IMB

Silicon-based IMBs with high sensitivity and multi-target capabilities have caught the attention of both industry and academia for future accurate and early disease diagnosis and treatment. Compared with conventional assays, such as ELISA and fluorescent detections, CMOS IMBs are advantageous in various aspects. Such advantages include low cost, large-scale, readily integration with signal extraction and processing front-end and back-end circuits monolithically for a lab-on-a-chip solution, etc. In this chapter, state-of-the-art silicon-based IMB assays, including electrochemical microarray biochips, GMR and MTJ microarray biochips and fully CMOS compatible oscillator microarray biochips, are reviewed and discussed. Multiple challenges and issues, as well as novel techniques in the research and production of silicon-based IMB, are also presented.

In summary, silicon-based IMB provides fertile ground for future innovations. More specifically, electrochemical microarray biochips are one type of widely interested assay, which are capable of mid-sensitivity multi-target detection. In future research, various issues need to be solved such as background noise and random currents in a biochemical sample solution medium during the shrinking of electrode feature size, post-CMOS fabrication technologies and parallel back-end signal processing and mode recognition. MTJ and GMR microarrays assays are proved with extremely high sensitivities and provide full support to multi-target (~100 targets) biochemical analysis applications. However, both biochips require extra fabrication flow beyond the CMOS process, preventing the monolithic integration with acquisition IC front-end, signal processing back-end and DSP for a SoC solution in order to further improve system performance and reduce costs. Efforts have to be made to fix the process compatibility issues. This can be achieved through chip-level integration and system-in-a-package (SiP) solutions, etc. The oscillator-based microarray biochip features high sensitivity, wide dynamic range, multi-target capabilities and full compatibility with CMOS technology. Thus, it is potentially a suitable candidate for future SoC solutions. Calibration techniques for phase noise of LC oscillators and frequency drifting due to PVT variations remains challenging and requires further innovations.

Acknowledgements

The authors would like to thank the support of the National Natural Science Foundation of China under grant 61101001 and Tsinghua University Initiative Scientific Research Programme.

Author details

Lei Zhang[1*], Cheng Zhu[1], Jinwen Geng[1], Xizeng Shi[2], Yunhua Gao[3], Zhijie Chang[4] and He Qian[1]

*Address all correspondence to: zhang.lei@tsinghua.edu.cn.

1 Institute of Microelectronics, Tsinghua University, Beijing, China

2 Dongguan Bosh Biotechnologies, LTD., China

3 Technical Institute of Physics and Chemistry of the Chinese Academy of Sciences (CAS)., China

4 Medical School, Tsinghua University, Beijing, China

References

[1] Cheng and L.J. Kricka. Biochip Technology. Harwood Academic Press; 2001.

[2] P.A. Serra, editor. Biosensors. InTech; 2010.

[3] T. G. Drummond, M. G. Hill, and J. K. Barton. Electrochemical DNA Sensors. Nature Biotechnology. 2003;21(10):1192-1199.

[4] M. Schienle, A. Frey, F. Hofmann, B. Holzapfl, C. Paulus, P. Schindler-Bauer, R. Thewes. A Fully Electronic DNA Sensor with 128 Positions and In-Pixel A/D Conversion. In: IEEE ISSCC Digest of Technical Papers; February; 2004. p. 220-221.

[5] M. Augustyniak, C. Paulus, R. Brederlow, N. Persike, G. Hartwich, D. Scmitt-Landsiedel, and R. Thewes. A 24×16 CMOS-Based Chronocoulometric DNA Microarray. In: IEEE ISSCC Digest of Technical Papers; February; 2006. p. 59-60.

[6] B.C. Jang, P.Y. Cao, A. Chevalier, A. Ellington, and A. Hassibi. A CMOS Fluorescent-based Biosensor Microarray. In: IEEE ISSCC Digest of Technical Papers; February; 2009. p. 436-437.

[7] L. Zhang, X.Q. He, Y. Wang, and Z.P. Yu. A Fully Integrated CMOS Nanoscale Biosensor Microarray. In: IEEE Custom Integrated Circuits Conference (CICC); September; 2011. p. 1-4.

[8] R. Thewes, C. Paulusa, M. Schienlea, F. Hofmanna, A. Freya, P. Schindler-Bauera, M. Atzesbergera, B. Holzapfla, T. Hanedera, and H.-C. Hankea. A CMOS Medium Density DNA Microarray with Electronic Readout. In: Material Research Society Symposium; 2005. p. D3.4.1-D3.4.11.

[9] J. Lian. A Fully Automated IVD System Based on MTJ arrays and Superparamagnetic Particles. Journal of Applied Physics. 2012;111(7):07B315.

[10] D.A. Hall, R.S. Gaster, K.A.A. Makinwa, S.X. Wang, B. Murmann. A 256 Pixel Magnetoresistive Biosensor Microarray in 0.18μm CMOS. IEEE Journal of Solid-State Circuits. 2013;48(5):1290-1301.

[11] C. Zhu, L. Zhang, X.Z. Shi, Y.H. Gao, Z.J. Chang, and H. Qian. A GMR Biosensing System with Sub-50ppm Sensitivity. In: IEEE International Conference on Electron Devices Solid-state Circuits (EDSSC); June; 2014.

[12] W. Wang, Y. Wang, L. Tu, T. Klein, Y. Feng, J.P. Wang. Surface Modification for Protein and DNA Immobilization onto GMR Biosensor. IEEE Transactions on Magnetics. 2013;49(1):296-299.

[13] H. Yang, B. Qu, B. Lei. Giant Magnetoresistive Biosensor for Myoglobin Immunoassay. In: IEEE Conference on Sensors; 2011. p. 805-808.

[14] B.M. de Boer, J.A.H.M. Kahlman, T.P.G.H. Jansen, H. Duric, J. Veen. An Integrated and Sensitive Detection Platform for Magneto-Resistive Biosensors. Biosensors and Bioelectronics. 2007;22(9):2366-2370.

[15] H. Wang, Y. Chen, A. Hassibi, A. Scherer, and A. Hajimiri. A Frequency-shift CMOS Magnetic Biosensor Array with Single-bead Sensitivity and No External Magnet. In: IEEE ISSCC Digest of Technical Papers; February; 2009. p. 438-439, 439a.

[16] H. Wang. Magnetic Sensors for Diagnostic Medicine. IEEE Microwave Magazine. 2013;14(5):110-130.

[17] H. Wang, C.-C. Weng, and A. Hajimiri. Phase noise and fundamental sensitivity of oscillator-based reactance sensors. IEEE Transactions on Microwave Theory and Techniques, 2013;61(5):2215-2229.

[18] R.M. Lequin. Enzyme Immunoassay (EIA)/Enzyme-Linked Immunosorbent Assay (ELISA). Clinical Chemistry. 2005;51(12):2415-2418.

[19] S. Weiss. Fluorescence Spectroscopy of Single Biomolecules. Science. 1999;283(5408): 1676-1683.

[20] S. Ayers, K.D. Gillis, M. Lindau, and B.A. Minch. Design of a CMOS Potentiostat Circuit for Electrochemical Detector Arrays. IEEE Transactions on Circuits and Systems I: Regular Papers. 2007;54(4):736-744.

[21] Basu, R.W. Robucci, and P.E. Hasler. A Low-power, Compact, Adaptive Logarithmic Transimpedance Amplifier Operating Over Seven Decades of Current. IEEE Transactions on Circuits and Systems I: Regular Papers. 2007;54(10):2167-2177.

[22] L. Zhang, Z.P. Yu, and X.Q. He. Design and Implementation of Ultralow Current-Mode Amplifier for Biosensor Applications. IEEE Transactions on Circuits and Systems II: Express Briefs. 2009;56(7):540-544.

[23] G. Ferrari, F. Gozzini, and M. Sampietro. A Current Sensitive Front-end Amplifier for Nano-Biosensors with a 2MHz BW. In: IEEE ISSCC Digest of Technical Papers; February; 2007. p. 164-165.

[24] B. Razavi. RF Microelectronics. 2nd ed. Prentice Hall; 2012.

[25] C. Brosseau and P. Talbot. Effective Magnetic Permeability of Ni and Co Micro- and Nanoparticles Embedded in a ZnO Matrix. Journal of Applied Physics. 2005;97(10): 104325-1-11.

[26] Y. Zheng and J.G. Tront. Biological Agent Sensing Integrated Circuit (BASIC): A New CMOS Magnetic Biosensor System. In: IEEE Sensors; October; 2012. p. 1-4.

Chapter 5

Supramolecular Materials for Optical and Electrochemical Biosensors

Tatiana Duque Martins,
Antonio Carlos Chaves Ribeiro, Flavio Colmati,
Geovany Albino de Souza,
Henrique Santiago de Camargo, Diogo Lopes Dias,
Paulo Alves da Costa Filho and
Diericon de Sousa Cordeiro

Additional information is available at the end of the chapter

http://dx.doi.org/10.5772/60710

Abstract

It is incontestable that the interactions and bonds that keep molecules united to generate unique supramolecular compounds, with individual properties, morphologies and behaviour, are of special dynamics and singular forces. Therefore, it is necessary to discuss and consider the types of interactions that may occur in a determined system, their dynamics and number, which directly act on the energetic balance that strengthen the union between participants and give rise to a supramolecule.

In this chapter, a number of such supramolecular systems that find application as any component of a biosensor are presented and discussed, considering intermolecular interaction forces that confer them shape, function and unique properties. To better understand their structural dynamics and the mechanisms through which they can be used in biosensing, a brief explanation on the interaction thermodynamics, types of intermolecular interactions that compete against each other and the energetic equilibrium that originate and stabilize supramolecular systems is given. To explain how this balance of forces can be extensively exploited to develop methods to produce supramolecular compounds, an overview on supramolecular strategies is presented and their contribution is explored in each example presented in this text, to evidence the importance of planning and developing methodologies of preparation, based on accurate information on structural characteristics and properties of the initial compounds. Thus, an approach to different procedures of preparation or adaptation methods applied to supramolecular compounds to fit a determined application, as well as newly developed materials, is the focus of the following discussions. For instance, supramolecules presenting interesting and singular characteristics that can be advantageous in biosensors, such as high quantum yield luminescence, high efficiency of energy conversion, which enable charge transfer and migration processes, adaptive and mimetic properties, among others, are discussed on preparation and function basis. A variety of supramolecules and supramolecule-based materials produced via distinct supramolecular strategies are discussed and examples of their applications are presented. Specifically, new and extraordinary supramolecular materials that present properties and characteristics that could not occur in ordinary organic/inorganic compounds, but notoriously enhance optical and electrochemical biosensor performance, will be presented and their properties, detailed. Also, the mechanisms of action of supramolecule-based optical and electrochemical biosensors and their dependence on singular properties of each type of supramolecular system are discussed to provide an understanding of the most prominent aspects that govern a compound behaviour in a device and guide the scientist through the choice of the appropriate material for biosensing applications in the wide field of available supramolecular compounds. In an attempt to provide accurate information to those who desire to understand the importance of supramolecular materials in biosensor applications, an overview on the recent progress on materials for optical, electrochemical and photoelectrochemical biosensors are presented, in order to inspire research on new and even more delicate and effective supramolecular materials for biosensing applications. Moreover, we discuss some of our recent results, which are provided as examples, that highlight the advantages of producing supramolecules based on peptides for applications in optical biosensors and introduce some interesting methods to anchor enzymes in electrodes to enhance device performances, as well as provide new technologies for distinct device production.

Keywords: Self-assembled materials, optical biosensors, electrochemical biosensors, fluorescence, cyclic voltammetry, electronic energy transfer, peptides

1. Introduction

1.1. Recognition mechanism — The fundamental concept of supramolecular chemistry

The property of molecular recognition is of great importance to describe most biological processes, such as DNA replication, enzyme activity, among several others, and consequently, it has played a significant role in supramolecular chemistry development. Among them, the implementation of molecular recognition as a method to control the production of supramolecular species, since it is based on the combination of small components, was a breakthrough.

Supramolecular control of synthesis of compounds is achieved by the development of methods that combine the design of the spontaneously generated complex molecular structures, based on self-assembly of chosen components, with pre-determined experimental conditions. As these self-assembled structures are obtained with impressive success, self-assembly property has been exploited in a distinguished number of organic and inorganic systems, which gave rise to the known *bottom-up* and *top-down* supramolecular built-up strategies.

The first step on producing a supramolecular compound is to have in mind the properties expected for this compound to present. It will both, to justify its applications and to enable the selection of the adequate experimental steps to produce it with high quality, yield and purity. Bottom-up strategies are preferred when a property found in a single molecule is desired in the supramolecular complex, since they tend to preserve such molecular characteristics. On the other hand, top-down strategies are preferred when the objective is to prepare representative small structures of major ones, presenting the desired structure and properties, being important on miniaturization processes.

1.2. Supramolecular strategies — Towards excellent structures for biosensors

1.2.1. Bottom-up strategies

In comparison to top-down strategies, bottom-up ones enable better chemical control of the produced supramolecular surface, better size control and higher degree of purity. They correspond to synthetic processes that involve the supramolecular growth from atoms, achieving molecules and further to aggregate with the desired dimensions and presenting the desired properties. Most of the bottom-up strategies occur in vapour and liquid phases. Among them are the following:

- Condensation methods, usually occurring in sealed chambers in which the precursor compound is vaporized and condensed in the substrate. The sublimation of the compound can be processed either by laser ablation or by thermal heating.
- Vapour phase nucleation methods, in which structure growth occurs usually in a vapour phase reactor. These methods present advantages, such as the use of precursors in solid, liquid and gas phase, mild conditions of preparation and homogeneous multi-component supramolecular structures production.
- Liquid phase nucleation methods, in which structures are prepared in a wide variety of conditions, depending on the desired supramolecular structure.
- Layer-by-layer methods of film preparation, which enable film production from large compounds such as polymers, proteins, even DNA or from small molecules and atoms. Figure 1 illustrates some bottom-up strategies to obtain supramolecules.

1.2.2. Top-down strategies

To particulate materials is the fundamental conception of any top-down preparation method that can be conceived. Especially, physical methods of fragmentation or disaggregation, in

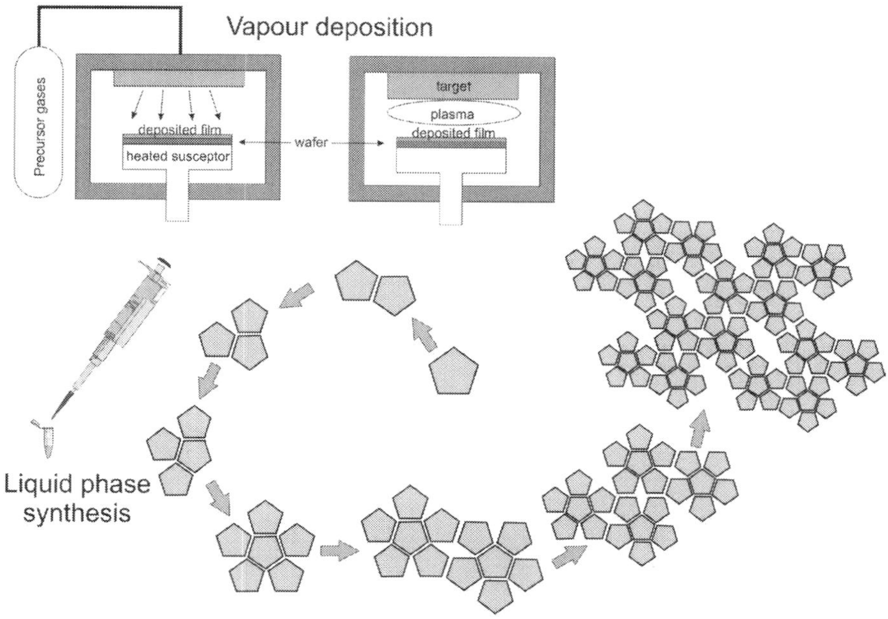

Figure 1. Schematic illustration of distinct bottom-up strategies to produce supramolecular compounds.

which the major structure is broken, nevertheless conserving its organization and physical-chemical properties, are based on the basic principle of top-down methods. Among the most employed top-down methods are the following:

- Laser ablation, in which supramolecular compounds of distinct nature are obtained by a surface irradiation process that promotes particle removal.
- Ball milling, in which highly energetic collisions between small balls in the milling chamber and the material repeatedly occur, resulting in a fine and uniform material dispersion and conserving all physicochemical characteristics of the initial material.
- Atomic force microscopy (AFM), which has been recently proved to be adequate to manipulate particles of a surface [1], due to an attraction/repulsion force balance between the surface and the cantilever of the microscope.
- Thermal treatment of a compound to weaken interaction forces that keep the major structure together and sonication of the material solution in an appropriate solvent are simple methods of physical separation that are employed to separate the aggregation units and generate the desired supramolecules. Figure 2 illustrates some top-down supramolecular strategies.

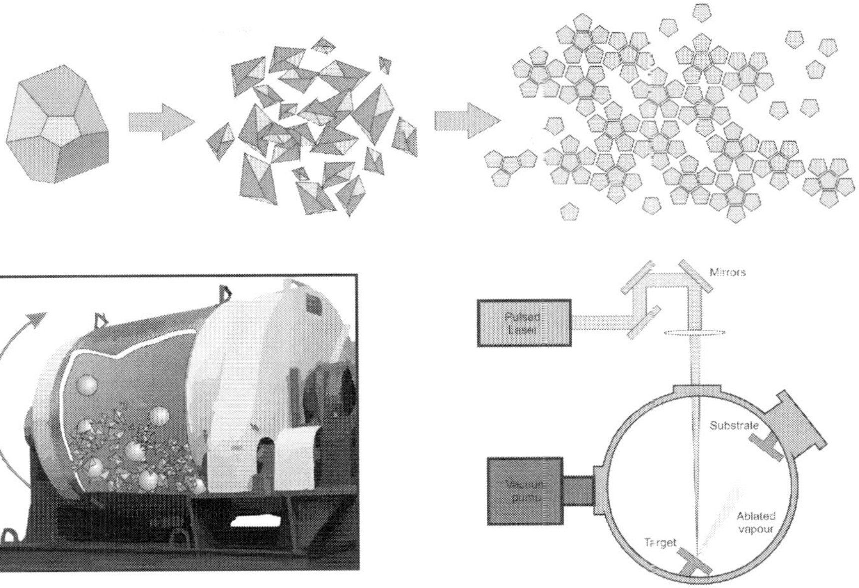

Figure 2. Scheme of supramolecular top-down strategies based on ball mills and ablation techniques.

2. Supramolecular interactions

Whatever strategy is considered adequate to build a super compound, the nature of interactions that govern the self-organization of primary compounds into the complex one is usually of the weak kind. This means that the units responsible for super compounds or supramolecules are kept together by non-covalent interactions, rather than covalent bonds.

It is clear that covalent bonds are, and always will be, the fundamental occurrence between atomic orbitals that originate molecules, structures, tissues and, therefore, materials and life. Nevertheless, the participation of weak forces on conferring identity and possibilities of change to every molecule formed cannot be neglected. In supramolecular chemistry, the modelling characteristics of compounds are derived from weak interaction forces that are exploited to confer them the abilities of formation and change, by accumulation of interaction forces. In fact, to generate a supramolecular compound, weak interaction forces work synergistically to create alternative forms of bonds that give rise to functional structures, in dynamic environments and mild conditions. As predicted by Steed and Atwood [2], very stable complexes can be formed based on non-covalent interactions that present, each one, a small contribution to the supramolecular compound stabilization, which, when allied to others, become important and overwhelms any environmental disturb that could lead to the compound dissociation.

Indeed, to have an idea of the binding forces and which supramolecular structures are energetically favourable, it is important to correctly identify the non-covalent bonds that dominate the formation processes and their comparative forces. In an attempt to better identify the most prominent processes and the equilibrium that is formed between components and the environment, the most important non-covalent interactions that are present in such complexes are as follows:

- **Charge interactions**

Due to their characteristic of making use of the strong effect that charged group may exert upon uncharged molecules, these interactions are favourable, since they promote a temporary completion of orbital charges. They involve positively or negatively charged entities affecting uncharged molecules or groups that present surface charge. Among them are the following:

- *Ion-ion interaction:* in compounds structured upon these interactions, there is a strong electrostatic interaction of an attractive character that enables the strong interaction that holds the isolated participants together, although at a distance from each other, which preserves their individual identities. These interactions present strength similar to those of covalent bonds with energetic values around 100-350 kJ mol^{-1}.

- *Ion–Dipole Interactions:* these interactions occur between an ion and a polar molecule, also susceptible to electrostatic effects. Since one of the components is uncharged, these interactions are weaker than the former, presenting binding energies of 50 – 200 kJ mol^{-1}.

- *Cation/Anion–π,π Interactions:* these are examples of electrostatic interactions that occur between charged moieties, mostly aromatic rings, although they are of the same nature of those observed between positively charged elements and carbon-carbon double bonds (C=C). They are of moderate intensity, presenting binding energies of 30-80 kJ mol^{-1} [3].

- *Hydrogen Bonding:* Linus Pauling described it [4] as a versatile interaction, presenting energy of moderate intensity (4 – 100 kJ mol^{-1}) and occurring between a compound that possesses a hydrogen atom adjacent to an electronegative or electron withdrawing group and the dipole portion of a neighbouring molecule.

- **Induced charge interactions**

These are interactions that occur whenever neutral groups or molecules approximate to each other, being perceived due to their distinct electron densities. They occur with small energetic changes and are more important when numerous in a given structure. They are modulated by the identity of participants and, therefore, by the singular characteristics of each molecular combination. They may present a wide range of lengths, angles and can be classified depending on the main characteristics of their participants:

- *π–π Interactions:* they involve π-orbitals of any compound and are common in aromatic rings, being very important in self-assembling processes. Presenting energies of 1-10 kJ mol^{-1}, their occurrence is justified by an electron density divergence between the participants.

- *Dipole–Dipole Interactions:* When a dipole is nearby another, they tend to align, due to their perception of each other, by attractive and/or repulsive interaction forces. They are of moderate intensity, presenting energies of 5-50 kJ mol^{-1}.
- *Van der Waals Interactions:* they are weak interactions of approximately 1-5 kJ mol^{-1} that arise from the perception of polarizable molecules on charged or possessing charge character compounds.

All these interactions occur in any supramolecular system, and their balance of forces and the role that each interaction plays in the configuration and stability of each system depends exclusively upon the preparation methods, the individual characteristics of the participants and the proposed molecular designs. Some of the above-mentioned interactions are illustrated in figure 3, in which water interacts with poly[5-methoxy-2-(3-sulfopropoxy)-1,4-phenylene-vinylene] (MPS-PPV) moieties through hydrogen bonds and the polymer moieties interact with other molecules through π- π interactions.

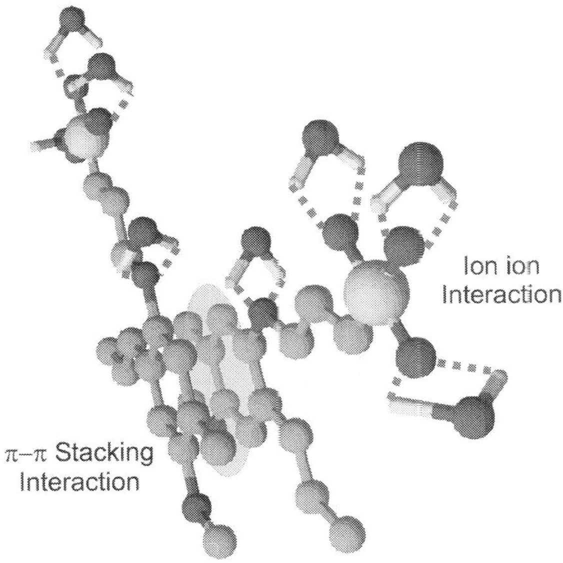

Figure 3. Examples of ion-ion and π- π interactions as they occur in an aqueous solution of MPS-PPV.

It is important to consider that all these interactions work synergistically in order to form a compound with some ideal characteristics that enhance the effects of the environment on their ability to act as host/guest system, such as the structural characteristics that are assumed when solvation is possible, participation in lattices, component changes, etc. Regarding these abilities, supramolecular chemistry has played a fundamental role in the biosensors development.

3. Optical and electrochemical biosensors

Since biosensors are devices that can perform a rigorous exploration of a system, under conditions that require the minimum possible human intervention, they can be constructed upon a variety of compounds, depending on their application and environmental conditions they will be subject to. The selection of the appropriate recognition element considers not only what the information needed is, but also the ease of construction of the devices employing such elements and their durability.

Once the recognition component is selected and, thus, the mechanism in which it interacts with the analyte is selected, it is important to find an appropriate mechanism to transform the resulting perturbation, or signal, into comprehensible information. The selection of the appropriate transducer determines the type of the biosensor to be constructed. Its selection enables a fast and precise data interpretation, which is an important requirement. Examples of biosensor components are displayed in figure 4.

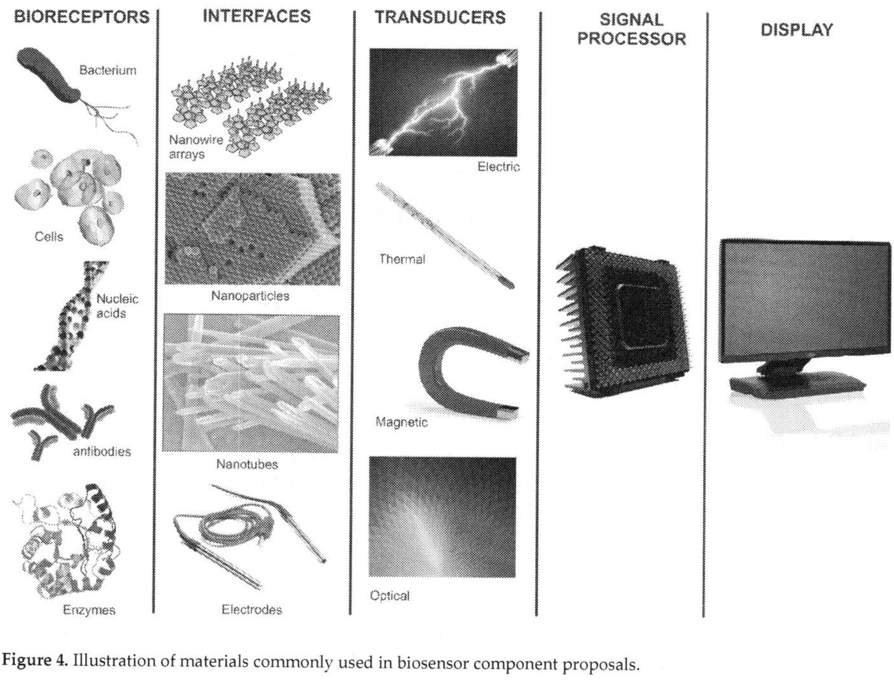

Figure 4. Illustration of materials commonly used in biosensor component proposals.

Transducers are available based on a variety of sensing techniques. In optical biosensors they include Fluorescence Spectroscopy, intensity/lifetime measurements of luminescence, light absorption or light reflectance in the ultraviolet (UV), visible, or near-infrared (NIR), spectral regions and Surface Plasmon Resonance (SPR). Also included are techniques based on microscopy, such as Atomic Force Microscopy (AFM), Fluorescence Confocal Microscopy, Raman Confocal Microscopy or Surface Enhanced Raman Spectroscopy (SERS). Also based on the transducer choice, electrochemical biosensors are classified into amperometric, conductometric, potentiometric and voltammetric categories. The variety of electrochemical and optical techniques that are exploited as transducers confer to the devices improvements on selectivity, sensibility, configuration versatility, ease of construction, possibility of miniaturization. The combination of optical to electrochemical transducers give rise to optical-electrochemical biosensors, which deliver either electrical or optical signals that are interpreted together or separately.

Independently of the chosen components, an efficient biosensor must effectively recognize the analyte and selectively interact with it, to deliver comprehensible data in short time, to avoid loss of information, mainly by analyte's degradation. This means that the guest-host ability of the biosensor is the key factor for a device to show high selectivity and sensibility and, thus, this ability may be modulated by the recognition of material choice. As depicted by Lehn, molecular recognition is related to the binding energy and the interaction ability between host and guest [5]. When it involves compounds with unique structures, it can occur in a specific way, and implies in the union of the spatial and functional characteristics of the molecular groups, that interact to each other and shapes the selectivity and sensibility characteristics of a compound. As predicted by Araki [6], selectivity and specificity are measures of the interaction of free energy between the binding moieties and, thus, the role of surface and electronic aspects is fundamental to modulate the free energy. In fact, shape, size, conformation changes of binding sites in host-guest systems, polarity, polarizability, charge density of groups and terminations are the most important features to be determined and considered when evaluating the selectivity of a host. Those are also the most important features when designing an efficient recognition element for a biosensor.

Indeed, with the rise of a variety of supramolecular structures and the number of studies that aim to describe their singular properties and performance in several applications, the diversity of compounds that showed perspective of use in biosensors arose significantly, and the resulting devices have enhanced its selectivity, sensibility and usage spectra.

4. Supramolecular compounds for optical biosensors

In optical biosensors, due to the wide range of compounds that can be employed in their construction, to result in selective recognition systems that converts the optical signal by several techniques, significant development have been achieved in the past twenty years. In particular, biosensors based on fluorescent responses greatly benefit from supramolecular compounds used as recognition systems and transducers, since they are able to significantly enhance charge transfer processes and this property can be exploited to label the analyte and, hence, enable detection via fluorescence imaging or enabling FRET, which can be efficiently detected using special biosensors [7-9].

A wide range of luminescent supramolecular compounds are available for biosensor construction, and the reason for this diversity is that supramolecular compounds can self-assemble. Self-assembling is the characteristic that distinguishes large molecules and supramolecules. Due to it, the alternatives for the design of a supramolecule are almost unlimited and production of supramolecules is becoming cheaper as they become comprehensible to the scientists [10]. Based on the variety of interactions that is possible in self-assembled structures and due to their lability, supramolecular materials can be designed based on organic and inorganic polymers, hybrid materials, charged molecules and macromolecules, crystals, liquid crystals, gels, metallic nanoparticles, some alternatives containing rare-earths [11], giving rise to new materials and nanomaterials presenting so many distinguished properties that confer them new applications and classifications, such as adaptive and self-healing materials.

In the past five years, much has been produced in the supramolecular and new materials research themes. New materials are often proposed to improve devices and technologies, but sometimes they carry so many innovations that new applications must be proposed. Due to that, the sensing development has achieved this great status and is still on its way up!

Earlier, the secondary interactions that predominate in supramolecular chemistry and their associated energies were discussed; nevertheless, it is important to reinforce that supramolecular compounds are formed by a balance of interaction forces that form a dynamic system, formed by distinct structural parts that struggle to remain in their possible less energetic conformation. In self-assembling processes, the major influence of one or a couple interaction types is often recognizable. When electrostatic interactions dominate, ionic self-assembling (ISA) is expected to occur. This type of self-assembling process gives rise to an enormous variety of supramolecules, each presenting its own characteristics [10, 12]. An example is the result of the combination of two dyes synthesized by Bohm et al. [13], 1,4-dihydroxyanthraquinone (AQ-OH) and 1,4-di- N –adamantyl-amino-anthraquinone (AQ-H), with the hyperbranched polyethylenimine attached to β –cyclodextrin (β -CD-PEI), which is a spontaneously formed host–guest complex with high hydrodynamic diameter. In their work, to demonstrate enlarged networks produced on supramolecular basis, due to host–guest interactions, they had measured the initial and final hydrodynamic diameters by dynamic light scattering, and they found that, after the assembling process, the hydrodynamic diameter presented a 40-fold increase.

Supramolecular compounds formed upon electrostatic interactions find many applications, including in optical and electrochemical biosensors. In their work, Wang et al. [14] produced a H_2O_2 non-enzymatic biosensor based on a self-assembled peptide nanofiber that was, then, metalized to give silver nanowires. This material was supported in graphene nanosheets to build the device, which presented high sensitivity, high selectivity and low detection limit. The use of this interesting designed peptide presented two important breakthroughs: it enabled the production of 1D peptide nanofibers by self-assembly and it rendered metal nanowires by metallization. The authors claim that this system may present potential applications in nanodevices, especially other biosensors, as well as in biomedicine and in Raman analysis.

Also, the self-assembling properties presented by polyelectrolytes have been widely explored to develop a series of multifunctional compounds. In their work, Habibi et al. [15] presented a self-assembled system based on anionic poly(styrene sulfonate) (PSS) and cationic poly(allylamine hydrochloride) (PAH) modified with S-protein to promote the loading and release of macromolecules from the interior of the obtained capsule. The permeability of this system was pH tuned and the encapsulation and release of macromolecules were promoted by controlled pH change. The functionality is based on the fact that S- proteins self-assemble into monomolecular arrays at different interfaces, which provide a regular arrangement of functional groups in the final structure, following a bottom-up self-assembly strategy to generate functional supramolecular structures and devices. As they showed that this system is biocompatible [16, 17], they suggest that its application in biosensing, in drug delivery and as micro-reactors, among several others, is highly probable.

Nanostructured materials have been studied aiming the biosensing application due to their specific physical-chemical properties and to their quantum-size effects when compared to bulk [18]. Specifically, highly luminescent compounds and charge transporter materials have also been widely employed to generate supramolecular structures for biosensing applications. This class of materials includes quantum dots, metallic or magnetic nanoparticles, nanostructured materials into one-dimensional or bi-dimensional structures such as nanowires, nanoribbons, nanospheres or nanosheets and may improve device sensitivities due to their large surface-to-volume ratios. Also, due to their small sizes, they can be delivered into living cells, enabling *in vivo* sensing applications. Among the devices that can be thought for *in vivo* applications, fluorescence-based biosensors have been developed, possessing several configurations, sizes and, thus, functions. Tunceroglu et al. [19] described a time-dependent fluorescence-based biosensor with the ability to evaluate drug resistance, by monitoring Bcr-Abl activity in patients diagnosed with chronic myelogenous leukemia (CML), and Gulyani et al. [20] developed a device that analyses the fluorescence enhancement when it binds to the target kinase with the determined conformation, becoming a Src merobody biosensor.

Rare-earth-based materials and quantum dots (QD) are of great interest due to their high quantum yield of luminescence at UV/Vis solar spectrum region [21, 22]. They also present narrow and tunable emission spectra and usually present good photostability, which substantially increase their hall of applications. Nguyen et al. [23] described a CdTe quantum dot–based biosensor, developed specifically to detect flu virus H5N1. They reported that the biosensors consisted of CdTe/CdS quantum dots of high quantum yield, chromatophores extracted from a bacteria and of β-subunit antibody, all connected to a protein, which was, then, connected to the H5N1 antibody. They recorded the changes in the QD photoluminescence spectra and collected images of the QD-labelled chromatophores in an optical microscope to evaluate the specificity of the biosensor [23].

Hybrid materials are often thought as interesting materials for biosensing, since nanoparticles, which present high surface-to-volume ratio, can be attached to biomolecules, resulting in a biologically active compound with improved characteristics. Semiconductor QDs are frequently a good alternative for this purpose. Boeneman et al. [24] developed a hybrid material based on CdSe/ZnS core/shell quantum dot conjugated to fluorescent proteins and to a light-harvesting compound. The activity of the resulting hybrid is based on the QDs electron-donor characteristic for Förster resonance energy transfer (FRET) and on the fact that the fluorescent proteins are FRET acceptors. Their results showed that QDs and fluorescent proteins can be conjugated in cellular environment, inclusively, which confer them the potential for live-cell imaging and biosensing applications. Brunetti et al. [25] also successfully produced a Cadmium-free QD, with *in vivo* and *in vitro* toxicity testing done.

Since Cadmium-based quantum dots may present a risk for living organisms, their substitution for more biocompatible elements has leveraged quantum dots research. Recently, biocompatible quantum dots have been developed, and for biosensors application and *in vivo* imaging, silicon QDs (Si QDs), which are expected to be less toxic than the common group II–VI based QDs, are good alternatives. [26]. Notwithstanding, an interesting cadmium-free quantum dots class based on carbon was rapidly suggested as biocompatible and environmentally friendly, with its multi-coloured luminescence, as presented by Sun et al. [27], being its most attractive characteristic. It is prepared by the supramolecular top-down strategy of laser ablation of a carbon target in water vapour, employing Argon as the carrier gas. The surface passivation is necessary and it is held by attaching organic biocompatible polymers such as $PEG_{1500}N$, as they purposed, since this polymer is hydrophilic and can be readily conjugated to proteins, antibodies and several others bioactive molecules.

Although proteins are very important in selective QD-based biosensor design, they are often too large, which causes sterical problems that restrict the biosensor activity. As an alternative, peptides have been thought of as substitutes, since they also can provide biological activity to the device with the advantage of size accommodation. They are also of facile synthesis, biocompatible, commercially available and, as a chemical advantage, they present a well-known structure. Also, their ability to self-assemble enables their use in strategically designed biosensors. In their work, Nagy et al. [28] presented biosensors based on peptide-functionalized quantum dots, exploring the Forster Resonance Energy Transfer that occurs between the peptide and the Quantum Dot. Choulier et al. [29] compared the sensibility of protein-based and peptide-based fluorescent biosensors and they found that peptides perform as good as protein receptors; nevertheless, the ability of forming complexes with the target provided a significant difference due to size interference. They employed a 3-hydroxychromone (3HC) derivative as dye in this system, since it presents a two-band fluorescence emission, whose ratio is susceptible to environmental changes. They observed that a quantitative target determination is more efficient in the peptide–target interface than in the protein–target interface, due to its smaller size and high flexibility [29]. Another important advantage of using peptides in biosensors, with no doubt, lies on their ability to self-assemble. Kim et al. [30] produced a hydrogel based on diphenylalanine dipeptide that encapsulates enzymatic bioreceptors along with fluorescent probes based on Cadmium quantum dots. They argued that enzymes and quantum dots (QDs) physically immobilized in a self-assembled hydrogel peptide result in efficient biosensors due to the three-dimension network formed of 70–90 nm diameter nanofibers. This system was tested for detection of glucose and phenols by means of photoluminescence quenching of the hybridized QDs.

Recently, our research group showed that fluorescent environmental biosensors can be produced with self-assembled peptides and they can be equally active in solution or solid phase, with high sensibility, reproducibility, durability and low costs of fabrication. Souza [31] registered a fluorescent system consisting of diphenylalanine self-assembled nanotubes doped with a coumarin dye derivative and supported in glass substrate. It was tested regarding efficiency, reproducibility and durability, with good results. The sensor was tested for water dissolved Oxygen (O_2) after respiration processes of living bodies in the environment. Coumarin fluorescence is modulated by O_2 presence in any environment, since its populated electronic excited state is resonant with O_2 triplet electronic state. An energy transfer process occurs to convert O_2 triplet into O_2 singlet, resulting in the coumarin fluorescence intensity decreases. Very low limits were detected, which suggest that this system can be used to explore cellular environments. Since its function is based on singlet O_2 generation and since it is constructed with biocompatible components, this system may also be used in photodynamic strategies for cancer treatment.

Other self-assembled structures find application in biosensors, such as those produced with carbon. An example is the work by Robinson et al. [32], which applied single-walled carbon nanotubes (SWNTs) for *in vivo* imaging of tumour cells in mice. Their system took advantage of the unique optical properties of SWNTs to perform 3D reconstructions of *in vivo* fluorescent imaging. They functionalized it with an amphiphilic surfactant to turn the SWNTs biocompatible and performed video imaging of tumours. The tumour was pointed within ~20 s after injection!

In fact, the use of carbon nanotubes in biosensors presents several advantages, mainly due to their photophysical properties that enable to explore phenomena at the NIR emission range, which is coincident to the tissue transparency window and guarantee *in vivo* security of exploration. They do not suffer bleaching or blinking and present a large Stokes Shift, turning them ideal materials for long-term sensing applications. Nevertheless, they cannot be easily regenerated due to their facile functionalization [33]. Some other carbon structures are being currently tested in biosensors construction. One of the most interesting is graphene oxide, due to its ability of charge transfer quenching. In their work, Shi et al. [34] reported a biosensor consisting of a peptide supported in 2D graphene oxide film to monitor, via fluorescence resonance energy transfer (FRET), the BoNT serotype-A protease activity The system was able to selectively detect it. Tao et al. [35] produced biosensors based on self-assembled graphene modified with DNA, whose sensitivity is such that they claim it could be a universal biosensor. It consists of a colorimetric label-free sensor able to detect distinct compounds, such as metal ions, DNA and small molecules as well. Several other optical and electrochemical biosensors using graphene were already reported [36, 37].

Strategies with good results are often based on combining materials. Metallic nanoparticles are interesting due to the possibility of modulating their properties depending on size. Combining it with carbon nanostructures have been widely explored for biosensing applications. Among the nanoparticles with application in sensing, those of gold are highlighted due to their high surface-volume ratio, stability, biocompatibility and their unique optical properties, such as their intense colour changes upon aggregation, enabling their use in optical biosensors. An example is the work by Kim et al. [38], in which these nanoparticles were assembled to distinct cathepsins to show selectivity to the B variety, which is related to distinct types of cancer, such as brain, breast, gastric, thyroid cancers, among others.

Metallic nanoparticles are also interesting in biosensing and therapies due to their property of magnetism. Since their magnetic properties can be modulated by controlling nanoparticle size and by coating them to avoid irreversible aggregation, biosensors based on distinct transduction methods such as electrochemical, optical, piezoelectric or exploiting magnetic fields, can be built. In a recent work, Wang et al. [39] presented a biosensor constructed upon magnetic nanoparticle composites, while another researcher Wang produced biosensors for rabbit IgG detection employing magnetic nanoparticle supported on SiO_2 surfaces [40].

In our research activities, Dias prepared a fluorescent self-assembled peptide nanotube combined to iron nanoparticles and to a highly fluorescent cadmium-based quantum dot via liquid synthesis. The resulting system presented high fluorescence quantum efficiency dependent of the dominant interactions between the system components, being tunable and occurring in a wide spectral range, from near UV to visible red-region, including the therapeutic window (621 nm), which indicated that this system is adequate for biosensing and therapeutic applications [41].

Polymers are also widely used in several types of devices, including biosensors, in which they are used to immobilize proteins, enzymes, DNA [42], as coatings and traps [43, 44] and as electron donor/acceptor in transducers [45]. They are also combined to metallic nanoparticles [46] and carbon nanotubes [47] to enhance the biosensor sensitivity and to diversify the analytical routine that can be conducted.

New supramolecular compounds have been developed in order to facilitate devices fabrication and to turn sampling, sensing and fabrication procedures as much sustainable as possible. In this path, non-covalent interactions are being stimulated in order to create new conceptions of compounds. It had its start on the behaviour characterization of nanoscaled materials and, now, it has been applied to achieve complex and ordered non-covalent systems. Due to that, some interesting characteristics have been achieved such as multi-functions, self-healing, adaptativity, recycling ability, self-assembling, characteristics that enable a variety of biomedical, technological and scientific applications. Adaptive materials are able to respond to externally imposed situations by implementing changes in structure or function. In supramolecular compounds, this is possible due to the lability of non-covalent interactions, which can quickly change upon environmental changes. In biosensor development, this characteristic can be of great importance [48]. Recently, Zhang et al. [49] reported a biosensor based on an adaptive material composed by an electro-active supramolecular ionic material of imidazolium di-cation and di-anionic dyes. These compounds self-assemble, resulting in a structure with unique optical and electrochemical properties and adaptive encapsulation properties in the presence of cationic dyes. It was tested to sense NADH at a low potential of -0.07 V.

Self-healing materials are, in a simple description, stretchable materials in which non-covalent interactions enable them to recover their form, even after severe stress. This characteristic turns them into convenient materials for almost any application one can imagine. When this physical property is associated to electrochemical sensibility or energy transfer, the material can be explored in sensing, biomedical, biophysical and photovoltaic applications. Kim et al. [50] reported a biosensor designed upon self-healing materials that are able to conveniently mould themselves into tissues of complex geometry. They were applied in sensors for cardiac muscles, while Ramuz et al. [51] described their optical and pressure biosensor as having activity similar to electronic skin, since it is suitable for covering large areas. Figure 5 illustrates the self-healing behaviour presented by some supramolecular compounds.

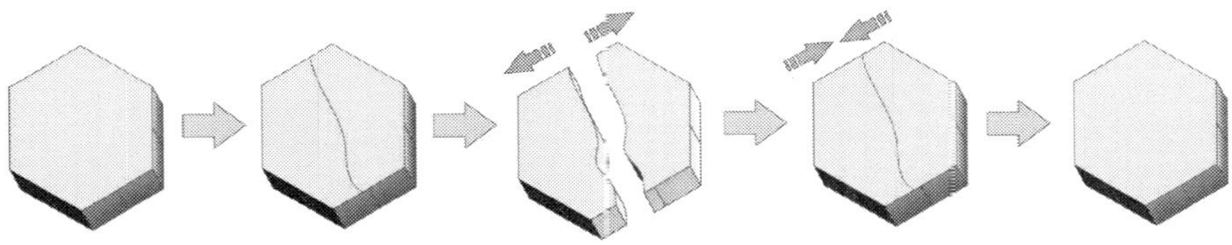

Figure 5. Schematic concept of self-healing activity of some supramolecular materials.

5. Supramolecular compounds for electrochemical biosensors

An electrochemical sensor is a device that provides an electrochemical response in the presence of an analyte such as amperometric, potentiometric, conductometric or impedimetric responses, which can be, thus, detected by electrochemical techniques [52]. Electrochemical biosensors consist of electrochemical sensors that use a biomolecule in its conception, either as recognition element or as transducer. For biosensing applications, it is common to employ enzymes as recognition element, adding simplicity to analysis, reproducibility, selectivity, low-cost, high sensitivity and short response time to the device. Durability is often observed in enzyme-based electrochemical biosensors. In such biosensors, the enzyme catalyses a chemical reaction, generating an electron flow that is identified by the device as dependent on analyte concentration. Therefore, the enzyme must be in contact with a transducer to conveniently enable the electron transfer to an electrode and generate a comprehensible signal.

The enzymes catalyse electrochemical reactions in biosensors by two main mechanisms as described below and illustrated in figure 6:

- Through substrate modification, producing an intermediate that can be reduced or oxidized at lower potentials, compared with those of the original substrate. In this system, the electron transfer is promoted by an electron carrier, which is immobilized in the electrode surface and in contact with the activated substrate, modified by an enzymatic reaction. This is a common process, exploited for instance, on the ethanol oxidation process, enabled by the enzymatic NADH/NAD+ reduction couple. In their work, Neto et al. [53] developed a supramolecular complex based on gold nanoparticle coupled to carbon nanotubes, whose function is to enhance NADH/NAD+ conversion rates, leading to lower potentials.

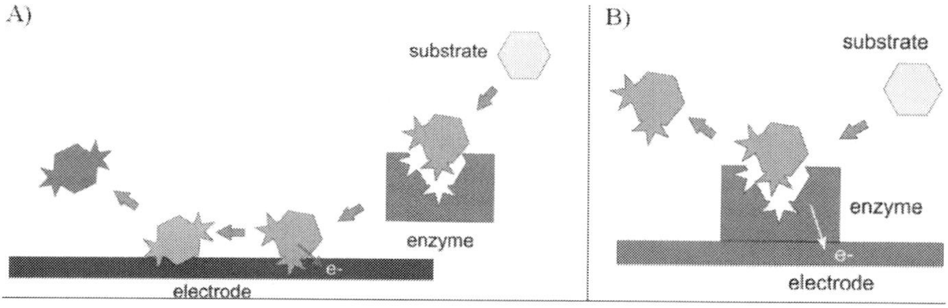

Figure 6. Scheme of the catalytic activity of an enzyme A) through electrochemical process of involving substrate modification and B) through enzyme-electrode interactions.

- Through enzyme immobilization in the electrode surface, enabling direct electron transfer to the electrode and eliminating the carrier.

Although many forms of attaching an enzyme to the electrode have been developed in the last years, it still remains the limitation of the bioelectrochemical devices. The most common method is the physical adsorption of an enzyme onto an electrode surface, nevertheless it usually results in enzymatic activity loss and, since the enzyme- electrode interaction is weak, the enzyme can be dragged by the electrolyte, which leads to a low efficient electron transfer [54] and may also result in biomolecule inactivation [55]. Supramolecular strategies of immobilization are at the forefront of the enzyme-based biosensors [56-60] in an attempt to repair the hazardous effects of binding-immobilization, such as reduction of enzymatic activity or inactivation and loss of electrode due to irreversible modifications on its surface. The common strategy of permanent modification is cross-linking of enzymes, which results in immobilization with no significant activity loss and stability, therefore, there are several strategies available for the fabrication and characterization of cross-linked enzymes [61]. Glutaraldehyde, is the most popular cross-linking agent for enzymes, however cross-linking conditions must be optimized and the best results are often obtained at glutaraldehyde concentrations of 0.1 to 5%. Even in supramolecular methods, this is popular, as in Alves et al. [62] work, which prepared a glucose biosensor by immobilizing glucose oxidase onto platinum electrode, executing the enzyme immobilization via cross-linking process, using a 2.5 % glutaraldehyde solution.

Arya et al. [63] proposed a biosensor for human influenza virus, which employed a self-assembled peptide interacting with the virus hemagglutinin antibody, in a microelectrode design that is claimed to be more sensitive than common macro-electrodes, due to its radial diffusion profile, as opposed to planar diffusion in macroelectrodes. Electrochemical impedance and cyclic voltammetry techniques were used to characterize the electrode and to estimate human Influenza virus hemagglutinin antibody concentration.

Among the enzymes that can be immobilized in biosensor electrodes, peroxidases are by far the most popular to detect distinct organic compounds, such as phenol, peroxides, poisons, hormones, gases, etc. [64]. Due to its powerful catalytic activity, through electron transfer mechanisms, peroxidase has found application in a variety of biosensors, mainly electrochemical ones, in which distinct types of responses can be recorded and treated to give accurate information on a specific analyte. Attar et al. [65] constructed a biosensor based on horseradish peroxidase to detect cyanides, which consist of extremely poisonous substances and is present in surface and ground-waters. In their proposal, peroxidase enzymatic activity is inhibited, and the decrease is inversely proportional to cyanide concentration increase. This process is followed by current density measurements, in an amperometric biosensor. Oliveira et al. [66] prepared a bi-enzymatic biosensor based on laccase and tyrosinase to detect carbamates, which are dangerous pesticides largely used to increment crop yields. The biosensor consisted of a hybrid film self-assembled onto a grapheme-doped carbon paste electrode, which presented good sensitivity and fast response.

Another strategy for biosensors development is to consider the electrode modification. Krzyczmonik et al. [67] produced a glucose oxidase biosensor in which the electrode was modified with poly(3,4-ethylenedioxythiophene) and polyacrylic acid doped with poly(4-lithium styrene-sulfonic acid), conferring good sensitivity and durability to the electrode.

Pure enzymes are highly selective, but they are costly and present low stability. As alternatives, crude enzyme preparations have been commercially available and scientifically exploited for devices application [61]. In our recent work [68], crude Brazilian zucchini (*Cucurbirta pepo*) extract was used as peroxidase source. The procedure to extract the enzyme was adapted from Fatibello and Vieira [69] and consisted of the adequate preparation of *C. pepo* aqueous suspension that is added to active carbon. To carry out the electrochemical experiments, carbon paste was prepared by with crude enzyme extract, carbon Vulcan XC 72R and mineral oil ca. Nujol®. The resulting emulsion was used to coat a carbon electrode. The bioelectrocatalytic property was characterized by cyclic voltammetry in the presence and absence of hydrogen peroxide. Figure 7A shows cyclic voltammetry data registered in this condition.

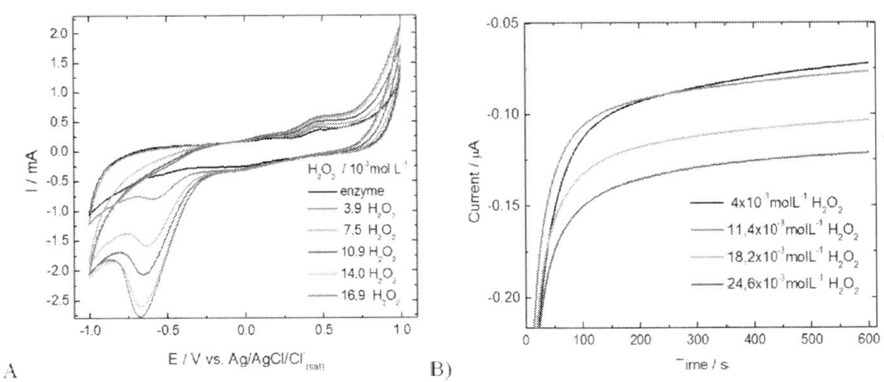

Figure 7. A) Cyclic Voltammetry of crude extract source peroxidase enzyme in buffer phosphate (pH = 6.5) in absence and in presence of H_2O_2 at distinct concentrations, v = 100m V s^{-1}, T = 25°C; B) Cronoamperometry of crude extract source of peroxidase enzyme catalysing the peroxide reduction reaction in buffer phosfate (pH = 5.5) with H_2O_2 at distinct concentrations. Electrode polarized at -0.5V vs. Ag/AgCl

In figure 7A, data shows no Faraday current in the positive scan until 0.45 V vs. Ag/AgCl/Cl-(sat). At 0.5 V, a small peak related to oxidation of some electroactive species is observed, above this potential the current increases due to enzyme oxidation process. Negative scan shows a current peak related to hydrogen peroxide reduction. The onset potential is -0.2 V and the peak current is 0.65 V. Moreover, with increase of H_2O_2 concentration in the electrolyte, the current peak increases due to the catalytically activated peroxide reduction, enabling electron transfer from the electrode to the electrochemical active species formed (reduced), which is H_2O. Since the electrode contains the crude enzymatic extract, cofactor and coenzymes are present and they enable the electron transfer. Regarding stability, cronoamperometry experiments were carried out in crude extract polarized at 0.5 V (half peak) during 10 minutes and the current decrease in different hydrogen peroxide concentrations was recorded. No current changes were observed during the experiment, as observed in Figure 7B. The crude peroxidase extract-based biosensor is easily prepared, presents low cost and is ideal to determine peroxide. Yet, in this device, reproducibility is a challenge.

Lima et al. [70] developed a biosensor based on the enzymatic extract of the Brazilian fruit pequi (Caryocar brasiliense), suitable for thiodicarb determination. In this case, the crude enzymatic extract is the source of polyphenol oxidase.

The development of an enzymatic biosensor for 3,4-dihydroxyphenyl ethylamine (dopamine) determination, based on functionalized multi-walled carbon nanotubes and horseradish peroxidase obtained from the crude extract of zucchini (Cucurbita pepo L.) as supramolecular active interface, was proposed by Ribeiro et al. [71]. In the procedure, C. pepo L. was homogenized in phosphate buffer (pH 6.5) and Polyclar SB-100 to produce the enzymatic source. In a similar procedure, Moccelini et al. [72] constructed a biosensor based on alfalfa sprout (*Medicago sativa*) as peroxidase source, also for thiodicarb determination. This enzyme was immobilized by self-assembled monolayers of l-cysteine on gold electrode.

Depending on the biosensor finality, the immobilization technique may vary, making use of a wide range of material classes, from organic polymers to metallic nanoparticles, biological materials, self-assembled materials, among others. This variety is stimulated by supramolecular approaches. Besides the above-mentioned surfaces with enzymes immobilized, carbon structures are widespread [52]. In their work, Puri et al. [47] proposed a biosensor for myoglobin detection, presenting high sensibility of 118% per decade, with high specificity. Their device consisted of a single-walled carbon nanotube (SWNT)-based label-free biosensor, containing poly(pyrrole-co-pyrrolepropylic acid) with pendant carboxyl groups electrochemically deposited on carbon structures to act as a conducting linker for immobilization, specifically of cardiac myoglobin antibodies. They performed the device characterization by source-drain current-voltage (I-V) and charge transfer studies. It presented a linear change in conductance in SWNT channel towards a wide range of myoglobin concentrations.

Lu et al. [58] proposed a graphene-functionalized self-assembled cyclodextrin for horseradish peroxidase immobilization by host-guest supramolecular strategy, to construct a biosensor for H_2O_2 determination. It was evaluated by means of electrochemical impedance spectroscopy and cyclic voltammetry, while Díez et al. [59] employed cyclodextrin combined with (3-aminopropyl)triethoxysilane-coated superparamagnetic Fe_3O_4 nanoparticles to construct amperometric biosensors towards catechol and xanthine, since the capped- magnetic nanoparticles are able to support the host–guest supramolecular immobilization of two different enzymes, tyrosinase and xanthine oxidase. Also, Ozturk et al. [73] developed a supramolecular version of an amperometric biosensor to determine xanthine in urine samples, consisting of xanthine oxidase immobilized into Fe_3O_4 nanoparticles-modified carbon paste. Its activity is based on the electron transfer properties between the electrode components and it was characterized by cyclic voltammetry and electrochemical impedance spectroscopy. Its activity was amplified due to an increase of the electroactive surface area of the electrode with the addition of Fe_3O_4 nanoparticles and to the resulting efficient electron transfer that occurs at the solution/electrode interface. A composition of these self-assembled materials with distinct characteristics can be also explored to result in better properties and innovative biosensor applications [56, 60].

6. Photoelectrochemical biosensors

The photon-to-electricity conversion process consists of a versatile phenomenon presented by many supramolecular structures and, nowadays, have been extremely advantageous to several technological and scientific areas of application. It is based on the charge separation and subsequent charge transfer that occurs in a photoactive material after its interaction with photons of light presenting the proper energy. The electron–hole pairs that rise in the material interface are responsible for several photo-physical-chemical processes that go into the materials towards its ground and stable electronic state. Due to the increasing demand for faster and reliable bioanalysis, supramolecular compounds have been thought of as facilitators in the process to integrate new and even more sensitive and analysis and imaging techniques, in such a way that gave rise to a new sensing area, which integrates the photoelectrochemistry to bioanalysis, the photoelectrochemical (PEC) bioanalysis [74, 75].

Research in this area is recent and in the past 5 years, a variety of contributions were registered. Freeman et al. [76] produced a hybrid system based on quantum dots as doping agents of nucleic acids for photoelectrochemical biosensor application. In their approach, the recognition and catalytic properties of nucleic acids were added to the photophysical properties of a quantum dot (QD), which enabled different photophysical mechanisms, such as fluorescence emission, electron transfer quenching, as well as fluorescence resonance energy transfer (FRET) and chemiluminescence resonance energy transfer (CRET), which were exploited in a multi-parameter biosensor construction. These processes are schematically represented in figure 8.

Moreover, Zhang et al. [77] constructed a bi-enzyme-based biosensor electrostatically interacting with multi-walled carbon nanotubes (MWCNT), along with a set of bilayers consisting of MWCNT-polyethyleneimine and MWCNT-DNA supported on glassy carbon electrode for selectively detect pesticides. Using apple samples, they reported remarkable detection results, with highly discriminative signals obtained for organophosphorus pesticides towards non-organophosphorus pesticides. They employed cyclic voltammetry and UV-Vis absorption as detection methods.

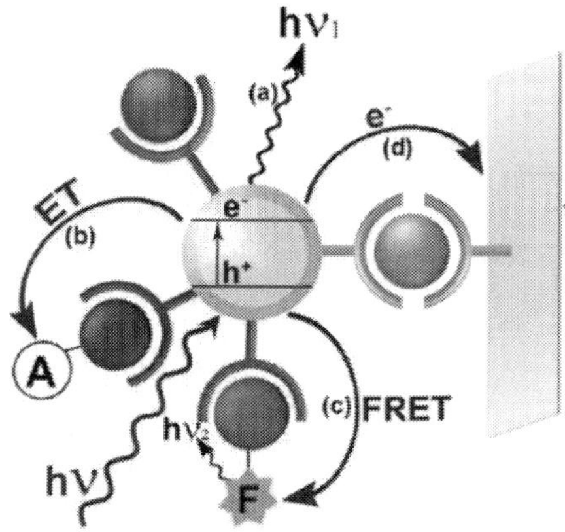

Figure 8. Scheme of QD-based sensing processes suitable for detection in Freeman's biosensor: (a) luminescence, (b) electron transfer, (c) FRET, (d) photocurrent generation. Reprinted with permission from ref [76] *Freeman R, Girsh J, Willner I. Nucleic Acid/Quantum Dots (Qds) Hybrid Systems for Optical and Photoelectrochemical Sensing. ACS Applied Material and Interfaces. 2013, 5 (8) 2815-2849.* Copyright 2015 American Chemical Society.

These multiple detection method devices are the most reliable perspectives for biosensing application. It opens several opportunities for device designs, conserving the integrity of organisms and the individuality of each assessed environment. The detection principle is that the change of photocurrent into photopotential could be driven by biological interactions between recognition elements of distinct classes and chemical constitutions and any analyte.

In a crude way, photoelectrochemical detection is the reverse of the electrochemiluminescence process, which is applied to photovoltaics, photocatalysis, and photosynthesis, since light excites a compound and the generated photocurrent results in the detected signal. Therefore, as the same principles are applied to these distinct applications, the developments achieved in photovoltaic device fabrication could be brought about to photoelectrochemical biosensor fabrication, in an attempt to achieve low cost fabrication, enabling their commercial production and practical use [78].

Due to the versatility of excitation sources and the variety of detection techniques that can be exploited in such biosensors, they are advantageous and attractive alternatives for bioanalysis. They are becoming more popular than conventional electrochemical methods, as they present higher sensitivity and selectivity and are more simple and cheaper than optical techniques, which usually need complicated and expensive equipment.

7. Conclusion: What about the future?

Supramolecular materials-based biosensors provide a new and powerful enhancement of biosensor functionalities and construction, which encompasses distinct applications in

http://dx.doi.org/10.5772/60710

diagnosis and therapeutics, as well as contribute to biological and biochemical research areas. Several electrochemical, photophysical and photochemical biosensing strategies are enabled by the unique physical-chemical properties such supramolecular compounds present, which is the reason for the rapid development of this research area.

From the achievements commented in this chapter, the share of supramolecular compounds in biosensing development is incontestable. Because of them, optical and electrochemical biosensors constitute potent and sensitive devices for *in vitro*, cellular environment and *in vivo* sensing, making use of innumerous processes that enable monitoring dynamic molecular events, living processes, degradation processes and, also, enable accurate imaging of organisms and biological processes with high precision and sensibility. The possibilities incremented by, for instance, the fluorescence spectroscopic development, giving rise to a variety of techniques that explore specific aspects of ground and excited, electronic and vibrational energy levels, open the perspective for many other materials for biosensor construction and to several other applications. Also, electrochemical sensors constructed on the supramolecular basis have enhanced the current molecular abilities of diagnosis, by enabling rapid and highly accurate responses delivery and it is thought, in the near future, to contribute to the integration of diagnostic and therapeutic techniques, enhancing medical care procedures and reliability. The possibility of combining detection techniques provide the proposal of versatile devices and enable a variety of designs and materials for biosensors, fitting clinical needs. With the analytical techniques development and the simplicity of the synthetic routines enabled by supramolecular chemistry in order to obtain pure and diverse compounds, improvements on key issues, such as sensitivity and signal-noise ratio have been fixed and the biosensors confidence have been exponentially enhanced. The perspective for the near future of biosensors application senses no limits and comprehends distinct areas such as drug delivery, therapeutics, biophysics, artificial tissue engineering, imaging, diagnosis, cellular and molecular levels exploration, as well as complete organisms, environmental studies, recycling processes development, energy harvesting and production and its limits depend almost exclusively on scientists' imagination.

Author details

Tatiana Duque Martins*, Antonio Carlos Chaves Ribeiro, Flavio Colmati, Geovany Albino de Souza, Henrique Santiago de Camargo, Diogo Lopes Dias, Paulo Alves da Costa Filho and Diericon de Sousa Cordeiro

*Address all correspondence to: tatiana@ufg.br

Chemistry Institute, P.O. Box , Campus II- Samambaia, Federal University of Goiás, Goiânia, Brazil

References

[1] Darwich S., Mougin K., Rao A., Gnecco E., Jayaraman S., Haidara H., Manipulation of Gold Nanoparticles with Atomic Force Microscopy in Dynamic Mode: Influence of Particle-Substrate Chemistry and Morphology, and of Operation Conditions. Nanotechnology. 2011; 2: 85–98.

[2] Steed J. W., Atwood J. L. editors. Supramolecular Chemistry. 2nd ed. Chichester: John Wiley & Sons; 2009. 970 p.

[3] Ma J. C., Dougherty D. The Cation-π Interaction. Chemical Reviews. 1997; 97: 1303–1324.

[4] Pauling L. The Nature of Chemical Bond and the Structure of Molecules and Crystals: An Introduction to Modern Structural Chemistry. Ithaca: Cornell University Press; 1939. 664 p.

[5] Lehn J. M. Cryptates: Inclusion Complexes of Macropolycyclic Receptor Molecules. Pure and Applied Chemistry. 1978; 50: 871–892.

[6] Araki K., Toma H. E. Química de Sistemas Supramoleculares Constituídos por Porfirinas e Complexos Metálicos. Química Nova. 2002; 25(6): 962–975.

[7] Dang Y. Q., Li H. W., Wu. Y. Construction of a Supramolecular Föster Resonance Energy Transfer System and Its Application Based on the Interaction Between Cy3-Labeled Melittin and Phosphocholine Encapsulated Quantum Dots. ACS Applied Materials and Interfaces. 2012; 4: 1267–1272.

[8] Ghadiali J. E., Lowe S. B., Stevens M. M. Quantum-Dot-Based FRET Detection of Histone Acetyltransferase Activity. Angewandte Chemie – International Edition. 2011; 50: 3417–3420.

[9] Biswas P., Cella L. N., Kang S. H., Mulchandani A., Yates M. V., Chen W. A Quantum-Dot Based Protein Module for *in Vivo* Monitoring of Protease Activity Through Fluorescence Resonance Energy Transfer. Chemical Communications Cambridge. 2011; 47: 5259–5261.

[10] Faul C. F. J., Antonietti M. Ionic Self-Assembling: Facile Synthesis of Supramolecular Compounds. Advanced Materials. 2003; 15(9): 673–683.

[11] Wang S., Wang L. Lanthanide-Doped Nanomaterials for Luminescence Detection and Imaging. Trends in Analytical Chemistry. 2014; 62: 123–134.

[12] Guan Y., Yu S. H., Antonietti M., Bottcher C., Faul C. F. J. Synthesis of Supramolecular Polymers by Ionic Self-Assembly of Oppositely Charged Dyes. Chemistry: An European Journal. 2005; 11: 1305–1311.

[13] Böhm I., Kreth S. K., Ritter H., Branscheid R., Kolb U. Switchable Supramolecular Crosslinking of Cyclodextrin-Modified Hyperbranched Polyethylenimine Via Anthraquinone Dyes. Macromolecular Chemistry and Physics. 2012; 213: 243–248.

[14] Wang J., Zhao X., Li J., Kuang X., Fan Y., Wei G., Su Z. Electrostatic Assembly of Peptide Nanofiber–Biomimetic Silver Nanowires onto Graphene for Electrochemical Sensors. ACS Macroletters. 2014; 3: 529–533.

[15] Habibi N., Pastorino L., Sandoval O. H., Ruggiero C. Polyelectrolyte Based Molecular Carriers: The Role of Self-Assembled Proteins in Permeability Properties. Journal of Biomaterials Applications. 2012; 28(2): 262–269.

[16] Pastorino L., Erokhina S., Soumetz F. C., Bianchini P., Konovalov O., Diaspro A., Ruggiero C., Erokhin V. Collagen Containing Microcapsules: Smart Containers for Disease Controlled Therapy. Journal Colloid Interface Science. 2011; 4: 56–62.

[17] Habibi N., Pastorino L., Soumetz F. C., Sbrana F., Raiteri R., Ruggiero C. Nanoengineered Polymeric S-layers Based Capsules with Targeting Activity. Colloids and Surfaces B: Interfaces. 2011; 88(1): 366–372.

[18] Sagadevan S., Periasamy M. Recent Trends in Nanobiosensors and Their Applications – A Review. Reviews on Advanced Materials Science. 2014; 36: 62–69.

[19] Tunceroglu A., Matsuda M., Birge R. B. Real-Time Fluorescent Resonance Energy Transfer Analysis to Monitor Drug Resistance in Chronic Myelogenous Leukemia. Molecular Cancer Therapeutics. 2010; 9: 3065–3073.

[20] Gulyani A., Vitriol E., Allen R., Wu J., Gremyachinskiy D., Lewis S., Dewar B., Graves L. M., Kay B. K., Kuhlman B., Elston T., Hahn K. M. A Biosensor Generated Via High-Throughput Screening Quantifies Cell Edge Src Dynamics. Nature Chemical Biology. 2011; 7: 437–444.

[21] Holzinger M., Le Goffand A., Cosnier S. Nanomaterials for Biosensing Applications: A Review. Frontiers in Chemistry. 2012; 2(article 63): 1–10.

[22] Liab J., Zhu J. J. Quantum Dots for Fluorescent Biosensing and Bio-imaging Applications. Analyst. 2013; 138: 2506–2515.

[23] Nguyen T. H., Ung T. D. T., Vu T. H., Tran T. K. C., Dong V. Q., Dinh D. K., Nguyen K. L. Fluorescence Biosensor Based on CdTe Quantum Dots for Specific Detection of H5N1 Avian Influenza Virus. Advances in Natural Sciences: Nanoscience and Nanotechnology. 2012; 3: 035014.

[24] Boeneman K., Delehanty J. B., Susumu K., Stewart M. H., Deschamps J. R., Medintz I. L. Quantum Dots and Fluorescent Protein FRET-Based Biosensors. Advances on Experimental Medicine and Biology. 2012; 733: 63–74.

[25] Brunetti V., Chibli H., Fiammengo R., Galeone A., Malvindi M. A., Vecchio G., Cingolani R., Nadeau J. L., Pompa P. P. InP/ZnS as a Safer Alternative to CdSe/Zns Core/

Shell Quantum Dots: *In Vitro* and *In Vivo* Toxicity Assessment. Nanoscale. 2012; 5: 307–317.

[26] Zhu J. J. Quantum Dots for Fluorescent Biosensing and Bio-imaging Applications. Analyst. 2013;138: 2506–2515.

[27] Sun Y. P., Zhou B., Lin Y., Wang W., Fernando K. A. S., Pathak P., Meziani M. J., Harruff B. A., Wang X., Wang H. F., Luo P. G., Yang H., Kose M. E., Chen B., Veca L. M., Xie S. Y. Quantum-Sized Carbon Dots for Bright and Colorful Photoluminescence. Journal of the American Chemical Society. 2006; 128: 7756–7757.

[28] Nagy A., Gemmill K. B., Delehanty J. B., Medintz I. L., Sapsford K. E. Peptide-Functionalized Quantum Dot Biosensors. IEEE Journal of Selected Topics in Quantum Electronics. 2014; 20(3): article 6900512, 1–12.

[29] Choulier L., Shvadchak V. V., Naidoo A., Klymchenko A. S., Mély Y., Altschuh D. A Peptide-Based Fluorescent Ratiometric Sensor for Quantitative Detection of Proteins. Analytical Biochemistry. 2010; 401: 188–195.

[30] Kim J. H., Lim S. Y., Nam D. Y., Ryu J., Ku S. H., Park C. B. Self-Assembled, Photoluminescent Peptide Hydrogel as a Versatile Platform for Enzyme-Based Optical Biosensors. Biosensors and Bioelectronics. 2011; 26: 1860–1865.

[31] Souza G. A. Caracterização Fotofísica e Morfológica de Estruturas Peptídicas Contendo Composto Fluorescente para Aplicação Ambiental. [dissertation]. Goiânia, Brazil. Federal University of Goiás; 2014. 110 p.

[32] Robinson J. T., Hong G., Liang Y., Zhang B., Yaghi O. K., Dai H. *In Vivo* Fluorescence Imaging in the Second Near-Infrared Window with Long Circulating Carbon Nanotubes Capable of Ultrahigh Tumor Uptake. Journal of American Chemical Society. 2012; 134: 10664–10669.

[33] Kruss S., Hilmer A. J., Zhang J., Reuel N. F., Mu B., Strano M. S. Carbon Nanotubes as Optical Biomedical Sensors. Advanced Drug Delivery Reviews. 2013; 65: 1933–1950.

[34] Shi J., Guo J., Bai G., Chan C., Liu X., Ye W., Hao J., Chen S., Yang M. A Grapheme Oxide Based Fluorescence Resonance Energy Transfer (FRET) Biosensor for Ultra Sensitive Detection of Botulinum Neurotoxin A (Bont/A) Enzymatic Activity. Biosensors and Bioelectronics. 2015; 65: 238–244.

[35] Tao Y., Lin Y., Re J., Qu X. Self-Assembled, Functionalized Graphene and DNA as a Universal Platform for Colorimetric Assays. Biomaterials. 2013; 34: 4810–4817.

[36] Ma H., Wu D., Cui Z., Li Y., Zhang Y., Du B., Wei Q. Graphene-Based Optical and Electrochemical Biosensors: A Review. Analytical Letters. 2012; 46(1): 1–17.

[37] Shao Y., Wang J., Wu H., Liu J., Aksay I. A., Lina Y. Graphene Based Electrochemical Sensors and Biosensors: A Review. Electroanalysis. 2010; 22(10): 1027–1036.

[38] Kim C. J., Lee D. I., Kim C., Lee K., Lee C. H., Ahn I. S. Gold Nanoparticles-Based Colorimetric Assay for Cathepsin B Activity and the Efficiency of Its Inhibitors. Analytical Chemistry. 2014; 86: 3825–3833.

[39] Wang J., Song D., Zhang H., Zhang J., Jin Y., Zhang H., Zhou H., Sun Y. Studies of Fe_3O_4/Ag/Au Composites for Immunoassay Based on Surface Plasmon Resonance Biosensor. Colloids and Surfaces B: Biointerfaces. 2013; 102: 165–170.

[40] Wang L., Sun Y., Wang J., Wang J., Yu A., Zhang H., Song D. Preparation of Surface Plasmon Resonance Biosensor Based on Magnetic Core/Shell Fe_3O_4/SiO_2 and Fe_3O_4/Ag/SiO_2 Nanoparticles. Colloids and Surfaces B: Biointerfaces. 2011; 84: 484–490.

[41] Dias D. L. Espectroscopia de Fluorescência Aplicada ao Estudo das Interações de Formação de Nanoestruturas de Peptídeos Contendo Maguemita. [dissertation]. Goiânia, Brazil. Federal University of Goiás; 2014. 92 p.

[42] Wang C., Tang Y., Liu Y., Guo Y. Water-Soluble Conjugated Polymer as a Platform for Adenosine Deaminase Sensing Based on Fluorescence Resonance Energy Transfer Technique. Analytical Chemistry. 2014; 86(13): 6433–6438.

[43] Wong L. S., Wong C. S. A New Method for Heavy Metals and Aluminum Detection Using Biopolymer-Based Optical Biosensor. IEEE Sensors Journal. 2015; 15(1): 471–475.

[44] Hosseini S., Ibrahim F., Djordjevic I., Rothan H. A., Yusof R., Van der Marel C., Benzinae A., Koole L. H. Synthesis and Characterization of Methacrylic Microspheres for Biomolecular Recognition: Ultrasensitive Biosensor for Dengue Virus Detection. European Polymer Journal. 2014; 60: 14–21.

[45] Wang C., Tang Y., Guo Y. Adenosine Deaminase Biosensor Combining Cationic Conjugated Polymer-Based FRET with Deoxyguanosine-Based Photoinduced Electron Transfer. ACS Applied Materials & Interfaces. 2014; 6(23): 21686–21691.

[46] Chen B., Liu C., Hayashi K. Selective Terpene Vapor Detection Using Molecularly Imprinted Polymer Coated Au Nanoparticle LSPR Sensor. IEEE Sensors Journal. 2014; 14(10): 3458–3464.

[47] Puri N., Niazi A., Biradar A. M., Mulchandani A., Rajesh. Conducting Polymer Functionalized Single-Walled Carbon Nanotube Based Chemiresistive Biosensor for the Detection of Human Cardiac Myoglobin. Applied Physics Letters. 2014; 105(15): 153701–153705.

[48] Rybtchinski B. Adaptive Supramolecular Nanomaterials Based on Strong Noncovalent Interactions. ACS Nano. 2011; 5 (9): 6791–6818.

[49] Zhang L., Qi H., Hao J., Yang L., Yu P., Mao L. Water-Stable, Adaptive, and Electroactive Supramolecular Ionic Material and Its Application in Biosensing. ACS Applied Materials and Interfaces. 2014; 6: 5988–5995.

[50] Kim D. H., Ghaffari R., Lu N., Wang S., Lee S. P., Keum H., D'Angelo R., Klinker L., Su Y., Lu C., Kim Y. S., Ameen A., Li Y., Zhang Y., Graff B., Hsu Y. Y., Liu Z., Ruskin J., Xu L., Lu C., Omenetto F. G., Huang Y., Mansour M., Slepian M. J., Rogers J. A. Electronic Sensor and Actuator Webs for Large-Area Complex Geometry Cardiac Mapping and Therapy. Proceedings of the National Academy of Sciences of the United States of America. 2012; 109: 19910–19915.

[51] Ramuz M., Tee B. C., Tok J. B., Bao Z. Transparent, Optical, Pressure-Sensitive Artificial Skin for Large-Area Stretchable Electronics. Advanced Materials. 2012; 24: 3223–3230.

[52] Bahadir E. B., Szgintürk M. K. Electrochemical Biosensors for Hormone Analyses. Biosensors and Bioelectronics. 2015; 68: 62–71.

[53] Neto S. A., Almeida T. S., Belnap D. M., Minteer S. D., Andrade A. R. Enhanced Reduced Nicotinamde Adenine Dinucleotide Eletrocatalysis onto Multi-Wall Carbon Nanotubes-Decorated Gold Nanoparticles and Their Use in Hybrid Biofuel Cell. Journal of Power Sources. 2015; 273: 1065–1072.

[54] Albareda-Sirvent M., Merkoçi A., Alegret S. Configurations Used in the Design of Screen-Printed Enzymatic Biosensors. A Review. Sensors and Actuators B. 2000; 69: 153–163.

[55] [55]Mendes R. K., Carvalhal R. F., Kubota L. T. Effects of Different Self-Assembled Monolayer on Enzyme Immobilization Procedures in Peroxidase-Based Biosensor Development. Journal of Electroanalytical Chemistry. 2008; 612(2): 164–172.

[56] Diez P., Piuleac C. G., Martinez-Ruiz P., Romano S., Gamella M., Villalonga R., Pingarrón J. M. Supramolecular Immobilization of Glucose Oxidase on Gold Coated with Cyclodextrin-Modified Cysteamine Core PAMAM G-4 Dendron/Pt Nanoparticles for Mediatorless Biosensor Design. Analytical and Bioanalytical Chemistry. 2013; 405(11): 3773–3781.

[57] Sioniewska A., Palys B. Supramolecular Polyaniline Hydrogel as a Support for Urease. Electrochimica Acta. 2014; 126 (Special Issue-SI): 90–97.

[58] Lu L. M., Qiu X. I., Zhang X. B., Shen G. L., Tan W., Yu R. Q. Supramolecular Assembly of Enzyme on Functionalized Graphene for Electrochemical Biosensing. Biosensors and Bioelectronics. 2013; 45: 102–107.

[59] Díez P., Villalonga R., Villalonga M. L., Pingarrón J. M. Supramolecular Immobilization of Redox Enzymes on Cyclodextrin-Coated Magnetic Nanoparticles for Biosensing Applications. Journal of Colloid and Interface Science. 2012; 386: 181–188.

[60] Villalonga R., Diez P., Eguilaz M., Martínez P., Pingarrón J. M. Supramolecular Immobilization of Xanthine Oxidase on Electropolymerized Matrix of Functionalized Hybrid Gold Nanoparticles/Single-Walled Carbon Nanotubes for the Preparation of

Electrochemical Biosensors. ACS Applied Materials & Interfaces. 2012; 4(8): 4312–4319.

[61] Roy J. J., Abraham T. E. Strategies in Making Cross-Linked Enzime Crystals. Chemical Reviews. 2004; 104(9): 3705–3727.

[62] Alves W. A., Fiorito P. A., Torresi S. I. C., Torresi R. M. Design of Molecular Wires Based on Supramolecular Structures for Application in Glucose Bionsensors. Biosensors and Bioelectronics. 2006; 22: 298–305.

[63] Arya S. K., Kongsuphol P., Wong C. C., Polla L. J., Park M. K. Label Free Biosensor for Sensitive Human Influenza Virus Hemagglutinin Specific Antibody Detection Using Coiled-Coil Peptide Modified Microelectrode Array Based Platform. Sensor and Actuators B. 2014; 194: 127–133.

[64] Barton S. C., Gallaway J., Atanassov P. Enzimatic Biofuel Cells for Implantable and Microscale Devices. Chemical Reviews. 2004; 104(9): 4867–4886.

[65] Attar A., Cubillana-Aguilera L., Naranjo-Rodríguez I., de Cisceros J. L., Palacios-Santander J. M., Amine A. Amperometric Inhibition Biosensors Based on Horseradish Perxidase and Gold Sononanoparticles Immobilized onto Different Electrodes for Cyanide Measurements. Bioelectrochemistry. 2015; 101: 84–91.

[66] Oliveira T. M. B. F., Barroso M. F., Morais S., Araújo M., Freire C., Lima-Neto P., Correia A., Oliveira M. P. P. P., Delerue-Matos C. Sensitive Bi-enzymatic Biosensor Base on Polyphenoloxidases-Gold Nanopartilcles-Chitonan Hybrid Film-Graphene Doped Carbon Paste Electrode for Carbamates Detection. Bioelectrochemistry. 2015; 98: 20–29.

[67] Krzyczmonik P., Socha E., Skrzyped S. Immobilization of Glucose Oxidase on Modified Electrode with Composite Layers Based on Poly(3, 4-Ethylenedioxythiophene). Bioelectrochemisty. 2015; 101: 8–13.

[68] YI L. H., Colmati F. Estudo da Bioeletroredução de Peróxido de Hidrogênio Catalisada por Extrato Bruto Fonte de Enzima Peroxidase. In: I Seminário de Iniciação em Desenvolvimento Tecnológico e Inovação/VIII Congresso de Pesquisa, Ensino e Extensão da UFG/63a Reunião Anual da SBPC, 10–15 July 2011. Goiânia, Brazil. editor. VIII Congresso de Pesquisa, Ensino e Extensão da UFG/63a Reunião Anual da SBPC; 2011.

[69] Fatibello-Filho O., Vieira I. C. Uso Analítico de Tecidos e de Extratos Brutos Vegetais como Fonte Enzimática. Química Nova. 2002; 25(3): 455–464.

[70] Lima F., Lucca B. G., Barbosa A. M. J., Ferreira V. S., Moccelini S. K., Franzoi A., Vieira I. Biosensor Based on Pequi Polyphenol Oxidase Immobilized on Chitosan Crosslinked with Cyanuric Chloride for Thiodicarb Determination. Enzyme and Microbial Technology. 2010; 47: 153–158.

[71] Ribeiro F. A. S., Tarley C. R. T., Borges K. B., Pereira A. C. Development of a Square Wave Voltammetric Method for Dopamine Determination Using a Biosensor Based

on Multiwall Carbon Nanotubes Paste and Crude Extract of Cucurbita Pepo L. Sensors and Actuators B. 2013; 185: 743–754.

[72] Moccelini S. K., Vieira I. C., Lima F., Lucca B. G., Barbosa A. M. J., Ferreira V. S. Determination of Thiodicarb Using a Biosensor Based on Alfalfa Sprout Peroxidase Immobilized in Self-Assembled Monolayers. Talanta. 2010; 82: 164–170.

[73] Öztürk F. O., Erden P. E., Kaçar C., Kiliç E. Amperometric Biosensor for Xanthine Determination Based on Fe_3O_4 Nanoparticles. Acta Chimica Slovenica. 2014, 61, 19–26.

[74] Zhao W. W., Xu J. J. Chen H. Y. Photoelectrochemical DNA Biosensors. Chemical Reviews. 2014; 114(15): 7421–7441.

[75] Zhao W. W., Xiong M., Li X. R., Xu J. J., Chen H. Y. Photoelectrochemical Bioanalysis: A Mini Review. Electrochemistry Communications. 2014; 38: 40–43.

[76] Freeman R., Girsh J., Willner I. Nucleic Acid/Quantum Dots (QDs) Hybrid Systems for Optical and Photoelectrochemical Sensing. ACS Applied Material and Interfaces. 2013; 5(8): 2815–2849.

[77] Zhang Y., Arugula M. A., Wales M., Wild J., Simonian A. L. A Novel Layer-by-Layer Assembled Multi-enzyme/CNT Biosensorfor Discriminative Detection Between Organophosphorus and Non-organophosphrus Pesticides. Biosensors and Bioelectronics. 2014; S0956-5663(14): 00624–00625.

[78] Yue Z., Lisdat F., Parak W. J., Hickey S. G., Tu L., Sabir N., Dorfs D., Bigall N. C. Quantum-Dot-Based Photoelectrochemical Sensors for Chemical and Biological Detection. ACS Applied Materials and Interfaces. 2013; 5: 2800–2814.

Chapter 6

Immunosensors

Cecilia Cristea, Anca Florea, Mihaela Tertiș and Robert Săndulescu

Additional information is available at the end of the chapter

http://dx.doi.org/10.5772/60524

Abstract

Immunosensors are solid-state devices in which the immunochemical reaction is coupled to a transducer. They form one of the most important classes of affinity biosensors based on the specific recognition of antigens by antibodies to form a stable complex, in a similar way to immunoassay. Depending on the type of *transducer* there are four types of immunosensor: electrochemical, optical, microgravimetric and thermometric. The most commonly used bioelements for the development of electrochemical immunosensors are antibodies (Ab), followed by aptamers (Apt) and, in the last five years, microRNA (miRNA). In order to perform an early diagnosis, a method that is able to measure peptides and proteins *directly* in a sample, without any sample pre-treatment or any separation, is preferred. This direct detection can be performed with methods making use of the specific interaction of proteins with Ab, Apt and miRNA. The recent developments made in the immunosensor field, regarding the incorporation of nanomaterials for increased sensitivity, multiplexing or microfluidic-based devices, may have potential for promising use in industry and clinical analysis. Some examples of assays for several commercially available biomarkers will be presented. The main application fields, beside biomedical analysis, are drug abuse control, food analysis and environmental analysis.

Keywords: Immunosensors, immunoassay, antibody, aptamer, clinical analysis, environmental monitoring

1. Introduction

1.1. Immunosensors and immunoassays

Immunosensors are solid-state devices in which the immunochemical reaction is coupled to a transducer. They form one of the most important classes of affinity biosensors based on the specific molecular recognition of antigens by antibodies to form a stable complex, in a similar way to immunoassay. In contrast to immunoassay, modern transducer technology enables the label-free detection and quantification of the immune complex [1]. The immune system possesses a great ability to distinguish self from non-self. Antibodies (Abs) with high specificity are synthesized by the organism in reasonable quantities after sensing the foreign species called antigens (Ag), as part of the immune defence system. The ability of organisms to recognize the presence of Ags and to respond rapidly by synthesizing Abs that exhibit high binding constants [2] was and continues to be exploited by scientists in order to create new specific analytical devices. Molecules generally designed as Abs include a number of classes and subclasses of immunoglobulins with high specificity towards various targets.

The high sensitivity and selectivity of the immunoreactions together with the availability of Abs or Apt for a broad range of molecules make immunochemical methods useful tools in a large variety of applications, including clinical analysis. The use of immunosensors instead of other immunochemical techniques simplifies the analysis considerably, making it rapid and reliable [3].

Lately, a plethora of biomolecules whose presence or level of expression is an indicator of some pathological condition (we refer here to biomarkers) are currently used in immunoassays. Those devices represent a convenient way of measuring the concentration of biomolecules in biological fluids (such as serum, urine, etc.) by means of immunological reactions.

Most clinical protein biomarker detection is done today by enzyme-linked immunosorbent assay (ELISA), but requirements for the relatively expensive test kits and the plate readers limit ELISA's usefulness for fast diagnostics [4]. LC-MS-based proteomics is currently used for biomarker discovery, but is currently too expensive and technically too complex for routine clinical diagnostics [5, 6]. Microarrays developed on a 96- or 384-well plate format, Ab-coated with colorimetric detection are also being used in clinical diagnostic. These arrays are simple and highly selective and allow multiplexed measurement of proteins [7, 8, 9, 10]. Several commercially available automatic or semiautomatic analysers for multiplexed protein measurements, employing fluorescence (Luminex, Myriad RBM), electrochemiluminescence (ECL) (Roche Diagnostics, Mesoscale Discovery), or surface plasmon resonance (Horiba Inc., BIO-RAD; Biacore Life Sciences, GE Healthcare) measurement technologies are currently used in hospital laboratories [11]. Assay kits are available for measurement of up to 10 selected target proteins per sample with detection limits (LOD) of 1-100 pg/mL^{-1} in serum [4]. These commercial instruments require specialized consumables, including sample well plates, chips and reagent kits [12], which are expensive and, thus, they have limited usefulness for point-of-care (POC) applications where resources are limited. Fluorescence-based detection strategies typically require laser sources and precise alignment of optical components, whereas electrochemical detection strategies offer robust, quantitative measurements using low-cost, simple instrumentation [11].

Depending on the type of *transducer* there are four types of immunosensor: electrochemical (also classified into potentiometric, amperometric and impedimetric), optical, microgravimetric and thermometric. All of these types can either be run as direct (non-labelled) or indirect (labelled) immunosensors. The most commonly used labels are enzymes such as peroxidase, glucose oxidase, alkaline phosphatase, catalase or luciferase. Some other labels such as electroactive compounds (ferrocene or In^{2+} salts), fluorescent reagents (rhodamine, fluorescein, Cy5, ruthenium diimine complexes, phosphorescent porphyrin dyes, etc.) and since recently metallic nanoparticle (gold or silver produced *in situ* electrochemically) are also used [3]. Although indirect immunosensors are highly sensitive, mainly due to the analytical label, the concept of direct sensing represents a reliable alternative in the development of immunoassay systems [13]. The appropriate immobilization of the recognition element at the surface of the transducer element is of paramount importance. Different immobilization techniques, such as direct adsorption onto the electrode surface, self-assembled monolayers (SAMs), polymer matrices or magnetic beads (MBs), etc., can be used. Direct adsorption is simple, but the proper orientation of the recognition element for the immunocomplex formation is poor and passivation of the electrode surface frequently occurs. A more ordered, oriented immobilization is obtained with SAMs that are generally formed on gold surfaces, using either thiolated Abs or DNA, or compounds containing a thiol group at one end and a carboxyl or amino group at the other end able to form a covalent bond with the Ab. SAMs can also be obtained on graphite surfaces using diazonium salt chemistry. Polymers with different functional groups, such as chitosan, polysulphone or polyaminobenzoic acid, may also be used for the immobilization of these recognition elements [14]. MBs have gained popularity in the development of immunosensors due to their good stability, small size and spherical geometry with a high load of recognition elements, fast-reaction kinetics, and the ease of surface modification with functional groups, DNA, enzymes, streptavidin, protein A or G, etc. [15].

The role of electrochemical immunosensors in analysis has considerably increased in the last years, due to their low cost, high sensitivity, potential for miniaturization and automation, and low power requirements. Important developments have been secured, especially in sensor design concerning the type of bioelement (Ab, Apt, Affibody, miRNA) and the immobilization technique.

2. Label-free versus labelled immunosensors

The immunoassay is a device based on biochemical detection and measurement of biomolecules in a test solution using a specific Ab or immunoglobulin. Immunoassays rely on the specificity of the immunoreaction, and more precisely on the ability of an Ab or Apt to recognize and bind to a specific target, usually an Ag. The measurable signal proportional to the Ag concentrations is obtained with the aid of a label. Some of the commonly employed labels include radioisotopes [16], chromophores [17], fluorophores [18] and enzymes [19]. Depending on the nature of the label, the immunoassay could be classified into radioimmunoassay, chromoimmunoassay, fluoroimmunoassay and enzyme immunoassay, respectively [20].

The enzyme immunoassay or enzyme-linked immunosorbent assay (ELISA) is the most commonly used technique employed as a diagnostic tool for clinical purposes. The working principle of ELISA consists in the immobilization of an unknown amount of Ag onto a solid surface, and then a specific Ab is added over the surface so that it can bind to the Ag. This antibody is usually labelled with an enzyme; the substrate of the enzyme is then added, which will be converted into a compound that can be quantified by means of optical methods. There are four types of ELISA: direct ELISA, indirect ELISA, sandwich ELISA and competitive ELISA [20].

The simplest protocol is the one used in the case of direct ELISA [21]. The Ag is adsorbed to a plastic plate, then an excess of another protein (normally bovine serum albumin) is added to block the free binding sites. An enzyme is linked to an Ab in a separate reaction and the formed enzyme-Ab complex is added to the well. The Ab interacts with its Ag, forming an Ag/Ab-enzyme complex, and the excess of unreacted Ab-enzyme is washed off. Afterwards the substrate of the enzyme is added, and is converted into a compound that gives a measurable signal directly linked to the Ag quantity.

Indirect ELISA is a two-step ELISA assay which involves the use of two Abs to bind the Ag: a primary capture Ab and a secondary signalling Ab. The primary Ab is incubated with the Ag followed by the incubation with the secondary Ab. However, this may lead to non-specific signals because of a cross-reaction that the secondary Ab may bring about. This technique has a high sensitivity because more than one labelled Ab is bound per Ag molecule.

Sandwich ELISA is a less common alternative, but is highly efficient in sample Ag detection. Moreover, many commercial ELISA pair sets are built on this sandwich ELISA.

The sandwich ELISA quantifies antigens between two layers of antibodies (i.e., capture and detection antibody). The Ag to be measured must contain at least two antigenic epitopes capable of binding to Ab, since at least two Abs act in the sandwich. Either monoclonal or polyclonal Abs can be used as the capture and detection Abs in sandwich ELISA systems. Monoclonal Abs recognize a single epitope that allows fine detection and quantification of small differences in Ag. A polyclonal is often used as the capture Ab to pull down as much of the Ag as possible. The advantage of sandwich ELISA is that the sample does not have to be purified before analysis, and the assay can be very sensitive (up to two to five times more sensitive than direct or indirect ELISA, but lower than ELISpot). The central event of competitive ELISA is a competitive binding process executed by original Ag (sample Ag) and add-in Ag. The procedures of competitive ELISA are different in some aspects compared with indirect, sandwich and direct ELISA assays. In competitive ELISA assays, the following successive steps are involved: 1) primary Ab (unlabelled) is incubated with sample Ag; 2) Abs-Ag complexes are then added to 96-well plates which are pre-coated with the same Ag; 3) unbound Ab is removed by washing the plate (the more Ags in the sample, the fewer Abs will be able to bind to the Ag in the well, hence "competition"); 4) the secondary Ab that is specific to the primary

http://dx.doi.org/10.5772/60524

Ab and conjugated with an enzyme is added; 5) a substrate is added, and the remaining enzymes elicit a chromogenic or fluorescent signal. For competitive ELISA, the higher the sample Ag concentration, the weaker the eventual signal is. The major advantage of a competitive ELISA is the ability to use crude or impure samples and still selectively bind any Ag that may be present.

In order to avoid confusion, the term 'immunoassay' is used for tests based on immunoreactions, while the term 'immunosensor' is specifically employed to describe whole instruments, i.e., immunoreaction-based biosensors. In the heterogeneous immunoassays, the Ab or Ag is immobilized at a solid substrate (e.g., microplate), while homogeneous immunoassays take place in solution [22]. Heterogeneous immunoassays are easier to design and construct, but they require a separation of the Ag and the Ab from the samples and their immobilization at a solid surface. In contrast, the homogeneous immunoassays usually involve the immobilization of the biomolecules on nano-/microbeads, thus allowing the integration of multiple liquid handling processes [23, 24]. The homogeneous immunoassays coupled with microfluidic devices are relatively inexpensive and efficient alternatives and can be used for the detection of complex samples, such as urine or blood, without any sample pre-treatment or large volume consumption. Microfluidic lab-on-a-chip technology also has the advantages of portability, integration and automation.

In the immunosensor design, the sensing element is formed by means of immobilization of Ags or Abs, and the binding event is transformed into a measurable signal by the transducer [25]. The performances of immunosensors are obviously related to the selectivity and affinity of the Ab-analyte binding reaction. The immunosensors can be either direct (where the immunochemical reaction is directly determined by measuring the physical changes induced by the formation of the complex) or indirect (where a sensitively detectable label is combined with the Ab or Ag of interest, Figure 1) [20]. The distinction has an entirely different meaning to the terms direct and indirect in the immunoassay fields, which are both tracer-related. They are distinguished by whether Ab binding is directly detected, for example, after an enzyme-labelled Ab is bound to an immobilized coating conjugate or an enzyme tracer to an immobilized Ab, or whether the detection only takes place after a secondary binding reaction, for example, if an enzyme-labelled secondary Ab is used to label the first Ab bound to an immobilized coating conjugate [21, 22, 23].

Immunoassays can be performed either in competitive or sandwich-type configurations. The competitive-type assay is based on the competition between the analyte in the sample which is unlabelled and a labelled analyte for a limited number of captured Abs. Two approaches can be employed: the first is based on immobilized Abs that react with free Ags in competition with labelled Ags; the second consists in immobilized Ags and labelled Abs. The sandwich-type ELISA-like assay is the most widely used format and employs a pair of matched, high-affinity Abs and appropriate labels, leading to an increased sensitivity and selectivity. The Ag of interest is captured between a pair of Abs and a labelled Ab is further used to obtain a detectable signal [26, 27, 28].

The reaction of proteins in solution with their complementary proteins immobilized on a surface leads to changes in, for example, refractive index, thickness and dielectric constant of

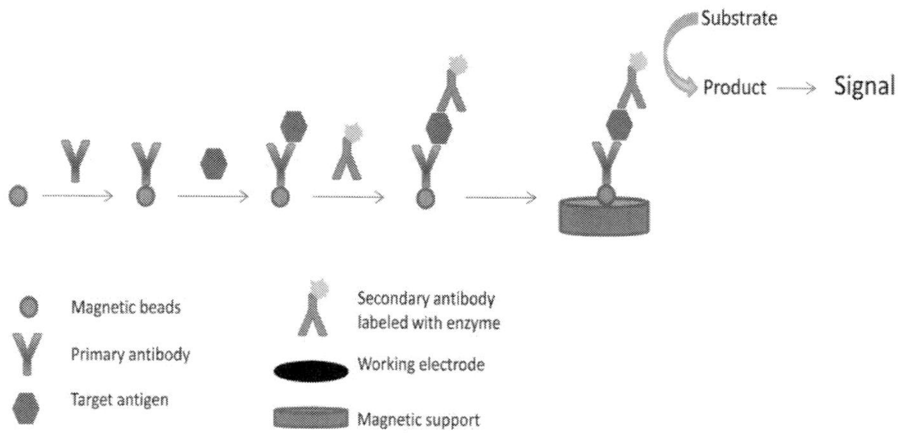

Figure 1. Scheme of the sandwich assay for protein biomarker detection using magnetic beads

the immobilized layer. With proteins immobilized on a piezoelectric material, a change in resonance frequency is detected which is proportional to the mass change on the surface. These properties are exploited in optical, electrochemical and piezoelectric immunosensors. Ideally, immunosensors are devices with fast response, a high specificity and sensitivity. It is also preferable that they are regenerable, which means that they can be reused immediately after dissociation of the Ab-Ag complex, e.g., by using a chaotropic reagent [29].

To be able to detect the interaction, one of the immunoagents has to be labelled. Various labels have been applied, of which radioisotopes were among the first because of their inherent sensitivity. Other frequently used labels are chemiluminescent compounds and enzymes (e.g., alkaline phosphatase, horseradish peroxidase), which convert an enzyme substrate into a measurable product (Figure 2).

Many different configurations have been described and the four most commonly used formats are shown in Figure 3. In a homogeneous immunoassay Abs (Figure 3a), Ags and labelled Ags are mixed. Free labelled Ags and those bound to an Ab, can be distinguished by a change of the label's activity upon binding. Usually, immunoassays are heterogeneous, which means that either the Ab or the Ag is immobilized on a solid carrier and an immunocomplex is formed upon contact with a solution containing the other immunoagent.

The unbound proteins are removed by washing and the response obtained from the labels is proportional to the amount of protein bound. In a sandwich immunoassay, Abs are immobilized and after addition of the sample containing the Ag, a labelled secondary Ab is added (Figure 3b). Besides these non-competitive formats, competitive configurations can also be applied. In a competitive assay, competition takes place between free and bound Ag for a limited amount of labelled Ab (Figure 3c) or between Ag (the sample) and labelled Ag for a limited amount of Ab (Figure 3d) [30]. Electrochemical detection of immune interaction can

http://dx.doi.org/10.5772/60524

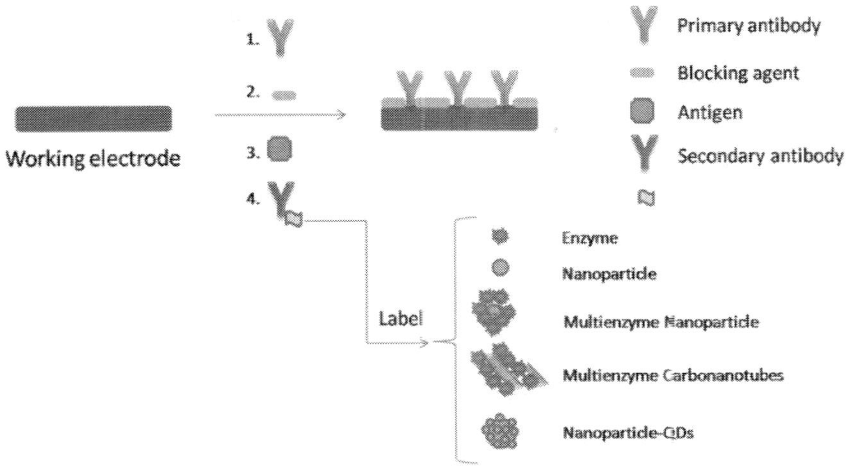

Figure 2. Bioconjugates for signal amplification strategies: enzymes, nanoparticles, multienzyme nanoparticles, multi-enzyme CNTs (carbon nanotubes), nanoparticle QDs (quantum dots)

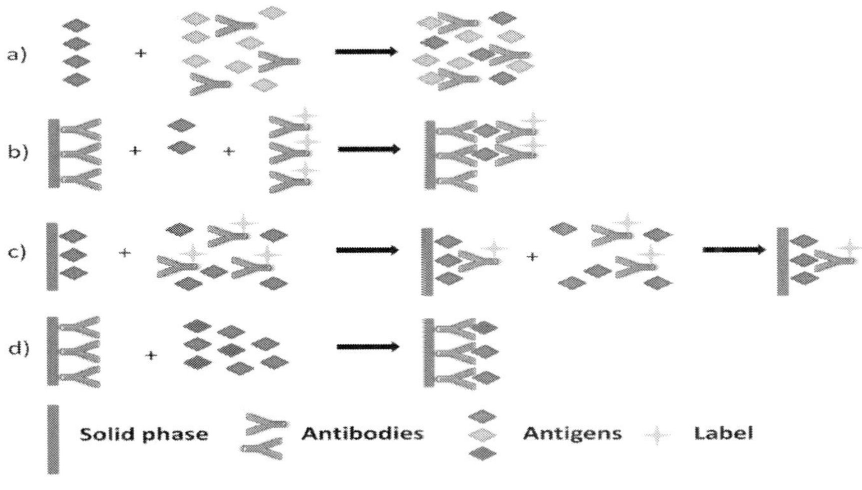

Figure 3. (a) Homogeneous competitive immunoassay; (b) Heterogeneous non-competitive immunoassay; (c) Heterogeneous competitive immunoassay; (d) Heterogeneous competitive immunometric assay

be performed both with and without labelling [31]. A frequently used format in electrochemical immunosensing is the amperometric immunosensor, where proteins are labelled with enzymes producing an electroactive product from an added substrate [32]. Direct detection without labelling can be performed by cyclic voltammetry, chronoamperometry, impedimetry, and by measuring the current during potential pulses (pulsed amperometric detection). These methods are able to detect the electrode change in capacitance and/or resistance, induced by the binding of protein. These immunosensors have been developed using various supporting materials. Those built on silicon, silanized metal, or polypyrrole are often regenerable, while those based on SAMs, on silver or gold, are not. Chaotropic reagents, used to accomplish disruption of the Ab-Ag bond, have been reported to affect the SAM [33, 34] or to induce a decrease in the sensitivity of the sensor [34, 35].

3. Types of bioelement used in the development of immunosensors

The most common bioelements used in the development of electrochemical immunosensors are Abs, followed by Apt, and in the last five years the miRNA. In order to perform an early diagnosis, a method that is able to measure peptides and proteins *directly* in a sample, without any sample pre-treatment or any separation, is preferred. This direct detection can be performed with methods making use of the specific interaction of proteins with Ab, Apt and miRNA.

Abs are proteins produced in animals by an immunological response to the presence of a foreign substance (with a molecular weight larger than 1.5 kDa), a so-called Ag, and which have a specific affinity for this Ag [36, 37, 38, 39, 40]. Abs may be monoclonal or polyclonal. Polyclonal Abs have a specific affinity for the Ag, but they are directed to different binding sites (epitopes) on the Ag with different affinities. In immunoassays, often the use of monoclonal Abs is preferred. These Abs are formed by using identical cells cultivated by the hybridoma technology, and have optimized characteristics such as high affinity or high specificity. To conclude, polyclonal Abs are isolated after immunization of animals, while monoclonal Abs require hybridomal selection. Abs are structurally very similar. Of the five classes of immunoglobulins (IgA, IgD, IgE, IgG and IgM), which differ in glycosylation and the number and position of disulphide bridges, IgG (150 kDa) are mainly used for immunoassays. An IgG consists of two heavy and two light chains, which are interconnected by disulphide bridges (Figure 4). All chains have a variable and a constant region. The variable regions of the heavy and light chain combine in one interaction site for the Ag, which is called the antigenic site. Thus, an IgG molecule has two identical binding locations for the Ag [41, 42]. Abs often demonstrate strong binding features and high selectivity; however they have some intrinsic limitations related to their molecular properties. The most commonly used Ab type, the IgG molecule, is a large, bivalent, multidomain protein, dependent on disulphide bonds and with a complex glycosylation pattern. This explains its relatively poor heat stability and a difficult and expensive manufacturing process. In addition, Abs use only a minor part of the molecule for Ag recognition [43].

A number of techniques are known today for the *in vitro* generation of a wide range of molecular recognition bioelements besides Abs, from which binders with high affinity and specificity can be selected. Besides the favourable molecular recognition properties of Abs, these alternative scaffold proteins often offer improved characteristics such as small size, high stability in rough conditions, absence of cysteine, high yield in bacterial expression and the option to use them for multispecific analysis.

Affibody molecules are a class of engineered affinity proteins with high potential for therapeutic, diagnostic and biotechnological applications. Affibody molecules are small (6.5 kDa), single-domain proteins that can be isolated for high affinity and specificity to any given protein target [43].

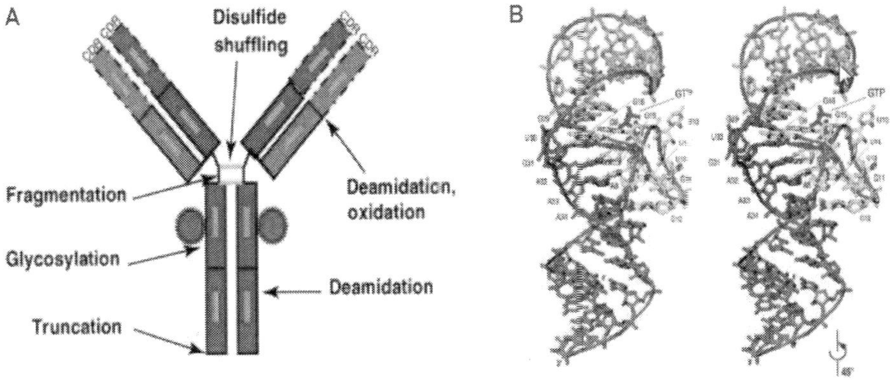

Figure 4. (A) Illustration of the structure of a typical monoclonal Ab along with the various sites of heterogeneity that are commonly observed [38]. (B) Class I GTP Apt structure [39]. "Reprinted from Trends. Biotechnol. (Cell Press). 27(12), Ratore AS, Follow-on protein products: Scientific issues, developments and challenges. A review, 698-705, copyright (2009), with permission from Elsevier". Figure 4b reprinted from an open access article according to Creative Commons Attribution 4.0 International License"

Affibody molecules have already been investigated as affinity probes in ELISA setup and protein microarrays. For example, in one such study, dimeric Affibody molecules with affinity for IgA, IgE, IgG, TNFα, Insulin and Taq polymerase were immobilized on thiol dextran microarray and an LOD of 70 fM was obtained [44]. Affibody molecules were also evaluated as capture agents in a sandwich array format with unlabelled target protein and monoclonal Abs for detection, demonstrating specificity in a complex serum or plasma sample. The use of Affibody molecules in sandwich assays for analysis of human samples might also be advantageous in a cross-reactivity perspective, since human anti-animal Abs are occasionally present which might lead to cross-linking of capture and detection Abs, resulting in elevated background signals, which are avoided when affibodies are used [44].

Apts, derived from the Latin word "aptus" (meaning to fit) are synthetic single-stranded DNA or RNA molecules that can bind with high affinity and specificity to their target molecules. These artificial "nucleic acid" ligands can be generated against proteins, small molecules, ions, whole cells, tissues and organs [45]. Those specific nucleic acids, selected from random sequence pools, have been shown to bind various targets that do not have a nucleic acid structure. The development of *in vitro* selection and amplification techniques has allowed the identification of specific Apts, which bind to the target molecules with high affinity. Many small organic molecules with molecular weights from 100 to 10,000 Da have been shown to be good targets for selection. Moreover, Apts can be selected against difficult target haptens, such as toxins or prions. The selected Apts can bind to their targets with high affinity and even discriminate between closely related targets. Apts can thus be considered a valid alternative to Abs or other biomimetic receptors, for the development of biosensors and other analytical methods [46]. The process to select Apts, known as *in vitro selection and amplification* or *SELEX* (from Systematic Evolution of Ligands by Exponential enrichment), was first reported in 1990, independently by two research teams [42].

It was demonstrated that the Apts could bind their targets by adaptive binding. Thus, small molecules usually become an integral part of a three-dimensional structure, while for proteins, the oligonucleotide combines with the protein into a larger molecular structure [46].

Apts offer several advantages over Abs. Being isolated through an *in vitro* procedure, they can be easily adapted to completely non-physiological conditions and against targets that would otherwise be invisible to the immune system. The conditions of their selection process can be adjusted to change their kinetic parameters (i.e., k_{on} and k_{off} rates). Other advantages include the ability of Apts to denature and refold easily, with better shelf life and ease of attachment and detachment of the bound molecule. Their resistance in harsh conditions (temperatures, pH) and their ability to distinguish between similar proteins (e.g., mucin class) are also appreciated by scientists developing aptasensors [47]. The advantage of having a recognition region that can repeatedly refolded as been effectively demonstrated recently in a microarray application [48], where denaturing conditions (e.g., with 7M urea) were used to regenerate free Apts that were refolded into their active conformations in the binding buffer after the measurement was completed. Such a procedure allows for multiple measurements with the same Apt.

Discovered in 1993 [49], microRNAs (miRNAs) are 22 nucleotide-long non-coding RNA molecules that regulate the expression of target genes at the post-transcriptional level by either translational repression or degradation of miRNAs [50]. It has been estimated that the human genome may encode over 1000 miRNAs, which are thought to regulate about 60 % of the mammalian genes. Altered expression patterns of miRNAs have been associated with many diseases, including several types of cancer, cardiovascular disorders, diabetes, rheumatic disease and neurological disorders [51]. Due to their unique miRNA expression signatures that have been demonstrated in many pathological conditions, these molecules have been proposed as novel biomarkers in diagnostics. The miRNA profiles in clinical samples will not only signify the presence of a disease, but can also help in the prediction of its course and help to select personalized treatments. miRNAs are currently detected in blood serum, plasma, urine, saliva, sweat and other bodily fluids. The levels of circulating miRNAs in serum were estimated to be in the 200 aM to 20 pM range [52]. The field of circulating miRNA analysis is in its early stages, but it has great potential due to the fact that miRNAs are relatively stable molecules and can be easily accessible in a non-invasive manner.

4. Immunosensors for biomedical and environmental analysis

There is a strong and continuous demand for fast, easy and cheap analytical devices for trace detection of compounds of clinical and environmental interest, as well as for pharmaceutical and food industries. The high selectivity and sensitivity of immunosensors make them the most appropriate for the detection and monitoring of the above mentioned domains. The recent developments made in the immunosensor field regarding the optimal orientation of Abs, incorporation of nanomaterials for increased sensitivity, multiplexing and microfluidic-based devices may have potential for promising use in industry and clinical analysis.

Some examples of assays for several commercially available biomarkers are presented in Table 1.

Target	Method	LOD	Reference
Mucin 1 (MUC1)	Ab-lectins sandwich optical immunoassays	1.2 U/mL 0.4 U/mL	[53]
MUC1	ZnO nanorods QCM sensor	1 U/mL	[54]
MUC1	Electrochemical aptasensors based on AuNP	EIS: 3.6 ng/mL DPV: 0.95 ng/mL	[55]
Total Prostate specific antigen (PSA) Free-PSA	White-light reflectance spectroscopy	0.20 ng/mL 0.15 ng/mL	[56]
PSA	3D label-free immunosensor based on crumpled graphene-Au nanocomposites	0.59 ng/ml	[57]
Carcinoembryonic antigen (CEA) Carcinoma antigen 125 (CA125)	Electrochemical 3D origami microfluidic paper based immunosensor	0.08 pg/mL 0.06 mU/mL	[58]
Myoglobin	SPR immunosensor	31.0 ng/mL	[59]
Troponin I (cTnI) C-reactive protein (CRP)	poly(dimethylsiloxane)-AuNP – QDs microfluidic electrochemical sensor	cTnI: 0.004 μg/L CRP: 0.22 μg/L	[60]
Thrombin	Sandwich-type aptasensor based on planar hall magnetoresistivetransducer and superparamagnetic labels	86 pM	[61]
Thrombin	Electrochemical aptasensor based on methylene blue and ferrocenne	170 pM	[62]
LDL cholesterol	Impedimetric immunosensor based on AuNP-AgCl/polyaniline composite film	0.34 pg/mL	[63]
LDL cholesterol	Impedimetric immunosensor based on reduced graphene platform	5 mg/dL	[64]
Creatinine	Electrochemical immunosensor based on polypyrrole immobilization platform	0.46 mg/dL	[65]
Influenza virus	Electrochemical DNA sensor based on avidin-biotin affinity reaction	$8.51 \cdot 10^{-14}$ M	[66]
Influenza A virus	Surface acoustic sensor based on Love waves	1 ng/mL	[67]
Legionella pneumophila	Electrochemical immunosensor based on ZnO nanorods	1 pg/mL	[68]
Neisseria gonorrhoeae	Electrochemical DNA sensor based on polyaniline films on In-SnO_2 plates	$0.5 \cdot 10^{-15}$ M	[69]
HIV	Capacitive immunosensor based on polytyramine films	$7.9 \cdot 10^{-8}$ pg/ml	[70]
Clenbuterol	ECL sensor with incorporated AuNP and QD	8.4 pg/mL	[71]
Ketamine	Impedimetric sensor based on SAM	0.41 pmol/L	[72]
Ciprofloxacin	Label free polypyrrole based immunosensor	10 pg/mL	[73]
Acetaminophen	Electrochemical immunosensor based on graphene oxide	0.17 μM	[74]

Target	Method	LOD	Reference
Ochratoxin A	Evanescent wave all fibersensor	3 nM	[75]
Estradiol Ehinylestradiol	MBs based electrochemical immunosensor	1 ng/ml 10 ng/mL	[76]
Trinitrotoluene	Fluorescence microfluidic immunosensor	0.01 ng/mL	[77]
Endosulfan Paraoxon	Electrochemical immunosensor based on SWCNT forrest	0.05 ppb 2 ppb	[78]
Aflatoxin B1	EQCM imunosensor based on SAM of 4-aminothiophenol	0.012 ng/mL	[79]
Escherichia coli O157:H7	Label free ITO based impedimetric immunosensor	1 CFU/mL	[80]
CEA	Electrochemical immunosensor based on MoS_2 nanosheets-Au composites and Ag nanosphere labels	0.27 pg/mL	[81]
Lymphoblastic leukemia antigen (CD10)	Label free QCM immunosensor	$2.4 \cdot 10^{-12}$ M	[82]
Carbohydrate antigen 72-4	Sandwich electrochemical immunosensor based on nanoporous Au film platform and polyaniline-Au asymmetric multicomponent NP labels	0.10 U/mL	[83]
Carbohydrate antigen 19-9	ECL immunosensor based on graphene oxide	0.0005 U/mL	[84]
Squamos cell carcinoma antigen	Electrochemical immunosensor based on montmorillonite-polyaniline-gold nanocomposites labels	0.3 pg/mL	[85]
α-fetoprotein-AFP	Electrochemical immunosensor based on graphene sheets-thionine platforms and ferroferric oxide-HRP labels	4 pg/mL	[86]
Human chorionic gonadotrophin-hCG	Electrochemical sensor based on SWCNT on SPE	0.01 pg/mL	[87]
Histamine	Electrochemical immunosensor in based on graphene sheets-HRP initiated deposition of an insulated polymer film	0.5 pg/mL	[88]
Bovine interferon gamma	Electrochemical impedance immunosensor based on TiO_2 nanorods	0.1 pg/mL	[89]
Heart fatty-acid binding protein	Capacitive interdigitated immunosensor based on SAMs	0.836 ng/mL	[90]
Zearalenona	Label free amperometric immunosensor with trimetallic nanorattles(Au core and AgPt shell)	1.7 pg/mL	[91]
Microcystin LR	Fluorescence immunosensor based on Au nano-crossed surface enhacement	-	[92]
Penicilin G	Competitive amperometric immunosensor	10^{-10} M	[93]
Neurotensin	Label-free ECLsandwich immunosensor based on sombrero model with graphene-hyaluronate-luminol composite	0.001 pg/mL	[94]
Melanoma adhesion molecule antigen-CD146	Electrochemical immunosesnor based on amination graphene and mesoporous nano-Co_3O_4 sheet combined with AuNPs	3.4 pg/ml	[95]
Cancer biomarker miRNA 21	Electrochemical immunosensor based on graphene	3.12 pmol	[96]

Table 1. Immunosensors reported in scientific journals in the last decade

4.1. Immunosensors as diagnostic tools

The most important feature of immunosensors, the specificity of Ab-Ag reactions, makes them attractive candidates for developing diagnostic tools. Other advantages immunosensors may offer are: decreased medical cost by reducing, e.g., the hospitalization period due to their rapid analysis response; ease of use and miniaturization, making them suitable for "in-home" or "POC" use, thus improving the compliance of patients in follow-up therapy; and rapid

http://dx.doi.org/10.5772/60524

response, which can have a determining role in life-threatening diseases such as myocardial infarct or infectious diseases. Different molecules such as proteins, nucleic acids, metabolites, hormones, whole cells and various pathogens are bioanalytical targets for the development of diagnostic devices.

4.1.1. *Immunosensors for detection of proteins*

4.1.1.1. *Cancer biomarkers*

The high prevalence of cancer worldwide, together with its high recurrence and high mortality, creates an urge for rapid, reliable and sensitive detection methods, focusing the scientists' attention on proteins that are frequently investigated as cancer biomarkers. Cancer biomarkers, like all other biomarkers, are generally defined as "a characteristic that is objectively measured and evaluated as an indicator of normal biological processes, pathogenic processes, or pharmacological responses to a therapeutic intervention" [97].

Their aberrant expression in biological fluids (serum, saliva) or cells and tissues can be an indicator of the presence of neoplastic processes. Because their abnormal levels appear from the first stages of cancer, they can play an important role in early stage detection, improving the survival rate of patients. Generally, more than one cancer biomarker is overexpressed for the same type of cancer. Therefore, in order to avoid false positive results, a panel for simultaneous detection of several significant cancer biomarkers is usually required.

Several strategies will be presented below for the detection of cancer protein biomarkers, such as Mucin 1 (MUC1, CA 15-3), prostate-specific antigen (PSA), cancer antigen CA125 and carcinoembryonic antigen (CEA), with an emphasis on electrochemical immunosensors reported in the last years.

Mucins are glycoproteins, either secreted or membrane bound, that have a specific pattern of expression altered during tumour progression, recently investigated as new potent biomarkers in epithelial cancers [98]. Among them, MUC1 has been more extensively studied, being involved in tumour proliferation and metastasis [99], as well as in chemoresistance and alteration of drug metabolism [100, 101]. Overexpression of MUC1 has been associated with breast, ovarian, pancreatic, liver, colon or lung cancer.

An Ab-lectin sandwich assay coupled with optical detection was proposed by Park *et al*. Abs against MUC1 were immobilized on a gold sensing surface and the protein was captured between Abs and lectins conjugated with fluoro-microbeads. Two different lectins were used: *Sambucus nigra* agglutinin and peanut agglutinin lectins. The fluoro-microbead guiding chip contained multiple sensing and fluidic channels and exhibited LODs of 1.2 and 0.4 U/mL, respectively, over the clinically important range of 1.25 to 25 U/mL [53]. Quartz crystal microbalance (QCM) sensors have gained increased attention lately due to their simple operating principles and the absence of labelling. A QCM sensor based on ZnO nanorods was developed by Wang et al. The label-free detection method exhibited a fast response time of 10 s and high sensitivity of 25.34±0.67 Hz/scale (1 U/mL) in the concentration range of CA15.3 from 0.5 to 26 U/mL [54]. Two simple aptasensors based on gold nanoparticle (AuNPs) modified screen-printed electrodes (SPEs) were reported for sensitive detection of MUC1. A

SAM of thiolated Apt was firstly formed on the surface of AuNP-modified disposable electrodes. The quantitative detection of MUC1 protein was achieved by electrochemical impedance spectroscopy (EIS), by measuring the changes in the charge transfer resistance after binding various concentrations of Ag, and by differential pulse voltammetry (DPV) using the signal of intercalated methylene blue (MB), which decreased after protein binding. LODs of 3.6 ng/mL by EIS and 0.95 ng/mL by DPV were obtained, in the linear range of 0-10 ng/mL [55]. Another Apt sandwich-based sensor for MUC 1 detection was proposed, using Abs as capture probes and MB labels as detection probes. Specific Abs were immobilized on the poly(*o*-aminobenzoic acid) polymeric film (PABA) through the interaction between the COOH groups activated by the EDC/NHS chemistry and NH_2 groups of the Abs (Figure 5) [102].

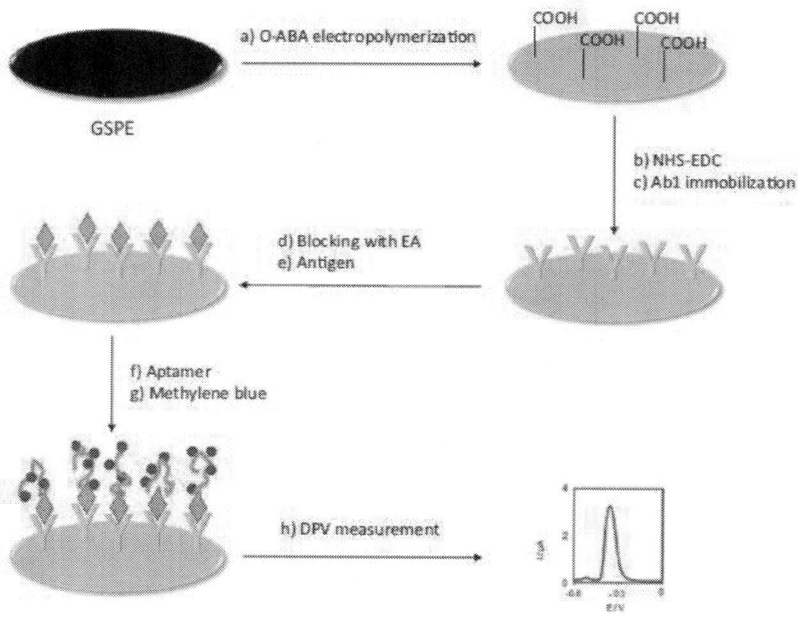

Figure 5. Preparation of sandwich aptasensor for MUC1 detection: (a) o-ABA polymerization; (b, c) MUC1 monoclonal mouse Ab immobilization onto PABA-modified graphite SPEs by carboxyl activation; (d) PABA free binding sites blocking with 10 mM ethanolamine; (e) incubation with human MUC1 of different concentrations; (f) specific binding of Apt with human MUC1 Ag (g) interaction of MB with G bases in Apt; (h) DPV measurements [102]. "Reprinted from J. Electroanal. Chem. 717-718, Taleat Z, Cristea C, Marrazza G, Mazloum-Ardakani M, Săndulescu R. Electrochemical immunoassay based on aptamer–protein interaction and functionalized polymer for cancer biomarker detection, 119-124, Copyright (2014), with permission from Elsevier"

After the reaction with Ag, the Apt binds to the immunocomplex which allows the accumulation of MB by interaction with the guanine bases of the Apt. DPV was then performed to measure the oxidation peak of MB. A linear response of the current with MUC1 concentration was observed in the range of 1-12 ng/mL with an LOD of 0.62 ppb [102]. Other strategies for

MUC1 detection include electrochemical aptasensors based on MB-labelled hairpin Apt, where the detection is based on the conformational changes of the labelled Apt after the binding of the protein [103] or on a dual signal amplification strategy, using poly(*o*-phenylenediamine)-AuNPs film for the immobilization of the captured Apt and AuNPs functionalized-silica/MWCNT core-shell nanocomposites as the tracing tag for the Apt detection [104].

PSA is produced by the prostate gland with normal levels ranging from 0 to 4 ng/mL, and it is overexpressed in prostate cancer. A label-free immunosensor based on white light reflectance spectroscopy for the determination of total- and free-PSA in human serum was reported. A two-site immunoassay was used, employing a biotinylated Ab in combination with streptavidin. High sensitivity was achieved with LOD of 0.2 and 0.15 ng/mL for total and free PSA, in the range of 0.5 to 100 ng/mL, and 0.5 to 20 ng/mL for total and free PSA, respectively [56]. Another interesting approach was used for the construction of a 3D label-free immunosensor for PSA-based crumpled graphene-gold nanocomposites (Figure 6). The graphene sheets decorated with AuNPs were prepared by aerosol spray pyrolysis. LOD of 0.59 ng/mL and a linear range of 0-10 ng/mL were achieved, due to the synergistic effect of graphene and Au, with increased biocompatibility, high level of bound Ag and fast electron transfer. The sensors also proved to be selective towards interferents such as alpha-fetoprotein, glucose, uric acid, bovine serum albumin and vitamin C [57].

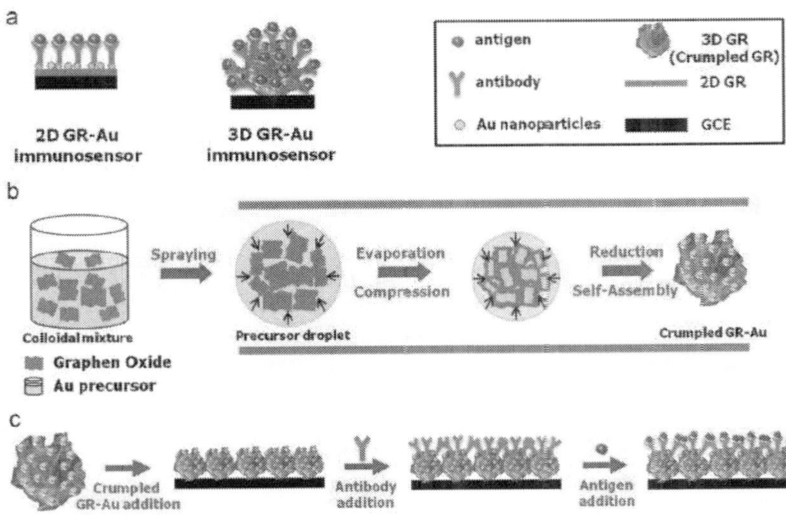

Figure 6. Illustration of (a) 2D GR-Au and 3D GR-Au electrodes, (b) the formation of crumpled GR-Au composites via aerosol spray pyrolysis and (c) fabricating step of 3D label-free PSA immunosensor using crumpled GR-Au composites [57]. "Reprinted from Biosens. Bioelectron. 63, Jang HD, Kim SK, Chang H, Choi J-W. 3D label-free prostate specific antigen (PSA) immunosensor based on graphene-gold composites, 546-551, Copyright (2015), with permission from Elsevier"

CA125 is a glycoprotein that exhibits abnormal levels in ovarian, breast, liver, lung and gastric cancer, serving as biomarker for the diagnosis and therapeutic monitoring of these diseases. Since its discovery in the ovarian epithelia by Bast *et al.* [105], CA 125 has served as a target for the development of several label-free [106] or labelled electrochemical or optical sensors [107, 108]. CEA is another tumour-associated Ag used as a biomarker for clinical diagnosis of ovarian, breast and colon carcinoma, and was also widely investigated as a target for the development of optical or electrochemical sensors [109, 110, 111].

Efforts have been made to improve cost effectiveness and miniaturize diagnostic devices, as well as in multiplexing detection of several cancer biomarkers, which would reduce analysis costs and increase diagnose accuracy. Special attention has been given to microfluidic devices that reduce the volume of used reagents, are easy to use, transport and store, and give a fast analytical response. Paper-based platforms have also been investigated as low-cost and simple to use materials. For example, a 3D origami microfluidic paper-based immunosensor for simultaneous detection of CA125 and CEA using metal ions as electroactive labels was developed (Figure 7).

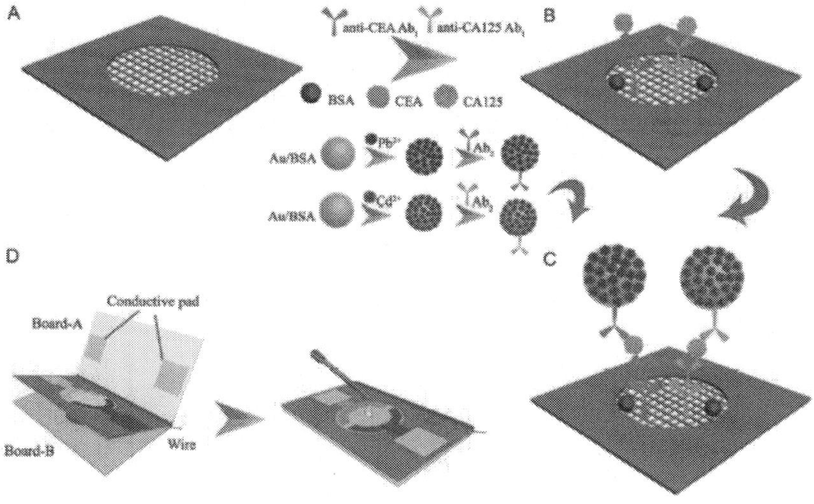

Figure 7. Origami EC immunodevice [58]. "Reprinted from Biosens. Bioelectron., 63, Ma C, Li W, Kong Q, Yang H, Bian Z, Song X, *et al*, 3D origami electrochemical immunodevice for sensitive point-of-care testing based on dual-signal amplification strategy, 7-13, Copyright (2015), with permission from Elsevier"

The support for the immobilization of captured Abs was constituted by a paper working electrode, modified with Au nanorods to ensure a large surface area and superior conductivity. For signal amplification, Au/BSA nanocarriers were used for loading Pd^{2+} and Cd^{2+}, which were conjugated with anti-CA125 and anti-CEA for a sandwich assay. DPV measurements were recorded and the quantification of the two different Ags was obtained, based on the different stripping peaks obtained for the oxidation of the two metal tracers. LOD of 0.08 pg/mL and 0.06 mU/mL were obtained and the developed microfluidic POC device was successfully applied for the quantification of the biomarkers in human serum [58].

An interdigitated electrode-based capacitive sensor was reported for multiple biomarker detection: CEA, epidermal growth factor receptor (hEGFR) and CA15-3, based on AuNPs amplification. CEA and epidermal growth factor receptor (hEGFR) could be successfully detected in the concentration range of 20-1000 pg/mL, while cancer antigen 15-3 (CA15-3) was detected in the range of 10-200 U/mL, with LODs of 20 pg/mL and 10 U/ml, respectively [112].

4.1.1.2. Cardiac biomarkers

Cardiovascular diseases include diseases of the heart and blood vessels and are a leading cause of death globally. The WHO estimated that deaths due to cardiovascular diseases in 2008 accounted for 30 % of worldwide deaths, and by 2030 the number of deaths per year will reach 23.3 million [113]. The main risk factors include smoking, alcohol abuse, hyperlipidaemia, high blood pressure, diabetes, physical inactivity and unhealthy diet. Myocardial infarct and stroke are acute events caused by obstructions of blood vessels or rupture of brain blood vessels. Rapid diagnosis in this case is crucial for the survival of patients. Beside signs and symptoms, current diagnosis of myocardial infarction is based on ECK and measurements of cardiac enzymes, while stroke is usually diagnosed using brain computed tomography, magnetic resonance imaging or carotid ultrasound. Other methods for diagnosis of cardiac conditions include echocardiography, chest X-ray, angiography or catheterization. Although ECK remains the main management method, some patients show normal ECK even in the presence of cardiopathies, and thus ECK may be inconclusive. Several markers exhibit elevated blood levels in the presence of heart conditions serving as diagnostic tools with important prognostic significance. As in the case of cancer, simultaneous multianalyte detection is preferred.

Myoglobin is released from injured muscle tissues and is a potential marker used in the diagnosis of myocardial infarction. However myoglobin is not specific for cardiac damage, and therefore it is usually used together with other biomarkers such as troponin or creatinine kinase MB subform (Ck-MB). Troponin I or T are proteins with important roles in muscle contraction that also detect cardiac damage as they are released in blood in elevated levels in case of cardiac muscle cell death. Ck-MB is an enzyme of the creatinine phosphokinase family specific to the heart cell; increased levels are present in case of heart muscle injuries. Serum levels of these markers rise rapidly within the first hours after a heart attack (myoglobin being the first to increase) reaching a maximum after several hours, depending on the size of the infarct [114]. Due to the fluctuation of the levels of cardiac markers, real-time monitoring is important for accurate measurements.

An SPR-based sensor for continuous monitoring of myoglobin levels in human serum was reported [59]. A recyclable Ab was immobilized at the surface of a label-free sensor that can quantify Ab-Ag reaction in real time. Two flow channels were used, one with specific Ab and the second, serving as control, with non-specific Ab. Buffered solutions and spiked serum samples of different concentration were then injected for the formation of immunocomplexes. The problems associated with the continuous sensing of myoglobin in human serum were overcome as follows: the non-specific binding of the analyte was avoided by pre-treating the sample with a mixture of detergents, specifically sodium dodecyl sulphate and P20, and adjusting the micelle size and net charge; and the aggregation of serum components was avoided by inactivating certain serum components by chelation of heavy metals. The sensor was able to detect the varying concentration of myoglobin over about 8 h with periodic one-point calibration every 3 h, with an LOD of 31.0 ng/mL, in the range from 31 ng/mL to 2000 ng/mL [58]. An electrochemical sensor for simultaneous detection of cardiac markers Troponin I (cTnI) and C-reactive protein (CRP), a known marker for inflammation and cardiac damage, has been reported based on a poly(dimethylsiloxane)-AuNP composite microfluidic system. Captured Abs for the two proteins were conjugated with CdTe and ZnSe QDs and square-wave anodic stripping voltammetry was used to detect the metal ions from the QDs in order to quantify the amount of bonded proteins. A linear range was obtained between 0.01 and 50 μg/L and 0.5 and 200 μg/L, with a very low LOD of approximately 0.004 μg/L and 0.22 μg/L, for cTnI and CRP, respectively [60]. Another target biomarker for cardiovascular disease is thrombin, a serine protease involved in blood coagulation, which acts as both a procoagulant (converting soluble fibrinogen in insoluble fibrin) and anticoagulant factor. It is also involved in the process of atherosclerosis and its cardiovascular complications [115]. The first Apt for thrombin, synthesized via SELEX in 1992, has a G-quadruplex structure and binds exosite I of thrombin [116]. A highly specific aptasensor was developed in a sandwich-type assay using planar Hall magnetoresistive transducer in combination with superparamagnetic labels. The sensor surface was modified with thiolated Apt to capture the thrombin added in the next step. Finally, a biotin-labelled secondary Apt was added to bind the thrombin-primary Apt complex and streptavidin-magnetic nanoparticles were used to measure the magnetoresistance signal. A linear response to thrombin concentration in the range of 86 pM-8.6 mM and an LOD as low as 86 pM was achieved using low sample volumes of only 2uL [61]. Interestingly, a simple Apt-target-Apt sensor was elaborated using the signal obtained after the electro-oxidation of intercalated MB and ferrocene for the quantification of thrombin. MB-tagged Apts were introduced by hybridization with complementary DNA sequences immobilized on the surface of the electrode. Thrombin was then added and its interaction with the Apt led to conformational changes of MB-Apt, driving the MB tag far from the electrode surface. A second Apt, tagged with ferrocene on both sides, was added to interact with thrombin, and the Apt's conformational changes allow the two Fc tags to be close to the electrode surface. These conformational changes resulted in a decrease in the oxidation peak current of MB (I_{MB}) and an increase in that of Fc (I_{Fc}), and the logarithmic value of I_{Fc}/I_{MB} was shown to be linear with the logarithm of thrombin concentration. An LOD of 170 pM was achieved, in the linear range of 1 nM to 600 nM [62]. Several label-free aptasensors for thrombin detection coupled with electrochemical [117, 118] or optical detection [119, 120] were reported.

4.1.2. Immunosensor for detection of metabolites

The detection of metabolites has gained increasing attention in recent years for the diagnosis of various diseases, metabolites being correlated with different pathologies. Glucose, for example, exhibits high levels in patients with diabetes and its monitoring plays a crucial role in the management of the disease. The devices used for the detection of glucose are mainly constituted by enzymatic sensors. The well-known glucometer accounts for 85 % of the market of commercially available biosensors [121]. Other metabolites of great interest are cholesterol and triglycerides, due to their involvement in atherosclerosis and in the pathology of cardiovascular diseases. Creatinine is a metabolite of creatine, eliminated by kidneys and excreted in urine, and it has become the preferred clinical marker for the diagnosis of renal filtration dysfunctions. The development of POC diagnostic devices could have an important impact on the improvement of the quality of life of patients suffering from metabolic disorders. Several POC devices for the detection of creatinine are already commercially available: IRMA TRUpoint (ITC, Edison, NJ), Radiometer ABL800 FLEX (Radiometer A/S, Bronshoj, Denmark), StatSensor (Nova Biomedical, Waltham, MA), and i-STAT (Abbott Diagnostics, East Windsor, NJ).

An interesting approach for detecting low-density lipoprotein (LDL) cholesterol was proposed based on immobilization of a large number of Abs on AuNP-silver chloride/polyaniline composite film. Abs specific to apolipoprotein 100 were chosen that recognized and bound apolipoprotein B present in the phospholipidic coats of the structure of LDL. EIS studies were performed to monitor the formation of Ab-Ag complex using $[Fe(CN)_6]^{4-/3-}$ redox probe. An increase in the charge transfer resistance was observed after the binding of LDL, due to its low conductivity coupled with the repulsion between negatively-charged phospholipid units and the negative charge of the redox probe. This simple, label-free immunosensor allowed for the detection of low concentrations of LDL of 0.34 pg/mL, in the linear range of 0-33.5 pg/mL [63]. With a similar approach, an impedimetric label-free sensor was developed using aminated reduced graphene oxides for the immobilization of anti-apolipoprotein B on the electrode surface. The aminated graphene oxides were synthesized chemically and deposited on the surface of the transducers through electrophoretic deposition. EDC/NHS was then used for covalent attachment of Abs via amide bond formation, followed by the addition of LDL solutions. The change in the resistance after the formation of Ab-Ag complexes allowed for the detection of higher concentrations of LDL of 5mg/dL [64]

Besides metal nanoparticles and carbon-based nanomaterials, conductive polymers have been widely applied, for many years, as immobilization platforms for the development of immunosensors, due to their low cost, ease of preparation, and most importantly good biocompatibility. A simple electrochemical sensor was developed for the detection of creatinine in whole blood obtained from patients with renal transplant. The surface of a 16-gold electrode array was covered with a film of polypyrrole via electropolymerization, by applying a cyclic square-wave electrical field. Creatinine was embedded in the film in the same step. Anti-creatinine Abs labelled with HRP were mixed with a low volume of blood sample containing creatinine for the formation of immunocomplexes and added to the array for competitive reaction with the immobilized Ag. The decreased current of the HRP-Ab is proportional to the level of creatinine in the samples in the linear range of 0 mg/dL to 11.3 mg/dL with an LOD of 0.46 mg/dL. The method has the advantage of short response time <5 min and low sample volume (40 uL) [65].

4.1.3. Immunosensors for detection of pathogens

Infectious diseases have increased rapidly in recent years, causing high fatality rates due to incorrect diagnosis, delay in treatment and various complications. Worldwide, infectious diseases account for nearly 40 % of the total 50 million annual estimated deaths, and represent the major cause of death in many developing countries [122].

Infections are caused by pathogenic microorganisms, such as bacteria, viruses, fungi or parasites. Most of the infections are highly contagious and can spread rapidly with the risk of causing epidemic or pandemic. Rapid and portable devices for the detection of pathogens are of paramount importance. Rapid screening for differentiation between viral and bacterial infection is extremely important for diagnosis and correct management of disease, limiting the overuse of antibiotics and reducing antibiotic resistance. A correct, rapid diagnosis may reduce the need for or period of hospitalization, with a high impact on medical costs. Current diagnosis is mostly based on cell culture and microscopic methods, serology or nucleic amplification testing, which are time-consuming. Most of the existing POC tests consist of immunoassays: agglutination, immunochromatographic and immunofiltration tests [123]. The development of portable devices that can be used at home or in doctors' offices may permit a rapid diagnosis of various infections, allowing quick establishment of treatment and fast recovery by the patients. POC devices also have the advantages of being used in biological fluids with no need for complex sample pre-treatment.

Most of the sensors reported in the literature for the detection of pathogens are based on electrochemical techniques, due to their simplicity, low cost, rapid response and suitability for developing in small portable devices. The majority of these devices employ nucleic acids as recognition elements for the pathogenic agent, based on hybridization processes with a complementary DNA strand immobilized onto transducers, intercalation of redox mediators in double-stranded DNA, or label-free detection based mainly on impedimetric measurements.

Influenza is the most common acute respiratory infection, accounting for 52,294 deaths in the USA in 2011 [113]. It is caused by influenza type A, B and C viruses; the type A virus being responsible for the annually recurring epidemics. Mutations can also occur (e.g., avian or swine flu), raising health concerns.

A simple electrochemical DNA sensor was proposed based on avidin-biotin reaction for the immobilization of DNA capture probes on the surface of glassy carbon electrodes (GCEs). The surface of the electrode was firstly functionalized with carboxylic groups using 4-carboxyphenyl diazonium salts. Avidin was then bound, after ethylene carbodiimide/*N*-hydroxysuccinimide activation. Afterwards biotinylated DNA probes were immobilized through avidin-biotin interaction followed by the hybridization with target DNA. The hybridization process between the probe (5-biotin-ATGAGTCTTCTAACCGAGGTCGAA-3) and the target DNA (5-TTCGACCTCGGTTAGAAGACTCAT-3, a fragment of influenza virus gene sequence) was investigated by CV. A dynamic range between 10^{-13} M and 10^{-10} M was achieved with an LOD of 8.51 10^{-14} M [66].

Another label-free electrochemical immunosensor was reported for swine flu influenza virus (SIV), based on a self-assembled SWCNT network modified with anti-SIV Abs. The change in resistance upon binding of the Ag permits LODs of 180 TCID 50/mL (50 % Tissue Culture Infective Dose) to be achieved [124].

Piezoelectric techniques have attracted interest lately. A surface acoustic sensor based on Love waves was recently developed for the detection of influenza virus type A. SiO_2coated piezoelectric $LiNbO_3$ wafers were used to bind Ags against haemagglutinin (HA) of influenza A H1N1 subtype. After a silanization process of the wafers using tri-ethoxy-silyl-butyraldehyde and tri-ethoxy-silyl-undecanal ethylene-glycol-acetal, and the reaction of the silan groups with EDC/NHS, the substrates were immersed in a solution containing anti-HA Ags. Phycoerythrin-conjugated Abs were immobilized for fluorescence microscopy analysis in order to check the successful binding of HA. Non-conjugated Abs were used for surface acoustic wave testing giving an LOD of 1ng/mL [67].

Legionella pneumophila is a bacterium that causes Legionnaires' disease and Pontiac fever, life-threatening diseases with high mortality. An electrochemical immunosensor based on ZnO nanorods was reported on its detection. The nanorods were prepared by hydrothermal method at low temperature, and used to immobilize primary Abs by electrostatic interactions. The Ag was then captured by the recognition of its peptidoglycan-associated lipoprotein. A sandwich ELISA-like assay was performed by adding secondary Abs labelled with HRP in the next step, and the electroactivity of the product of the enzymatic reaction was investigated. A dynamic range of 1-5000 pg/mL was obtained in this case with an LOD of 1 pg/mL [68].

Sexually-transmitted diseases (STDs) are viral or bacterial infections transmitted from one human to another through unprotected sexual contact. The most common viral STDs are HIV/AIDS, hepatitis B, herpes and human papillomavirus, while for bacterial infections the highest numbers of cases are diagnosed for gonorrhoea, chlamydia and syphilis. The incidence of these diseases is worryingly high; approximately 20 million people will contract an STD each year in the USA, and half of the infections occur in the age range 15-24 [125].

A simple device was developed for the detection of *Neisseria gonorrhoea* by electropolymerization of polyaniline films on indium-tin-oxide plates. Avidin was then immobilized on the films after EDC/NHS activation to bind biotinylated ssDNA. The hybridization process with the target DNA and the amount of bound DNA was investigated by DPV and EIS measurements. DPV measurements have been performed using two methods: the first is based on monitoring of the oxidation peak of guanine in the capture probe, which decreased after the binding of complementary DNA; the second approach used the oxidation peak of intercalated MB for the monitoring of the hybridization process. The MB peak disappears after hybridization with the complementary oligomer, due to steric inhibition of MB packing between the double helix of the hybrid. EIS measurements confirmed the hybridization: an increase in the impedance was observed due to electrostatic repulsion between DNA and the redox probe. An LOD of 0.5 10^{-15} M was obtained [69].

Human immunodeficiency virus (HIV) causes acquired immunodeficiency syndrome (AIDS) and is transmitted mainly by unprotected sexual intercourse or use of unclean syringes [126].

A rapid, label-free capacitive immunosensor has been developed for real-time analysis of p24 Ag, a HIV-1 capsid protein that is detectable in serum shortly after infection. Silicon wafers were nanopatterned with gold via photolithography. A polytyramine film was then electro-deposited on the surface of the transducer, which chemisorbed AuNPs through the amine groups. Anti-HIV-1 p24 monoclonal Abs were then immobilized to capture the Ag from injected buffered or plasma samples, and the capacitance change was investigated. A linear relationship was observed from 2.4 10^{-6} to 2.4 10^{-3} pg/mL with an LOD of 7.9 10^{-8} pg/ml [70].

A simple electrochemical ELISA-type sensor for the detection of HIV-1 and HIV-2 Abs was recently reported. HIV-1 gp41 and HIV-2 gp36 Ags were immobilized at the surface of the transducers followed by the addition of anti-HIV Abs and IgG labelled with alkaline phosphatase. The substrate *p*-aminophenyl phosphate was added to produce the redox-active product, *p*-aminophenol. It was observed that its oxidation peak increased linearly over a wide Ab concentration range (0.001-1μg/mL), with an LOD of 1 ng/mL (6.7 pM) for both HIV-1 and HIV-2 [127].

Hepatitis is an inflammatory condition of the liver affecting many people worldwide, caused by various types of hepatitis viruses (A-E). Hepatitis B and C are chronic diseases causing cirrhosis and liver cancer [128].

An optical immunosensor for the detection of hepatitis C virus Abs was developed using the photoimmobilization method [129]. For this purpose, a photoactivatable electrogenerated polymer film was deposited on the surface of conductive fibre optics in order to link the bioreceptor to the fibre tip through light mediation, and a solution containing HCV-E2 envelope protein Ag was added and illuminated with UV light. Due to the photochemical reaction, a thin layer of the Ag is covalently attached to the polypyrrole-benzophenone-modified surface that is able to bind anti-E2 Abs from samples. Chemiluminescence measurements allowed anti-E2 Ig detection at a low titer [129].

A disposable impedimetric sensor was reported for the detection of *Streptococcus*, based on gold SPEs modified with polytyramine film. Avidin was bound through the polymer amine groups, and the biotinylated Ab was then immobilized. The sensor was tested in (50 % (v/v) human saliva, resulting in a detection range of 100 to 105 cells per 10μL and 100 to 104 cells per 10 μL of bacteria, for cumulative and single-shot incubations, respectively [130].

4.2. Immunosensors for drug safety

Therapeutic monitoring is important for drug adjustment to achieve optimum efficiency and minimal toxicity of the drug. Ideally, the concentrations of drugs should be measured at the site of the action - the receptor - but due to its inaccessibility, drug concentrations are measured in bodily fluids such as serum, plasma, saliva, urine or cerebrospinal fluids [131]. The most commonly used techniques for establishing drug levels for drug monitoring such as HPLC, GC-MS, LC-MA, radioimmunoassay, chemiluminescence, etc., are time- and reagent-consuming, require qualified personnel and complex sample pre-treatment, and are expensive. Therefore, there is an urge to develop new, minimally invasive devices for drug detection that could be easily used in homes, laboratories and clinics. Special attention should be given to targeting drugs with a narrow therapeutic range, which may be easily over- and under-dosed.

A simple ECL sensor was developed for the detection of clenbuterol, a drug used for pulmonary diseases, incorporating AuNPs and QDs for enhanced signals. The AuNP/ovalbumin-clenbuterol/anti-clenbuterol-QDs sensor was characterized using EIS and ECL emission achieving LOD of 0.0084 ng/mL in the range of 0.02-50 ng/mL [71]. An impedimetric sensor based on 3-mercaptopropionic acid SAMs was reported for the sensitive and selective detection of ketamine, an anaesthetic drug. The immobilization of Abs via SAM and the recognition of the target Ag were evaluated by the change in the charge transfer resistance, allowing for the detection of ketamine levels of 0.41 pmol/L [72]. EIS has also been employed for the detection of ciprofloxacin by a label-free immunosensor, which is widely used for the treatment of pulmonary, urinary or digestive infections. Polypyrrole-NHS electrogenerated films were used to immobilize anti-ciprofloxacin Abs, with an LOD of 10 pg/mL being obtained [73].

Developing new methods for drug quantification is also important in the quality control of drug formulations in the pharmaceutical industry. A label-free immunosensor based on the graphite oxide immobilization platform was proposed to quantify acetaminophen from pharmaceutical formulations. Anti-acetaminophen Abs were linked via amide groups by NHS/EDC chemistry on graphene oxide-modified SPEs. The preparation steps were investigated by several electrochemical techniques such as SWV, EIS and EQCM, and the LOD found by SWV measurements was 0.17 μM [74].

Another important social and health issue is **drug abuse and doping** [132]. The development of fast, sensitive and specific screening tests is gaining increasing interest, mainly in sport, with special attention being given to the development of portable on-site detection methods for doping agents. Several devices for the detection of such drugs from saliva are already commercially available, for example, a gold particle-based lateral in-flow immunoassay named Oratect is available for the detection of marijuana (THC), cocaine, amphetamines, opiates, methamphetamines including MDMA (ecstasy) and either phencyclidines or benzodiazepines. Some of the most commonly used drugs for sport doping are beta-blockers, steroidal hormones, growth hormones, theophylline and derivatives, peptides and methamphetamines.

A simple and sensitive electrochemical sensor was developed for the simultaneous detection of two illicit drugs: morphine and methamphetamine. The simple approach employed SAM of 3-mercaptopropionic acid on a gold electrode that served to bind Abs specific for the targeted drugs. CV and EIS were employed as electrochemical techniques, revealing a decrease in the current together with an increase in the impedance after Ab binding and immunocomplex formation, due to the hindrance of electron transfer from the redox probe to the surface of electrodes. The sensors were able to detect morphine and methamphetamine in the linear range of 4 to 80 pg/L and 20-200 pg/L, respectively, with an LOD of 0.27 pg/L and 10.1 pg/L. The sensor is stable, reusable and may also be used in serum samples [133]. Another anti-doping device has been reported for the detection of human growth hormone, which is usually associated with anabolic steroids used to improve sporting performance. An SPR platform-based sandwich assay was developed using avidin-biotin coupling for the proper orientation of primary Abs, and the obtained sensor allowed the simultaneous detection of both hCG isoforms (20K and 22K) in concentrations as low as 0.9 ng/mL [134].

4.3. Immunosensors for food safety

Detection of food contaminants is a major issue, since these contaminants may have serious consequences for food safety and human health, as well as for medical and economical costs. The rapid and sensitive detection of food contaminants is important to prevent intoxication and provide efficient treatment [135]. Diseases caused by food contaminants may vary widely from several million cases to 81 million cases in the USA, with bacterial foodborne outbreaks accounting for 91 % of the total outbreaks [122].

The most common foodborne pathogens include *Escherichia coliO157:H7, Salmonella typhimurium, Listeria monocytogenes, Staphylococcus aureus, Clostridium perfringens, Enterobacter sakazakii, Campylobacter jejuni and Yersinia enterocolitica*. The most common methods for their identification, such as the conventional cell culture and microscopy, as well as the more recently applied methods such as ELISA and PCR, are time-consuming, and therefore immunosensors, through their advantages, are promising tools. An electrochemical immunosensor for the simultaneous detection of *Escherichia coli* O157:H7 (E. coli O157:H7) and *Enterobacter sakazakii (E. sakazakii)* was reported using carbon screen-printed low-density arrays modified with MWCNTs/sodium alginate (SA)/carboxymethyl chitosan (CMC) composite films. The film served for the immobilization of HRP-labelled Abs against the two pathogens. The array exhibited a linear response from 10^4 to 10^{10} cfu/mL with LOD of 4.57 10^4 cfu/mL for *E.coli* and 3.27$10^4$ cfu/mL for *E. sakazakii* [136].

Salmonella is another dangerous foodborne pathogen causing infections with various clinical symptoms depending on the serotype, ranging from diarrhoea, fever and abdominal cramps to more severe illnesses such as Reiter's syndrome that can lead to chronic arthritis. About 1.4 million salmonellosis cases occur annually in the United States, resulting in about 16,000 hospitalizations with nearly 600 deaths [137].

An impedimetric immunosensor was developed for the detection of *Salmonella thyphurium* in milk samples. Gold electrodes were functionalized with: single 11-amino-1-undecanethiol (MUAM); a mixture of MUAM and 6-mercapto-1-hexanol (MCH); and a mixture of MUAM, MCH in the presence of triethylamine (TEA) to prevent formation of interplane hydrogen bonds among amine-terminated thiols. Specific Abs were then immobilized via SAM that served to capture the Ag from samples, resulting in changes of the resistivity of the solution. Signal amplification was obtained by performing the measurements in the sample media, which offered several advantages such as: higher response in shorter time due to simultaneous proliferation of the viable bacteria cells in the tested samples, insensitivity to the presence of dead cells, and elimination of the centrifugation and washing steps that are normally used to isolate bacterial cells. The results showed the importance of MCH in SAMs while the addition of TEA had rather negative effects. For a detection time of 2 h, the MUAM-MCH-based sensor provided reliable analytical signals for a concentration of three orders of magnitude lower than the infectious dosage of *S. typhimurium* [138].

Listeria monocytogenes can be transmitted to humans through consumption of contaminated food, such as smoked fish, poultry, meat and dairy products, and ready-to-eat foods [139, 140]. An impedance immunosensor for the detection of *Listeria monocytogenes* using TiO_2 nanowire bundle microelectrodes as the Ab immobilization platform was reported. The TiO_2 nanowire bundle was prepared through a hydrothermal reaction of alkali with TiO_2 powder and connected to the gold microelectrodes by mask welding. Impedance changes were investigated after the formation of nanowire-Ab-Ag complexes and correlated to bacterial number. The immunosensor was able to specifically detect 10 cfu/mL *Listeria monocytogenes* [141]. Ochratoxin A (OTA) is a thermostable mycotoxin produced by species of *Aspergillus* or *Penicillium* fungi, which contaminates a high variety of food stuffs, such as cereals, dried fruits, coffee, grapes, wine and beer, with nephrotoxic and carcinogenic effects on mammalians [142].

An Apt-based biosensing approach based on an evanescent wave all-fibre (EWA) platform with embedded dethiobiotin-modified fibres was recently reported In a simple target-capturing step using Apt-modified magnetic microbeads, signal probes (represented by Apt-complementary oligonucleotide) conjugated with streptavidin are released and further detected by an EWA biosensor via facial dethiobiotin streptavidin recognition. A reusable, sensitive sensor was thus obtained that can maintain its response after being reused 300 times, with an LOD of 3 nM in the linear range of 6 nM to 500 nM [75].

OTA was also detected in wine samples using two indirect competitive ELISA-type immunoassays coupled with electrochemical methods. pAb and mAb were investigated for the capture of the Ag and different enzymatic labels, horseradish peroxidase (HRP) and alkaline phosphatase (ALP) were used and compared. The results showed that although similar performance was observed in buffer solution for the HRP-Ab-based strategy and the ALP-conjugated Ab sensor, when it comes to wine samples, the signal for the ALP-based sensor was affected by interferences. LODs of 0.7 ng/mL and 0.3 ng/mL were obtained for HRP- and ALP-labelled sensors [143].

4.4. Immunosensors for environmental control

Environmental air pollution and water control monitoring have become areas of intense interest in the research field due to the serious consequences for public health. Gas sensors are generally used for the detection of indoor and outdoor air pollutants such as carbon monoxide, formaldehyde, ethylene oxide, nitric oxide, hydrogen sulphide or chloride. Many contaminants infiltrate the soil, arriving in wastewaters and drinking water. These contaminants belong to various classes such as hormones, antibiotics, surfactants, endocrine disruptors, human and veterinary pharmaceuticals, X-ray contrast media, pesticides and metabolites or explosives, and are typically detected at concentrations ranging from 1 ng/L to 1 μg/L, reaching in some cases as much as 100 μg/L [144]. Water pollution is harmful to human, animal and water life, resulting in death of aquatic animals, disruption of food chains and diseases. The effects of consuming contaminating water can range from mild effects (diarrhoea, vomiting and nausea) to more severe diseases or even death. Pesticides, for example organochlorines, organophosphates, carbamates, pyretroides and atrazine, can cause neurological effects, asthma, allergies and endocrine disruption, and even cancer and reproduction problems.

Other contaminants like nitroaromatic explosives, found in soil and groundwater near military zones, are toxic and carcinogenic to living organisms, causing aplastic anaemia, toxic hepatitis, hepatomegaly, cataract and skin irritation.

Standard methods for water quality analysis involve intensive sampling regimes and multistep sample preparation, requiring manual inputs which prohibit their integration into continuous monitoring systems, resulting in the need for developing new monitoring methods that can be easily integrated into water flow systems [145] such as electrochemical immunosensors.

Kanso's group [76] reported on the development of electrochemical immunosensors based on magnetic beads (MBs) for the detection of oestradiol and ethinyl oestradiol in wastewaters. Oestradiol and ethinyl oestradiol are water contaminants found at levels of nanomolar range, which act as endocrine-disrupting chemicals, interfering with the endocrine systems and causing breast and testis cancers in humans. MBs were chemically modified with four different types of synthetic oestrogen derivative that contained carboxylic or amino groups. The assay was based on the competition between free oestrogen in the sample and immobilized oestrogen using a primary Ab labelled with ALP. Using SWV, an LOD of 1 and 10 ng/mL was obtained [76]. An interesting approach for a high-throughput microfluidic immunosensor for the detection of explosives in water has been presented by Adams et al. Immunoaffinity and fluorescence detection assays were coupled on a microfluidic device containing 39 parallel microchannels with immobilized Abs against TNT. With a total volume throughput of less than 6 mL/min, the assay is performed 60 times faster than the usual immunoassay, achieving an LOD of 0.01 ng/mL for TNT [77].

An electrochemical immunosensor for the simultaneous detection of pesticides, endosulphan and paraoxon was reported. Microcontact printing was used to pattern an SWCNT forest on the surface of transducers using diazonium salt chemistry. Redox probes of 1,1′-di(amino-methyl) ferrocene (FDMA) and pyrroloquinoline quinone (PQQ) were linked to the nanotubes, followed by Ab immobilization. Pesticides were quantified based on the current change of the redox probes upon the formation of immunocomplexes: the electrochemical signal of FDMA allowed the detection of endosulphan in the range of 0.05 to 100 ppb with an LOD of 0.05 ppb, while PQQ signal allowed paraoxon detection in the range from 2 to 2500 ppb with an LOD of 2 ppb [78].

5. Conclusions and perspectives

New approaches using magnetic nanoparticles, quantum dots, carbon nanotubes, noble metal nanoparticles or hybrid nanomaterials, either as labels or immobilization platforms, have increased the number of electrochemical immunosensors and their applications. These materials offer important advantages, such as good biocompatibility, which maintains bioreceptor activity, a high surface-to-volume ratio, which increases the amount of immobilized bioreceptor, and unique conductivity and optical proprieties. Their high selectivity and sensitivity, as well as the possibility to miniaturize these systems, have made the use of electrochemical immunosensors for *in vivo* applications possible.

http://dx.doi.org/10.5772/60524

Although numerous developments have been made in the field of immunosensors, novel techniques are still needed to improve the selectivity, sensitivity and simplicity of these devices, ready to meet the exigent requirements of clinical diagnosis or industry. The need for miniaturization and integration on an electronic platform able to detect the biomarkers or pathogens in real time are other challenges in this active area of research. In the past decade, remarkable progress in nanotechnology techniques has been made, together with new strategies using nanoparticles and nanostructured surfaces for electrochemical detection of a large variety of proteins. Low-cost, reliable, portable, multiplexed protein detection devices, especially those coupled with microfluidics, could ensure accurate detection of panels of cancer biomarkers in blood, urine or saliva. The application of immunosensors in drug abuse or doping as well as in the quantification of pathogens was achieved using various immobilization nanoplatforms for immobilization, and also using multienzyme labelling on metal nanoparticles, magnetic beads and carbon nanotubes or graphene, allowing for the ultrasensitive detection of target analytes.

The integration of electrochemical immunoassays into microfluidic platforms will create a versatile platform for the fabrication of devices for clinical diagnostics. The development and deployment of these systems could ultimately lead to more rapid clinical decision-making, reducing patient stress and health care costs.

Author details

Cecilia Cristea, Anca Florea, Mihaela Tertiș and Robert Săndulescu*

*Address all correspondence to: rsandulescu@umfcluj.ro

Analytical Chemistry Department, "Iuliu Haţieganu" University of Medicine and Pharmacy Cluj-Napoca, Romania

References

[1] Luppa PB, Sokoll LJ, Chan DW. Immunosensors-principles and applications to clinical chemistry. Clin. Chim. Acta. 2001;314(1-2):1-26.

[2] Ullah MF, Aatif M. The footprints of cancer development: Cancer biomarkers. Cancer Treat. Rev. 2009;35(3):193-200.

[3] Liang S-L, Chan DW. Enzymes and related proteins as cancer biomarkers: A proteomic approach. Clin. Chim. Acta. 2007;381:93-97.

[4] Findlay JW, Smith WC, Lee JW, Nordblom GD, Das I, De Silva BS, Khan MN, Bowsher RR. Validation of immunoassays for bioanalysis: A pharmaceutical industry perspective. J. Pharmac. Biomed. Anal. 2000;21:1249-1273

[5] Hawkridge AM, Muddiman DC. Mass spectrometry- based biomarker discovery: Toward a global proteome index of individuality. Annu. Rev. Anal. Chem. 2009;2:265-277.

[6] Hanash SM, Pitteri SJ, Faca VM. Mining the plasma proteome for cancer biomarkers. Nature. 2008;452:571-579.

[7] Wilson DS, Nock S. Recent developments in protein microarray technology. Angew. Chem., Int. Ed. 2003;42:495-500.

[8] Lee HJ, Wark AW, Corn RM. Microarray methods for protein biomarker detection. Analyst. 2008;133:975-983.

[9] Bensmail H, Haoudi A. Postgenomics: Proteomics and bioinformatics in cancer research. J. Biomed. Biotechnol. 2003;2003:217-230.

[10] Rasooly A, Jacobson J. Development of biosensors for cancer clinical testing. Biosens. Bioelectron. 2006;21:1851-1858.

[11] Chikkaveeraiah BV, Bhirde AA, Morgan NY, Eden HS, Chen X. Electrochemical immunosensors for detection of cancer protein biomarkers. ACS Nano. 2012;6(8): 6546-6561.

[12] Healy DA, Hayes CJ, Leonard P, McKenna L, O'Kennedy R. Biosensor developments: Application to prostate-specific antigen detection. Trends Biotechnol. 2007;25:125-131.

[13] Oswald B, Lehmann F, Simon I, Terpetschnig E, Wolfbeis OS. Red laser-induced fluorescence energy transfer in an immunosystem. Anal. Biochem. 2000;280(2):272-277.

[14] Diaconu I, Cristea C, Hârceagă V, Marrazza G, Berindan-Neagoe I, Săndulescu R. Electrochemical immunosensors applied in diagnostic and monitoring of breast and ovarian cancer. Clin. Chim. Acta. 2013;425:128-138.

[15] Taleat Z, Ravalli A, Mazloum-Ardakani M, Marrazza G. CA125 Immunosensor based on poly-anthranilic acid modified screen-printed electrodes. Electroanal. 2012;25(1):269-277.

[16] Mulvaney SP, Mattoussi H, Whitman LJ. Incorporating fluorescent dyes and quantum dots into magnetic microbeads for immunoassays. BioTechniques. 2004;36(4): 602-609.

[17] Hong B, Kang KA. Biosensor compatible, nanogold particle fluorescence enhancer for optical immunoassay. Biosens. Bioelectron. 2006;21(7):1333-1338.

[18] Singh AK, Kilpatrick PK, Carbonell RG. Application of antibody and fluorophore-derivatized liposomes to heterogeneous immunoassays for D-dimer. Biotechnol. Prog. 1996;12(2):272-280.

[19] Pei X, Zhang B, Tang J, Liu B, Lai W, Tang D. Sandwich-type immunosensors and immunoassays exploiting nanostructure labels: A review. Anal. Chim. Acta. 2013;758:1-18.

[20] Mistry KK, Layek K, Mahapatra A, Roy Chaudhuri C, Saha H. A review on amperometric-type immunosensors based on screen-printed electrodes. Analyst. 2014;13(10): 2289-2311.

[21] Sino Biological Inc. Elisa Enciclopedia [Internet]. Available from: http://www.elisa-antibody.com [Accessed: 20.01.2015]

[22] Jiang H, Weng X, Li D. Microfluidic whole-blood immunoassays. Microfluid. Nanofluid. 2011;10(5):941-946.

[23] Zhang B, Tang D, Liu B, Chen H, Cui Y, Chen G. GoldMag nanocomposite-functionalized graphene sensing platform for one-step electrochemical immunoassay of alpha-fetoprotein. Biosens. Bioelectron. 2011;28(1):174-180.

[24] Li Q, Tang D, Tang J, Su B, Chen G, Wei M. Magneto-controlled electrochemical immunosensor for direct detection of squamous cell carcinoma antigen by using serum as supporting electrolyte. Biosens. Bioelectron. 2011;27(1):153-159.

[25] Marco M, Gee S, Hammock B. Immunochemical techniques for environmental analysis I. Immunosensors. TrAC Trends Anal. Chem. 1995;14(7):341-350.

[26] Rongen HAH, Bult A, van Bennekom WP. Liposomes and immunoassays. J. Immunol. Methods. 1997;204:105-133.

[27] Skládal P. Advances in electrochemical immunosensors. Electroanal. 1997;9:737-745.

[28] Berggren C, Johansson G. Capacitance measurements of antibody-antigen interactions in a flow system. Anal. Chem. 1997;69:3651-3657.

[29] Salvi G, De Los Rios P, Vendruscolo M. Effective interactions between chaotropic agents and proteins. Proteins. 2005;61(3):492-499.

[30] Liu M, Rechnitz GA, Li K, Li QX. Capacitive immunosensing of polycylic aromatic hydrocarbon and protein conjugates. Anal. Lett. 1998 31:2025-2038.

[31] Knichel M, Heiduschka P, Beck W, Jung G, Göpel W. Utilization of a self-assembled peptide monolayer for an impedimetric immunosensor. Sensor. Actuat. B-Chem. 1995;B28:85-94.

[32] Rickert J, Göpel W, Beck W, Jung G, Heiduschka P. A mixed self-assembled monolayer for an impedimetric immunosensor. Biosens. Bioelectron. 1996;11:757-768.

[33] Gosling JP. Enzyme immunoassay. Immunoassay. 1996;:287-308.

[34] Gosling JP. Enzyme immunoassay: With and without separation. In: Price CP, Newman DJ, editors. Principles and Practice of Immunoassay. 2nd ed. London: Stockton Press; 1997, pp. 351-388.

[35] Sparreboom A, Rongen HAH, van Bennekom WP. Assays for interferons and interleukins in biological matrices. Anal. Chim. Acta. 1994;295(1-2):1-26.

[36] Porstmann T, Kiessig ST. Enzyme immunoassay techniques. J. Immunol. Methods. 1992;150(1-2):5-21.

[37] Rongen HAH, Hoetelmans RMW, Bult A, van Bennekom W P. Chemiluminescence and immunoassays. J. Pharm. Biomed. Appl. 1994;12(4):433-462.

[38] Ratore AS. Follow-on protein products: scientific issues, developments and challenges. A review. Trends. Biotechnol. (Cell Press). 2009;27(12):698-705.

[39] Carothers JM, Davis JH, Chou JJ, Szostak JW. Solution structure of an informationally complex high-affinity RNA aptamer to GTP. RNA (Cold Spring Harbor Laboratory Press). 2006;12(4):567-579.

[40] Meir A, Marks RS, Stojanovic MN. Aptameric biosensors. In: Marks RS, Cullen DC, Karube I, Lowe CR, Weetall HH, editors. Handbook of Biosensors and Biochips. 1st ed. New York: John Wiley & Sons; 2007.

[41] Tombelli S, Minunni M, Mascini M. Analytical applications of aptamers. Biosens. Bioelectron. 2005;20:2424-2434.

[42] Ellington AD, Szostak JW. In vitro selection of RNA molecules that bind specific ligands. Nature. 1990;346:818-822.

[43] Löfblom J, Feldwisch J, Tolmachev V, Carlsson J, Ståhl S, Frejd FY. Affibody molecules: Engineered proteins for therapeutic, diagnostic and biotechnological applications. FEBS Lett. 2010;584(12):2670-2680.

[44] Renberg B, Nordin J, Merca A, Uhlén M, Feldwisch J, Nygren PA, Karlström AE. Affibody molecules in protein capture microarrays: evaluation of multidomain ligands and different detection formats. J. Proteome Res. 2007;6:171-179.

[45] Andersson M, Rönnmark J, Areström I, Nygren PA, Ahlborg N. Inclusion of a non-immunoglobulin binding protein in two-site ELISA for quantification of human serum proteins without interference by heterophilic serum antibodies. J. Immunol. Methods. 2003;283:225-234.

[46] Hermann T, Patel DJ. Adaptive recognition by nucleic acid aptamers. Science. 2000;287:820-825.

[47] Florea A, Ravalli A, Cristea C, Sandulescu R, Marrazza G. An optimized multiplexed bioassay for sensitive Mucin1 detection in serum samples. Electroanal. Forthcoming. DOI: 2015, accepted.

[48] Collett JR, Cho EH, Ellington AD. Production and processing of aptamer microarrays. Methods. 2005;37:4-15.

[49] Lee RC, Feinbaum RL, Ambros V. The C. elegans heterochronic gene lin-4encodes small RNAs with antisense complementarity to lin-14. Cell. 1993;75:843-854.

[50] Ambros V. The functions of animal microRNAs. Nature. 2004;431(7006):350-355.

[51] Labib M, Berezovski MV. Electrochemical sensing of microRNAs: Avenues and paradigms. Biosens. Bioelectron. 2015;68:83-94.

[52] Keller A, Leidinger P, Lange J, Borries A, Schroers H, Scheffler M et al. Multiple sclerosis: microRNA expression profiles accurately differentiate patients with relapsing-remitting disease from healthy controls. PloS One. 2009;4(10):e7440.

[53] Park YM, Kim SJ, Kim K, Han YD, Yang SS, Yoon HC. Lectin-based optical sensing for quantitative analysis of cancer antigen CA15-3 as a breast cancer marker. Sensor. Actuat. B-Chem. 2013;186:571-579.

[54] Wang X, Yu H, Lu D, Zhang J, Deng W. Label free detection of the breast cancer biomarker CA15.3 using ZnO nanorods coated quartz crystal microbalance. Sensor. Actuat. B-Chem. 2014;195:630-634.

[55] Florea A, Taleat Z, Cristea C, Mazloum-Ardakani M, Săndulescu R. Label free MUC1 aptasensors based on electrodeposition of gold nanoparticles on screen printed electrodes. Electrochem. Comm. 2013;33:127-130.

[56] Koukouvinos G, Petrou PS, Misiakos K, Drygiannakis D, Raptis I, Goustouridis D et al. A label-free flow-through immunosensor for determination of total- and free-PSA in human serum samples based on white-light reflectance spectroscopy. Sensor. Actuat. B-Chem. 2014;209:1041-1048. DOI: 10.1016/j.snb.2014.11.104.

[57] Jang HD, Kim SK, Chang H, Choi J-W. 3D label-free prostate specific antigen (PSA) immunosensor based on graphene-gold composites. Biosens. Bioelectron. 2015;63:546-551.

[58] Ma C, Li W, Kong Q, Yang H, Bian Z, Song X et al. 3D origami electrochemical immunodevice for sensitive point-of-care testing based on dual-signal amplification strategy. Biosens. Bioelectron. 2015;63:7-13.

[59] Kim D-H, Seo S-M, Cho H-M, Hong S-J, Lim D-S, Paek S-H. Continuous immunosensing of myoglobin in human serum as potential companion diagnostics technique. Biosens. Bioelectron. 2014;62:234-241.

[60] Zhou F, Lu M, Wang W, Bian Z, Zhang J, Zhu J-J. Electrochemical immunosensor for simultaneous detection of dual cardiac markers based on a poly(dimethylsiloxane)-gold nanoparticles composite microfluidic chip: A proof of principle. Clin. Chem. 2010;56:1701-1707.

[61] Sinha B, Ramulu TS, Kim KW, Venu R, Lee JJ, Kim CG. Planar Hall magneto resistive aptasensor for thrombin detection. Biosens. Bioelectron. 2014;59:140-144.

[62] Yu P, Zhou J, Wu L, Xiong E, Zhang X, Chen J. A ratiometric electrochemical aptasensor for sensitive detection of protein based on aptamer-target-aptamer sandwich structure. J. Electroanal. Chem. 2014;732:61-65.

[63] Yan W, Chen X, Li X, Feng X, Zhu JJ. Fabrication of a label-free electrochemical immunosensor of low-density lipoprotein. J. Phys. Chem. B. 2008;112:1275-1281.

[64] Ali MA, Reza KK, Srivastava S, Agrawal VV, John R, Malhotra BD. Lipid-lipid interactions in aminated reduced graphene oxide interface for biosensing application. Langmuir. 2014;30:4192-4201.

[65] Wei F, Cheng S, Korin Y, Reed EF, Gjertson D, Ho C et al. Serum creatinine detection by a conducting-polymer-based electrochemical sensor to identify allograft dysfunction. Anal. Chem. 2012;84(18):7933-7937.

[66] Chung D-J, Kim K-C, Choi S-H. Electrochemical DNA biosensor based on avid in biotin conjugation for influenza virus (type A) detection. Appl. Surf. Sci. 2011;257(22): 9390-9396.

[67] Jiang Y, Tan CY, Tan SY, Wong MSF, Chen YF, Zhang L et al. SAW sensor for Influenza A virus detection enabled with efficient surface functionalization. Sensor. Actuat. B-Chem. 2015;209:78-84.

[68] Park J, You X, Jang Y, Nam Y, Kim MJ, Min NK et al. ZnO nanorod matrix based electrochemical immunosensors for sensitivity enhanced detection of Legionella pneumophilla. Sensor. Actuat. B-Chem. 2014;200:173-180.

[69] Singh R, Prasad R, Sumana G, Arora K, Sood S, Gupta RK et al. STD sensor based on nucleic acid functionalized nanostructured polyaniline. Biosens. Bioelectron. 2009;24(7):2232-2238.

[70] Teeparuksapun K, Hedstro M, Wong EY, Tang S, Hewlett IK, Mattiasson B. Ultrasensitive detection of HIV-1 p24 antigen using nanofunctionalized surfaces in a capacitive immunosensor. Anal. Chem. 2010;82:8406-8411.

[71] Yan P, Tang Q, Deng A, Li J. Ultrasensitive detection of clenbuterol by quantum dots based electrochemiluminescent immunosensor using gold nanoparticles as substrate and electron transport accelerator. Sensor. Actuat. B-Chem. 2014;191:508-515.

[72] Chen Y, Yang Y, Tu Y. An electrochemical impedimetric immunosensor for ultrasensitive detection of ketamine hydrochloride. Sensor. Actuat. B-Chem. 2013;183:150-156.

[73] Ionescu RE, Jaffrezic-Renault N, Bouffier L, Gondran C, Cosnier S, Pinacho DG et al. Impedimetric immunosensor for the specific label free detection of ciprofloxacin antibiotic. Biosens. Bioelectron. 2007;23(4):549-555.

[74] Tertiş M, Hosu O, Fritea L, Farcau C, Cernat A, Săndulescu R et al. A novel label-free immunosensor based on activated graphene oxide for acetaminophen detection. Electroanal. 2015; 27(3):638-647.

[75] Wang R, Xiang Y, Zhou X, Liu L, Shi H. A reusable aptamer-based evanescent wave all fiber biosensor for highly sensitive detection of Ochratoxin A. Biosens. Bioelectron. 2015;66:11-18.

[76] Kanso H, Barthelmebs L, Inguimbert N, Noguer T. Immunosensors for estradiol and ethinylestradiol based on new synthetic estrogen derivatives: Application to wastewater analysis. Anal. Chem. 2013;85:2397-2404.

[77] Adams AA, Charles PT, Deschamps JR, Kusterbeck AW. Demonstration of submersible high-throughput microfluidic immunosensors for underwater explosives detection. Anal. Chem. 2011;83:8411-8419.

[78] Liu G, Guo W, Song D. A multianalyte electrochemical immunosensor based on patterned carbon nanotubes modified substrates for detection of pesticides. Biosens. Bioelectron. 2014;52:360-366.

[79] Chauhan R, Solanki PR, Singh J, Mukherjee I, Basu T, Malhotra BD. A novel electrochemical piezoelectric label free immunosensor for aflatoxin B1 detection in groundnut. Food Control. 2015;52:60-70.

[80] Barreiros dos Santos M, Azevedo S, Agusil JP, Prieto-Simón B, Sporer C et al. Label-free ITO-based immunosensor for the detection of very low concentrations of pathogenic bacteria. Bioelectrochemistry. 2015;101:146-152.

[81] Wang X, Chu C, Shen L, Deng W, Yan M, Ge S et al. An ultrasensitive electrochemical immunosensor based on the catalytical activity of MoS2-Au composite using Ag nanospheres as labels. Sensor. Actuat. B-Chem. 2015;206:30-36.

[82] Yan Z, Yang M, Wang Z, Zhang F, Xia J, Shi G et al. A label-free immunosensor for detecting common acute lymphoblastic leukemia antigen (CD10) based on gold nanoparticles by quartz crystal microbalance. Sensor. Actuat. B-Chem. 2015;210:248-253.

[83] Fan H, Guo Z, Gao L, Zhang Y, Fan D, Ji G et al. Ultrasensitive electrochemical immunosensor for carbohydrate antigen 72-4 based on dual signal amplification strategy of nanoporous gold and polyaniline–Au asymmetric multicomponent nanoparticles. Biosens. Bioelectron. 2015;6:51-56.

[84] Sha Y, Guo Z, Chen B, Wang S, Ge G, Qiu B et al. A one-step electrochemiluminescence immunosensor preparation for ultrasensitive detection of carbohydrate antigen 19-9 based on multi-functionalized graphene oxide. Biosens. Bioelectron. 2015;66:468-473.

[85] Jia H, Gao P, Ma H, Li Y, Gao J, Du B et al. Ultrasensitive electrochemical immunosensor for squamous cell carcinoma antigen detection using lamellar montmorillonite-gold nanostructures as signal amplification. Talanta. 2015;132:803-808.

[86] Wang H, Li X, Mao K, Li Y, Du B, Zhang Y et al. Electrochemical immunosensor for α-fetoprotein detection using ferroferric oxide and horseradish peroxidase as signal amplification labels. Anal. Biochem. 2014;465:121-126.

[87] Teixeira S, Conlan RS, Guy OJ, Goreti M, Sales F. Novel single-wall carbon nanotube screen-printed electrode as an immunosensor for human chorionic gonadotropin. Electrochim. Acta. 2014;136:323-329.

[88] Yang M, Zhang J, Chen X. Competitive electrochemical immunosensor for the detection of histamine based on horseradish peroxidase initiated deposition of insulating film. J. Electroanal. Chem. 2015;736:82-92.

[89] Yang Z, Jian Z, Chen X, Li J, Qin P, Zhao J et al. Electrochemical impedance immunosensor for sub-picogram level detection of bovine interferon gamma based on cylinder-shaped TiO_2 nanorods. Biosens. Bioelectron. 2015;63:190-195.

[90] Mihailescu CM, Stan D, Iosub R, Moldovan C, Savin M. A sensitive capacitive immunosensor for direct detection of human heart fatty acid-binding protein (h-FABP). Talanta. 2015;132:37-43.

[91] Liu L, Chao Y, Cao W, Wang Y, Luo C, Pang X et al. A label-free amperometric immunosensor for detection of zearalenone based on trimetallic Au-core/AgPt-shell nanorattles and mesoporous carbon. Anal. Chim. Acta. 2014;847:29-36.

[92] Li Y, Sun J, Wu L, Ji J, Sun X, Qian X. Surface-enhanced fluorescence immunosensor using Au nano-crosses for the detection of microcystin-LR. Biosens. Bioelectron. 2014;62:255-260.

[93] Merola G, Martini E, Tomassetti M, Campanella L. New immunosensor for β-lactam antibiotics determination in river waste waters. Sensor. Actuat. B-Chem. 2014;119:301-313.

[94] Guo Z, Du S, Chen B, Sha Y, Qiu B, Jiang X et al. A sandwich-type label-free electrochemiluminescence immunosensor for neurotensin based on sombrero model with graphene-hyaluronate-luminol composite. Electrochim. Acta. 2014;135:519-525.

[95] Ren X, Yan T, Zhang Y, Wu D, Ma H, Li H, Du B, Wei Q. Nanosheet Au/Co_3O_4-based ultrasensitive nonenzymatic immunosensor for melanoma adhesion molecule antigen. Biosens. Bioelectron. 2014;58:345-350.

[96] Kilic T, Erdem A, Erac Y, Seydibeyoglu MO, Okur S, Ozsoz M. Electrochemical detection of a cancer biomarker mir-21 in cell lysates using graphene modified sensors. Electroanal. 2014;26:1-16.

[97] Atkinson AJ, Colburn WA, DeGruttola VG, DeMets DL, Downing GJ, Hoth DF et al. Biomarkers and surrogate endpoints: Preferred definitions and conceptual framework. Clin. Pharmacol. Ther. 2001;69:89-95.

[98] Jonckeere N, Skrypek N, Van Seuningen I. Mucins and tumour resistance to chemotherapeutic drugs. Biochim. Biophys. Acta - Reviews on Cancer. 2014;1846(1):142-151.

[99] Vlad AM, Diaconu I, Gantt KR. MUC1 in endometriosis and ovarian cancer. Immunol. Res. 2006;36:229-236.

[100] Leteutre E, Gouyer LV, Rousseau K, Moreau O, Barbat A, Swallow D et al. Differential mucin expression in colon carcinoma HT-29 clones with variable resistance to 5-fluorouracil and methotrexate. Biol. Cell. 2004;96:145-151.

[101] Nath S, Daneshvar K, Roy LD, Grover P, Kidiyoor A, Mosley L et al. MUC1 induces drug resistance in pancreatic cancer cells via upregulation of multidrug resistance genes. Oncogenesis. 2013;2:e51. DOI: 10.1038/oncsis.2013.16

[102] Taleat Z, Cristea C, Marrazza G, Mazloum-Ardakani M, Săndulescu R. Electrochemical immunoassay based on aptamer–protein interaction and functionalized polymer for cancer biomarker detection. J. Electroanal. Chem. 2014;717-718:119-124.

[103] Ma F, Ho C, Cheng AKH, Yu HZ. Immobilization of redox-labeled hairpin DNA aptamers on gold: Electrochemical quantitation of epithelial tumour marker mucin 1. Electrochim. Acta. 2013;110:139-145.

[104] Chen X, Zhang Q, Qian C, Hao N, Xu L, Yao C. Electrochemical aptasensor for mucin 1 based on dual signal amplification of poly(o-phenylenediamine) carrier and functionalized carbon nanotubes tracing tag. Biosens. Bioelectron. 2015;64:485-492.

[105] Bast RC, Feeney JM, Lazarus H, Colvin RB, Knapp RC. Reactivity of a monoclonal antibody with human ovarian carcinoma. J. Clin. Invest. 1981;68:1331-1337.

[106] Ravalli A, Pilon dos Santos G, Ferroni M, Faglia G, Yamanaka H, Marrazza G. New label free CA125 detection based on gold nanostructured screen-printed electrode. Sensor. Actuat. B-Chem. 2013;179:194-200.

[107] Li J, Xu Q, Fu C, Zhang Y. Dramatically enhanced electrochemiluminescence assay for CA125 based on dendrimer multiply labeled luminol on Fe_3O_4 nanoparticles. Sensor. Actuat. B: Chem. 2013;185:146-153.

[108] Wu L, Yan F, Ju H. An amperometric immunosensor for separation-free immunoassay of CA125 based on its covalent immobilization coupled with thionine on carbon nanofiber. J. Immunol. Methods. 2007;322:12-19.

[109] Gao J, Guo Z, Su F, Gao L, Pang X, Cao W et al. Ultrasensitive electrochemical immunoassay for CEA through host-guest interaction of β-cyclodextrin functionalized graphene and Cu@Ag core–shell nanoparticles with adamantine-modified antibody. Biosens. Bioelectron. 2015;63:465-471.

[110] Huang J, Tian J, Zhao Y, Zhao S. Ag/Au nanoparticles coated graphene electrochemical sensor for ultrasensitive analysis of carcinoembryonic antigen in clinical immunoassay. Sensors and Actuators B: Chem. 2015;206:570-576.

[111] Wu Z, Li H, Liu Z. An aptasensor for carcinoembryonic antigen based on upconversion fluorescence resonance energy transfer. Sensors and Actuators B: Chem. 2015;206:531-537.

[112] Altintas Z, Kallempudi SS, Gurbuz Y. Gold nanoparticle modified capacitive sensor platform for multiple marker detection. Talanta. 2014;118:270-276.

[113] World Health Organization. Cardiovascular Diseases (CVDs) [Internet]. 2014 [Updated: January 2015]. Available from: www.who.int [Accessed: 10.01.2015].

[114] de Winter RJ, Koster RW, Sturk A, Sanders GT. Value of myoglobin, troponin T, and CK-MB mass in ruling out an acute myocardial infarction in the emergency room. Circulation. 1995;92(12):3401-3407.

[115] Tracy RP. Thrombin, inflammation, and cardiovascular disease an epidemiologic perspective. CHEST. 2003;124:49S–57S.

[116] Bock LC, Griffin LC, Latham LA, Vermaas EH, Toole JJ. Selection of single-stranded DNA molecules that bind and inhibit human thrombin. Nature. 1992;355(6360): 564-566.

[117] Jiang B, Wang M, Li C, Xie J. Label-free and amplified aptasensor for thrombin detection based on background reduction and direct electron transfer of hemin. Biosens. Bioelectron. 2013;43:289-292.

[118] Xie S, Yuan R, Chai Y, Bai L, Yuan Y, Wang Y. Label-free electrochemical aptasensor for sensitive thrombin detection using layer-by-layer self-assembled multilayers with toluidine blue-graphene composites and gold nanoparticles. Talanta. 2012;98:7-13.

[119] Xu N, Wang Q, Lei J, Liu L, Ju H. Label-free triple-helix aptamer as sensing platform for "signal-on" fluorescent detection of thrombin. Talanta. 2015;132:387-391.

[120] Wang K, Liao J, Yang X, Zhao M, Chen M, Yao W et al. A label-free aptasensor for highly sensitive detection of ATP and thrombin based on metal-enhanced PicoGreen fluorescence. Biosens. Bioelectron. 2015;63:172-177.

[121] Wang J. Electrochemical Glucose Biosensors. Chemical Reviews. 2008;108(2):814-825.

[122] Ivnitski D, Abdel-Hamid I, Atanasov P, Wilkins E. Biosensors for detection of pathogenic bacteria. Biosens. Bioelectron. 1999;14:599-624.

[123] Clerc O, Greub G. Routine use of point-of-care tests: Usefulness and application in clinical microbiology. Clin. Microbiol. Infect. 2010;16(8):1054-1061.

[124] Lee D, Chander Y, Goyal SM, Cui T. Carbon nanotube electric immunoassay for the detection of swine influenza virus H1N. Biosens. Bioelectron. 2011;26(8):3482-3487.

[125] Jones RE, Lopez KH. Chapter 17 – Sexually transmitted diseases, reference module in biomedical sciences. In: Stanley D, Cameron B, Tyndall M, editors. Human Reproductive Biology. 4th ed. London: Elsevier Academic Press; 2014. pp. 323-347.

[126] Sankoh O, Arthur SS, Nyide B, Weston M. The history and impact of HIV & AIDS. A decade of INDEPTH research. HIV & AIDS Rev. 2014;13(3):78-84.

[127] Bhimji A, Zaragoza AA, Live LS, Kelley SO. Electrochemical enzyme-linked immunosorbent assay featuring proximal reagent generation: Detection of human immunodeficiency virus antibodies in clinical samples. Anal. Chem. 2013;85:6813-6819.

[128] Robotis JF, Vassilaki N. Viral hepatitis. Reference Module in Biomedical Sciences. 2014. Update of Robotis JF, Boleti H. Viral Hepatitis, xPharm: The Comprehensive Pharmacology Reference; 2007, pp. 1-10.

[129] Konry T, Novoa A, Shemer-Avni Y, Hanuka N, Cosnier S, Lepellec A et al. Optical fiber immunosensor based on a poly(pyrrole-benzophenone) film for the detection of antibodies to viral antigen. Anal. Chem. 2005;77(6):1771-1779.

[130] Ahmed A, Rushworth JV, Wright JD, Millner PA. Novel impedimetric immunosensor for detection of pathogenic bacteria Streptococcus pyogenes in human saliva. Anal. Chem. 2013;85:12118-12125.

[131] Dhillon S, Gill K. Basic pharmacokinetics. In: Dhillon S, Kostrzewski A, editors. Clinical Pharmacokinetics Pharmaceutical Press. 1st ed. RPS Publishing; 2006, pp. 2-42.

[132] Pidd K, Roche AM. How effective is drug testing as a workplace safety strategy? A systematic review of the evidence. Accident Anal. Prev. 2014;71:154-165.

[133] Ya Y, Jinyin P, Weijie H, Yifeng T. An approach for the preparation of highly sensitive electrochemical impedimetric immunosensors for the detection of illicit drugs. J. Electroanal. Chem. 2014;726:1-6.

[134] De Juan-Franco E, Rodríguez-Frade JM, Mellado M, Lechuga LM. Implementation of a SPR immunosensor for the simultaneous detection of the 22K and 20K hGH isoforms in human serum samples. Talanta. 2013;114:268-275.

[135] Yang L, Bashir R. Electrical/electrochemical impedance for rapid detection of foodborne pathogenic bacteria. Biotechnol. Adv. 2008;26:135-150.

[136] Dou W, Tang W, Zhao G. A disposable electrochemical immunosensor arrays using 4-chanel screen-printed carbon electrode for simultaneous detection of Escherichia coli O157:H7 and Enterobacter sakazakii. Electrochim. Acta. 2013;97:79-85.

[137] Lee KM, Runyon M, Herrman TJ, Phillips R, Hsieh J. Review of Salmonella detection and identification methods: Aspects of rapid emergency response and food safety. Food Control. 2015;47:264-276.

[138] Mantzila AG, Maipa V, Prodromidis MI. Development of a faradic impedimetric immunosensor for the detection of Salmonella typhimurium in milk. Anal. Chem. 2008;80:1169-1175.

[139] Spanu C, Scarano C, Ibba M, Spanu V, De Santis EPL. Occurrence and traceability of Listeria monocytogenes strains isolated from sheep's milk cheese-making plants environment. Food Control. 2015;47:318-325.

[140] Muhterem-Uyar M, Dalmasso M, Bolocan AS, Hernandez M, Kapetanakou AE et al. Environmental sampling for Listeria monocytogenes control in food processing facilities reveals three contamination scenarios. Food Control. 2015;51:94-107.

[141] Wang R, Ruan C, Kanayeva D, Lassiter K, Li Y. TiO2 nanowire bundle microelectrode based impedance immunosensor for rapid and sensitive detection of Listeria monocytogenes. Nano Lett. 2008;8(9):2625-2631.

[142] Vettorazzi A, González-Peñas E, López de Cerain A. Ochratoxin A kinetics: A review of analytical methods and studies in rat model. Food Chem. Toxicol. 2014;72:273-288.

[143] Prieto-Simon B, Campas M, Marty J, Noguer T. Novel highly-performing immunosensor-based strategy for ochratoxin A detection in wine samples. Biosens. Bioelectron. 2008;23:995-1002.

[144] Pal A, He Y, Jekel M, Reinhard M, Yew-Hoong Gin K. Emerging contaminants of public health significance as water quality indicator compounds in the urban water cycle. Environ. Int. 2014;71:46-62.

[145] Gowen A, Marini F, Tsuchisaka Y, De Luca S, Bevilacqua M, O'Donnell C et al. On the feasibility of near infrared spectroscopy to detect contaminants in water using single salt solutions as model systems. Talanta. 2015;131:609-618.

Chapter 7

Cell Concentration Systems for Enhanced Biosensor Sensitivity

Óscar Castillo-Fernandez, Naroa Uria,
Francesc X. Muñoz and Andrey Bratov

Additional information is available at the end of the chapter

http://dx.doi.org/10.5772/61088

Abstract

Concentration processes for analytical systems based on different types of biosensors are very important for many applications. The sample conditioning is oriented to enhance the sensitivity or directly to make the detection or analysis possible. Processes that may be used for concentration and conditioning of original samples are very diverse, depending on applications that may range from clinical diagnostics to industrial processes control, and there are different strategies to achieve the final goal.

This chapter presents an overview of the most important and relevant microscale techniques to produce an effective separation and concentration of cells, mostly bacterial cells. The main focus is put on the separation mechanisms as a tool for biosensing toward the development of complete devices in a lab-on-a-chip format by integrating the concentration, sample conditioning, and detection subsystems.

Keywords: Cell concentration, microfluidics, hydrodynamics, electrophoresis, dielectrophoresis

1. Introduction

The sensitivity and the detection limit of biosensors have natural limitations. It is possible to enhance these analytical parameters and the performance of the system by increasing cell concentration, reducing the sample volume in a preconcentration stage. The utilization of any concentration technique is crucial for the diagnostic and detection of particular cells [1–3] or metabolites [4] in clinical, food, or water environment [5].

Along with sensitivity and detection limits, another important aspect of any detection method is related to the signal-to-noise ratio and false positives, which determine the specificity of the system. This defines what percentage of the measured signal is related to our specific bacteria of interest. To enhance this signal-to-noise ratio, it is useful to apply a selective preconcentration methodology adding some specificity to detect only the cell of interest. The most selective methods to isolate specific bacteria from the complex mixture presenting a real sample are immunological systems based on the utilization of specific markers or identifiers.

Cell detection for clinical diagnostics sometimes requires detection of low abundance cells circulating in complex samples such as blood, e.g., specific detection of cancer cells from tissue or determination of small quantities of cells in noncellular samples as urine. These are only some examples of different types of bioanalysis that require the preconcentration of samples [1, 3].

Bacteria detection, especially in environments affecting the public health risks, such as food, water, and clinical, requires an efficient concentration process to guarantee low detection limits corresponding to official law regulations. This means that the objective sometimes implies the detection of few bacteria in large water volumes. The best option to solve this problem and improve detection limits of a biosensor detection method is the sample treatment by retaining all bacteria present in the sample, decreasing its volume and, in this way, increasing the concentration of bacterial cells [1].

This section will mainly be focused on the concentration of cells for biosensor applications, but we will try to present a complete view of the physical and chemical phenomena that could be potentially used for this purpose. We will provide important details of each evaluated methodology focusing on the main characteristics of the concentration system effectiveness: the capacity of sample volume processing, the velocity of processing or extraction time, and the recovery, concentration, purity, and detection factor.

The modern appeal for miniaturization requires the use of microtechnologies for sensor fabrication as well as the application of microfluidic systems. These may be regarded as crucial tools for the integration of concentration steps in sensor-based bioanalytical systems [6]. It is clear that speaking about large volumes may cause confusion in terms of relative volume amounts. In this chapter, we will focus on those concentration systems that could process volumes up to 60 mL.

The main concept of concentration is the separation. Thus, many different techniques based on various physical principles (electric, magnetic, gravitational, acoustic, hydrodynamic, affinity chemistry, etc.) have been proposed for cell and protein separation within microfluidic devices [6]. From the point of view of the flow conditions, the preconcentration techniques on chip can be classified as static and dynamic [4]. The static techniques are those used to stop or trap cells, blocking them in particular places[7], while dynamic techniques are those that separate them from the main carrier flow to lateral streams by focusing or deviating the cells, bringing them to specific places under continuous flow conditions[8].

The most interesting static concentrators are those that trap the cells at particular places where the sensing would be carried out. These systems enhance the sensitivity of the biosensor technique, in this case, the detection limit, as they perform the measurements in a volume limited by the sensor area. The dynamic concentration techniques are very interesting for their use in sequential systems or devices, build up in a modular architectural design [9].

However, along with certain advantages, both methodologies possess some drawbacks, mainly due to the relationship between the trapping forces and the dragging forces of the fluid movement. This fact imposes limitations in terms of velocity and sample processing capacity as well as proper characteristics such as efficiency, purity, and recovery factor. These limitations may be improved by using large channel dimensions to generate relatively low flow velocities, which implies high dead volumes and a low concentration ratio (samples/volume unit). Thus, it is necessary to find a compromise between design limitations and concentration volume requirements for a measurement or analysis.

In this chapter, from a general point of view, we present the main methods with the objectives to show the most interesting technical solutions and to give some numbers characterizing their performances in order to give an effective perspective in this field.

2. Filtration-based processes

Fibrous filters are routinely used in laboratory protocols for filtration of water and wastewater samples. Typical wastewater filtration methods include the use of paper filters or glass fiber filters as a mechanical trapping surface. After the trapping process, particle recovery should be performed by diffusion, interception, gravitational settling, or impaction [10].

These mechanical filtration methods remain the most easy, cheap, and effective for very large volumes. In particular, the tangential or cross-flow filtration has demonstrated a good performance for volumes over 200 mL [11]. However, they present several drawbacks if its integration into microfluidic devices is required.

Nevertheless, a filter-based bacteria preconcentrator embedded in a microfluidic system with a miniaturized sensor based on electrical impedance spectroscopy method was presented by Jiang and coworkers [12]. The design of a multistage filtering was comprised by the first layer of a silicon chip having a large array of holes with a diameter of 10 μm and the second layer of a nanoporous filter paper with submicron pore size. This direct integration of a paper filter is very interesting since the system is able to process 60 mL *Escherichia coli* samples with a detection limit of 10^2 cell/mL.

A different perspective in terms of the reutilization of the system and release capacity represents the system reported by Bao and Lu [13]. They created an array of microscale beads by pinching the polydimethylsiloxane (PDMS) channel using a pneumatic valve, which allows the formation of a conglomeration of beads that are used as a filter. The device was tested using *E. coli* cells, showing an increase in the concentration of two orders of magnitude.

Therefore, different filtration-based concentration methods may be used for the detection of low concentration cells. However, as it has been said, their integration in miniaturized systems for the sensor fabrication is still a challenge.

3. Specific interaction of cells with functionalized surfaces or particles

This approach is based on the adsorption or binding of the microbial cells to a specific material surfaces treated with special reagents. The main idea is to generate large surfaces with affinity to certain bacteria cells or class of bacteria. Basically, this methodology is considered as a static concentration process; however, the utilization of beads may be performed in dynamic processes.

3.1. Treated surfaces

This concentration concept is fully applicable to sequential systems in which trapped cells are released from the surface and resuspend in a smaller volume giving sufficient concentration to be addressed by the detection system. In this kind of concentration process, it is important to take into account the relationship between fluidic phenomena considerations and the surface reactivity, in terms of its capability to capture and retain the target cells. This technique may be modeled [14], studying the capacity of the system to transport cells or bacteria in the sample solution to the treated surface (k_0) and the probability that an association event will occur during the time that the target cells are in proximity of the surface with capture molecules. Another critical point is the relationship between the adhesion force (F_A) experienced by the target cell at the treated surface and the shear force (F_S) generated by the flow stream that is tangentially directed trying to tear off the cell.

Cells can be trapped at particular places using different specific and nonspecific affinities. Specific cell surface binding exploits selective interactions between a ligand and a target-specific receptor, for example, a membrane protein. In this case, antigen/antibody is the most extended strategy, although they present certain disadvantages such as difficulty of surface immobilization or instability. To overcome these problems, new strategies have been proposed [1], which make use of aptamers, peptides, or toxins to selectively bind and concentrate a specific microorganism or cell.

A very important application of this separation technique is aimed on the separation and concentration of rare cells or low concentration samples. Thus, we may find many different options of isolation of circulating tumor cells (CTCs) from blood [15–21]. For example, Nagrath and colleagues [15] presented in 2007 a device with 78,000 microposts coated with EpCAM antibody, increasing the interaction surface to 970 mm^2. The EpCAM antibody was demonstrated to trap different cells, such as lung, breast, prostate, and bladder cancer cells, in a whole blood. The optimization of the microfluidic design and micropost distribution permitted to minimize the shear stress at flow rates of 1–2 mL/h. This system permits to process maximum sample volumes of 5.1 mL, reaching a best capture of the 65% of the target cells (meaning ≈65 cells/mL) with a final sample containing around 50% of CTC along with blood cells. The micropost array, forming obstacles for the cells within the flow, may be further optimized in terms of the particular post diameter and distances between posts to enhance the collision probability of cells with particular sizes, thus enhancing the purity of the concentration up to 90% [18, 21, 22].

These capture concepts also require the development of the cell release method to recover thus immobilized cells for detection. The most relevant systems use digestive enzymes like trypsin to break the bond between the ligands and the surface [15, 18, 21] or biopolymers to reduce the possibility of damaging the recovered cells [23]. Adams et al. [16] presented an integrated system where the cells are trapped and released to perform the sample concentration. The following measurement of recovered cells was carried out by a conductivity sensor.

Another interesting approach is the concentration of cells by reactive binding directly on a sensor surface. Recently, Souiri et al. [24] studied and characterized the immobilization of antibodies on an indium-tin oxide (ITO) electrode and the capture of *Legionella pneumophila*. The immobilization of the specific antibody for *L. pneumophila* was carried out by a covalent binding to a silane self-assembled monolayer (SAM). Surface characterization by AFM, confocal microscopy, and impedance spectroscopy demonstrated the deposition of different layers at the ITO surface used as an impedimetric electrode sensor for characterization. Dulay and colleagues [25] also used this silane SAM technique with a specific antibody for an electrochemical ELISA detection of *Francisella tularensis*.

Surface chemistry and different physicochemical factors, such as hydrophobicity, hydrophilicity, steric hindrance, or roughness between others, play an important role in bacterial adhesion to the surfaces [26]. Although the major part of the research in the bacterial adhesion to solid surfaces is aimed on the prevention of biofilm formation, their results may be useful for an opposite strategy [27, 28].

3.2. Treated magnetic particles

A particular case of the implementation of functionalized surfaces is the utilization of microbeads modified with antibodies to recognize certain groups of bacteria. This is a good strategy to increase the interaction surface. Besides, it offers an easy way for the manipulation of the captured cells. Thus, concentrated samples may be carried by a flow, fixing the beads in particular places or deflecting them under flow conditions by different forces.

In particular, preconcentration techniques with magnetic particles are very popular in microfluidic devices for biological applications, implementing and integrating on a chip normal bench protocols to include high specificity attributed to the utilization of immunological interaction. The use of magnetic beads is very widespread due to the facility of separation and concentration by using a magnet. The utilization of this technique was implemented in a system to extract *E. coli* from such complex samples as soil [29].

Verbarg et al. [30] presented a very promising spinning magnetic trap for automated assay systems, which gives the possibility to perform a simple protocol assays typical for any laboratory that includes catch, clean, and release steps. The system called "MagTrap" consists of a mobile disk with magnetic rods disposed radially forming rotating magnet arrays that permit to entrap or to move the magnetic beads within the channel during processing. They demonstrated the capabilities of the system presenting a 98% of capture efficiency with a release capacity of 80%. The experiment was carried out processing from 0.1 to 1 mL sample volumes with a flow rate of 10 μL/min. The system was used for the detection of *E. coli*, *Salmonella*, *Listeria*, and *Shigella* integrated with flow microcytometer [31].

In terms of sensor sensitivity, magnetic particles may be used to increase the interaction between the sample and the sensing transducer, enhancing the sensitivity to detection levels of around 10 cfu/mL [32–34].

Some interesting examples may be presented concerning the use of microbeads for direct concentration and posterior off-chip detection. Beyor et al. [35] presented an immobilized bed of magnetic beads functionalized with specific antibodies. They processed 10, 20, and 100 μL samples with different concentrations of *E. coli* with a mean flow rate of 0.24 μL/min. Analysis was performed by using off-chip polymerase reaction and capillary electrophoresis to evaluate the trapping efficiency. The system presents a capacity of 70% capture yield from a pure sample of 10^3 *E. coli* in 100 μL PBS, while a 40% capture efficiency was achieved in a complex mixture of *E. coli* and *S. aureus* in a 1:100 ratio.

Recently, Lee and colleagues [36] presented a new device that implements the same idea of the surface coated with magnetic beads. They used specific geometry of magnets and nanomagnetic beads of around 150 nm functionalized with specific antibodies for *Salmonella* to trap bacteria in a large volume container (see Figure 1). The release of particles was implemented by removing the magnets and then flushing the particles with the captured bacteria. With this system, a complete assay of 10 mL of a real food sample may be accomplished in less than 3 min, under flow rates of 25 mL/min. The concentrated sample of released cells is then analyzed by ATP luminescence measurements, demonstrating a detection limit of 10 cfu/mL.

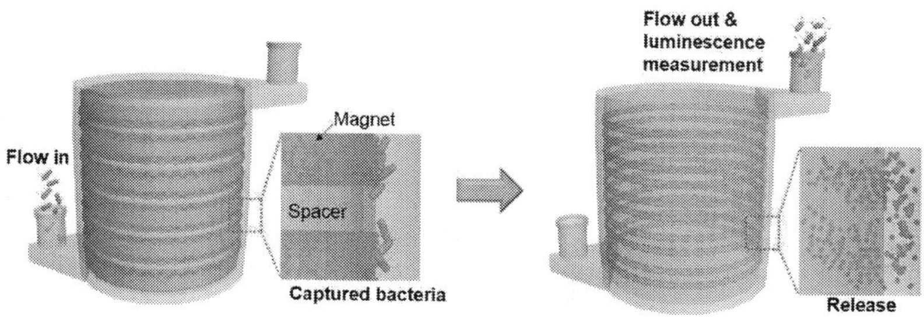

Figure 1. Schematic illustrations of 3D immunomagnetic flow assay. The magnet-spacer assembly was placed in the center opening during the capture and rinsing of the magnetic particles and the antibody–Salmonella complexes, which were trapped. After removing the magnet assembly from the center opening, the concentrated AbMNC–Salmonella solution was collected using a disposable syringe. Reprinted with permission from Lee et al. [36]. Copyright 2014, American Chemical Society.

As mentioned at the beginning of the section, trapping processes have to take into account dragging forces. With some devices, it is required to reduce the flow rates below 0.1 mL/min

[31, 35] to allow effective reaction binding, which increases the time needed to process the sample volume. However, the device presented by Lee and colleagues [36] is able to perform at high flow rates of 25 mL/min, but it presents a dead volume (≈3 mL) comparable to the initial sample volume(10 mL). Thus, working with microfluidic systems, it is necessary to consider possible dead volumes of devices, which may affect the final cell concentration even for systems with high trapping coefficients.

Most of the cited works employ immunochemical methods for cell entrapment. However, in the case of bacteria concentration, we think that this entrapment concept should be extended more, focusing on different bacteria cell capture strategies involving hydrophobic interactions that are related to the adhesion involved in microbial infections [37], biofilm formation of specific species in the presence of calcium ions [38],bacteria interactions with some known polysaccharides [39], or antibacterial peptides [40, 41].

4. Electro capture

This process is based on the capture of charged or polarizable cells or other micro-objects due to the interaction of the material with the external electric field. These phenomena generally depend on the frequency and strength of the electric field. Some bacteria are typically charged due to the presence of ionizable groups in its outer membrane and also are polarizable under an electric field. Within electro capture-based methods, we may distinguish two different groups of techniques: electrophoresis and dielectrophoresis, which can be used to perform static or dynamic separations.

4.1. Electrophoresis

Electrophoresis is defined as the motion of a charged particle within a uniform electric field. Due to such electric field of magnitude *E*, a particle with a charge *Q* will experience a Coulomb force.

Some bacteria have a net negative charge on the cell wall, although the magnitude of this charge varies from strains depending on the molecular structure of their membranes as well as on the pH of the medium [42, 43].

The velocity of a particle under applied electric field force is defined by its intrinsic mobility in a given buffer solution. This specific mobility of a particle (μ_e) could be expressed by

$$\mu_e = \frac{v}{E} \propto \frac{\xi}{\eta}, \tag{1}$$

where v is the linear velocity of the particle and E is the electric field applied. The mobility of the particle is then independent of the electric field applied and is proportional to the zeta potential (ξ) (which depends on the ionic strength of the fluid) and on the medium dynamic viscosity (η).

The static electrophoretic devices retain the particles at particular places due to the specific net charges on the surface of some bacteria. For example, Balasubramanian et al. [44] used two glass slides covered with chromium (15 nm) and gold (35 nm) with a spacer to create a microcontainer 60 mm wide, 40 mm long, and 150 μm high, for static concentration. The glass covers work as big electrodes where the bacteria may be attached depending on their net charge. A water sample with a concentration of 10^6 cfu per mL is injected into the microcontainer passing during few minutes at a very low flow rate (2–6 mL/h) while applying 1 V between the electrodes. The efficiency of the trapping container, which was measured with a negative control in agar plates, resulted in 95% of capture efficiency for *Salmonella*, *E. coli*, and *Pseudomonas* under different flow rates.

An important improvement of this idea was presented by Podszun and colleagues [45], who developed a more complex device for free-flow electrophoresis adapted for bacteria enrichment of *Enterococcus faecalis*. The system incorporated a gel container where the bacteria may be trapped and released by electrophoretic control. With this device, Podszun presented trapping efficiencies of 80% and a concentration factor of 25 over 200 μL sample. The maximum flow rate achieved was 6 μL/min.

This electrical separation concept permits to perform trapping and release steps, crucial for a concentration process. However, it presents important drawbacks related to the necessity of high voltages (>1 V) that may induce electrolysis, high temperatures associated with the current pass, and cell damage. Also, the compromise between trapping forces and flow rates should be achieved.

4.2. Nanogap: Exclusion-Enrichment Effect (EEE)

The generation of exclusion-enrichment zone effect by using nanogaps or nanopores to generate a trap for charged particles has been demonstrated to be a very promising tool for analytes static concentration [4]. The translation of this method to the preconcentration of charged particles or cells may be regarded as a good option for biosensing.

The exclusion-enrichment effect is produced when an electric field is applied along a nanochannel. Figure 2 represents a simplified scheme of this effect. If dimensions of the nanochannel are comparable to the Debye length of the electrical double layer formed at the nanochannel walls—electrolyte solution interface, then an overlap in electrical double layers occurs, resulting in a higher concentration of positive ions (for a negatively charged surface) inside the nanochannel. This effect produces unbalanced ionic flows that provide ionic enrichment at the cathode side of the nanochannel, resulting in formation of an ionic exclusion zone at the anode side [46].

Wang and colleagues [47] recently presented a preconcentrator microsystem consisting of two microchannels connected by 205 nm high nanochannels fabricated by electric breakdown of a 25-μm-thick polydimethylsiloxane (PDMS) membrane using high electric shock. Figure 2 shows the final concentration of *E. coli* produced by the ionic-enrichment effect and the electroosmotic flow. The detection of the concentrated bacteria was performed by fluorescence measurements. Using this methodology, they estimated to reach a 10^4-fold concentration factor within a few minutes. The main drawback of this method is similar to the classical electrophoresis, as it also requires high voltages. Even so, the capture capacity of this method seems very promising and useful.

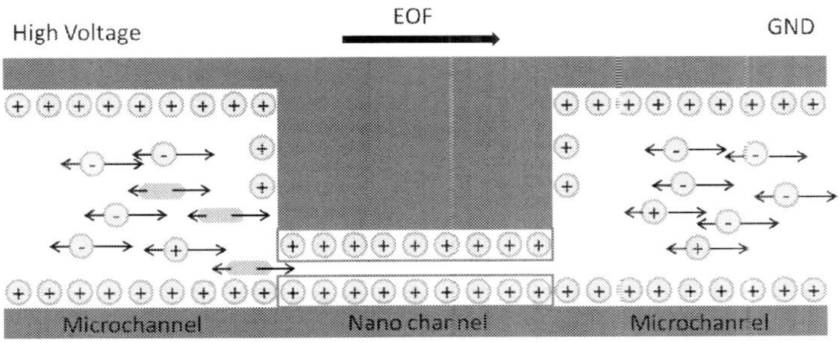

Figure 2. Shows a scheme of the distribution of charges inside and outside the nanochannel. The green ellipsoids indicate the concentration of *E. coli* cells at the cathode side.

4.3. Dielectrophoresis

Dielectrophoresis uses the effect of electrical polarization of particles under the influence of nonuniform electric fields to induce a translational motion. The inner elements of the particle under the influence of an electric field, together with the free charges and surrounding media, are polarized, forming an induced dipole. The positive and the negative extremes of the dipole are influenced by forces of different magnitude because of the nonuniformity of the electric field. These forces depend on the strength and frequency of the applied field, as well as on the conductivity of the supporting electrolyte. Thus, cells can be polarized in a highly conductive electrolyte suspension [48, 49] or in diluted solutions.

The dielectrophoretic (DEP) model for cells is normally described by a spherical particle approach [50], and the DEP force, F_{DEP}, may expressed by

$$F_{DEP} = 2\pi r^3 \varepsilon_0 \varepsilon_m \operatorname{Re}\left[K(\omega)\right] \nabla \left|E_{rms}\right|^2, \quad (2)$$

where r is the radius of the particle, $\varepsilon_0 \varepsilon_m$ are the permittivities of the free space and suspending medium, and E_{rms} is the root-mean-square electric field. The factor $K_{(\omega)}$ is the Clausius–Mossotti factor (CM), which depends on the relationship between the particle and the medium complex permittivity, that is,

$$K(\omega) = \frac{\varepsilon_p^* - \varepsilon_m^*}{\varepsilon_p^* + 2\varepsilon_m^*}, \quad (3)$$

where ε^* is the dielectric complex permittivity, for the particle and medium, expressed by

$$\varepsilon^* = \varepsilon_0 \varepsilon - \frac{j\sigma}{\omega}, \tag{4}$$

where j is ($\sqrt{-1}$), σ is the conductivity of particle or medium, and ω is the frequency in radians.

According to the simplified model proposed by Huang et al. in 1992 [51], cells can be presented using a simple spherical concentric multishell model, composed of different spheres, shells, contained from one to another with interfaces separating the different dielectric layers, as presented in Figure 3.

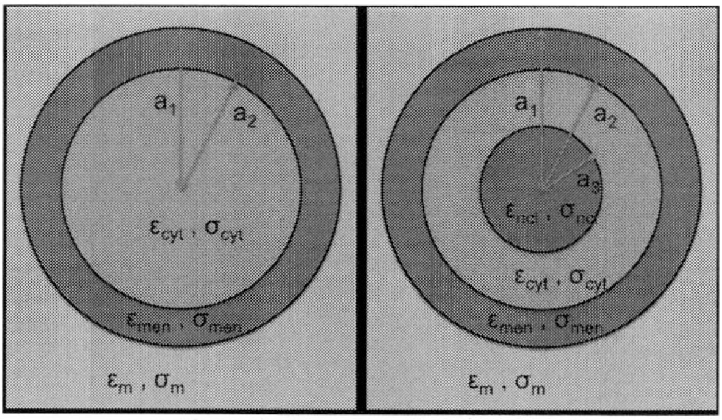

Figure 3. Schematic representation of the multishell model for a nonnucleated cell (left) and a nucleated cell (right). m indicates the medium; mem, cell membrane; cyt, cytoplasm; ncl, nucleus, a_1, a_2, and a_3, radius of the shells.

The complex permittivity of the cell is calculated by a geometrical relationship between shell geometries, and it defines the sign of the dielectrophoretic forces. When the factor is positive, the particles are attracted to the places where maximums of the electric field distribution appear, and we talk about positive dielectrophoresis (pDEP). On the other hand, when the CM is negative, the particles are repelled from the maximum of the electric field, and we talk about negative dielectrophoresis (nDEP). Figure 4 presents the evolution of the CM factor with the frequency of the applied electric field for a nonnucleated cell of 6-μm size with typical electric parameters [52] in different conductivity mediums. It is important to note the evolution of the pDEP with the solution conductivity, which disappears in high conductivity mediums. Moreover, the maximum value of CM factor for pDEP is over 0.8, whereas that for nDEP is –0.5, indicating that forces generated by negative DEP are weaker than positive DEP forces.

Despite that the bacteria multishell dielectrophoretic model [51] may be simplified to a sphere, the majority of bacteria have different morphological shapes, and it is necessary to modify the calculation of the CM factor taking into account their geometry, such as the ellipsoidal bacterial cells [53].

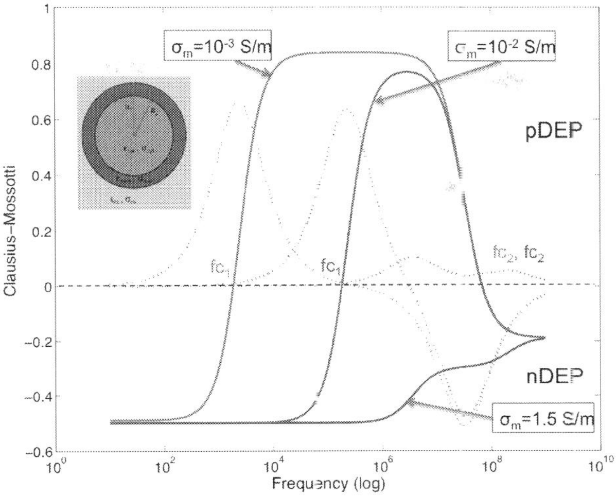

Figure 4. The solid lines represent the variation of the real part of the CM factor, and the dotted lines represent the evolution of the imaginary part of CM. The conductivity of the medium, σ_m, affects to the cutting frequencies that define the sign of the DEP force. fc1 indicates the transition from nDEP to pDEP, and fc2 indicates the transition from pDEP to nDEP. The color indicates the medium conductivity case. In high conductivity medium (blue line), the cell is under negative DEP for all frequencies.

As we said previously, a nonuniform electric field is required for produce a DEP effect. These nonuniform electric fields may be generated by electrodes of different geometrical arrangement, the most common being coplanar electrode arrays, in which the current distribution forms a maximum of electric field between the electrodes close to the electrode plain. This electrode configuration opens the possibility to exploit two different functions offered by an array of interdigitated electrodes (IDE), namely, simultaneous trapping and impedimetric detection. Ferrier et al. [54] implemented an nDEP system for yeast cells to elevate the cells over electrodes, measuring at the same time the capacitance changes associated to the cell distance from the electrodes. Dastider et al. [55] used pDEP for *E. coli* trapping to enhance impedance sensitivity and to lower detection limit down to 10^2 cfu/mL.

Another option for nonuniform electric field generation is the utilization of pore surfaces [56, 57]. Cho et al. [57] presented a system that implements a porous SU8 polymer membrane with the optimized pore shape designed to generate a maximum electric field gradient inside the pore. In this work, they optimized conditions for *E. coli* trapping in terms of frequency, media conductivity, and flow rates. The optimal parameters were found to be 300 kHz, 0.2 S/m, and flow rates below 50 μL/min. A complete concentration experiment performed with a 500-μL sample with a concentration of ≈9 × 10^3 cells/mL at 100 μL/min flow rate in 5 min consisting in capture and release of bacteria cells resulted in a trapping efficiency of more than 60% and a recovery rate of 93% in approximately 150 μL.

When it is required to work with physiological medium, normally highly conductive buffers, it is necessary to use negative DEP because, as it was shown earlier, the CM values under these medium conditions are negative for frequencies below 10^8 Hz. This implies certain limitations because the forces generated by the negative DEP are weaker than of the positive DEP [58].

A very interesting method of the *E. coli* deviation and concentration by negative DEP was implemented by Park and colleagues [59]. The complex fluidic system presented in Figure 5 involves simultaneous parallel laminar flow of two different solutions: a highly conductive sample volume (where the separation is carried out by nDEP) and a low conductive water solution (for trapping the cells by pDEP). The separation is carried out by deflecting the bacterial cells moving them from the sample stream to the low conductivity buffer stream, as it is shown in Figure 5(A). This buffer stream delivers cells to an IDE structure where the cells are trapped by a positive DEP (Figure 5(B)).

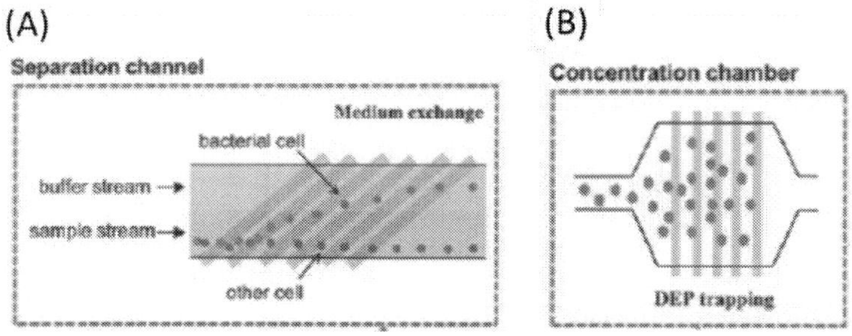

Figure 5. Scheme of the microfluidic device presented by Park et al. [59]. In scheme (A), the deviation of bacterial cells from the sample stream is represented, to be incorporated to the buffer stream that brings bacteria to the concentration chamber (B). Adapted from Park et al. [59], with permission of the Royal Society of Chemistry.

This method showed a separation efficiency of 95% at 800 μL/hand. The capture ratio in the final chamber with the volume of 0.5 μL was reported to be of 100%. The concentration ratios were unknown, and the measurements were performed by taking instant images that were analyzed by ImageJ (Image processing software). The flow rate of sample processing (≈13 μL/min) is quite low, but the importance of this system lies in the use of nDEP for separation process. The pDEP trapping permits to achieve separation at higher velocities, thanks to higher values of the CM factor as, for example, in the device presented by Cho and colleagues [57], with a retention capacity over 60% at flow rates 100 μL/min. These results were obtained by the positive DEP in a low conductivity buffer (0.2 S/m).

4.3.1. Combination of DEP with functionalized surfaces or particles for enhanced capture strategies

Dielectrophoresis is a commonly used technique for low abundance sample enrichment in many different applications, from the simple cells concentration to the differentiation of living and dead bacteria or to enhance the efficiency of bacteria binding to immune-treated surfaces [60].

The capacity of the DEP forces to attract particles to particular places gives a possibility to combine this method with cell interaction with a specially treated surface [61–63] as discussed previously. Yang and colleagues [61] used a microchannel system with interdigitated electrodes on the surface of SiO_2, where monoclonal antibodies were attached. Binding of antibodies was carried out by using a biotin/streptavidin strategy. Combining DEP with specific binding, it was possible to collect 90% of *Listeria monocytogenes* under flow conditions of 0.2 μL/min. The capture efficiency was calculated by counting colonies on plates and for concentrations from ≈3 10^7 cells/mL to ≈2.5 10^6 cells/mL was estimated to be 90%.

Koo and colleagues [63] presented the increase in the capture rate of *L. monocytogenes* by using a biotinylated Hsp60, a protein receptor responsible for cell adhesion during infection processes. The capture efficiency of the Hsp60 in the DEP system in comparison with static flow conditions was increased by 60%.

Another electric effect known as AC electro-osmotic flow could be useful to enhance the trapping efficiency of functionalized surfaces [64]. This phenomenon is related to the generation of an electro-osmotic flow at the surface by the application of AC electric fields over coplanar electrodes [52]. The net fluid flow depends directly on the electrode geometry. Using a very specific geometry (asymmetric geometry with a conical electrodes), Vaidyanathan et al. [64] were able to generated a lateral flow and microvortices that generate local shear forces that prevent weak interactions at the surface, reducing the nonspecific attachment of cells. This system permitted to capture breast cancer cells with 87% of efficiency.

Due to weakness of dielectrophoretic forces in comparison with dragging forces, the utilization of surface-modified microbeads is more advantageous as the effect of electric field is higher for a bigger particle. This is the case of the device implemented by Hu et al. [65], in which pDEP deviation under flow conditions was successfully used for separation of specific *E. coli*. In this system, target cells were attached to polystyrene microbeads modified with monoclonal antibody using a streptavidin–biotin method. The application of the positive DEP caused the deviation of modified beads from the main stream flow facilitated by an angled electrode. This method that permitted to recollect 95% of injected beads was reported under 300 μL/h flow conditions.

5. Hydrodynamic mechanism

We would like to finish this chapter, introducing technologies permitting to perform sample separation by hydrodynamics phenomena [66, 67], such as an inertial flow dynamics and Dean flows [68] or hydrodynamic focusing [69, 70]. These concepts are very interesting for sample separation and concentration in terms of sample processing capabilities. While the systems based on surface binding, magnetic interaction, electrical trapping, acoustic separation, and optical effects generate forces that are weaker in comparison to the dragging forces at flow conditions, hydrodynamic mechanisms implement forces and effects produced at relatively high flow rates.

A very useful effect at high flow rates is the formation of fluid vortices, in which the cells are trapped or focused. The generation of microvortices in microfluidic channels may be implemented using spiral channels or asymmetrical geometries, called Dean vortices, as represented in Figure 6(A) for a vortex formation on a channel curve. Under inertial conditions in curved channels, the faster fluid at the central section tends to move outward by inertial effect, while the slower parts of the fluid near to the channel walls tend to recirculate to the internal part of the curve, generating two rotating vortices. On the other hand, vortices can also be created in linear channels by a stepwise change of the channel width. Figure 6(B) shows the creation of vortices, geometry, and sizes that depend on the Reynolds number, a value proportional to the fluid velocity.

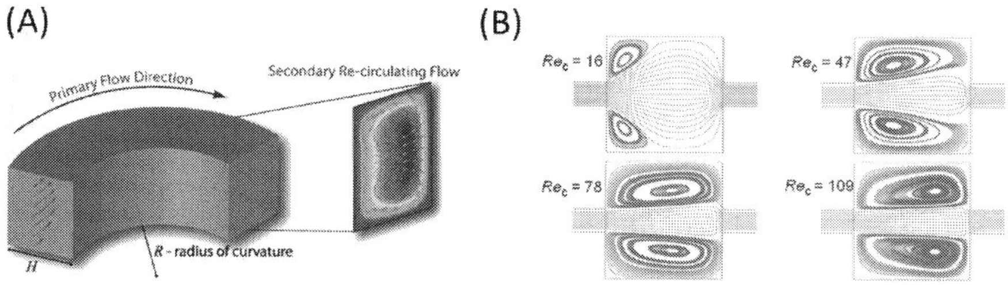

Figure 6. (A) An example of a Dean flow [68] under inertial conditions. Faster moving fluid near the channel center tends to continue outward, and to conserve mass, more stagnant fluid near the walls recirculates inward. This creates two vortices perpendicular to the primary flow direction. (B) the representation of different vortices produced by the change on the channel width at different Reynolds number (Re proportional to fluid velocity) [67]. (A) Reproduced from Di Carlo [68], with permission of the Royal Society of Chemistry. (B) Adapted from Park et al. [67], with permission of the Royal Society of Chemistry.

Using a system with a consecutive succession of width variations in long channels, Mach et al. [71] implemented a microfluidic device performing as a two-step concentration system able to capture cells at the created vortices, releasing then afterward by reducing those vortices lowering the fluid velocity. With this system, they treated a 10-mL sample of a diluted blood (5% v/v) with a ratio of 10^9 blood cells to 500 cancer cells, processing the sample with ~5 mL/min flow rates, concentrating the cancer cells into a final volume of 200 μL in less than 3 min with a purity of 40%. Moon et al. [72] used a similar system with multiorifices combined with an angled electrode DEP system to perform cell separation with high purification. The sample flow rated decay to 126 μL/min, but the purity ratios increased up to 90%.

Inertial systems are mostly used for separation based on a particle size; however, these hydrodynamic-based systems can perform a separation based on inner physical properties of cells, such as its deformability. This kind of sorting can be determinant for the diagnosis of illnesses that affect to the cell physical properties, such as cancer or diabetes [66]. Using an inertial system with a final aperture, Hur et al. [73] demonstrated that at the aperture cells suffer a lateral displacement depending on their deformability. According to this lateral displacement, the cells were collected in different reservoirs. With this method, it was possible to generate an enrichment by factor 5 with a recovery ratio of 96%, with flow rates between tens to hundreds of microliters per minute.

6. Conclusions

We have tried to show that the application of concentration techniques may result in considerable improvement of biosensor-based analytical systems, helping to increase the sensitivity and selectivity and reduce the overall analysis time. Some examples of the most interesting and potentially useful strategies were presented. However, there are other mechanisms to handle and concentrate cell samples in microfluidic environments, such as acoustic [74], CD centrifugal microfluidic technologies [75, 76], or optical [77] methods that we have not treated in this review.

Among different existing concentration mechanisms, one should implement the most convenient, satisfying the requirements of the detection sensor used. There are many possibilities if we are looking for the concentration of large sample volumes with a very high concentration ratio. However, in many cases, a very specific system is required with high rates of purity or recovery efficiency.

	Device	Cell	Processed Volume	Initial Concentration / Detection	Final concentration	Trapping Efficiency / Separation Efficiency	Purity	Flow rates
Hydrdynamic	Hur [73]	CTC Breast cancer cells	-	1:100 (CTC:RBC) ≈2·10³ CTC/ml	≈5 times c/ml	-	96%	≈ 100 μl/min
	Mach [71]	CTC	10 ml	1:10⁷ (CTC:RBC) 500 CTC/ml	≈ 100 CTC/200μL ≈ 500 CTC/ml	20%	40%	5 ml/min
Dielectrophoresis	Yang [61]	*Listeria Monocytogenes*	-	-	-	90%	*Pure samples*	0.2 μL/min
	Park [59]	*E. Coli*	-	10⁸ c/ml	-	94%	*Pure samples*	13 μl/min
	Cho [57]	*E. Coli*	500 μL	≈ 9.3·10³ c/ml	≈ 4·10⁶ c/ml	60%	*Pure samples*	100 μL/min
Filter	Filter [12]	*E. Coli*	60 ml	10⁸ c/ml	-	-	*Pure samples*	-
Surface	Nagrath [15]	CTC	5.1 ml	10³ CTC/ml in whole blood	-	65%	50%	≈ 30 μl/min
Magnetic	Beyor [35]	*E. Coli*	100 μL	≈ *10⁴ c/ml*	-	70% (Pure samples) 60% (mixtures 1:1) 40% (mixtures 1:100)	-	0.24 μL/min
	Lee [36]	*Salmonella*	10 ml	10 cfu/ml	-	-	*Pure samples*	25 ml/min
Electrophoresis	Podszum [45]	*E. Faecalis*	200 – 100 μL	10² c/ml	Calculated factor = x 25	80%	*Pure samples*	3 - 6 μL/min
	Wang [47]	*E. Coli*	-	10⁶ c/ml	Calculated factor = x 10⁴	-	*Pure samples*	

Table 1. Summary of some experimental parameters of different reported methods of cell concentration and separation.

Table 1 summarizes experimental parameters of different reported methods of cells concentration and separation, including processed sample volumes, initial and final concentration, trapping efficiency, purity of the sample after separation, and respective flow rates. Some of the table data marked in cursive are our own estimations based on the direct data presented by the authors and were introduced to provide the possibility to perform direct comparison of different methods. However, due to the differences in the methodologies and measuring setups and sometimes poor details on the release and final concentration obtained, this comparison should be performed with care.

In general, to cover the real necessities of industry and the public health system in micrototal analysis systems and for the wide application of concentration techniques for biosensor applications, three main problems should be resolved:

- integration of known mechanisms into complete microfluidic systems
- capacity of sample processing and velocities of the most specific methodologies
- recovery performances and the purity of the recovered analyte

It seems that to enhance the performance of these systems, it is reasonable to look for a combination of different capture and separation methods combining them effectively with detection routines. This will open a future option to implement efficient concentration and sensing on a chip for real-life applications, providing high levels of integration and velocity, high volumes of sample processing, and high ratios of concentration and purity.

Acknowledgements

This work was carried out thanks to the financial support from ENIAC-JU, by the SILVER-ENIAC project. Authors also acknowledge financial support from Spanish Ministry of Economy and Competitiveness (projects IPT-2011-1055-900000 and CTQ2014-54553-C3-1-R).

Author details

Óscar Castillo-Fernandez, Naroa Uria, Francesc X. Muñoz and Andrey Bratov*

*Address all correspondence to: andrei.bratov@imb-cnm.csic.es

BioMEMsgroup, Instituto de Microelectrónica de Barcelona, Centro Nacional de Microelectrónica IMB-CNM (CSIC) Campus UAB, Cerdanyola del Vallès, Barcelona, Spain

References

[1] Dharmasiri, U., et al., Microsystems for the capture of low-abundance cells. Annual Review of Analytical Chemistry, 2010. 3(1): 409–431. DOI: 10.1146/annurev.anchem.111808.073610.

[2] Bhagat, A., et al., Microfluidics for cell separation. Medical & Biological Engineering & Computing, 2010. 48(10): 999–1014. DOI: 10.1007/s11517-010-0611-4.

[3] Mach, A.J., O.B. Adeyiga, and D. Di Carlo, Microfluidic sample preparation for diagnostic cytopathology. Lab on a Chip, 2013. 13(6): 1011–1026. DOI: 10.1039/C2LC41104K.

[4] Lin, C.-C., J.-L. Hsu, and G.-B. Lee, Sample preconcentration in microfluidic devices. Microfluidics and Nanofluidics, 2011. 10(3): 481–511. DOI: 10.1007/s10404-010-0661-9.

[5] Jokerst, J.C., J.M. Emory, and C.S. Henry, Advances in microfluidics for environmental analysis. Analyst, 2012. 137(1): 24–34. DOI: 10.1039/C1AN15368D.

[6] Gossett, D., et al., Label-free cell separation and sorting in microfluidic systems. Analytical and Bioanalytical Chemistry, 2010. 397(8): 3249–3267. DOI: 10.1007/s00216-010-3721-9.

[7] Nilsson, J., et al., Review of cell and particle trapping in microfluidic systems. Analytica Chimica Acta, 2009. 649(2): 141–157. DOI: 10.1016/j.aca.2009.07.017.

[8] Lenshof, A. and T. Laurell, Continuous separation of cells and particles in microfluidic systems. Chemical Society Reviews, 2010. 39(3): 1203–1217. DOI: 10.1039/B915999C.

[9] El-Ali, J., P.K. Sorger, and K.F. Jensen, Cells on chips. Nature, 2006. 442(7101): 403–411.

[10] Logan, B.E., T.A. Hilbert, and R.G. Arnold, Removal of bacteria in laboratory filters: models and experiments. Water Research, 1993. 27(6): 955–962. DOI: 10.1016/0043-1354(93)90059-Q.

[11] Li, X., et al., Rapid Sample Processing for Detection of Food-Borne Pathogens via Cross-Flow Microfiltration. Applied and Environmental Microbiology, 2013. 79(22): 7048–7054. DOI: 10.1128/aem.02587-13.

[12] Jiang, J., et al., Smartphone based portable bacteria pre-concentrating microfluidic sensor and impedance sensing system. Sensors and Actuators B: Chemical, 2014. 193(0): 653–659. DOI: 10.1016/j.snb.2013.11.103.

[13] Bao, N. and C. Lu, A microfluidic device for physical trapping and electrical lysis of bacterial cells. Applied Physics Letters, 2008. 92(21). DOI: 10.1063/1.2937088.

[14] Chang, K.-C. and D.A. Hammer, The forward rate of binding of surface-tethered reactants: effect of relative motion between two surfaces. Biophysical Journal, 1999. 76(3): 1280–1292. DOI: 10.1016/S0006-3495(99)77291-7.

[15] Nagrath, S., et al., Isolation of rare circulating tumour cells in cancer patients by microchip technology. Nature, 2007. 450(7173): 1235–1239. DOI: 10.1038/nature06385.

[16] Adams, A.A., et al., Highly efficient circulating tumor cell isolation from whole blood and label-free enumeration using polymer-based microfluidics with an integrated

conductivity sensor. Journal of the American Chemical Society, 2008. 130(27): 8633–8641. DOI: 10.1021/ja8015022.

[17] Wan, Y., et al., Velocity effect on aptamer-based circulating tumor cell isolation in microfluidic devices. Journal of Physical Chemistry B, 2011. 115(47): 13891–13896. DOI: 10.1021/jp205511m.

[18] Kirby, B.J., et al., Functional characterization of circulating tumor cells with a prostate-cancer-specific microfluidic device. PLoS One, 2012. 7(4): e35976. DOI: 10.1371/journal.pone.0035976.

[19] Zhao, W., et al., Bioinspired multivalent DNA network for capture and release of cells. Proceedings of the National Academy of Sciences, 2012. 109(48): 19626–19631. DOI: 10.1073/pnas.1211234109.

[20] Galletti, G., et al., Isolation of breast cancer and gastric cancer circulating tumor cells by use of an anti HER2-based microfluidic device. Lab on a Chip, 2013. DOI: 10.1039/c3lc51039e.

[21] Sheng, W., et al., Capture, release and culture of circulating tumor cells from pancreatic cancer patients using an enhanced mixing chip. Lab on a Chip, 2014. 14(1): 89–98. DOI: 10.1039/C3LC51017D.

[22] Gleghorn, J.P., et al., Capture of circulating tumor cells from whole blood of prostate cancer patients using geometrically enhanced differential immunocapture (GEDI) and a prostate-specific antibody. Lab on a Chip, 2010. 10(1): 27–29. DOI: 10.1039/B917959C.

[23] Shah, A.M., et al., Biopolymer system for cell recovery from microfluidic cell capture devices. Analytical Chemistry, 2012. 84(8): 3682–3688. DOI: 10.1021/ac300190j.

[24] Souiri, M., et al., AFM, CLSM and EIS characterization of the immobilization of antibodies on indium–tin oxide electrode and their capture of *Legionella pneumophila*. Talanta, 2014. 118(0): 224–230. DOI: 10.1016/j.talanta.2013.09.049.

[25] Dulay, S.B., et al., Automated microfluidically controlled electrochemical biosensor for the rapid and highly sensitive detection of *Francisella tularensis*. Biosensors and Bioelectronics, 2014. 59(0): 342–349. DOI: 10.1016/j.bios.2014.03.024.

[26] Cunliffe, D., et al., Bacterial adhesion at synthetic surfaces. Applied and Environmental Microbiology, 1999. 65(11): 4995–5002.

[27] Xie, G.-J., et al., Photo-hydrogen production by *Rhodopseudomonas faecalis* RLD-53 immobilized on the surface of modified activated carbon fibers. RSC Advances, 2012. 2(6): 2225–2228. DOI: 10.1039/C2RA01075E.

[28] Hwang, G., et al., Adhesion of nano-sized particles to the surface of bacteria: mechanistic study with the extended DLVO theory. Colloids and Surfaces B: Biointerfaces, 2012. 97(0): 138–144. DOI: 10.1016/j.colsurfb.2012.04.031.

[29] Sen, A., T. Harvey, and J. Clausen, A microsystem for extraction, capture and detection of E-Coli O157:H7. Biomedical Microdevices, 2011. 13(4) 705–715. DOI: 10.1007/s10544-011-9540-8.

[30] Verbarg, J., et al., Spinning magnetic trap for automated microfluidic assay systems. Lab on a Chip, 2012. 12(10): 1793–1799. DOI: 10.1039/C2LC21189K.

[31] Verbarg, J., et al., Catch and release: integrated system for multiplexed detection of bacteria. Analytical Chemistry, 2013. 85(10): 4944–4950. DOI: 10.1021/ac303801v.

[32] Liu, R.H., et al., Self-contained, fully integrated biochip for sample preparation, polymerase chain reaction amplification, and DNA microarray detection. Analytical Chemistry, 2004. 76(7): 1824–1831. DOI: 10.1021/ac0353029.

[33] Chan, K.Y., et al., Ultrasensitive detection of *E. coli* O157:H7 with biofunctional magnetic bead concentration via nanoporous membrane based electrochemical immunosensor. Biosensors and Bioelectronics, 2013. 41(0): 532–537. DOI: 10.1016/j.bios.2012.09.016.

[34] Barallat, J., et al., Chronoamperometric magneto immunosensor for myeloperoxidase detection in human plasma based on a magnetic switch produced by 3D laser sintering. Analytical Chemistry, 2013. 85(19): 9049–9056. DOI: 10.1021/ac401549d.

[35] Beyor, N., et al., Immunomagnetic bead-based cell concentration microdevice for dilute pathogen detection. Biomedical Microdevices, 2008. 10(6): 909–917. DOI: 10.1007/s10544-008-9206-3.

[36] Lee, W., et al., Ultrarapid detection of pathogenic bacteria using a 3D immunomagnetic flow assay. Analytical Chemistry, 2014. 86(13): 6683–6688. DOI: 10.1021/ac501436d.

[37] Doyle, R.J., Contribution of the hydrophobic effect to microbial infection. Microbes and Infection, 2000. 2(4): 391–400. DOI: 10.1016/S1286-4579(00)00328-2.

[38] Cruz, L.F., P.A. Cobine, and L. De La Fuente, Calcium increases *Xylella fastidiosa* surface attachment, biofilm formation, and twitching motility. Applied and Environmental Microbiology, 2012. 78(5): 1321–1331. DOI: 10.1128/aem.06501-11.

[39] Walker, S.L., J.A. Redman, and M. Elimelech, Role of cell surface lipopolysaccharides in *Escherichia coli* K12 adhesion and transport. Langmuir, 2004. 20(18): 7736–7746. DOI: 10.1021/la049511f.

[40] Kulagina, N.V., et al., Antimicrobial peptides for detection of bacteria in biosensor assays. Analytical Chemistry, 2005. 77(19): 6504–6508. DOI: 10.1021/ac050639r.

[41] Mannoor, M.S., et al., Electrical detection of pathogenic bacteria via immobilized antimicrobial peptides. Proceedings of the National Academy of Sciences, 2010. 107(45): 19207–19212. DOI: 10.1073/pnas.1003768107.

[42] Corpe, W.A., Attachment of Marine Bacteria to Solid Surfaces. Vol. 73. New York: Academic Press, 1970.

[43] Dickson, J.S. and M. Koohmaraie, Cell surface charge characteristics and their relationship to bacterial attachment to meat surfaces. Applied and Environmental Microbiology, 1989. 55(4): 832–836.

[44] Balasubramanian, A.K., et al., A microfluidic device for continuous capture and concentration of microorganisms from potable water. Lab on a Chip, 2007. 7(10): 1315–1321. DOI: 10.1039/B706559K.

[45] Podszun, S., et al., Enrichment of viable bacteria in a micro-volume by free-flow electrophoresis. Lab on a Chip, 2012. 12(3): 451–457. DOI: 10.1039/C1LC20575G.

[46] Pu, Q., et al., Ion-enrichment and ion-depletion effect of nanochannel structures. Nano Letters, 2004. 4(6): 1099–1103. DOI: 10.1021/nl0494811.

[47] Wang, Z., et al., Rapid detection and quantification of bacteria using an integrated micro/nanofluidic device. Sensors and Actuators B: Chemical, 2013. 178(0): 683–688. DOI: 10.1016/j.snb.2013.01.017.

[48] Markx, G.H., et al., Dielectrophoretic characterization and separation of micro-organisms. Microbiology, 1994. 140(3): 585–591. DOI: 10.1099/00221287-140-3-585.

[49] Markx, G.H. and C.L. Davey, The dielectric properties of biological cells at radiofrequencies: applications in biotechnology. Enzyme and Microbial Technology, 1999. 25(3–5): 161–171.

[50] Pethig, R., Dielectrophoresis: status of the theory, technology, and applications. Biomicrofluidics, 2010. 4(2): 022811–35.

[51] Huang, Y., et al., Differences in the AC electrodynamics of viable and non-viable yeast cells determined through combined dielectrophoresis and electrorotation studies. Physics in Medicine and Biology, 1992. 37(7): 1499.

[52] Morgan, H. and N.G. Green, AC Electrokinetics: colloids and nanoparticles. Herts: Research Studies Press, 2003.

[53] Castellarnau, M., et al., Dielectrophoresis as a tool to characterize and differentiate isogenic mutants of *Escherichia coli*. Biophysical Journal, 2006. 91(10): 3937–3945. DOI: 10.1529/biophysj.106.088534.

[54] Ferrier, G.A., et al., A microwave interferometric system for simultaneous actuation and detection of single biological cells. Lab on a Chip, 2009. 9(23): 3406–3412. DOI: 10.1039/B908974H.

[55] Dastider, S.G., et al., A micromachined impedance biosensor for accurate and rapid detection of *E. coli* O157:H7. RSC Advances, 2013. 3(48): 26297–26306. DOI: 10.1039/C3RA44724C.

[56] Kovarik, M.L., and S.C. Jacobson, Integrated nanopore/microchannel devices for ac electrokinetic trapping of particles. Analytical Chemistry, 2008. 80(3): 657–664. DOI: 10.1021/ac701759f.

[57] Cho, Y.-K., et al., Bacteria concentration using a membrane type insulator-based dielectrophoresis in a plastic chip. Electrophoresis. 2009. 30(18): 3153–3159. DOI: 10.1002/elps.200900179.

[58] Han, K.-H. and A.B. Frazier, Lateral-driven continuous dielectrophoretic microseparators for blood cells suspended in a highly conductive medium. Lab on a Chip, 2008. 8(7): 1079–1086.

[59] Park, S., et al., Continuous dielectrophoretic bacterial separation and concentration from physiological media of high conductivity. Lab on a Chip, 2011.

[60] Yang, L., A review of multifunctions of dielectrophoresis in biosensors and biochips for bacteria detection. Analytical Letters, 2012. 45(2–3): 187–201. DOI: 10.1080/00032719.2011.633182.

[61] Yang, L., et al., A multifunctional micro-fluidic system for dielectrophoretic concentration coupled with immuno-capture of low numbers of *Listeria monocytogenes*. Lab on a Chip, 2006. 6(7): 896–905. DOI: 10.1039/B607061M.

[62] Suehiro, J., et al., Selective detection of bacteria by a dielectrophoretic impedance measurement method using an antibody-immobilized electrode chip. Sensors and Actuators B: Chemical, 2006. 119(1): 319–326. DOI: 10.1016/j.snb.2005.12.027.

[63] Koo, O.K., et al., Targeted capture of pathogenic bacteria using a mammalian cell receptor coupled with dielectrophoresis on a biochip. Analytical Chemistry, 2009. 81(8): 3094–3101. DOI: 10.1021/ac9000833.

[64] Vaidyanathan, R., et al., Tunable "nano-shearing": a physical mechanism to displace nonspecific cell adhesion during rare cell detection Analytical Chemistry, 2014. DOI: 10.1021/ac4032516.

[65] Hu, X., et al., Marker-specific sorting of rare cells using dielectrophoresis. Proceedings of the National Academy of Sciences of the United States of America, 2005. 102(44): 15757–15761. DOI: 10.1073/pnas.0507719102.

[66] Karimi, A., S. Yazdi, and A.M. Ardekani, Hydrodynamic mechanisms of cell and particle trapping in microfluidics. Biomicrofluidics, 2013. 7(2): 021501. DOI: 10.1063/1.4799787.

[67] Park, J.-S., S.-H. Song, and H.-I. Jung, Continuous focusing of microparticles using inertial lift force and vorticity via multi-orifice microfluidic channels. Lab on a Chip, 2009. 9(7): 939–948. DOI: 10.1039/B813952K.

[68] Di Carlo, D., Inertial microfluidics. Lab on a Chip, 2009. 9(21): 3038–3046.

[69] Wu, Z., et al., Soft inertial microfluidics for high throughput separation of bacteria from human blood cells. Lab on a Chip, 2009. 9(9): 1193–1199. DOI: 10.1039/B817611F.

[70] Geislinger, T.M., et al., Separation of blood cells using hydrodynamic lift. Applied Physics Letters, 2012. 100(18): 183701. DOI: 10.1063/1.4709614.

[71] Mach, A.J., et al., Automated cellular sample preparation using a Centrifuge-on-a-Chip. Lab on a Chip, 2011. 11(17): 2827–2834. DOI: 10.1039/C1LC20330D.

[72] Moon, H.-S., et al., Continuous separation of breast cancer cells from blood samples using multi-orifice flow fractionation (MOFF) and dielectrophoresis (DEP). Lab on a Chip, 2011. 11(6): 1118–1125.

[73] Hur, S.C., et al., Deformability-based cell classification and enrichment using inertial microfluidics. Lab on a Chip, 2011. 11(5): 912–920. DOI: 10.1039/C0LC00595A.

[74] Ai, Y., C.K. Sanders, and B.L. Marrone, Separation of *Escherichia coli* bacteria from peripheral blood mononuclear cells using standing surface acoustic waves. Analytical Chemistry, 2013. 85(19): 9126–9134. DOI: 10.1021/ac4017715.

[75] Martinez-Duarte, R., et al., The integration of 3D carbon-electrode dielectrophoresis on a CD-like centrifugal microfluidic platform. Lab on a Chip, 2010. 10(8): 1030–1043. DOI: 10.1039/B925456K.

[76] Ren, Y., L.-C. Chow, and W.-F. Leung, Cell culture using centrifugal microfluidic platform with demonstration on *Pichia pastoris*. Biomedical Microdevices, 2013. 15(2): 321–337. DOI: 10.1007/s10544-012-9735-7.

[77] Yang, A.H.J., et al., Optical manipulation of nanoparticles and biomolecules in sub-wavelength slot waveguides. Nature, 2009. 457(7225): 71–75. DOI: 10.1038/nature07593.

Chapter 8

Mechanical Sensing of Living Systems — From Statics to Dynamics

F. Argoul, B. Audit and A. Arneodo

Additional information is available at the end of the chapter

http://dx.doi.org/10.5772/60883

Abstract

Living systems are fascinating sensing machines that outmatch all artificial machines. Our aim is to put a focus on the dynamics of mechanosensing in cellular systems through concepts and experimental approaches that have been developed during the past decades. By recognizing that a cellular system is not simply the intricate assembly of active and passive macromolecular actors but that it can also manifest scale-invariant and/or highly nonlinear global dynamics, biophysicists have opened a new domain of investigation of living systems. In this chapter, we review methods and techniques that have been implemented to decipher the cascade of temporal events which enable a cell to sense a mechanical stimulus and to elaborate a response to adapt or to counteract this perturbation. We mainly describe intrusive (mechanical probes) and nonintrusive (optical devices) experimental methods that have proved to be efficient for real-time characterization of stationary and nonstationary cellular dynamics. Finally, we discuss whether thermal fluctuations, which are inherent to living systems, are a source of coordination (*e.g.*, synchronization) or randomization of the global dynamics of a cell.

Keywords: Mechanosensing, biosensing, mechanotransduction, cytoskeleton, multiscale, scale-invariant, nonlinear dynamics, focal adhesion, cellular rheology, coherent dynamics

INTECH
open science | open minds

1. Introduction

The concept of mechanical sensing dates back to the 19th century with the emergent theory of tone sensing proposed by H. Helmholtz and J. Muller [1, 2]. All living organisms have the ability to sense mechanical stress and/or hydrostatic pressure, either locally or globally. One of the most studied example is the touch perception of metazoa [3-6]. The first model of mechanosensing proposed by physiologists was constructed on the concept of mechanical receptors [3, 4, 7], *i.e.*, mechanical machineries that are able to transform a mechanical information in another signal that will afterward be interpreted by the cell and potentially transformed into an adaptative response [8]. This concept of receptor or transducer is inspired from mechanical engineering methods. The term mechanotransduction [5] has been more recently introduced to explain how a single cell transforms a mechanical stress through signaling pathways down to nuclear molecular processes.

Mechanosensing [9-11] is a complex mechanism that involves not only a whole range of molecular actors with nanometer-scale sensibility but also a dynamical integration and regulation of these molecular actors that allow a much larger scale (μm to mm) response in amplitude, with strength and duration adapted to the perturbation. Our physical models of cellular nanomechanics still rarely consider active viscoelastic systems [12] and despite a recent increase of the rate of publications devoted to nanobiomechanics of cells, the concept of cellular biodynamics is still in its middle age. This relative slow progression comes from the necessity to introduce concepts of active matter [13-18] into biology. All cells interact physically with their surrounding tissue and they can establish their response on various timescales. To get a full understanding of cellular biodynamics, we actually need to master the most fundamental concepts of atomic and statistical physics, submicron-scale hydrodynamics and out-of-equilibrium nonlinear dynamics and to associate nanotechnologies with optogenetic, microfluidic tools and molecular and cellular biology methods to achieve a complete mechanogenetic characterization of living cells. Actually, a cell is able to combine biochemical submolecular and supramolecular active or passive interactions with micron-size mechanical and electromagnetic informations that we still have much difficulty to reproduce, with either our experimental devices or our huge computer machines.

There are three aspects that we would like to put forward in this chapter. The first one is the importance of molecular machines of living systems, also called molecular motors, which drive all the cell movements thanks to ATP consumption. These molecular machines are not fully deterministic motors but are in part driven by thermal fluctuations. The second aspect that seems important to us is the fact that fluctuations are predominant in biological systems, and that the fluctuation dissipation theorem [19] must not be taken as granted in all situations. This means that standard approaches in statistical physics and corresponding mathematical models must be used with caution. The third aspect is the importance of multiscale properties in mechanosensing: short-, middle-, and long-distance interactions contribute to the dynamics of cellular systems and more widely of living systems [20-29]. Because they are nonlinear and nonstationary, these interactions produce a global dynamics that each element could not achieve alone. However, the nature of these interactions is still the subject of current debate. We will illustrate different approaches that have been used so far to address this issue.

The most striking example of the interplay of mechanics and cell dynamics is illustrated in cell migration and adhesion. Actually, the mobility of a cell is a subtle combination of two counteracting mechanisms: on the one hand, adhesion which tends to immobilize the cell and, on the other hand, protrusion/retraction mechanisms which modify the cell shape and assist its movement. These processes also underlie the ability of a cell to deform under a mechanical perturbation. The cytoskeletal dynamics and mechanics are univocally linked to cellular tension in cell adhesion [30-33]. Cellular movement by membrane protrusion and formation of new adhesions at the cell front cannot occur without a tight link of the cell cortex to the whole cytoskeleton (CSK), allowing the settling of traction forces that drive the cell forward in motion in synchrony with the disassembly of the rear fibers. External membrane protrusions are important components of the ability of a cell to migrate or interact with other neighboring cells [34-36]. These protrusions can be viewed as local instabilities of the cell cortex. They are not independent of the internal dynamics of the CSK: the microtubule plus ends associate with F-actin via plus-tip proteins and act as a scaffolding complex [37] that recruits further down other protein effectors involved in the actin network remodeling [38].

When cells are placed in adherent conditions, they rapidly develop integrin-mediated adhesion complexes that link the extracellular matrix (ECM) to the actin CSK. These trans-membrane proteins are associated with a complex of proteins (vinculin, actin, paxillin, tensin, etc.) which allows very fast and reversible connectivity of the intracellular CSK to the outer membrane complexes [39, 40]. The integrin-based molecular complexes concentrate in small domains with different size and shapes that focalize the cell traction force on the ECM. In addition to their function as adhesion sites, matrix adhesion foci also participate in the adhesion-dependent signaling pathways via tyrosine kinases, tyrosine phosphatases, etc. Focal adhesion (FA) centers function as both adhesion and signal transduction hubs that communicate the external stresses of the ECM to the cell interior [39]. The maturation of these FA complexes cannot exist if the cell does not have a contractile machinery, *e.g.*, canceling the cell contractility by inhibiting the Rho GTPase or tyrosine kinase activity aborts this maturation. The higher the cell tension, the larger and more mature are the FA complexes. The FA complexes can therefore be considered as mechanosensing intermediates coupling the internal cell traction forces with the ECM ridigity [41-44].

The actin network can be viewed as a fluid-gel structure which plays both a passive (viscous) and an active (ATP-driven) role in the spatiotemporal dynamics of the cells. This network is dynamically intertwined with microtubules and interfilament networks in such a way that the leading edge of the cell undergoes retrograde flow away from this edge simultaneously to the cell migration. This retrograde flow occurs in two steps: (i) on short timescales, a fast flow in the most peripheral region of the cell, the lamellipodium (a 1-4 μ m width extension filled with a dense network of branching F-actin filaments); and (ii) on longer timescales, a slower centripetal flow over the broad (more than 10 μ m width) lamellum [45]. The fast lamellipodium flow principally involves an F-actin network. The slower lamellum flow involves all the CSK filaments (actin stress fibers, microtubules, and intermediate filaments) and a relatively more sparse actin network. These two types of flow are each characterized by a specific organization of the cytoskeletal network and a different turnover rate. In addition, they are driven by distinct forces, namely actin assembly/disassembly in lamellipodium and actomyosin contraction in lamellum [46]. This indicates that the whole dynamics of a cell during protrusion, traction, and migration is a highly correlated, multiscale (in time and in space) process that entails long-range and short-range mechanisms that can only be tackled using multiscale experimental concepts and methods.

2. Physical probes to capture the mechanical response of living cells

In this section, we concentrate on the nano and micromechanical tools which have been designed in the past two decades to record in real time the mechanics and rheology of a living cell, with the specific purpose to understand its mechanosensing properties. These methods can be classified into two groups: (i) the methods that introduce mechanical tracers inside the cell and follow their spatiotemporal dynamics and (ii) the methods remaining external to the cell and that bring a mechanical device (nano- or microscale in size) close to the cell to follow its response in real time. In each case, a few examples will be described as regard to their ability and efficiency in extracting characteristic temporal and/or spatial scales in the dynamics of cell adaptation to a mechanical stress.

Cellular rheology from the outer membrane Rheological properties of cells, their deformability under stress are key features of their ability to sense their environment. Recent studies of the microrheology of the intracellular medium have highlighted the fact that this viscoelastic medium is complex and cannot be modeled by the association of a finite set of elastic and viscous elements as usually done in mechanical engineering [25, 47]. Actually, the viscoelastic complex modulus of the cell medium exhibits a weak power-law behavior over a wide frequency range. Using magnetic twisting cytometry (MTC) coupled to an optical detection of the motion of a bead coupled to membrane RGD receptor, Fabry *et al.* [20, 21] succeeded in probing the cell surface dynamics in the frequency range from 0.01 Hz to 1 kHz. During the bead displacement on the cell surface (forced by a twisting magnetic field), the cell responds with an opposing torque that reflects the cell mechanical strength. The ratio of the complex torque T to the complex bead displacement $\tilde{d}$ in Fourier space is defined as the elastic modulus $\tilde{G}(f)$:

$$\tilde{G}(f) = f_g \frac{\tilde{T}(f)}{\tilde{d}(f)}, \tag{1}$$

where the proportionality geometrical factor f_g depends on the shape, thickness of the cell, and the degree of embedding of the bead in the cell cortex. $\tilde{G} = G' + iG''$, where G' is the shear modulus and G'' the loss modulus. The range of stress and deformation used in this study was limited to the linear response regime for the cell. These authors found for five types of adherent cell models that both G' and G'' increase with excitation frequency as a weak power-law over the whole frequency range. These power-law dependence of G' and G'' on frequency was also observed by other groups [22, 24, 48-50] and with other methods, such as atomic force microscopy [51, 52]. Except for a small additive viscous term that emerged only at high frequencies, mechanical responses collected from the cell surface did not appear to be tied to any specific frequency and in that respect was considered as (time) scale-invariant.

When a power-law behavior emerges in the rheological response of a cell, a wide range of frequencies is required to bring the experimental demonstration of the existence of scale invariance in the cell dynamical response to stress. Actually, a limited range of frequencies could still be parameterized by a combination of a small number of viscoelastic elements, as an exponential crossover between two regimes. The most impressive result of the above studies is the fact that all the curves captured from different cells of various types could be collapsed to single master curves typical of a soft glassy material (SGM) [25, 53], demonstrating the universality of this behavior [24, 54]. This universality law can be written as

$$\tilde{G}(f) = G'(f) + iG''(f) = G_0\left(i\frac{f}{f_0}\right)^{x-1} + 2\pi if\mu, \tag{2}$$

where x is a unifying parameter, G_0 and f_0 are cell-type-dependent scaling factors for stiffness and frequency, μ is an additive Newtonian viscosity term that is negligible for frequencies lower than 30 Hz. This equation tells us that (below 30 Hz) the phase angle ϕ of $\tilde{G}$,

$$\phi = \tan^{-1}\left(\frac{G''}{G'}\right) = \frac{\pi}{2}(x-1), \tag{3}$$

is independent of the forcing frequency. This unifying parameter x depends on the cell state; x decreases to 1 when the cell approaches an ideal elastic material (for instance, by increasing its contraction) whereas x increases toward 2 (limit of a Newtonian viscous fluid) when the cell prestress is diminished (*e.g.*, by disrupting the actin CSK).

The common and generic features of SGMs are due to the fact that they are composed of numerous discrete elements which are interconnected in a random way via weak interactions. These materials are out-of-equilibrium metastable systems, very much like living cells. However, soft glassy dynamics as proposed by soft glass rheology theory [53] is not the only mechanism that can lead to scale-free mechanical behavior as expressed by power-law stress relaxation. Power-law behavior can also be produced by models containing a large number of viscoelastic compartments with a particular distribution of characteristic relaxation times $P(\tau) \sim \tau^{-x}$ that must be related to intracellular processes.

With the same MTC device, it was also possible to track the spontaneous motions (without magnetic twisting) of small beads linked to cell membrane integrin receptors on adherent cells. Bursac *et al.* [48] observed that these motions were intermittent with periods of confinement (stalling) punctuated by directed movement (hopping). Plotting the mean-square displacement of the beads versus time revealed that they were subdiffusing (stalling) at short times but superdiffusing at longer times (hoping) and that intermittent motions reflecting nanoscale CSK rearrangements depended on both the approach to kinetic arrest and energy release due to ATP hydrolysis. The percentage of hopping events in the bead motion was shown to be higher than in soft glassy systems where only thermal energy can push the system out of a microenergy well. Thanks to ATP-driven motors and polymerization/depolymerization cycles, the active properties of the cell CSK and cortex provide alternative ways to visit different microconfigurations and adjust to the ECM changes, with an effective local temperature which increases with the parameter x. Therefore, the exponent x tells us the extent to which the cell behaves as a fluidic system ($x \sim 2$) or as an elastic solid ($x \sim 1$).

Cytoplasm rheology Microbeads engulfed inside living cells were used as tracers of the internal cellular activity [55-57]. Two different regimes of transport were observed: on the one hand, the passive fluctuations (local movements of the tracers) which characterize the local viscosity of the cytoplasm and, on the other hand, the active trajectories which are driven by molecular motors such as kinesins and dyneins along microtubules. The same type of experiments has also been performed more recently by nanoparticule tracing and manipulation inside A7 melanoma, MCF-10A and MCF-7 cells [58] with optical tweezers, leading again to the conclusion that the elastic modulus follows a power-law: $|G(f)| \sim f^{\beta}$, with $\beta = x - 1 = 0.15$, in agreement with previous measurements on the cell exterior [20, 21]. They also noted that the measured cytoplasmic modulus is approximately of a few Pa, much lower than previously estimated for its actin cortex [20]. Thus, these tweezers measurements confirmed the rubber-like elastic properties of the cytoplasm of these cells in two-dimensional (2D) adherent conditions.

Fluctuation dissipation theorem and cell rheology The equilibrium fluctuation-dissipation theorem (FDT) [19] assumes that the response of a system to a small perturbation is hampered by spontaneous fluctuations at equilibrium (damping term). Understanding the nature and the amplitude of nonequilibrium forces driving the dynamics of cells out of equilibrium is a very important challenge for statistical physicists and biologists. There are many evidences that FDT fails in living cells [59-62] as well as in active gel systems [63], and it is therefore important to estimate the critical timescale at which active-force-driven fluctuations are predominant over thermal fluctuations. Using micron-size silica beads attached to the wall of a living adherent myoblast cell, Bohec *et al.* [64] have recently identified a crossover time ($\tau \sim 1$s) between thermally controlled fluctuations and active-force-driven fluctuations. This seems to corroborate that the short-time behavior has equilibrium-like properties, from which the subdiffusive nature of viscoelasticity emerges, while the long-time behavior is strongly governed by active nonequilibrium forces. Bohec *et al.* [64] also provided a quantitative estimation of the power dissipated by the active forces into the system, which turns out to be three orders of magnitude smaller than the chemical power injected into the underlying motors by the ATPase machinery.

Long-range cell deformation capture with an atomic force microscope As emerging in the late 1990s from scanning tunneling microscopy (STM) technologies, atomic force microscopy (AFM) was early recognized to provide a unique opportunity to investigate the structure, morphology, micromechanical properties, and biochemical signaling activity of cells under physiological environment, and this with high temporal and spatial resolutions [65, 66]. The principle of AFM is to bring directly in soft (or hard) contact a sharp-tip cantilever probe over a cell surface and to capture with piconewton sensitivity the interaction force of the tip with the cell surface. AFM is a very powerful technique that has been used to detect single biomolecules (receptors, lipids) on single cell surface without the need for fixation or staining. AFM has such sensitivity that it can be used to measure interaction between and within single biomolecules [67-69]. Beyond its preliminary application for imaging the topography of biological objects [70-72], AFM has become a multitask scanning probe versatile tool (antigen recognition, molecular and membrane flexibility, single molecule, gel, cell and tissue elasticity, electric current, conductance, near-field electromagnetic field) [73, 74]. AFM force spectroscopy can be applied to probe the elastic properties of a cell, either adherent or confined in a narrow chamber [75-78]. Its unique ability to detect and to map the cellular elasticity of living cells with a few tens of nanometers' resolution definitely outmatches the performance of other techniques such as magnetic or optical tweezers. However, it has as a main limitation that it cannot probe cell internal structure without crossing before the cell cortex. This difficulty has been partly overcome recently, thanks to a singular space-scale analysis of force-distance curves to disentangle the viscoelastic moduli of the cell cortex and of the underlying CSK [79]. Interestingly, 2D mapping of mean elastic modulus on a large variety of cells [75, 76, 80-83] was reconstructed, revealing for the first time intracellular interplay of mechanical forces in living cells.

The purpose of this paragraph is not to make a detailed review of AFM or to advertise its latest technological development which can be found in an increasing number of published reports [84-88], but rather to (i) pinpoint the few approaches which were focused on dynamical characteristics of living cells during large deformation and (ii) propose new research directions to perform real-time capture of the cell dynamics when the cell is not in a stationary phase. Unlike AFM-based microrheology measurements [51, 52] discussed in the previous paragraph which were limited to very small deformations, we consider now much larger deformations (more than 1/10 of the cell size) and their temporal and/or frequency decomposition [89]. To perform large deformation cell study, a new experimental strategy has been recently proposed that consists in exciting the cantilever and recording the cell response over a band of frequencies rather than at a single frequency [90]. Note that broadband excitation of the cantilever can also be achieved by thermal excitation [91]. When the probed object is in a stationary regime, power spectral analysis of cantilever fluctuations, based on Fourier analysis, is the best way to understand how the interaction of the cantilever tip is changed when coming in contact with the sample surface When the sampled surface is not stationary, it is no longer possible to perform a simple spectral analysis which only displays an averaged decomposition of the signal in frequency domain. The lower frequency part of the power spectrum is biased by the cell dynamical adaptation to the cantilever stress. To circumvent this difficulty, time-frequency analysis based on the wavelet transform has recently been proposed [92-95]. The continuous wavelet transform (WT) performs the spectral analysis of the signal on a compact window (given by the wavelet) and allows, therefore, to follow how the cell mechanics changes during its strain-to-stress response. Such a study has been recently performed on HOPG surfaces [96, 97] and on living myoblasts [98].

Intracellular stress measurements External forces that are transferred across integrins in FAs and channeled through the CSK can alter signaling activities deep inside the cell [99]. Another evidence for long-distance force transfer was provided by intracellular stress tomography measurements [89]. The long-distance force transmission mediated by intermediate filaments was observed in response to fluid flow-induced shear stress applied to the apical surface of endothelial cells [100]. Using the MTC technique, Laurent *et al.* [101] also found from alveolar epithelial cells, that the submembranous "cortical" CSK, which is mainly composed of actin, is less stiff and more responsive to external forces than the "deep" subcortical CSK, which also includes intermediate filaments and microtubules. Based on these findings, these authors concluded that "mechanical deformation is transmitted globally throughout the network, whereas the cell surface is able to 'sense' very local deformation forces." To analyze the distribution and dynamics of traction stress within individual FAs, Plotnikov *et al.* [102] applied high-resolution traction force microscopy (TFM) [103, 104] to mouse embryonic fibroblasts expressing enhanced green fluorescent protein (eGFP)-paxillin as FA marker. These fibroblasts were plated on fibronectin-coupled elastic polyacrylamide supports (PAA) of known rigidity embedded with a mixture of red and far-red fluorescent beads. Cell-induced ECM deformation was visualized by spinning disk confocal microscopy, and traction fields were reconstructed at 0.7 μm resolution with Fourier transform traction cytometry [104]. FAs were found to exhibit tugging traction fluctuations on a wide range of ECM rigidities, but the choice of tugging versus stable traction states was shown to be regulated by both tension and a specific signaling pathway. These experiments suggest that strengthening the molecular clutch via the FAK/phosphopaxillin/vinculin pathway broadens the range of rigidities over which dynamic ECM rigidity sampling operates. The requirement for tugging focal adhesion traction in durotaxis suggests that tugging is a means of repeatedly sensing the local ECM rigidity landscape over time. Individual FAs within a single cell sense dynamically the sample rigidity by applying fluctuating pulling forces to the ECM and behave therefore as dynamical sensors to guide durotaxis.

3. Biochemical sensors based on fluorescence methods for capturing cell dynamics: from the nano to the microscale

3.1. Fluorescence-based nanomechanical sensors of intra- and intermolecular dynamics

In the late nineties, combined progress in the biology of fluorescent proteins, miniaturization of optical systems, and nanotechnologies have provided a tremendous asset throughout the investigation of the kinetic properties of macromolecules in living cells [105]. The chemical interaction between two molecular complexes of a metabolic pathway is conditioned by their ability to come in contact, which is often assisted by ATP driven molecular motors. Transport of protein actors in a randomly crowded space such as the cellular cytoplasm differs markedly from a batch reactor. A common form of biochemical regulation is allostery, where an effector molecule binds to a regulatory site and favors a global conformational change that alters further down the structure and function of the active site. Mechanical forces regulate receptor-ligand binding conformation through control of allosteric conformational changes [106]. This general idea of mechanical regulation of active site functions through allosteric-like regulation of a distal site is termed mechanochemistry and is well accepted for motor proteins [107, 108]. We focus here on cytoskeletal proteins, since they are directly involved in mechanosensing pathways; however, the approaches discussed below could be generalized to a wide variety of biochemical interactions.

To dissect how mechanical stress impacts the structure of cytoskeletal proteins, molecular labels have been designed by physicists to provide a fluorescence signal that could report on the molecular strain. Whereas many proteins have been shown *in vitro* or predicted by numerical simulations to undergo conformational changes in response to external mechanical stress, we had to wait until the 2000s to get the demonstration of these changes *in vivo*. Most of these molecular sensors use the Forster resonance energy transfer [109] (FRET). FRET is a technique that can measure the proximity or spatial distance between a donor and an acceptor molecule. It has been widely used to detect protein conformational changes, thanks to molecular constructions with various fluorescent proteins [110]. The principle of stress FRET sensors was elaborated by combining two mutants of a fluorescent protein [111] with either a stable α -helix linker [112, 113], a spectrin linker [114], or a spider silk domain as linker [115]. These force sensors share a common mechanism for interpreting force: tension in the host induces strain in the linker, leading to increased distance between the donor and acceptor. The dynamic range of these sensors is limited by the nearly linear relationship between FRET efficiency and strain [113].

Recently, using the high flexibility of the vinculin linker domain, a sensor based on force transmission through FAs was developed [115, 116]. When the head integrin domain Vh binds to talin, it recruits vinculin to FA, whereas on the other side the tail integrin domain Vt binds to F-actin and paxillin. This intermediate flexible vinculin linker plays an important role in the transmission of adhesion strength from the FAs to the actin CSK. This calibrated biosensor has piconewton (pN) sensitivity, and the tension across vinculin in stable FAs was estimated to 2.5 pN. It was also demonstrated that higher tension across vinculin favors adhesion assembly and enlargement, and conversely that low tension vinculin favors disassembly or sliding of FAs at the trailing edge of migrating cells. Finally, this study [115] revealed that FA stabilization under force requires both vinculin recruitment and force transmission, and surprisingly, that these processes can be controlled independently.

Another type of strain sensors was elaborated from proximity imaging microscopy (PRIM) combined with GFP dimers [117] and further called PRI-based strain sensor module (PriSSM). If two GFP molecules are brought into physical contact, changes in the ratio of fluorescence emitted when excited with 395 nm and 475 nm light occur. Proximity imaging exploits these changes to reveal homotypic protein-protein interactions *in vivo*. Unlike FRET, PRIM involves only 2 types of fluorescent excitation spectra corresponding to monomeric and dimeric GFP, so that an estimated excitation ratio should simply reflect a mixing ratio of the monomer and the dimer. By combining the GFP-based PRIM technique and myosin-actin as the model system, Iwai *et al.* [117] used this genetically encoded fluorescent sensor to visualize the interaction between myosin II and F-actin in *Dictyostelium* cells. Both spectroscopic and microscopic studies suggested that the fraction of PriSSM-myosin bound to F-actin is low in normal cells.

We have just given few examples of application of FRET to probe cellular internal molecular structures and their transformation under mechanical stress and association with molecular partners. If one can use this method to identify the mechanical organizing centers in a mechanotransduction pathway, the range of forces estimated *in vivo* can be biased by several limitations. FRET is sensitive to changes in distance between 0.5 and 2 times the Forster radius (*i.e.*, between 2 and 10 nm) and to the fluorescence lifetimes of the donor (free or engaged in FRET). Indeed, the dynamic range accessible to this technique is rather low because the interactions which are probed by FRET are typically in the range of a few ns (the fluorescence lifetime of the donor) and difficult to discriminate from background thermal or shot noise. The donor fluorescence lifetime decreases due to energy transfer in the excited state. In adhesion, molecules interact with their closer neighbors but also with other partners on much longer distances which can reach several tens of nm, and thus may be missed by FRET. Conversely, molecules that do not actually participate to a mechanotransduction complex can nevertheless be in juxtaposition and show FRET. FRET experiments suffer from additional artifacts: the method is aimed at probing the interaction of two partners but it is quite impossible to separate the fluorescence responses coming from multiple donor and acceptor interactions even if extensive controls for every FRET pair studied have been performed *a priori*. Finally, none of these methods can determine the number of molecules of a certain component and stoichiometry within an adhesion. A partial correction of these limitations has recently been proposed by coupling FRET biosensors with fluorescence lifetime imaging microscopy [118-120]. Note that this method also provides a millisecond temporal resolution.

3.2. Spanning short- to long-range interactions and transport with fluorescence correlation spectroscopy

Analyzing the fluorescence fluctuation signals offered a simple, high-resolution, quantitative method to probe the intracellular dynamics that other fluorescence imaging techniques could not afford. From a single fluctuation temporal signal it is possible to get several informations over a wide range of frequencies, such as molecular densities, interaction rate and stochiometry, intra- and extracellular transport (diffusion, advection, etc.). These fluctuation signals should therefore be a very good candidate to capture the multiscale properties of cells in space and in time. Fluorescence correlation spectroscopy (FCS) was originally developed [121] to measure diffusion coefficients and chemical rate constants of biomolecules in solution. It has also been applied successfully to characterize the nature of transport processes of colloidal particles in complex flows [122]. This method uses a focused laser beam to define a very small focal volume ($\sim$1 femtoliter) from which the fluctuations of fluorescence intensity are recorded. These fluctuations are analyzed in the nanosecond-to-hour temporal range, and can therefore give information about many different processes including transport, exchange and binding interactions, fluorescence bleaching or blinking. The characteristic times of these different processes are uncovered by computing autocorrelation functions (ACF). Modeling of these ACFs allows the estimation of diffusion, transport, and reaction rates, but it can also be generalized to cross-correlation analysis to quantify molecular interactions if two fluorophores are used simultaneously in the confocal volume [123-125].

In experiments where the fluorescence signal is too weak or lacks contrast, thereby preventing separation of features from background signals, the spatial and temporal fluorescence cross-correlation functions allow to recover enough contrast thanks to their temporal fluctuations. This method was recently applied by Chiu *et al.* [126] to capture actin flow as well as F-actin dynamics and location of F-actin bundles in human breast adenocarcinoma cells (MDA-MB-231) grown in three-dimensional collagen gels. By recording simultaneously the collagen signal using confocal reflection microscopy, they also showed that collagen fibers move in concert with the actin-bundle flow. This experimental study is an impressive demonstration of the impact of fluctuations on internal cell dynamics, and of the power of fluorescence cross-correlation methods to discriminate different molecular entities inside living cells to reach a quantitative model of the intracellular architecture that resembles a random obstacle network for diffusion proteins [125]. Correlation of fluorescence amplitude fluctuations with two colors was also applied to detect the presence of molecular complexes in FAs [127]. In addition to their participation in the structural linking of the ECM to actin filaments, these complexes also serve as signaling "hubs" that regulate many cellular processes, including their own assembly and turnover, migration, gene expression, apoptosis, and proliferation. Capturing the dynamics of assembly and disassembly of these complexes is therefore of major importance to understand how a cell senses its environment. Indeed, there is not a single irreversible event of adhesion but rather a fully orchestrated sequence of adhesion events, which may take from seconds to many minutes to be established. This is typically a multifrequency behavior and applying FCS was a definite step toward deciphering the intricate mechanisms of adhesion. Adhesion complexes and interacting protein actors comprise more than 100 different molecules; some are stably associated and others only transiently [128]. When staining FAK, paxillin (Pax) and vinculin (Vn), with enhanced green fluorescent protein (EGFP) and mCherry, respectively, Digman *et al.* [127] evidenced a heterogeneity in the dynamics and aggregation state of paxillin in different regions across the cell. Taking diffusion as the main mode of transport for adhesion molecules through the cytoplasm, they showed that exchange (binding-unbinding) kinetics with a broad range of rates ($0.1\text{-}10^{-1}$ s) dominate in the vicinity of adhesion zones. They observed large clusters and complexes exchanging rather slowly in the vicinity of the disassembling adhesion regions, whereas small aggregates (largely monomers) were observed exchanging rapidly in assembling adhesion zones.

In a very recent paper, Baum *et al.* [125] used FCS to link protein mobility and cellular structure in single cells at high resolution. They mapped the mobility of inert monomers, trimers, and pentamers of the GFP domain on multiple length and timescales in the cytoplasm and nucleus via parallelized FCS measurements. From the perspective of these proteins that cover the range of size of most enzymes, they showed that the cellular interior appears as a porous medium made up by randomly distributed obstacles that reorganize in response to intra- and extracellular cues for small molecules and acts as a viscous medium on large polymeric molecules.

3.3. Confining fluorescence measurements with near field optical probes to improve sensitivity

Reflection interference contrast microscopy (RICM) [129, 130] has been used since the seventies for imaging the internal structure of cells adhering on solid surfaces. Due to a lack of quantitative interpretation of these images, this method was early abandoned. This technique relies on reflections from an incident beam passing through materials of different refractive indices. The interference of these reflected beams is either constructive or destructive, depending on the thickness and index of the layer of both the liquid medium and the cell in contact with the glass coverslip. More recently, thanks to fast progress in data acquisition and storage and improved modeling of the reflection signals, RICM was applied to a variety of biological situations, such as adhesion of vesicles and cells [131]. It has the practical advantage of not requiring any staining or labeling of the sample, and can be implemented with relative ease and very little investment on a standard inverted microscope. It can also be combined with several other microscopy techniques such as fluorescence or other scanning probe microscopies (AFM, optical or magnetic tweezers [130, 132]). Reflected light imaging has also been coupled to fluorescence excitation in total internal reflection fluorescence (TIRF) microscopy [133-135] to capture the cellular structures involved in FA complexes. As compared to transmission microscopy, this planar confinement (evanescent field) of light not only provides a higher signal-to-noise ratio but also minimizes photodamage to the cellular material [136]. Interestingly, the fact that RICM can be performed without staining the cellular sample was exploited to capture the spontaneous fluctuations (called Fluctuation Contrast RICM or Dynamical RICM) of a soft interface to identify the organization of specific ligand-receptor bonds in cellular adhesion [137, 138].

More recently, surface plasmon microscopy has been proposed for imaging internal structures of cells without staining [139 -144]. This microscopy offers also the possibility to recover both the amplitude and the phase of the reflected field and in some situations to retrieve the index of the layer in contact with gold without needing to know its thickness [145-148]. This microscopy combines total internal reflection of light with surface plasmon resonance excitation to achieve high contrast and high resolution images. Lately, surface plasmon resonance imaging ellipsometry (SPRIE) has been applied to capture cell-matrix adhesion dynamics and strength [149].

3.4. Beyond fluorescence methods: quantitative phase microscopies for living cell data capture

In the fifties, phase contrast (PC) and differential interference contrast (DIC) microscopies [150] have revolutionized the biologist view of living systems, by inferring their morphometric features without the need for exogenous contrast agents [151]. However, both PC and DIC remain qualitative in terms of optical path-length measurement, since the relationship between the incoming light power and the optical phase of the image field is generally nonlinear. Quantifying the optical phase shifts associated with biological structures was expected to give access to important information about morphology and dynamics at the nanometer scale [152-154]. However, imaging large field of view samples required time-consuming raster scanning. Full-filed phase measurement techniques were also developed [155, 156], providing simultaneous information from a large number of points on the sample. Fourier phase microscopy (FPM) [157], digital holographic microscopy (DHM) [158] and quantitative phase microscopy (QPM) [159-164] have recently been implemented to provide quantitative phase images of biological samples with remarkable sensitivity and stability over extended periods of time. Thanks to its sub-nanometer path-length stability over long periods and efficient algorithms to retrieve the phase maps from fringe patterns [161, 165], QPM is well suited for studying a wide range of temporal scales. This technique has been applied to capture red blood cell fluctuations (spontaneous flickering), which manifest as submicron motions characterized by membrane displacements in the millisecond (or less) timescale. Amin *et al.* [166] showed that the frequency behavior of the complex modulus $G(f)$ of healthy red blood cells is similar to that obtained in SGMs. Over the frequency range 5 - 50 Hz, the storage and dissipation moduli approach power-law behavior, $G' \propto f^{0.5\pm0.02}$ and $G'' \propto f^{0.7\pm0.05}$, where the errors indicate cell-to-cell variations (N=13). As already discussed for MTC, the intermediate exponent of 0.7 tells us that normal (discoid) red blood cell membranes behave neither as purely elastic nor as purely viscous media, but as viscoelastic gels. For red blood cells switching to echinocyte and spherocyte shapes, the G'' exponent decreases consistently, indicating stronger confinement of the membrane viscous motions. Finally, above 35 Hz, the viscous modulus becomes dominant, *i.e.*, the cell transits toward a dissipation-dominated regime, which has been ascribed by the authors to the culture medium viscosity.

4. From fluctuations to deterministic behavior

4.1. Emergence of coherent dynamics in cellular systems

So far, most cellular models have been established at specific scales, those which focus on molecular mechanisms are not suited to pave macroscopic scales and inversely. Establishing a connection between the discrete stochastic microscopic and the continuous deterministic macroscopic descriptions of the same biological phenomenon is likely to give new clues toward the understanding of mechanotransduction and mechanosensing processes. The scale invariance properties of the cell rheology revealed by MTC [24, 54] suggest that for very small mechanical deformations, no characteristic timescale emerges. Even if fluctuations have been shown to play an essential role in many biological systems, *e.g.*, Brownian rachets [167-169], does that mean that the molecular motors dynamics is not cooperative or synchronized? Progressive molecular motors such as myosin, kinesin and dynein, RNA and DNA polymerases, and chaperonins are macromolecules which hydrolyze ATP while moving unidirectionally along a linear macromolecular track. These progressive transport processes cannot operate without the presence of thermal fluctuations, their directionality resulting from the rectification (asymmetrization) of the Brownian motion [170].

A typical mechanism that crawling cells use to probe their environments is called protrusion, which is a thin (sharp or flat) actin gel extension that the cells generate to move and invade their environment. These protrusions result from many dynamical multicale processes namely polymerization/depolymerization of cytoskeleton filaments (actin, microtubules, and inter-filaments), progressive molecular motors, and FA complexes recruitment [171]. These outer cellular extensions are called filopodia and lamellipodia depending on the shape and dynamics of the protrusion; they also vary with the presence of intra- and extracellular factors [172]. Protrusions grow and shrink in a random manner around the cell on a few minutes' timescales over micrometers. When protrusions are temporarily stabilized, adhesion mechanisms are triggered and the cell can develop traction forces on its ECM. If the cell is polarized, an imbalance between the protrusions at the cell ends may lead to a directional motion. Filopodia stochastic dynamics was shown to play a key role in turning the nerve growth cone to face the chemical signal of a specific partner cell [173-175]. In a recent experimental work, Caballero *et al.* [176] have illustrated the key role of fluctuating protrusions on ratchet-like structures in driving NIH3T3 cell migration. They have shown that stochasticity affects the short- and long-term cell trajectories. They confirmed with a theoretical model that an asymmetry in the protrusion fluctuations is sufficient for predicting the long-term motion, which can be described as a biased persistent random walk. Depending on the type of cells and their environment (ECM stiffness, culture medium), fairly nondeterministic ruffling- and bubbling-like shape dynamics [80, 177] or low-dimensional "periodic" and coherent dynamics [178] can be observed. When placed in conditions for adhesion and spreading on fibronectin-coated glass plates, mouse embryonic fibroblasts (MEFs) show two modes of spreading [178, 179]: on the one hand, anisotropic spreading extensions supported by randomly emerging filopodia [180] and, on the other hand, deterministic spreading extensions that otherwise are rather smooth and continuous. In the latter case, these smooth extensions were shown to be periodically interrupted with a period of about 24 s [178] and to depend on the rigidity of the ECM, integrin binding and myosin light chain kinase (MLCK) activation. Giannone *et al.* [178] suggested a local cytoskeletal signal transport via the actin cytoskeleton from the tips of the lamellipodia to the back where contraction can be activated to start a new cycle. However, in that situation the oscillating signal remained local and did not synchronize over the whole cell cortex. In different situations, global cortical oscillations in spreading cells were observed and attributed to a cyclic depolymerization of microtubules [181], to Arp2/3 complex [182], or to calcium oscillations [183, 184]. These studies question the nature of the transition from stochastic and dissipative [185, 186] local protrusions or membrane pearling [187, 188] processes to global periodic morphological protrusions. This accumulation of experimental evidences of the impact of fluctuations on cell dynamics together with our improved ability to quantify and to model them are pushing the whole cellular biology community to revisit our traditional models of cell shape and dynamics.

4.2. From mechanotransduction to mechanogenetics: is there a genomic signature of the cell dynamics?

We have seen above that the cell mechanosensing mechanisms involve many length and temporal scales; they are definitely out-of-equilibrium processes which can manifest as stochastic in some situations or low-dimensional periodic dynamics in other situations. Cellular systems have a unique property that no physical/chemical system can reproduce. Depending on the external perturbation, they have the ability to evolve as they synthesize some cytoskeletal elements and/or biochemical activators, which may drastically change gene expression and their mechanical phenotype. Cancer stem cells are vivid examples of very drastic transformations [189]. The mechanical environment of a cell has a direct impact on its genetic expression and reciprocally the interplay between the cell mechanics and its geometrical constraints is conditioned by the gene expression level of all the cytoskeletal and adhesion proteins. Large-scale cellular mechanosensing leads to an adaptative response of cell migration to stiffness gradients [11, 190]. This two-way communication initially termed as mechanotransduction could also be called mechanogenetics of a cell to enlighten the interplay of genomic and mechanical functions. Recent advances in cellular biology have put forward mechanical forces as major actuators in cell signaling in addition to biochemical pathways [191]. Within the cell, the cytoskeleton provides a physical continuity from the ECM down to the interior of the nucleus, enabling direct mechanical links between the cellular microenvironment and chromosome organization. Sensed mechanical signals influence information processing through complex cellular signaling and transcriptional networks that may or may not be specifically force dependent [192]. In many cases, these responses feedback to remodel the cytoskeleton and/or nuclear architecture and consequently modify also the mechanosensitive structures that were initially involved in the response. It has been shown that both integrin-mediated and cadherin-mediated adhesion foci enlarge and strengthen in response to tension in the range of a few tens of seconds [193]. On longer timescales, signaling pathways are activated over minutes (*e.g.* the small GTPase RhoA), which stimulates the formation of actin stress fibers [32], whereas gene expression pathways that operate over hours or days (*e.g.* the induction of vinculin through serum response factor [194]) change the composition and structure of FAs and of the CSK. Although it seems reasonable to assume that cell mechanics and motility require coordinated protein biosynthesis, nowadays the links between cytoskeletal actin dynamics and correlated gene activities are still poorly understood. Olson *et al.* [194] recently filled this gap by discovering that globular G-actin polymerization can modulate myocardin-related transcription factor (MRTF) cofactors, thereby inducing the nuclear transcription serum response factor (SRF) and subsequently impacting the expression of genes encoding structural and regulatory effectors of actin dynamics. In cancer, the genome architecture is often impacted directly by mutations and/or translocations or chromatin rearrangements, but the influence of the cellular microenvironment may also change the spatiotemporal program of replication and gene expression [195-199]. Recent large-scale sequencing efforts have also helped scientists to delineate the enormous complexity of cancer and the degree to which signaling, drug resistance and genomic alterations vary from patient to patient and even within one patient [200].

Abbreviations

ACF: autocorrelation functions

AFM: atomic force microscopy

ATP: adenosine triphosphate

CSK: cytoskeleton

DHM: digital holographic microscopy

DIC: differential interference contrast

DNA: deoxyribonucleic acid

ECM: extracellular matrix

EGFP: enhanced green fluorescent protein

FA: focal adhesion

FAK: focal adhesion kinase

FCS: fluorescence correlation spectroscopy

FDT: fluctuation dissipation theorem

FPM: Fourier phase microscopy

FRET: Forster resonance energy transfer

GFP: green fluorescent protein

GTPase: guanosine triphosphatase

HOPG: highly ordered pyrolytic graphite

MCF7 (-10): Michigan Cancer Foundation-7 (-10). a model of breast cancer cell

MEF: mouse embryonic fibroblast

MLCK: myosin light chain kinase

MRTF: myocardin-related transcription factor

MTC: magnetic twisting cytometry

PAA: polyacrylamide

PC: phase contrast

PRIM: proximity imaging microscopy

PriSSM: PRI-based strain sensor modulus

QPM: quantitative phase microscopy

RICM: reflection interference contrast microscopy

RNA: ribonucleic acid

SGM: soft glassy material

SPRIE: surface plasmon resonance imaging ellipsometry

SRF: serum response factor

STM: scanning tunneling microscopy

TFM: traction force microscopy

TIRF: total internal reflection fluorescence

WT: wavelet transform

Acknowledgements

This work was supported by the Agence National de la Recherche (ANR 10 BLANC 1615 and ANR-11 IDEX-0007-02 with the PRES-University of Lyon) and INSERM (Plan Cancer 2012 01-84862).

Author details

F. Argoul*, B. Audit and A. Arneodo

*Address all correspondence to: francoise.argoul@ens-lyon.fr

Laboratoire de Physique, CNRS, Ecole Normale Supérieure de Lyon, Lyon, France

References

[1] H. L. F. Helmholtz. *On the Sensation of Tone as a Physiological Basis for the Theory of Music* (Longmans, Green and Co, London, 1875).

[2] J. Müller. *Elements of Physiology* (Taylor and Walton, London, 18?8).

[3] E. D. Adrian & K. Umrath. The impulse discharge from the pacinian corpuscle. *J Physiol* 68, 139-154 (1929).

[4] W. R. Loewenstein & M. Mendelson. Components of receptor adaptation in a pacinian corpuscule. *J Physiol* 177, 377-397 (1965).

[5] M. Chalfie. Neurosensory mechanotransduction. *Natur Rev Mol Cell Biol* 10, 44-52 (2009).

[6] E. A. Lumpkin, K. L. Marshall & A. M. Nelson. The cell biology of touch. *J Cell Biol* 191, 237-248 (2010).

[7] W. R. Loewenstein. Excitation and changes in adaptation by stretch of mechanoreceptors. *J Physiol* 133, 588-602 (1956).

[8] O. P. Hamill & B. Martinac. Molecular basis of mechanotransduction in living cells. *Physiol Rev* 81, 685-740 (2001).

[9] T. Luo, K. Mohan, P. A. Iglesias & D. N. Robinson. Molecular mechanisms of cellular mechanosensing. *Natur Mater* 12, 1064-1071 (2013).

[10] H. B. Schiller & R. Fässler. Mechanosensitivity and compositional dynamics of cell-matrix adhesions. *EMBO Rep* 14, 509-519 (2013).

[11] B. Ladoux & A. Nicolas. Physically based principles of cell adhesion mechanosensitivity in tissues. *Rep Progress Phys* 75, 116601 (2012).

[12] A. C. Callan-Jones & F. Jülicher. Hydrodynamics of active permeating gels. *New J Phys* 13, 093027 (2011).

[13] F. Julicher, K. Kruse, J. Prost & J. Joanny. Active behavior of the cytoskeleton. *Phys Rep* 449, 3-28 (2007).

[14] G. H. Koenderink, Z. Dogic, F. Nakamura, P. M. Bendix, F. C. MacKintosh, J. H. Hartwig, T. P. Stossel & D. A. Weitz. An active biopolymer network controlled by molecular motors. *Proc Natl Acad Sci USA* 106, 15192-15197 (2009).

[15] J.-F. Joanny & S. Ramaswamy. A drop of active matter. *J Fluid Mech* 705, 46-57 (2012).

[16] M. C. Marchetti, J. F. Joanny, S. Ramaswamy, T. B. Liverpool, J. Prost, M. Rao & R. A. Simha. Hydrodynamics of soft active matter. *Rev Modern Phys* 85, 1143-1189 (2013).

[17] P. Marcq. Spatio-temporal dynamics of an active, polar, viscoelastic ring. *Eur Physic J E* 37, 1-8 (2014).

[18] H. Berthoumieux, J.-L. Maître, C.-P. Heisenberg, E. K. Paluch, F. Jülicher & G. Salbreux. Active elastic thin shell theory for cellular deformations. *New J Phys* 16, 065005 (2014).

[19] J. Weber. Fluctuation dissipation theorem. *Phys Rev* 101, 1620-1626 (1956).

[20] B. Fabry, G. Maksym, J. Butler, M. Glogauer, D. Navajas & J. Fredberg. Scaling the microrheology of living cells. *Phys Rev Lett* 87, 148102 (2001).

[21] B. Fabry, G. N. Maksym, J. P. Butler, M. Glogauer, D. Navajas, N. A. Taback, E. J. Millet & J. J. Fredberg. Time scale and other invariants of integrative mechanical behavior in living cells. *Phys Rev E* 68, 041914 (2003).

[22] G. Lenormand, E. Millet, B. Fabry, J. P. Butler & J. J. Fredberg. Linearity and time-scale invariance of the creep function in living cells. *J Royal Soc, Interface* 1, 91-97 (2004).

[23] C. T. Lim, E. H. Zhou, A. Li, S. R. K. Vedula & H. X. Fu. Experimental techniques for single cell and single molecule biomechanics. *Mater Sci Engin: C* 26, 1278-1288 (2006).

[24] X. Trepat, G. Lenormand & J. J. Fredberg. Universality in cell mechanics. *Soft Matter* 4, 1750-1759 (2008).

[25] B. D. Hoffman & J. C. Crocker. Cell mechanics: dissecting the physical responses of cells to force. *Ann Rev Biomed Engin* 11, 259-288 (2009).

[26] M. A. Stolarska, Y. Kim & H. G. Othmer. Multi-scale models of cell and tissue dynamics. *Philo Transact Royal Soc. Series A* 367, 3525-3553 (2009).

[27] D. Mitrossilis, J. Fouchard, A. Guiroy, N. Desprat, N. Rodriguez, B. Fabry & A. Asnacios. Single-cell response to stiffness exhibits muscle-like behavior. *Proc Natl Acad Sci USA* 106, 18243-18248 (2009).

[28] D. Mitrossilis, J. Fouchard, D. Pereira, F. Postic, A. Richert, M. Saint-Jean & A. Asnacios. Real-time single-cell response to stiffness. *Proc Natl Acad Sci USA* 107, 16518-16523 (2010).

[29] A. K. Miri, H. K. Heris, L. Mongeau & F. Javid. Nanoscale viscoelasticity of extracellular matrix proteins in soft tissues: a multiscale approach. *J Mechan Beh Biomed Mater* 30, 196-204 (2014).

[30] J. Small, T. Stradal, E. Vignal & K. Rottner. The lamellipodium where motility begins. *Trends Cell Biol* 12, 112-120 (2002).

[31] T. D. Pollard & G. G. Borisy. Cellular motility driven by assembly and disassembly of actin filaments. *Cell* 112, 453-465 (2003)

[32] J. T. Parsons, A. R. Horwitz & M. A. Schwartz. Cell adhesion: integrating cytoskeletal dynamics and cellular tension. *Natur Rev Mol Cell Biol* 11, 633-643 (2010).

[33] A. J. Ridley. Life at the leading edge. *Cell* 145, 1012-1022 (2011).

[34] A. Pierres, A.-M. Benoliel, D. Touchard & P. Bongrand. How cells tiptoe on adhesive surfaces before sticking. *Biophys J* 94, 4114-4122 (2008).

[35] E. S. Welf & J. M. Haugh. Stochastic dynamics of membrane protrusion mediated by the DOCK180/Rac pathway in migrating cells. *Cell Mol Bioengin* 3, 30-39 (2010).

[36] K. Lam Hui, C. Wang, B. Grooman, J. Wayt & A. Upadhyaya. Membrane dynamics correlate with formation of signaling clusters during cell spreading. *Biophys J* 102, 1524-1533 (2012).

[37] E. Sackmann & A.-S. Smith. Physics of cell adhesion: some lessons from cell-mimetic systems. *Soft Matter* 10, 1644-1659 (2014).

[38] A. E. Carlsson. Actin dynamics: from nanoscale to microscale. *Ann Rev Biophy* 39, 91-110 (2010).

[39] S. K. Mitra, D. A. Hanson & D. D. Schlaepfer. Focal adhesion kinase: in command and control of cell motility. *Natur Rev Mol Cell Biol* 6, 56-68 (2005).

[40] P. W. Oakes & M. L. Gardel. Stressing the limits of focal adhesion mechanosensitivity. *Curr Opin Cell Biol* 30, 68-73 (2014).

[41] D. Riveline, E. Zamir, N. Q. Balaban, U. S. Schwarz, T. Ishizaki, S. Narumiya, Z. Kam, B. Geiger & A. D. Bershadsky. Focal contacts as mechanosensors: externally applied local mechanical force induces growth of focal contacts by an mDia1-dependent and ROCK-independent mechanism. *J Cell Biol* 153, 1175-1186 (2001).

[42] A. Bershadsky, M. Kozlov & B. Geiger. Adhesion-mediated mechanosensitivity: a time to experiment, and a time to theorize. *Curr Opin Cell Biol* 18, 472-481 (2006).

[43] A. D. Bershadsky, C. Ballestrem, L. Carramusa, Y. Zilberman, B. Gilquin, S. Khochbin, A. Y. Alexandrova, A. B. Verkhovsky, T. Shemesh & M. M. Kozlov. Assembly and mechanosensory function of focal adhesions: experiments and models. *Eur J Cell Biol* 85, 165-173 (2006).

[44] B. Geiger, J. P. Spatz & A. D. Bershadsky. Environmental sensing through focal adhesions. *Natur Rev Mol Cell Biol* 10, 21-33 (2009).

[45] P. Vallotton, G. Danuser, S. Bohnet, J.-J. Meister & A. B. Verkhovsky. Tracking retrograde flow in keratocytes: news from the front. *Mol Biol Cell* 16, 1223-1231 (2005).

[46] A. Ponti, M. Machacek, S. L. Gupton, C. M. Waterman-Storer & G. Danuser. Two distinct actin networks drive the protrusion of migrating cells. *Science* 305, 1782-1787 (2004).

[47] W. N. Findley, J. S. Lai & K. Onaran. *Creep and Relaxation of Nonlinear Viscoelastic Materials* (Dover Publications Inc., 1990).

[48] P. Bursac, G. Lenormand, B. Fabry, M. Oliver, D. A. Weitz, V. Viasnoff, J. P. Butler & J. J. Fredberg. Cytoskeletal remodelling and slow dynamics in the living cell. *Natur Mater* 4, 557-561 (2005).

[49] D. Icard-Arcizet, O. Cardoso, A. Richert & S. Hénon. Cell stiffening in response to external stress is correlated to actin recruitment. *Biophys J* 94, 2906-2913 (2008).

[50] A. Asnacios, S. Hénon, J. Browaeys & F. Gallet. Microrheology of living cells at different times and length scales. In *Cell Mechanics: From Scale-Based Models to Multiscale Modeling*, 5-28. Chapman & Hall/CRC, Boca Raton (2010).

[51] J. Alcaraz, L. Buscemi, M. Grabulosa, X. Trepat, B. Fabry, R. Farré & D. Navajas. Microrheology of human lung epithelial cells measured by atomic force microscopy. *Biophys J* 84, 2071-2079 (2003).

[52] E. Moeendarbary, L. Valon, M. Fritzsche, A. R. Harris, D. A. Moulding, A. J. Thrasher, E. Stride, L. Mahadevan & G. T. Charras. The cytoplasm of living cells behaves as a poroelastic material. *Natur Mater* 12, 253-261 (2013).

[53] P. Sollich. Rheological constitutive equation for model of soft glassy materials. *Physic Rev E* 58, 738-759 (1998).

[54] E. H. Zhou, X. Trepat, C. Y. Park, G. Lenormand, M. N. Oliver, S. M. Mijailovich, C. Hardin, D. A. Weitz, J. P. Butler & J. J. Fredberg. Universal behavior of the osmotically compressed cell and its analogy to the colloidal glass transition. *Proc Natl Acad Sci USA* 106, 10632-10637 (2009).

[55] D. Heinrich & E. Sackmann. Active mechanical stabilization of the viscoplastic intracellular space of Dictyostelia cells by microtubule-actin crosstalk. *Acta Biomaterialia* 2, 619-631 (2006).

[56] D. Arcizet, B. Meier, E. Sackmann, J. O. Rädler & D. Heinrich. Temporal analysis of active and passive transport in living cells. *Physic Rev Lett* 101, 248103 (2008).

[57] M. Otten, A. Nandi, D. Arcizet, M. Gorelashvili, B. Lindner & D. Heinrich. Local motion analysis reveals impact of the dynamic cytoskeleton on intracellular subdiffusion. *Biophysic J* 102, 758-767 (2012).

[58] M. Guo, A. J. Ehrlicher, S. Mahammad, H. Fabich, M. H. Jensen, J. R. Moore, J. J. Fredberg, R. D. Goldman & D. A. Weitz. The role of vimentin intermediate filaments in cortical and cytoplasmic mechanics. *Biophysic Journal* 105, 1562-1568 (2013).

[59] A. W. C. Lau, B. D. Hoffman, A. Davies, J. C. Crocker & T. C. Lubensky. Microrheology, stress fluctuations, and active behavior of living cells. *Physic Rev Lett* 91, 198101 (2003).

[60] D. Mizuno, C. Tardin, C. F. Schmidt & F. C. Mackintosh. Nonequilibrium mechanics of active cytoskeletal networks. *Science (New York, N.Y.)* 315, 370-3 (2007).

[61] C. Wilhelm. Out-of-equilibrium microrheology inside living cells. *Physic Rev Lett* 101, 028101 (2008).

[62] R. H. Pritchard, Y. Y. S. Huang & E. M. Terentjev. Mechanics of biological networks: from the cell cytoskeleton to connective tissue. *Soft Matter* 10, 1864-1884 (2014).

[63] A. Basu, J. F. Joanny, F. Jülicher & J. Prost. Thermal and non-thermal fluctuations in active polar gels. *Eur Physic J E* 27, 149-60 (2008).

[64] P. Bohec, F. Gallet, C. Maes, S. Safaverdi, P. Visco & F. van Wijland. Probing active forces via a fluctuation-dissipation relation: application to living cells. *EPL (Europhys Lett)* 102, 50005 (2013).

[65] G. Binnig & C. F. Quate. Atomic force microscope. *Physic Rev Lett* 56, 930-933 (1986).

[66] D. J. Müller & Y. F. Dufrêne. Atomic force microscopy as a multifunctional molecular toolbox in nanobiotechnology. *Natur Nanotechnol* 3, 261-269 (2008).

[67] A. F. Oberhauser, P. K. Hansma, M. Carrion-Vazquez & J. M. Fernandez. Stepwise unfolding of titin under force-clamp atomic force microscopy. *Proc Natl Acad Sci USA* 98, 468-472 (2001).

[68] M. Rief, M. Gautel, F. Oesterhelt, J. M. Fernandez & H. E. Gaub. Reversible unfolding of individual titin immunoglobulin domains by AFM. *Science* 276, 1109-1112 (1997).

[69] A. E. X. Brown, R. I. Litvinov, D. E. Discher & J. W. Weisel. Forced unfolding of coiled-coils in fibrinogen by single-molecule AFM. *Biophysic J* 92, L39-L41 (2007).

[70] P. Milani, G. Chevereau, C. Vaillant, B. Audit, Z. Haftek-Terreau, M. Marilley, P. Bouvet, F. Argoul & A. Arneodo. Nucleosome positioning by genomic excluding-energy barriers. *Proc Natl Acad Sci USA* 106, 22257-22262 (2009).

[71] F. Montel, H. Menoni, M. Castelnovo, J. Bednar, S. Dimitrov, D. Angelov & C. Faivre-Moskalenko. The dynamics of individual nucleosomes controls the chromatin condensation pathway: direct atomic force microscopy visualization of variant chromatin. *Biophysic* J 97, 544-553 (2009).

[72] J. Moukhtar, C. Faivre-Moskalenko, P. Milani, B. Audit, C. Vaillant, E. Fontaine, F. Mongelard, G. Lavorel, P. St-Jean, P. Bouvet, F. Argoul & A. Arneodo. Effect of genomic long-range correlations on DNA persistence length: from theory to single molecule experiments. *J Physic Chem* B 114, 5125-5143 (2010).

[73] P. Hinterdorfer. Molecular recognition studies using the atomic force microscope. *Meth Cell Biol* 68, 115-139 (2002).

[74] Z. Deng, V. Lulevich, F. T. Liu & G. Y. Liu. Applications of atomic force microscopy in biophysical chemistry of cells. *J Physic Chem B* 114, 5971-5982 (2010).

[75] M. Radmacher, M. Fritz, C. M. Kacher, J. P. Cleveland & P. K. Hansma. Measuring the viscoelastic properties of human platelets with the atomic force microscope. *Biophysic J* 70, 556-567 (1996).

[76] M. Prass, K. Jacobson, A. Mogilner & M. Radmacher. Direct measurement of the lamellipodial protrusive force in a migrating cell. *J Cell Biol* 174, 767-772 (2006).

[77] C. Rotsch, F. Braet, E. Wisse & M. Radmacher. AFM imaging and elasticity measurements on living rat liver macrophages. *Cell Biol Int* 21, 685-696 (1997).

[78] M. J. Rosenbluth, W. A. Lam & D. A. Fletcher. Supplemental material for: force microscopy of non adherent cells. *Biophysic J* 90, 4-6 (2006).

[79] S. Digiuni, A. Berne-Dedieu, C. Martinez-Torres, J. Szecsi, M. Bendahmane, A. Arneodo & F. Argoul. Single cell wall nonlinear mechanics revealed by a multi-scale analysis of AFM force-indentation curves. *Biophysic J.* 108, 2235-2248 (2015).

[80] C. Rotsch, K. Jacobson & M. Radmacher. Dimensional and mechanical dynamics of active and stable edges in motile fibroblasts investigated by using atomic force microscopy. *Proc Natl Acad Sci USA* 96, 921-926 (1999).

[81] R. Matzke, K. Jacobson & M. Radmacher. Direct, high-resolution measurement of furrow stiffening during division of adherent cells. *Natur Cell Biol* 3, 607-610 (2001).

[82] S. E. Cross, Y.-S. Jin, J. Rao & J. K. Gimzewski. Nanomechanical analysis of cells from cancer patients. *Natur Nanotechnol* 2, 780-783 (2007).

[83] P. Milani, M. Gholamirad, J. Traas, A. Arnéodo, A. Boudaoud, F. Argoul & O. Hamant. In vivo analysis of local wall stiffness at the shoot apical meristem in Arabidopsis using atomic force microscopy. *Plant J* 67, 1116-1123 (2011).

[84] J. Helenius, C.-P. Heisenberg, H. E. Gaub & D. J. Muller. Single-cell force spectroscopy. *J Cell Sci* 121, 1785-1791 (2008).

[85] B. N. Johnson & R. Mutharasan. Biosensing using dynamic-mode cantilever sensors: A review. *Biosens Bioelectron* 32, 1-18 (2012).

[86] Y. Suzuki, N. Sakai, A. Yoshida, Y. Uekusa, A. Yagi, Y. Imaoka, S. Ito, K. Karaki & K. Takeyasu. High-speed atomic force microscopy combined with inverted optical microscopy for studying cellular events. *Scient Rep* 3, 2131 (2013).

[87] A. Pietuch & A. Janshoff. Mechanics of spreading cells probed by atomic force microscopy. *Open Biol* 3, 130084 (2013).

[88] A. M. Whited & P. S.-H. Park. Atomic force microscopy: a multifaceted tool to study membrane proteins and their interactions with ligands. *Biochim Biophys Acta* 1838, 56-68 (2014).

[89] N. Wang & Z. Suo. Long-distance propagation of forces in a cell. *Biochem Biophysic Res Commun* 328, 1133-1138 (2005).

[90] B. J. Rodriguez, C. Callahan, S. V. Kalinin & R. Proksch. Dual-frequency resonance-tracking atomic force microscopy. *Nanotechnology* 18, 475504 (2007).

[91] D. O. Koralek, W. F. Heinz, M. D. Antonik, A. Baik & J. H. Hoh. Probing deep interaction potentials with white-noise-driven atomic force microscope cantilevers. *Appl Phys Lett* 76, 2952 (2000).

[92] J. Morlet, G. Arenss, E. Fourgeau & D. Giard. Wave propagation and sampling theory-Part I: Complex signal and scattering in multilayered media. *Geophysics* 47, 203-221 (1982).

[93] J. Morlet, G. Arensz, E. Fourgeau & D. Giard. Wave propagation and sampling theory-Part II: Sampling theory and complex waves. *Geophysics* 47, 222-236 (1982).

[94] R. A. Carmona, W. L. Hwang & B. Torresani. *Practical Time-Frequency Analysis* (Academic Press, San Diego, 1998).

[95] Y. Meyer (ed). *Wavelets and their Applications*, (Springer, Berlin, 1992).

[96] G. Malegori & G. Ferrini. Wavelet transforms to probe long- and short-range forces by thermally excited dynamic force spectroscopy. *Nanotechnology* 22, 195702 (2011).

[97] V. Pukhova, F. Banfi & G. Ferrini. Complex force dynamics in atomic force microscopy resolved by wavelet transforms. *Nanotechnology* 24, 505716 (2013).

[98] L. Streppa, C. Martinez-Torres, L. Berguiga, F. Ratti, E. Goillot, A. Devin, L. Schaeffer, A. Arneodo & F. Argoul. Myoblasts are committed to maintaining a minimal actomyosin driven cytoskeletal tension for differentiating into myotubes: an AFM study. *Phys Biol*, submitted (2015).

[99] D. E. Ingber. Tensegrity: the architectural basis of cellular mechanotransduction. *Ann Rev Physiol* 59, 575-599 (1997).

[100] S. R. Heidemann, S. Kaech, R. E. Buxbaum & A. Matus. Direct observations of the mechanical behaviors of the cytoskeleton in living fibroblasts. *J Cell Biol* 145, 109-122 (1999).

[101] V. M. Laurent, R. Fodil, P. Cañadas, S. Féréol, B. Louis, E. Planus & D. Isabey. Partitioning of cortical and deep cytoskeleton responses from transient magnetic bead twisting. *Anna Biomed Engin* 31, 1263-1278 (2003).

[102] S. V. Plotnikov, A. M. Pasapera, B. Sabass & C. M. Waterman. Force fluctuations within focal adhesions mediate ECM-rigidity sensing to guide directed cell migration. *Cell* 151, 1513-1527 (2012).

[103] M. L. Gardel, K. E. Kasza, C. P. Brangwynne, J. Liu & D. A. Weitz. Chapter 19: Mechanical response of cytoskeletal networks. *Meth Cell Biol* 89, 487-519 (2008).

[104] B. Sabass, M. L. Gardel, C. M. Waterman & U. S. Schwarz. High resolution traction force microscopy based on experimental and computational advances. *Biophysic J* 94, 207-220 (2008).

[105] J. Lippincott-Schwartz, E. Snapp & A. Kenworthy. Studying protein dynamics in living cells. *Natur Rev Mol Cell Biol* 2, 444-456 (2001).

[106] W. E. Thomas. Mechanochemistry of receptor-ligand bonds. *Curr Opin Struct Biol* 19, 50-5 (2009).

[107] R. D. Astumian & M. Bier. Mechanochemical coupling of the motion of molecular motors to ATP hydrolysis. *Biophysic J* 70, 637-653 (1996).

[108] D. Keller & C. Bustamante. The mechanochemistry of molecular motors. *Biophysic J* 78, 541-556 (2000).

[109] T. Förster. Expermentelle und theoretische untersuchung des zwischengmolekularen übergangs von elektronenanregungsenergie. *Naturforsch* 4A, 321-327 (1949).

[110] I. T. Li, E. Pham & K. Truong. Protein biosensors based on the principle of fluorescence resonance energy transfer for monitoring cellular dynamics. *Biotechnol Lett* 28, 1971-1982 (2006).

[111] A. Miyawaki & R. Y. Tsien. Monitoring protein conformations and interactions by fluorescence resonance energy transfer between mutants of green fluorescent protein. *Meth Enzymol* 327, 472-500 (2000).

[112] G. Y. Chen, T. Thundat, E. A. Wachter & R. J. Warmack. Adsorption induced surface stress and its effects on resonance frequency of microcantilevers. *J Appl Phys* 77, 3618-3622 (1995).

[113] F. Meng, T. M. Suchyna & F. Sachs. A fluorescence energy transfer-based mechanical stress sensor for specific proteins in situ. *FEBS J* 275, 3072-3087 (2008).

[114] F. Meng, T. M. Suchyna, E. Lazakovitch, R. M. Gronostajski & F. Sachs. Real time FRET based detection of mechanical stress in cytoskeletal and extracellular matrix proteins. *Cell Mol Bioengin* 4, 148-159 (2011).

[115] C. Grashoff, B. D. Hoffman, M. D. Brenner, R. Zhou, M. Parsons, M. T. Yang, M. A. McLean, S. G. Sligar, C. S. Chen, T. Ha & M. A. Schwartz. Measuring mechanical tension across vinculin reveals regulation of focal adhesion dynamics. *Nature* 466, 263-266 (2010).

[116] H. Chen & D. C. Chan. Emerging functions of mammalian mitochondrial fusion and fission. *Human Mol Genet* 14, R283-R289 (2005).

[117] S. Iwai & T. Q. P. Uyeda. Visualizing myosin-actin interaction with a genetically-encoded fluorescent strain sensor. *Proc Natl Acad Sci USA* 105, 16882-16887 (2008).

[118] S. Padilla-Parra, N. Audugé, M. Coppey-Moisan & M. Tramier. Quantitative FRET analysis by fast acquisition time domain FLIM at high spatial resolution in living cells. *Biophysic J* 95, 2976-2988 (2008).

[119] S. Padilla-Parra, N. Audugé, H. Lalucque, J. C. Mevel, M. Coppey-Moisan & M. Tramier. Quantitative comparison of different fluorescent protein couples for fast FRET-FLIM acquisition. *Biophysic J* 97, 2368-2376 (2009).

[120] E. Hinde, M. A. Digman, C. Welch & K. M. Hahn. Biosensor FRET detection by the phasor approach to fluorescence lifetime imaging microscopy (FLIM). *Microsc Res Techniq* 75, 271-281 (2013).

[121] D. Magde, E. L. Elson & W. W. Webb. Fluorescence correlation spectroscopy. II. An experimental realization. *Biopolymers* 13, 29-61 (1974).

[122] C. Ybert, F. Nadal, R. Salomé, F. Argoul & L. Bourdieu. Electrically induced microflows probed by fluorescence correlation spectroscopy. *Eur Physic J E* 16, 259-266 (2005).

[123] A. I. Bachir, K. E. Kubow & A. R. Horwitz. Fluorescence fluctuations approaches to the study of adhesion and signaling. *Meth Enzymol* 519, 167-201 (2013).

[124] K. Bacia, S. A. Kim & P. Schwille. Fluorescence cross-correlation spectroscopy in living cells. *Natur Methods* 3, 83-89 (2006).

[125] M. Baum, F. Erdel, M. Wachsmuth & K. Rippe. Retrieving the intracellular topology from multi-scale protein mobility mapping in living cells. *Natur Commun* 5, 4494 (2014).

[126] C. L. Chiu, M. A. Digman & E. Gratton. Measuring actin flow in 3D cell protrusions. *Biophysic J* 105, 1746-1755 (2013).

[127] M. A. Digman, P. W. Wiseman, C. Choi, A. R. Horwitz & E. Gratton. Stoichiometry of molecular complexes at adhesions in living cells. *Proc Natl Acad Sci USA* 106, 2170-2175 (2009).

[128] R. Zaidel-Bar, S. Itzkovitz, A. Ma'ayan, R. Iyengar & B. Geiger. Functional atlas of the integrin adhesome. *Natur Cell Biol* 9, 858-867 (2009).

[129] J. Bereiter-Hahn, C. H. Fox & B. Thorell. Quantitative reflection contrast microscopy of living cells. *J Cell Biol* 82, 767-779 (1979).

[130] L. Limozin & K. Sengupta. Quantitative reflection interference contrast microscopy (RICM) in soft matter and cell adhesion. *Eur J Chemic Phys Physic Chem* 10, 2752-2768 (2009).

[131] H. Verschueren. Interference reflection microscopy in cell biology: methodology and applications. *J Cell Sci* 75, 279-301 (1985).

[132] A.-S. Smith, K. Sengupta, S. Goennenwein, U. Seifert & E. Sackmann. Force-induced growth of adhesion domains is controlled by receptor mobility. *Proc Natl Acad Sci USA* 105, 6906-6911 (2008).

[133] B. D. Axelrod. Total internal reflection fluorescence microscopy in cell biology. *Methods Enzymol* 361, 1-33 (2003).

[134] T. Bretschneider, S. Diez, K. Anderson, J. Heuser, M. Clarke, A. Müller-Taubenberger, J. Köhler & G. Gerisch. Dynamic actin patterns and Arp2/3 assembly at the substrate-attached surface of motile cells. *Curr Biol* 14, 1-10 (2004).

[135] M. E. Berginski, E. A. Vitriol, K. M. Hahn & S. M. Gomez. High-resolution quantification of focal adhesion spatiotemporal dynamics in living cells. *PLoS ONE* 6 (2011).

[136] M. Cebecauer, J. Humpolíčková & J. Rossy. Advanced imaging of cellular signaling events. *Methods Enzymol* 505, 273-289 (2012).

[137] A.-S. Smith. Physics challenged by cells. *Natur Phys* 6, 726-729 (2010).

[138] A. Zidovska & E. Sackmann. Brownian motion of nucleated cell envelopes impedes adhesion. *Physic Rev Lett* 96, 048103 (2006).

[139] M. G. Somekh. Surface plasmon fluorescence microscopy: an analysis. *J Microsc* 206, 120-131 (2002).

[140] L. Berguiga, S. Zhang, F. Argoul & J. Elezgaray. High-resolution surface-plasmon imaging in air and in water: V(z) curve and operating conditions. *Optics Lett* 32, 509-511 (2007).

[141] M. Mahadi Abdul Jamil, M. Youseffi, P. C. Twigg, S. T. Britland, S. Liu, C. W. See, J. Zhang, M. G. Somekh & M. C. T. Denyer. High resolution imaging of bio-molecular

binding studies using a widefield surface plasmon microscope. *Sensors Actuators B: Chemic* 129, 566-574 (2008).

[142] M. G. Somekh, G. Stabler, S. Liu, J. Zhang & C. W. See. Wide field high resolution surface plasmon interference microscopy. *Optics Lett* 34, 3110-3112 (2009).

[143] L. Berguiga, T. Roland, K. Monier, J. Elezgaray & F. Argoul. Amplitude and phase images of cellular structures with a scanning surface plasmon microscope. *Optics Expr* 19, 6571-6586 (2011).

[144] E. Boyer-Provera, A. Rossi, L. Oriol, C. Dumontet, A. Plesa, L. Berguiga, J. Elezgaray, A. Arneodo & F. Argoul. Wavelet-based decomposition of high resolution surface plasmon microscopy V (Z) curves at visible and near infrared wavelengths. *Optics Expr* 21, 7456-7477 (2013).

[145] J. Elezgaray, L. Berguiga & F. Argoul. Plasmon-based tomographic microscopy. *J Optic Soc Am A* 31, 155-161 (2014).

[146] L. Berguiga, E. Boyer-Provera, C. Martinez-Torres, J. Elezgaray, A. Arneodo & F. Argoul. Guided wave microscopy: mastering the inverse problem. *Optics Lett* 38, 4269-4272 (2013).

[147] V. Yashunsky, V. Lirtsman, M. Golosovsky, D. Davidov & B. Aroeti. Real-time monitoring of epithelial cell-cell and cell-substrate interactions by infrared surface plasmon spectroscopy. *Biophysic J* 99, 4028-36 (2010).

[148] V. Yashunsky, T. Marciano, V. Lirtsman, M. Golosovsky, D. Davidov & B. Aroeti. Real-time sensing of cell morphology by infrared waveguide spectroscopy. *PLoS ONE* 7, e48454 (2012).

[149] S.-H. Kim, W. Chegal, J. Doh, H. M. Cho & D. W. Moon. Study of cell-matrix adhesion dynamics using surface plasmon resonance imaging ellipsometry. *Biophysic J* 100, 1819-1828 (2011).

[150] F. Zernike. How I discovered phase contrast. *Science* 121, 345-349 (1955).

[151] D. J. Stephens & V. J. Allan. Light microscopy techniques for live cell imaging. *Science (New York, N.Y.)* 300, 82-86 (2003).

[152] C. G. Rylander, D. P. Davé, T. Akkin, T. E. Milner, K. R. Diller & A. J. Welch. Quantitative phase-contrast imaging of cells with phase-sensitive optical coherence microscopy. *Optics Lett* 29, 1509-1511 (2004).

[153] M. A. Choma, A. K. Ellerbee, C. Yang, T. L. Creazzo & J. A. Izatt. Spectral-domain phase microscopy. *Optics Lett* 30, 1162-1164 (2005).

[154] C. Joo, T. Akkin, B. Cense, B. H. Park & J. F. de Boer. Spectral-domain optical coherence phase microscopy for quantitative phase-contrast imaging. *Optics Lett* 30, 2131-2133 (2005).

[155] G. A. Dunn, D. Zicha & P. E. Fraylich. Rapid, microtubule-dependent fluctuations of the cell margin. *J Cell Sci* 110, 3091-3098 (1997).

[156] C. J. Mann, L. Yu, C.-M. Lo & M. K. Kim. High-resolution quantitative phase-contrast microscopy by digital holography. *Optics Expr* 13, 8693-8698 (2005).

[157] G. Popescu, L. P. Deflores, J. C. Vaughan, K. Badizadegan, H. Iwai, R. R. Dasari & M. S. Feld. Fourier phase microscopy for investigation of biological structures and dynamics. *Optics Lett* 29, 2503-2505 (2004).

[158] B. Rappaz, P. Marquet, E. Cuche, Y. Emery, C. Depeursinge & P. Magistretti. Measurement of the integral refractive index and dynamic cell morphometry of living cells with digital holographic microscopy. *Optics Expr* 13, 9361-9373 (2005).

[159] V. P. Tychinskii. Coherent phase microscopy of intracellular processes. *Physics-Uspekhi* 44, 617-629 (2001).

[160] V. P. Tychinskii. Coherent phase microscopy of intracellular processes. *UFN* 171, 649-662 (2001).

[161] G. Popescu, T. Ikeda, R. R. Dasari & M. S. Feld. Diffraction phase microscopy for quantifying cell structure and dynamics. *Optics Lett* 31, 775-777 (2006).

[162] P. Bon, G. Maucort & B. Wattellier. Quadriwave lateral shearing interferometry for quantitative phase microscopy of living cells. *Optics Expr* 17, 468-470 (2009).

[163] G. Popescu. *Quantitative Phase Imaging of Cells and Tissues* (McGraw Hill, New York, 2011).

[164] P. Bon, J. Savatier, M. Merlin, B. Wattellier & S. Monneret. Optical detection and measurement of living cell morphometric features with single-shot quantitative phase microscopy. *J Biomed Optics* 17, 076004 (2012).

[165] C. Martinez-Torres, L. Berguiga, L. Streppa, E. Boyer-Provera, L. Schaeffer, J. Elezgaray, A. Arneodo & F. Argoul. Diffraction phase microscopy: retrieving phase contours on living cells with a wavelet-based space-scale analysis. *J Biomed Optics* 19, 036007 (2014).

[166] M. S. Amin, Y. Park, N. Lue, R. R. Dasari, K. Badizadegan, M. S. Feld & G. Popescu. Microrheology of red blood cell membranes using dynamic scattering microscopy. *Optics Expr* 15, 17001-17009 (2007).

[167] M. Bier. Brownian ratchets in physics and biology. *Contemp Phys* 38, 371-379 (1997).

[168] D. Dan, A. M. Jayannavar & G. I. Menon. A biologically inspired ratchet model of two coupled Brownian motors. *Physica A* 318, 40-47 (2003).

[169] K. Kruse & D. Riveline. Spontaneous mechanical oscillations. Implications for developing organisms. *Curr Topics Develop Biol* 95, 67-91 (2011).

[170] M. O. Magnasco. Forced thermal ratchets. *Physic Rev Lett* 71, 1477-1481 (1993).

[171] T. D. Pollard, L. Blanchoin & R. D. Mullins. Molecular mechanisms controlling actin filament dynamics in nonmuscle cells. *Ann Rev Biophys Biomol Struct* 29, 545-576 (2000).

[172] K. Keren, Z. Pincus, G. M. Allen, E. L. Barnhart, G. Marriott, A. Mogilner & J. A. Theriot. Mechanism of shape determination in motile cells. *Nature* 453, 475-480 (2008).

[173] J. Q. Zheng, J. J. Wan & M. M. Poo. Essential role of filopodia in chemotropic turning of nerve growth cone induced by a glutamate gradient. *J Neurosci* 16, 1140-1149 (1996).

[174] D. J. Odde & H. M. Buettner. Autocorrelation function and power spectrum of two-state random processes used in neurite guidance. *Biophys J* 75, 1189-1196 (1998).

[175] T. Betz, D. Koch, B. Stuhrmann, A. Ehrlicher & J. Käs. Statistical analysis of neuronal growth: edge dynamics and the effect of a focused laser on growth cone motility. *New J Phys* 9, 426 (2007).

[176] D. Caballero, R. Voituriez & D. Riveline. Protrusion fluctuations direct cell motion. *Biophys J* 107, 34-42 (2014).

[177] P. W. Oakes, D. C. Patel, N. A. Morin, D. P. Zitterbart, B. Fabry, J. S. Reichner & J. X. Tang. Neutrophil morphology and migration are affected by substrate elasticity. *Blood* 114, 1387-1396 (2009).

[178] G. Giannone, B. J. Dubin-Thaler, H.-G. Döbereiner, N. Kieffer, A. R. Bresnick & M. P. Sheetz. Periodic lamellipodial contractions correlate with rearward actin waves. *Cell* 116, 431-443 (2004).

[179] B. J. Dubin-Thaler, G. Giannone, H.-G. Döbereiner & M. P. Sheetz. Nanometer analysis of cell spreading on matrix-coated surfaces reveals two distinct cell states and STEPs. *Biophys J* 86, 1794-806 (2004).

[180] G. L. Ryan, N. Watanabe & D. Vavylonis. A review of models of fluctuating protrusion and retraction patterns at the leading edge of motile cells. *Cytoskeleton* 69, 195-206 (2012).

[181] O. J. Pletjushkina, Z. Rajfur, P. Pomorski, T. N. Oliver, J. M. Vasiliev & K. A. Jacobson. Induction of cortical oscillations in spreading cells by depolymerization of microtubules. *Cell Mot Cytoskeleton* 48, 235-244 (2001).

[182] G. L. Ryan, H. M. Petroccia, N. Watanabe & D. Vavylonis. Excitable actin dynamics in lamellipodial protrusion and retraction. *Biophys J* 102, 1493-1502 (2012).

[183] M. Kapustina, T. C. Elston & K. Jacobson. Compression and dilation of the membrane-cortex layer generates rapid changes in cell shape. *J Cell Biol* 200, 95-108 (2013).

[184] G. Salbreux, J. F. Joanny, J. Prost & P. Pullarkat. Shape oscillations of non-adhering fibroblast cells. *Phys Biol* 4, 268-84 (2007).

[185] J. Étienne & A. Duperray. Initial dynamics of cell spreading are governed by dissipation in the actin cortex. *Biophys J* 101, 611-621 (2011).

[186] E. Bernitt, C. G. Koh, N. Gov & H.-G. Döbereiner. Dynamics of actin waves on patterned substrates: a quantitative analysis of circular dorsal ruffles. *PloS One* 10, e0115857 (2015).

[187] D. Heinrich, M. Ecke, M. Jasnin, U. Engel & G. Gerisch. Reversible membrane pearling in live cells upon destruction of the actin cortex. *Biophys J* 106, 1079-1091 (2014).

[188] N. Nisenholz, K. Rajendran, Q. Dang, H. Chen, R. Kemkemer, R. Krishnan & A. Zemel. Active mechanics and dynamics of cell spreading on elastic substrates. *Soft Matter* 10, 7234-7246 (2014).

[189] K. Polyak & R. A. Weinberg. Transitions between epithelial and mesenchymal states: acquisition of malignant and stem cell traits. *Natur Rev Can* 9, 265-273 (2009).

[190] L. Trichet, J. Le Digabel, R. J. Hawkins, S. R. K. Vedula, M. Gupta, C. Ribrault, P. Hersen, R. Voituriez & B. Ladoux. Evidence of a large-scale mechanosensing mechanism for cellular adaptation to substrate stiffness. *Proc Natl Acad Sci USA* 109, 6933-6938 (2012).

[191] R. P. Martins, J. D. Finan, F. Guilak & D. A. Lee. Mechanical regulation of nuclear structure and function. *Ann Rev Biomed Engin* 14, 431-455 (2012).

[192] B. D. Hoffman, C. Grashoff & M. A. Schwartz. Dynamic molecular processes mediate cellular mechanotransduction. *Nature* 475, 316-323 (2011).

[193] M. A. Schwartz & D. W. DeSimone. Cell adhesion receptors in mechanotransduction. *Curr Opin Cell Biol* 20, 551-556 (2008).

[194] E. N. Olson & A. Nordheim. Linking actin dynamics and gene transcription to drive cellular motile functions. *Natur Rev Mol Cell Biol* 11, 353-365 (2010).

[195] M. J. Bissell & W. C. Hines. Why don't we get more cancer? A proposed role of the microenvironment in restraining cancer progression. *Natur Med* 17, 320-329 (2011).

[196] F. Michor, J. Liphardt, M. Ferrari & J. Widom. What does physics have to do with cancer? *Natur Rev Cancer* 11, 657-670 (2011).

[197] B. Schuster-Böckler & B. Lehner. Chromatin organization is a major influence on regional mutation rates in human cancer cells. *Nature* 488, 504-508 (2012).

[198] M. J. M. Magbanua & J. W. Park. Advances in genomic characterization of circulating tumor cells. *Can Metast Rev* 33, 757-69 (2014).

[199] P. Polak, R. Karlić, A. Koren, R. Thurman, R. Sandstrom, M. S. Lawrence, A. Reynolds, E. Rynes, K. Vlahoviček, J. A. Stamatoyannopoulos & S. R. Sunyaev. Cell-of-origin chromatin organization shapes the mutational landscape of cancer. *Nature* 518, 360-364 (2015).

http://dx.doi.org/10.5772/60883

[200] N. Gjorevski, E. Boghaert & C. M. Nelson. Regulation of epithelial-mesenhymal transition by transmission of mechanical stress through epithelial tissues. *Can Microenviron* 5, 29-38 (2012).

Chapter 9

Impedimetric Sensors for Bacteria Detection

Sergi Brosel-Oliu, Naroa Uria,
Natalia Abramova and Andrey Bratov

Additional information is available at the end of the chapter

http://dx.doi.org/10.5772/60741

Abstract

The application of electrochemical biosensors based on impedance detection has grown during the past years due to their high sensitivity and rapid response, making this technique extremely useful to detect biological interactions with biosensor platforms. This chapter is focused on the use of electrochemical impedance spectroscopy (EIS) for bacterial detection in two ways. On one hand bacteria presence may be determined by the detection of metabolites produced by bacterial growth involving the media conductivity changes. On the other hand, faster and more selective bacterial detection may be achieved by the immobilization of bacteria on a sensor surface using biorecognition elements (antibodies, antimicrobial peptides, aptamers, etc.) and registering changes produced in the charge transfer resistance (faradic process) or interfacial impedance (nonfaradic process). Here we discuss different types of impedimetric biosensors for microbiological applications, making stress on their most important parameters, such as detection limits, detection times, selectivity, and sensitivity. The aim of the paper was to give a critical review of recent publications in the field and mark the future trends.

Keywords: Bacteria detection, impedance, biosensors, interdigitated electrode array

1. Introduction

Food- and water-borne bacterial outbreaks remain a major cause for disease and mortality throughout the world [1, 2]. The rapid detection of these pathogenic microorganisms is critical for the prevention of these outbursts [3]. The identification and quantification of microorganisms has become a key point in biodefense, food safety, diagnostics, and drug discovery researches. The detection of pathogens and indicator microorganisms in water and food samples plays a vital role in public and environmental health. Globally, there are nearly 1.7 billion cases of diarrheal disease every year, and it is responsible for killing around 760,000 children every year (http://www.who.int/mediacentre/factsheets/fs330/en/).

To date, the detection and identification of pathogens rely mainly on classical culturing techniques, which require several handling steps in most cases, or on advanced "rapid" techniques in microbiology, such as biochemical kits, enzyme-linked immunosorbent assay (ELISA), and polymerase chain reaction (PCR) assays [4, 5].These methods are laborious and time consuming and lack the ability to detect microorganisms in "real time" or outside the laboratory environment [4, 6]. Over the past decade, there has been an immense effort to develop new bioassays and biosensors for the detection of food- and water-borne pathogens [7, 8]. Various biosensors for rapid identification of bacteria in food and water have been reported [2], while the most popular are optical biosensors. These biosensors offer several advantages, including speed, selectivity, sensitivity, and reproducibility of the measurement [2]. To date, the most successful optical-based biosensors are based on surface plasmon resonance (SPR) [8–10], whereby biomolecular binding events cause a change in the refractive index that is recognized by a shift in the SPR signal. However, the widespread application of these technologies for bacteria detection is limited mainly by the labor, high cost (US$10,000–150,000), and complexity of the SPR biosensor system.

Electrochemical biosensors based on impedance technique [11] have proved to be a promising method for pathogenic bacteria detection [12, 13] due to their portability, rapidity, sensitivity, low cost, ease of miniaturization, and label-free operation, and more importantly, they can be used for on-the-spot detection. There is a lot of literature about impedance microbiology, which is based on impedance changes that occurs in culture mediums due to bacterial growth as changes in conductance, due to charged ions and compounds resulting from biological metabolism, or due to bacteria cell adhesion to the electrode surface in interfacial capacitance. It must be noted that traditional impedance microbiology is not a selective method. Some selectivity may be achieved by using selective culture mediums. However, as it is presented in this chapter, functionalization of the electrodes with high-affinity recognition elements, such as antibodies, aptamers, proteins, etc., that selectively bind target cells permit to considerably enhance the selectivity of the method. Along with this, the separation of the target cells from the rest of the sample microorganisms and their preconcentration, as discussed in the chapter "Cell Concentration Systems for Enhanced Biosensor Sensitivity" of this book, may help to reduce the detection limits and raise the selectivity of the method.

Impedance biosensors register changes in the electrical properties at their surface (either capacitance or resistance), affected by interactions between biorecognition element attached to its surface and analyte present in a sample solution. Faradic impedance measurements in the presence of a redox pair in a test solution may be performed on planar metal electrodes. However, to enhance the sensitivity of the measurements and to miniaturize the final sensor element, an impedimetric transducer with two planar interdigitated electrodes called *interdigitated electrode array* (IDEA) [11, 14, 15] was introduced which consists of a series of parallel planar electrodes in which alternating electrodes are connected together, forming a set of interdigitated electrode fingers. This sensor design permits to perform a label-free detection of bacteria utilizing different biorecognition elements.

All important aspects of pathogen detection using electrochemical impedance spectrometry (EIS) will be presented in more detail in the following sections of this review.

2. EIS for bacteria detection

The electrochemical technique of impedance has been used in microbiology for detecting and quantifying bacteria during last decades. The integration of impedance technique with biosensor technology in the past few years has allowed the development of impedance biosensors, reducing assay times and detection limits.

One of the positive features of the impedance technique is its simplicity. The impedance Z is determined by applying a voltage perturbation of a sinusoidal wave of small amplitude and detecting the current response. Then impedance extends the concept of pure ohmic resistance to alternate current (AC) circuits. The impedance is the quotient of the voltage–time function $V(t)$ and the resulting current–time function $I(t)$:

$$Z(t) = \frac{V(t)}{I(t)} = \frac{V_0 \sin(2\pi ft)}{V_0 \sin(2\pi ft + \varphi)} \quad (1)$$

where V_0 and I_0 are the maximum voltage and current signals, f is the frequency, t is the time, and φ is the phase shift between the voltage–time and the current–time functions.

The impedance is a complex value because the current can differ in terms of amplitude and also a phase shift compared to the voltage–time function. Thus, impedance, as shown in Figure 1, can be described by the modulus $|Z|$ and the phase shift φ as well as the real part Z_{re} and the imaginary Z_{im} of the impedance. Therefore, for evaluating data graphically, the most popular formats are the Bode and the Nyquist plots. In the Bode plots, log $|Z|$ and φ are represented as a function of log f, while Nyquist plot data are represented as the real component of impedance (Z_{re}) on the x axis and imaginary component (Z_{im}) on the y axis [16, 17].

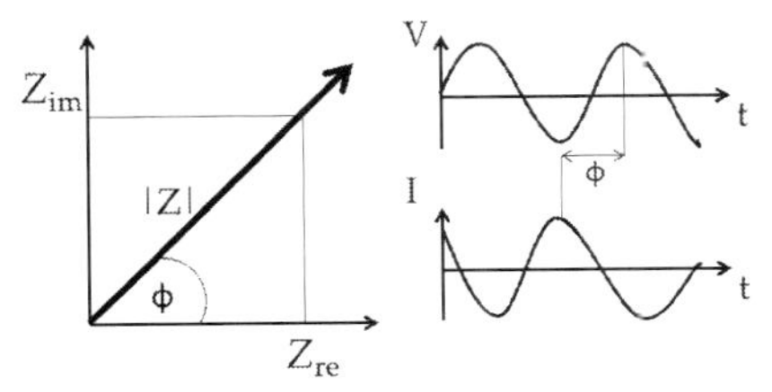

Figure 1. Complex impedance plane diagram and relation of voltage, current, and phase.

EIS studies the response of an electrochemical cell to a voltage at different frequencies. Thus, the impedance spectrum obtained allows the characterization of a complex electrode system composed of surfaces, layers, and membranes where electrical charge transfer and ion diffusion process take place. The most difficult part of the EIS is the correct interpretation of spectra that are analyzed using an equivalent circuit (EC), which consists of resistances and capacitances combined in parallel or serially, as required. Since an electrochemical cell is a complex system, an EC with components representing different physicochemical parameters and processes should be selected, reflecting the electrochemical cell's physical characteristics. However, it must be noted that typically more than one circuit model can fit obtained experimental data. Monitoring the variation of impedance elements as a function of the system properties (e.g., solution composition), it is possible to correlate total impedance changes to individual EC components and thus to confirm correct selection of the EC.

The weight with which individual EC components give their input into the total impedance depends on the applied frequency. This means that in some cases, it is possible to simplify the measurements by working in a limited range of frequencies or just one selected frequency where the relative changes of the component under interest are the largest.

Basically, for EIS performed on a metal electrode in an electrolyte solution in the presence of electroactive compounds, the elements of the EC are well known from general electrochemistry and include ohmic resistance of electrolyte (the bulk medium resistance), double-layer capacitance, charge transfer resistance, and the Warburg impedance, as is presented in Figure 2. For more complex experimental systems, additional components such as dielectric capacitor, polarization resistance, constant-phase element, interfacial impedance, coating capacitance, stray capacitance, and virtual inductors may be required to include. The measured impedance depends on all the individual contributions and distribution of this elements within the EC [12]. However, the impedimetric response in real systems is very complex, and some of the processes cannot be presented in the EC by simple (capacitor, resistance) elements. In this case, some additional EIS elements, such as constant phase element (CPE) or Warburg impedance, with known frequency response are introduced [16].

Resuming, EIS is a very powerful tool as it permits to elucidate physical and chemical phenomena occurring in an electrochemical system, thus allowing to obtain information on changes produced by the interaction of analytes of interest, such as proteins, antibodies, or whole microorganisms, with an impedimetric sensor surface [11].

2.1. Faradic impedance

Impedimetric detection can be achieved either in a direct manner in an anylyte solution or in the presence of an additional redox probe used as a marker. In the presence of electron mediator as $Fe(CN_6)^{3-/4-}$ (ferricyanide/ferrocyanide) or $Ru(NH_3)_6^{3+/2+}$ (hexaammineruthenium III/II ions), the impedance is termed faradic impedance. The use of electron mediators requires a plentiful supply of redox species to guarantee that impedance does not become limited by the charge transfer process between electrolyte and electrode surface. In faradic impedance measurements, the main parameter is the charge transfer resistance that depends on the interface blocking by surface products of biochemical reactions and thus may be used to measure concentration dependencies.

The behavior of simple impedance biosensors systems in faradic processes is typically interpreted by a Randles EC presented in Figure 2.

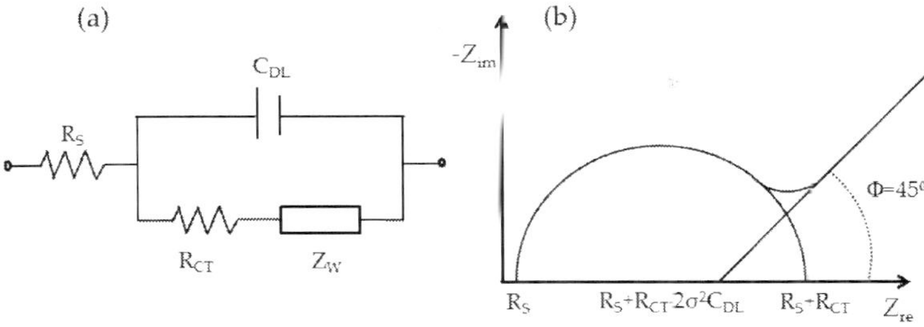

Figure 2. The Randles EC (a) and the Nyquist plot (b) of its frequency response.

The Randles EC, presented in Figure 2a, consists of solution resistance (R_s), double-layer capacitance (C_{dl}), electron transfer resistance (R_{et}) (also called charge-transfer resistance (R_{ct})), and Warburg impedance (Z_w). R_s is inserted as a series element in the EC because all the current passes through the uncompensated solution, while the parallel elements are introduced because the total current through electrodes is the sum of distinct contribution from faradic process and double-layer charging. R_s and Z_w represent bulk properties of the electrolyte and diffusion of the redox probe, while C_{dl} and R_{ct} depend on dielectric and insulating features at the electrode/electrolyte interface. The attachment of bacteria on the electrode surface would retard the interfacial electron transfer process blocking partially the surface and increase the electron transfer resistance [18].

The Nyquist plot (Figure 2b) is the best way to visualize and determinate the Randles EC elements. The semicircle observed at high frequencies corresponds to the electron transfer limited process and linear part at lower frequencies represents the diffusion limited process. The intercept of semicircle at high frequencies with the Z_{re} axis is equal to R_s, while extrapolation of semicircle to lower frequencies into another intercept with Z_{re} axis is equivalent to R_s + R_{ct}. The double-layer capacitance C_{dl} can be calculated from the frequency at the maximum of the semicircle. The Warburg impedance can be determined by extrapolating the 45° line observed in Figure 2 to the real axis. In some analytical applications, the Warburg impedance is often neglected by choosing a frequency range where no 45° line is observed in the Nyquist plot and bulk impedance is dominant.

2.2. Nonfaradic impedance

In the case when a redox pair is absent in the electrolyte solution, the impedance is termed nonfaradic [19] and depends on the conductivity of the supporting electrolyte and impedimetric electrode interfacial properties (interfacial capacitance or surface conductivity). Figure 3 shows the basic elements of EC in the case of nonfaradic process:

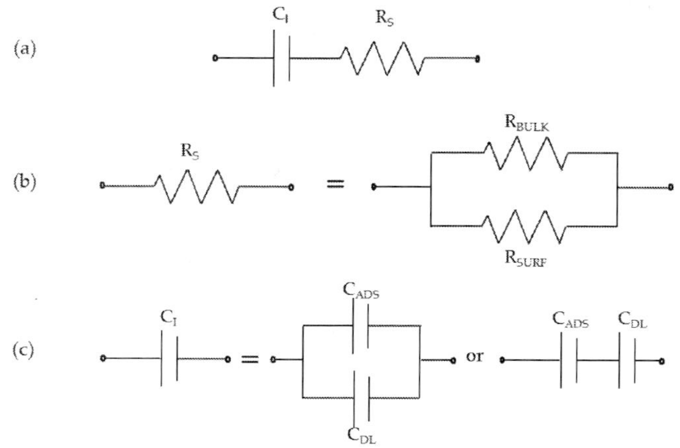

Figure 3. Typical electrical components on ECs characterizing nonfaradic impedance. (a) General circuit elements; (b) the resistance R_s in some cases may depend on bulk solution resistance and surface resistance; (c) adsorption of bacteria cells on the sensor surface results in additional capacitance that may be in parallel or in series with the electrolyte double-layer capacitance depending on the bacteria amount.

In the absence of a redox pair or if its charge transfer rate on the electrode is very slow, no faradic process occurs, and subsequent electron transfer is not produced. In these cases, the interfacial capacitance changes are often studied [20]. These capacitance changes occur when the dielectric constant or the thickness of the interfacial capacitance layer on the transducer surface change their values due to surface chemical reactions [17]. The formation of biochemical reaction products may be represented by an additional capacitor that depending on the process may be included in parallel or in series with the double-layer capacitor (Figure 3c).

It must be noted that a lot of published works refer to changes in capacitance registered by impedance spectroscopy as variations produced in the electrical double-layer capacitance. However, the double-layer capacitance, defined as an outer capacitance at the solid/liquid interface, depends basically on ionic species concentration, while interfacial capacitance depends on the presence of adsorbed species or interfacial layer formation on the electrode surface. On the other hand, R_s, which represents the solution resistance in the case of parallel electrode arrangement, may be constituted as a parallel combination of solution bulk resistance and surface resistance in the case of in-plane electrodes, for example, IDEAs [14].

All these show once again the importance of accurate interpretation of impedance data that should be based on a correct EC choice with the components that unambiguously reflect real physicochemical processes at the electrode surface.

The use of different impedimetric sensors designs, the advances in microfabrication technologies resulting in miniaturization and integration of sensors into a chip format, and better understanding of biochemical interfacial phenomena helping the analysis of impedance components using ECs should help us to improve the biosensor detection systems serving to reduce the assay time and improve the bacteria detection limits [18].

3. Impedimetric detection of metabolites produced by bacterial cells

Metabolism refers to all the biochemical reactions that occur in a cell or organism. By metabolic pathways, bacteria convert large molecules, such as polysaccharides, lipids, nucleic acids, and proteins, into smaller units as monosaccharides, fatty acids, nucleotides, and amino acids, respectively, to release energy. Consequently, this conversion of large organic substrate molecules in the medium into charged, small and more mobile ionic metabolites, which can include lactic acid, acetic acid, carbon dioxide, ammonia, bicarbonate, and urea, results in a change of the ionic composition of the growth media. In this way, these changes can be measured and related to bacterial concentration for determination of microbial growth.

Different electrochemical transduction techniques have been used for the detection of products of microbial metabolism. Amperometric technique has been reported by the use of mediators [21], which are reduced by the microorganism as a consequence of substrate metabolism; however, no examples of direct metabolite detection are found in the literature, probably due to the electroactive interference produced by the sample matrix, which can cause the transducer to generate a false current reading [22]. Potentiometric methods have been developed to detect changes resulting from metabolite accumulation of hydrogen ions [23]. Nevertheless, these electrochemical methods show some disadvantages such as insufficient sensitivity, selectivity, and sample matrix effect [22]. In addition, these methods require the use of a reference electrode, which complicates the system miniaturization and prevents its use in a small volume samples. Among different electrochemical techniques, the most extended transduction method is based on measurements of electrical impedance changes in the medium resulting from the bacterial growth.

The correlation between microbial growth and impedance was first defined by Stewart in 1899 [24]. However, it was starting from 1970s when much attention and efforts were put in this research [25–30] to monitor bacterial activity detecting changes in electrical impedance caused by growing bacterial culture. Impedance technique was shown to be useful for the estimation of microbial biomass [31], detection of microbial metabolism, and determination of the physiological state of bacteria [32–37]. The advantages of this approach are high sensitivity, relative simplicity, and comparatively low cost of the required experimental equipment [25]. In 1992, the impedance method was approved by the Association of Official Analytical Chemists (AOAC) International as a first action method for screening *Salmonella* in food samples [38]. Finally, in 1996, AOAC approved it as a final action method for the detection of *Salmonella* in food [39].

Impedance microbiology is one of the most successful of all the recently introduced rapid methods. Several analytical systems have been developed for bacteria detection, such as Bactometer (Bio Merieux, Nuertingen, Germany), Malthus systems (Malthus Instruments Ltd., Crawley, UK), rapid automated bacterial impedance technique (RABIT) (Don Whitley Scientific Ltd., Shipley, UK), and BacTrac (Sy-Lab, Purkersdorf, Austria) [18, 40–42]. They have been validated against other conventional methods, such as the most probable number method (MPN) [36] or microbial colony counts [43], showing a sensitivity comparable to these standard methods. Existing commercial instruments are widely used for different applications. For example, all these systems have been reported to detect and make quantitative estimations and differentiation of bacteria, such as *Escherichia coli* or *Salmonella,* among others typically found in food [36, 44–48]. They have been also shown to be useful for the evaluation of different mediums for selective bacterial growth [47, 49, 50]. The use of impedance technique for bacteria determination is summarized in Table 1.

Target Microorganism	Growth Medium	Electrodes (Measure frequence)	Detection Limit	Detection Time (h)	Ref.
Enterobacteriaceae Family	BHI[1] + 0.1% yeast extract	-(8-Channel Mathus-Meter Bactometer 32)	10^4 cell·cm^{-2}	8-9	[37]
-	BHI[1]	Gold plated and stainless-steel (Bactometer, 2kHz)	10^5 cell·mL^{-1}	2,6	[28]
E. coli	TSB[2]	Stainless steel (Bactometer, 2kHz)	10^5 cell·mL^{-1}	5-6	[29]
Listeria innocua	Tris-Gly[3] buffer + dextrose	Interdigitated platinum electrodes (11.43 kHz)	10^5-10^7 cell·mL^{-1}	2	[51]
Listeria innocua Listeria monocytogenes E.coli	Tris-Gly[3] buffer + dextrose	Interdigitated platinum electrodes (11.43 kHz)	10^7-10^8 cell·mL^{-1}	2	[52]
Coliforms (*E.coli*)	SM[4]	-(Bactometer, 2kHz)	10^4 cell·mL^{-1}	5	[42]
E.coli	YPLT[5]	Interdigitated electrodes	8-8·10^8 cell·mL^{-1}	14.7-0.8	[53]
Bacillus lactis	SM[4]	-(Bac Trac)	10^8 cell·mL^{-1}	6	[48]

[1] Brain Heart Infusion broth; [2] Trypticase Soy Broth; [3] Buffer Tris-glicine; [4] Specific Medium; [5] Low conductivity Yeast-Peptone-Lactose-TMAO medium

Table 1. The use of impedance technique for bacteria determination

Impedance microbiological techniques can be used to monitor bacteria viability during growth. Since only live bacteria cells present metabolic activity and are able to produce changes in the conductivity of the medium, impedance microbiology is used for differentiating live and dead cells [51–53]. Kinetics monitoring may give additional information since the impedance growth curves under different conditions are found to be characteristic for different bacteria species. Most applications of the traditional microbiological impedance technique for the detection of bacteria were reviewed by Silley and Forsythe in 1996 [41] and Wawerla et al. in 1999 [54].

Impedance changes associated with metabolic activity of microbial cells are often expressed as the ratio of the reference impedance (medium without bacteria) to the sum of the reference and the sample, resulting in the parameter known as normalized impedance change (NIC):

$$\mathrm{NIC}(\%) = \frac{Z_{\mathrm{REF}}}{Z_{\mathrm{REF}} + Z_{\mathrm{SAMPLE}}}, \tag{2}$$

which is related to microbial growth.

The typical impedance growth curve is presented in Figure 4, where the measured impedance values are graphically plotted in relation to the incubation times.

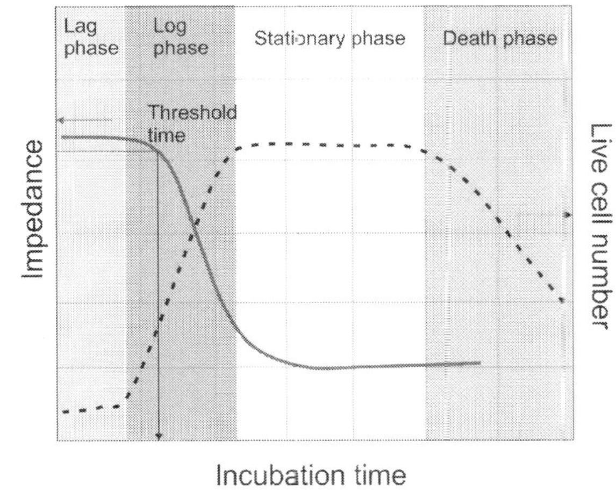

Figure 4. Microbial impedance growth curve with typical bacterial growth phases showing impedance changes (solid line) and live cell number (dashed line) in time.

To provide detectable changes in the measured impedance, a minimum concentration of microorganisms is needed in the medium. However, these microorganisms replicate and in time reach numbers sufficient to cause a detectable impedance change. This concentration of bacterial cells is defined as a threshold concentration. Thus, the threshold concentration, also called *detection limit*, refers to the lowest concentration of microorganisms that must be present to detect the change measured by impedance. The threshold concentration depends, in part, on how the detectable impedance change is defined [28], which is normally referred to the changes occurring in a control sample of a sterile broth (see Eq. (2)).

The time required for the organisms to grow to threshold concentrations is called *detection time*. The detection time depends on three main parameters:

- the initial concentration of microorganisms;
- lag phase (the initial period in which cellular metabolism is accelerated, cells are increasing in size, but the bacteria are not able to replicate);
- the generation time of the population of microorganisms (time it takes bacterial to double its concentration)

From this, it follows that if a population of organisms has similar generation times growing in a concrete medium, the detection time can be used to estimate initial concentrations [28, 29].

Metabolic activity measurements can be performed in either direct or indirect ways. In direct measurements, impedance electrodes immersed in the growth medium detect changes of the bacterial metabolism taking place in the bulk of a growth media. Indirect technique, however, detects CO_2 produced by microorganisms [51]. In this indirect impedance technique, the CO_2 produced due to bacteria biological activity reacts with potassium hydroxide (KOH) solution in a separate chamber. The formation of carbonates causes decrease in the solution conductivity. This technique was first described by Owens et al. [55], and it has been adapted for rapid automated bacterial impedance apparatus in other works [32]. Using this approach, a recent work of Johnson et al. [56] studies the viability of indirect impedance method using a commercial system to study microbial growth in complex food matrices. The ability of the system to detect different microorganisms in different food matrices was clearly demonstrated [56].

In direct measurements, the impedance changes may be produced by two primary sources: microbial metabolism, which alter the conductivity of the medium, and electrode interfacial impedance, changes in the surface properties of the electrodes affecting the capacitance of the electrode/electrolyte interface due to bacteria presence [26, 29, 57]. Thus, the growth of microorganisms usually results in an increase in both conductance and capacitance, causing a decrease in impedance [58]. To account for this, the detection of microbial metabolism by impedance systems is typically conducted by measuring relative or absolute changes in impedance at different frequencies at regular time intervals during the growth of bacteria at a given temperature. From the frequency dependence of the impedance using an appropriate EC, the system conductance/resistance and capacitance may be determined.

Different studies have analyzed the relationship between microbial growth and relative changes in both the capacitive and the resistive parts of impedance showing that both components are indicative of bacterial growth. In low conductivity media, the change in the conductance of the media clearly correlates with bacterial growth, whereas in more conductive media, the relative changes in conductance are smaller in comparison to impedance changes caused by polarization interfacial capacitance, the effect that can also be useful for monitoring bacterial growth [59].

As it was noted, both components can be studied separately by measuring impedance in a different frequency ranges. Various works have demonstrated the predominance at low frequency of electrode surface impedance, while impedance at high frequencies is associated with media conductance effect [60, 61]. However, the frequency range in which certain EC components give the main input into the overall impedance may vary depending on the dimensions of the electrodes and their separation. The differentiation of the impedance effects into electrode/electrolyte interfacial capacitance and medium resistance changes at different frequencies has led to the development of impedance-splitting methods for the detection of bacteria [18].

However, in most of publications on the impedance microbiology, only the conductance of the medium is measured. In this case, impedance always decreases with time, indicating that the microorganisms are consuming growth media substrates of low conductivity metabolizing them into ionic products of higher conductivity [59].

Thus, altogether, changes registered in the microbiological impedance are dependent on bacteria species, the number of microorganisms and properties of the medium in which they are growing in, the frequency of the applied signal, the surface properties and geometry of the measuring electrodes, and the temperature [31].

Since impedance microbiology relies on determining the changes in electrical impedance of a culture medium resulting from the bacterial growth, it depends largely on the design of the growth medium since the culture medium not only supports the bacteria growth but also provides high, noninterfering signals to the overall impedance or its components. Therefore, medium, besides providing the optimum growth, activity, products yield, and morphology of the microorganisms, should contain substrates with low conductivity contribution to the overall conductivity of the medium. In this case, the sensitivity of the impedance detection strongly depends on designing an appropriate culture medium [62]. For this reason, in recent years, much of the research in the field has been directed to find or design selective media with low conductivity [49, 52, 62]. Another important feature of the culture media is its selectivity that gives the main priority for growing only for specific bacteria. In the case of nonselective detection, growth media changes in impedance may be induced by the growth of different bacteria, not only the target microorganism one wants to determine. To solve the problem, selectivity can be provided by using specific culture media designed for certain microbial type by using specific inhibitors such as antibiotics [47].

Recently, Lopez Rodriguez and co-workers [62] designed a specific medium to impedance monitoring of *Streptomyces* strain M7. The importance of this bacterium is its capacity to grow in the presence of organochlorine pesticides used by them as a carbon source and using the amino acid asparagine as a nitrogen source. Thus, the presence of these pesticides makes this medium specific for bacteria growth. On the other hand, by monitoring *Streptomyces* bacteria growth, it is possible to determine the presence of these pesticides and quantify them. Hence, this system based on impedimetric biosensor has a potential use to detect these dangerous compounds. Since amino acids frequently used in culture media interfere in impedance determinations, the work presented an optimized design regarding the culture medium capabilities on the impedance response of *Streptomyces* M7 activity. Finally, a specific medium was designed using $(NH_4)_2SO_4$ instead of asparagine as nitrogen source [62].

Much attention has also been paid to miniaturization of impedimetric sensor systems by using microelectronics lithographic techniques to fabricate microelectrodes in order to improve sensitivities or add functionalities. Interdigitated array microelectrodes has been demonstrated as a promise in impedance measurements for monitoring the growth of bacteria since they present advantages in terms of the fast establishment of the steady-state signal, the increased signal-to-noise ratio, and the use of small sample solution volumes [63, 64].

To enhance and facilitate the impedance sensing, microfluidic flow cells can be added to the interdigitated microelectrodes to achieve a fully integrated microchip. This brings different benefits such as high-detection sensitivity, small volume handling, low contamination during bacterial growth, ability to concentrate cells, and rapid detection of small number of cells [64].

With the aim to study whether impedance measurements in the microscale could provide information about the metabolic activity of bacteria, Gomez *et al.* described a microfabricated biochip with integrated fluidic paths and electrodes for impedance spectroscopy of nanoliter volumes of bacterial suspensions [65]. Later experiments analyzed the use of this microsystem to detect metabolic activity of small concentrations of different bacteria (*Listeria* and *E. coli*) in two different media (Luria Bertani broth with high ionic content and a low-conductivity medium denominated Tris–Gly–Dext). Thus, they demonstrated the capacity of the system to detect viability of small numbers (around 10^5 cfu mL^{-1}, resulting in about five cells in the 5.27 nL chamber of the chip) in a couple of hours [52].

Varshney *et al.* [64] reported double interdigitated array microelectrodes-based flow system to detect viable cells of *E. coli* O157:H7 selecting a frequency of 1 MHz to monitor the change in the impedance values. They attribute the change in impedance to a decrease in the resistance of the medium due to an increased concentration of highly charged ions corresponding to the growth of the bacterium in a low conductive medium. Thus, they found that the system may be successfully employed for the *E. coli* O157:H7 detection in a range from 8.0 to 8.2 10^8 cfu mL^{-1} after 14.7 and 0.8 h of cultivation time, respectively.

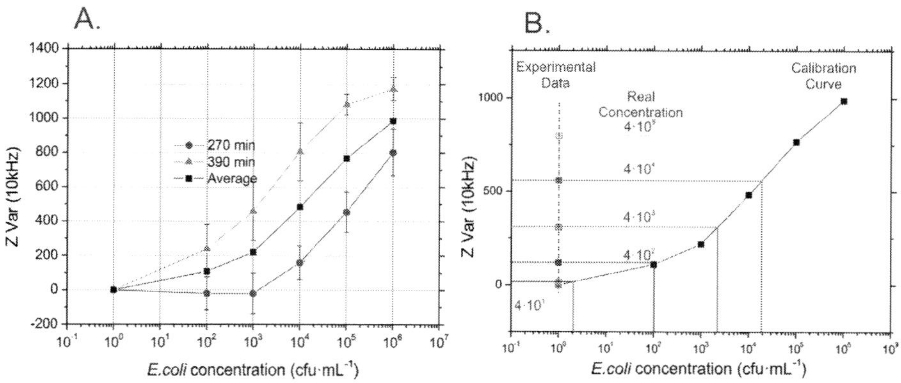

Figure 5. (A) Calibration curves showing changes in sensor impedance at different concentrations of *E. coli* measured at a fixed frequency of 10 kHz in a growth medium at 270 (●) and 390 (▲) min, and the average of these two calibrations (■). (B) Determination of *E. coli* concentration in milk samples after 6 h of incubation. Black squares represent calibration plot used to determine *E. coli* in spiked milk samples. Squares on the left side are the results of impedance changes measured in samples with "unknown" *E. coli* concentration.

It should be noted that most of the bacteria metabolic products are of acidic nature, so that the produced conductivity changes due to ionic products are accompanied by pH changes [25] of the growth media. On this effect, measurements of pH to control the bacteria growth are based [66].

Experiments performed in our research group [67] with *E. coli* in Luria Bertani (LB) bacterial growth medium supplemented with glucose performed with a miniature pH sensor–reference electrode pair showed subsequent pH changes in the pH 7–5 range with the bacteria incubation time.

In order to miniaturize the system, an IDEA [11] was used instead of the pH sensor and the reference electrode. Experimental results presented in Figure 5A show that changes of *E. coli* concentration in the growth medium provoke changes of impedance measured at a fixed frequency of 10 kHz. To avoid the undesirable effect of the bacteria attachment to the interdigitated electrode surface that might provoke additional changes in the impedance magnitude, the developed method consisted of a single measurement after a certain incubation time of *E. coli,* when a drop of the growth media was placed on a sensor surface. Impedance changes were found to depend on the number of microorganisms in a sample in the concentration range 10^2 to 10^6 cfu mL^{-1}. However, as the kinetics of the impedance changes greatly depends on the bacterial concentration, it was not possible to obtain a single calibration curve in a wide concentration range at one specifically fixed incubation time. To resolve this, a novel calibration method was proposed [67] by measuring the sensor response at 270 and 390 min of incubation and taking a mean value.

Thus, this method of measuring impedance using interdigitated microelectrodes at a fix frequency to control the bacterial growth was used for *E. coli* determination in real samples. Our impedimetric biosensor was tested on quantification of *Escherichia coli* bacterium in milk, demonstrating the capacity of the method to detect concentrations in a range between 10^2 and 10^6 cfu mL^{-1} with 6%–12% error margin in only 6 h (Figure 5B).

Summarizing, we may conclude that impedance microbiology, being a useful and well established bacteria control method, finds itself in a new stage of development based on the application of modern technologies oriented on a chip-based method. Advances in microfabrication are allowing the transfer of impedance microbiology to microdevices increasing signal and therefore sensitivity, minimizing sample volume, and reducing assay time [18].

4. Detection based on impedance measurements produced by surface changes of the electrodes

Most of bacteria cells are electrically charged, so due to bacterial cells, immobilization on the electrode surface of impedimetric biosensor variations in electrical impedance may be produced. The bacterial attachment also implies a reduction of the effective electrode area that may affect the charge transfer resistance in faradic impedance measurements. This means that bacterial cells attached to the sensor surface may produce variations in interfacial impedance due to changes in surface conductivity produced by their electrical charge or the surface layer capacitance. Direct label-free impedance method of bacteria detection has gained much interest permitting to reduce substantially the detection time compared with growth-based impedance methods because this methodology is not dependent on cells replication in a culture medium or the production of metabolites [18, 68]. Different kind of electrodes, especially IDEAs, also named as interdigitated array microelectrodes (IDAM), differing in their geometry and immobilization strategies can be used as impedimetric transducers for bacteria detection.

IDEA transducers present promising advantages compared to other impedimetric biosensors as rapid detection kinetics, increase of signal-to-noise ratio, fast establishment of a steady-state response, potential low cost, and ease of miniaturization. Moreover, IDEA eliminates the requirement of a reference electrode compared to three or four electrode systems or potentiometric and amperometric devices. IDEA devices consist of a pair of comblike metal electrodes formed on a planar insulating substrate, in which a series of parallel microband electrodes are connected together by a common bus, forming a set of interdigitating electrode fingers. At present, IDEAs are widely used as impedimetric biosensors for bacterial detection [12].

Parameter designs and materials employed for electrode fabrication are important as they affect the sensitivity and operation of an IDEA. The selection of materials for electrode fabrication depends on the future application, chosen surface modification method, ionic species involved, production costs, and fabrication process. The number of electrode fingers, the spacing between each pairs, and the width, length, or height will determinate the sensitivity of the sensor [69]. Several studies show the importance of geometry for microbiological applications [70]. Bratov et al. [15] developed a transducer for biosensor applications based on a three-dimensional interdigitated electrode array (3D-IDEA) with electrode digits separated by an insulating barrier. This sensor presented considerable improvement in sensitivity compared with a standard planar IDEA design, resulting in a viable option for integrated biosensing applications.

4.1. Nonspecific immobilization on electrodes surface

The majority of detection systems in impedimetric biosensors involve a biorecognition element directly immobilized on the electrode surface to react and attach bacterial cells. However, some studies [12] pointed out that when antibodies or another biomolecules with affinity against bacteria are immobilized on the surface of electrode, the functional area of the electrode is not optimally utilized. Moreover, these biosensors show lack of reproducibility as it is difficult to repeatedly achieve the same surface density of biorecognition molecules on the sensor surface. Different strategies of the detection of bacteria without the direct immobilization of biodetection molecules are discussed here.

Varshney and Li [12] suggested the use of biofunctionalized microbeads or nanoparticles as an indirect impedance measurement. The same authors developed a biosensor based on an IDEA coupled with magnetic nanoparticle–antibody conjugates for the rapid and specific detection of *E. coli* O157:H7 in ground beef samples [71]. Nanoparticles were prepared previously by immobilizing biotin-labeled polyclonal anti-*E. coli* antibodies onto streptavidin-coated magnetic nanoparticles to carry out the preselection and preconcentration of bacteria.

http://dx.doi.org/10.5772/60741

After the separation of bacteria immobilized on nanoparticle–antibody conjugates, they were resuspended into a low conductivity mannitol 0.1 M solution, and the concentrated sample was uniformly spread on the surface of the sensor. The frequency response of the impedance in a 10-Hz to 1-MHz range showed that the bulk resistance and the surface capacitance were responsible of the impedance changes caused by presence of *E. coli* on the surface of IDEA. The detection limit of *E. coli* O157:H7 was 7.4×10^4 in pure cultures and 8×10^5 cfu mL^{-1} in ground beef samples, while the detection time was 35 min. The same methodology was used in a microfabricated flow cell to detect *E. coli* O157:H7 [72]. In this case, a volume of 60 nL was used, and the detection limit was as low as 1.6×10^2 and 1.3×10^3 cells in pure cultures and beef samples, respectively. This detection limit, being recalculated for a 5 mL volume, corresponds to 8.4×10^4 cfu mL^{-1}.

Recent studies of Kanayeva et al. [73] used a preconcentration technique for *Listeria monocytogenes* detection, a food-borne pathogenic bacteria. Immunomagnetic nanoparticles were functionalized with anti-*L. monocytogenes* antibodies via biotin–streptavidin bonds to capture *Listeria* in a sample during 2-h immunoreaction. To collect the complex of nanoparticles with bacteria, a magnetic separator was used and, after a washing step, *L. monocytogenes* was removed from the samples and injected in a microfluidic chip. The impedance change produced by bacteria was measured by an IDEA in the microfluidic chip in a phosphate-buffered solution. An equivalent concentration of 10^3 cfu mL^{-1} of the original sample was detected without interferences by other bacteria as *Listeria innocua*, *E. coli* K12, *E. coli* O157:H7, *Salmonella typhimurium*, and *Staphylococcus aureus*. EC analysis indicated that impedance change was mainly produced by decrease in the medium resistance. Results obtained in milk, lettuce, and ground beef samples showed that the sample matrix effect affects the detection limit that was between 10^4 and 10^5 cfu mL^{-1}. The required detection time was around 3 h.

Advantages of using microbeads and nanoparticles are based in the separation and concentration of a specific strain of bacteria from the native sample previous to registration process, which permits to reduce the background noise caused by nontarget compounds or other bacteria. Furthermore, in some cases, the surface of the electrodes can be used multiple times because recognition elements are not attached to the sensor surface. However, the increase in number of the detection process steps from the initial sample treatment to the final bacteria detection could result in the sensitivity and reproducibility loss of a biosensor device.

In this field of detection without chemical attachment of a biorecognition element to the sensor surface, our group has recently developed a modified $TaSi_2$ IDEA on a SiO_2 substrate to study interactions with bacteria present in a sample solution [74]. Bacteria immobilized on the sensor affect the surface charge and produce changes in the superficial impedance. In the studied case, the sensor surface was chemically modified by a layer-by-layer method [75] with oppositely charged polyelectrolyte layers by alternating polyethyleneimine (PEI) and poly(sodium 4-styrenesulfonate) (PSS). *E. coli* ATCC 10536 was employed for sensitivity and time evaluation. Bacteria were immobilized on the IDEA with a PEI–PSS–PEI multilayer, taking into account the ability of PEI to react chemically with outer membrane compounds of gram-negative bacteria. Results obtained showed a detection limit of the sensor as low as 10^1 cfu mL^{-1} and response time around 20 min.

Despite the satisfactory results obtained, the main drawback of this technique was the nonspecificity of this methodology because all other negatively charged particles that might be present in the sample would adhere to the sensor surface as well. Moreover, the reuse of thus fabricated sensors is very complicated due to a very strong adhesion of PEI polyelectrolyte to silicon dioxide sensor surface. However, in combination with preconcentration and separation techniques mentioned above, this device [74] may be advantageous due to its higher sensitivity.

4.2. Bacteria detection by biorecognition elements on the sensor surface

Most of the studied impedimetric biosensors have been functionalized by the immobilization of biorecognition elements on their surface. A biorecognition element is a biomolecule (antibody, protein, peptide, etc.) with specific affinity that selectively reacts with a specific target analyte [17, 76]. The detection process involves the formation of a complex between the sensing recognition biomolecule and the specific analyte (proteins, nucleic acids, antibodies, antigens, microorganisms, or whole cells). Generally, the electrical properties of the sensor surface are altered by the two components, the biorecognition element in a first phase and the specific target in a second phase.

Different strategies are used to promote the immobilization of the biorecognition element on impedimetric biosensors [7, 17, 77]:

- Bioaffinity layers (avidin–biotin system)
- Thiol containing self-assembled monolayers (SAMs) on gold
- Langmuir–Blodgett films
- Chemical grafting through silanization strategies
- Thin polymers
- Polyelectrolyte films (layer by layer)

The choice of the appropriate immobilization technique depends on the biomolecule nature, reproducibility, cost, and difficulty of immobilization. Immobilized biomolecules have to maintain their active structure, function, high sensitivity and selectivity, fast reaction kinetics, and high stability and not to be desorbed during the use of biosensor. More information on immobilization strategies may be found in a specific review [78].

Previously in Section 2, the main differences between faradic and nonfaradic impedance has been reported. In next sections, impedimetric biosensors functionalized with biorecognition elements attached on electrode surface will be discussed, taking into account whether faradic or nonfaradic processes are responsible for sensitivity.

4.2.1. Impedimetric immunosensors

Immunochemistry is a well-studied and developed area, so the implementation of this technique to different kinds of electrochemical impedimetric biosensors has been widely used in the past years. The major advantage of immunosensors is the specificity and sensitivity of biomolecular interactions between the antibody and the antigen (from a little target molecule to bacteria). Moreover, the advances in production techniques of monoclonal antibodies, genetic engineering, and recombinant antibodies have improved binding-ability and stability on biosensor surfaces [79]. However, the main drawback of antibodies that prevents their widespread use in biosensors is the lack of stability, as many of them lose their activity quite rapidly. A few examples of reported impedimetric immunobiosensors for microbiological applications are discussed later in this section, stressing the impact of sensor geometry and electrodes configurations and their materials, antibodies, and bacteria species detected.

A high number of impedimetric immunobiosensors reported in literature are based on faradic impedance measurements. One of the earliest works on electrochemical impedance spectroscopic biosensors is by Ruan and Yang in 2002 [80], who reported an immunosensor based on a planar IDEA with indium tin-oxide (ITO) electrodes. Anti-*E. coli* antibodies were immobilized using an epoxysilane layer for chemical anchoring of antibodies to capture *E. coli* cells. Impedance was measured in the presence of a redox couple, $Fe(CN_6)^{3-/4-}$, in that way attached bacteria partially block the metal electrode surface, which allows to register the increase of electron transfer resistance (R^{et}) with increasing concentration of bacteria. The biosensor could detect the *E. coli* bacteria with a detection limit of 6×10^3 cells/mL and a linear response in the R_{et} between 6×10^4 and 6×10^7 cells/mL. Lately, Yang et al. [81] developed another immunosensor for detecting the same bacteria using the same antibodies, but in this case, anti-*E. coli* O157:H7 antibodies were immobilized on an ITO interdigitated array sensor surface through hydroxyl groups of ITO electrode and carboxylic groups of antibodies. As in previous studies conducted by Ruan and colleagues [80], faradic EIS was measured using a redox probe. The R_{et} increased on the immobilization of antibodies and bacterial cells that behaved as insulators, allowing to obtain a correlation between the electron transfer resistance and bacterial concentration between 10^5 and 10^8 cfu mL^{-1}, but a detection limit of 10^6 cfu mL^{-1} was quite high.

It should be noted that faradic impedance measurements with a redox probe do not use any possible advantages presented by IDEAs, as the charge transfer resistance is not dependent on the electrode geometry and is affected only by the total electrode area [11]. For these measurements, simple plane electrodes can be used as well.

Although *E. coli* is the most popular model bacterial system, many studies are focused on detecting other bacteria. For example, Mantzila et al. [82] developed a faradic impedimetric immunosensor for the detection of *S. typhimurium* in milk samples. Polyclonal antibodies anti-*Salmonella* were cross-linked in gold electrodes in presence of glutaraldehyde and different mixed SAMs. High selectivity was obtained in front *E. coli* bacteria in milk samples, while the detection limit for *Salmonella* was indicated at a concentration level three orders of magnitude lower than the infectious dosage that is around 10^5 cfu mL^{-1} [83].

The most important challenge in microbial sensor development is the reduction of the detection limit and detection time. One of the latest publications [84] report *E. coli* O157:H7 bacteria immobilization on a gold electrode with anti-*E. coli* antibodies through the SAM of mercaptohexadecanoic acid. EIS was used for detecting pathogenic bacteria, while SPR was used to monitor the antibody immobilization. Rct values obtained with a redox probe were used to monitor changes produced by bacterial interaction with antibody. To our knowledge, this work reported the lowest detection limit of 2 cfu mL^{-1}.

On the other hand, other studies were focused on nonfaradic impedance measurements. Radke and co-workers [85] used an IDEA chip with gold electrodes modified by a 3-mercaptomethyldimethylethoxysilane and a heterobifunctional cross-linker to immobilize antibodies. The reported optimum width and spacing were 3 and 4 μm, respectively. The impedance across the interdigitated electrode was measured after immersing the functionalized biosensor in a peptone solution with *E. coli* O157:H7, and the resulting impedance change caused by bacterial attachment was monitored. In this case, the impedance was measured in a frequency range of 100 Hz to 10 MHz, and main differences due to bacteria concentration were obtained at low frequencies where the impedance depends mainly on interfacial capacitance. The biosensor was able to discriminate between different concentrations from 10^5 to 10^7 cfu mL^{-1} at a frequency of 1 kHz. The same electrode system being used in pure cultures and food samples (romaine lettuce) showed reduced one order of magnitude detection limit (10^4 cfu mL^{-1}) [5].

Tan et al. [86] developed a PDMS microfluidic immunosensor integrated with specific antibodies on an alumina nanoporus membrane for the rapid detection of *S. aureus* and *E. coli* O157:H7 with EIS. In this case, antibodies were immobilized on the membrane using self-assembled (3-glycidoxypropyl)trimethoxysilane (GPMS). For bacteria detection, a frequency range from 1 to 100 Hz was applied for both bacteria, and changes produced by bacterial attachment were obtained around 100 Hz. Cross-bacteria experiments showed a high specificity of anti-*E. coli* and anti-*S. aureus* antibodies utilized, and detection within 2 h showed a detection limit of 10^2 cfu mL^{-1}. Thus, the combination of a microfluidic chamber and different substrates such as an alumina nanoporous membrane for the immobilization of antibodies offered new approaches for the immunodetection of bacteria.

One of the main problems of immunosensors is the difficulty to reuse the biosensor once bacteria are attached. Single-use disposable sensors are attractive; however, their production cost should be very low, and this typically is not the case. Hence, it is required to find some treatment to remove the sensor surface coating in order to use it several times. Dweik et al. [87] established a cleaning protocol for a biosensor based on a gold interdigitated microelectrodes for the detection of viable *E. coli* O157:H7 using anti-*E. coli* IgG antibodies. They assured that a 30-min treatment with acetone, followed by a wash with isopropanol and distilled water, and exposure to plasma for 2 min with a power of 48 W permit to restore the sensor surface to its initial state and thus to reuse each device at least for five times.

Without any doubt, antibodies are the most widely used bioreceptors in biosensor research and development. However, as it was mentioned, the main problem of the stability of antibodies after the immobilization on a sensor surface remains a challenge as well as short shelf lifetime and decrease of binding efficiency over time [19]. Furthermore, antibodies production and purification costs are an added difficulty. For this reason, there is a permanent search for other biorecognition elements as bioreceptors for biosensing.

4.2.2. Aptamers

Aptamers are short series of single-stranded DNA or RNA oligonucleotides obtained artificially via in vitro process called systematic evolution of ligands exponential enrichment

(SELEX) [88]. Aptamers have been used for biosensing applications due to their ability to bind with high selectivity to a specific target molecule. These artificial nucleic acid ligands can be generated against amino acids, proteins, drugs, and other molecules, and they can be applied for the detection of various targets molecules and even whole cells or organisms [12]. The high specificity and affinity to target molecules, the ease of synthesis, and immobilization without compromising their biological activity allows their use as biorecognition elements for bacterial detection. The majority of aptasensors are focused on the detection of protein targets, but recently appeared publications devoted to direct bacterial detection. During the SELEX process, whole microorganisms can be employed as target during aptamer synthesis because of the interesting membrane proteins bound specifically to aptamer [89]. One of principal advantages of this method of aptamer synthesis is the ability to target and specifically differentiate microbial strains without having previous knowledge of the membrane molecules or structural changes present in that particular microorganism [90]. Compared to antibodies, aptamers can be chemically modified and labeled more easily facilitating the functionalization of solid surfaces and nanoparticles and can be used in real samples, which is especially useful for environmental and food control applications. The major disadvantage is probably that DNA and RNA structures are highly sensitive to nuclease degradation, but in biosensing applications, the presence of nucleases is not very common.

The majority of assays with aptamers in impedimetric biosensing applications have been reported in terms of faradic measurements. Labib and co-workers reported impedimetric sensors for bacteria viability and typing [91, 92]. In particular, they developed DNA aptamers against *Salmonella enteritidis* pathogen and used it in a mixture of related pathogens including *S. typhimurium, E. coli, S. aureus, Pseudomonas aeruginosa,* and *Citrobacter freundii* to confirm specificity of the aptamers. The integration of aptamer onto an impedimetric biosensor was conducted via self-assembling onto gold nanoparticle-modified screen-printed carbon electrode. The aptasensor was incubated for 1 h in different aliquots with increasing concentration of *S. enteritidis* in phosphate-buffered saline. The binding between the target bacteria and the respective aptamers blocked the charge transfer resistance (R_{et}) from a solution-based redox probe to the electrode surface. The obtained detection limit was 600 cfu mL^{-1}, while R_{et} changes produced by other bacterial species were very low compared with *S. enteritidis* target bacteria. This work presented a significant proof of concept for the first aptamer-based impedimetric sensor for typing bacteria [92].

More recent studies has been focused on developing electrochemical impedimetric biosensors for Salmonella detection using a specific ssDNA aptamer [93]. In this case, the biosensor was based on a glassy carbon electrode modified with graphene oxide and gold nanoparticles. Nanoparticles were used for signal amplification and better biocompatibility to detect biological molecules. The modified electrode was incubated in the presence of Salmonella, and its faradic impedance was measured. The optimal incubation time was determined to be 35 min, while the detection limit obtained was as low as 3 cfu mL^{-1}. Furthermore, the specificity was also compared with different strains of bacteria as *L. monocytogenes, B. subtilis, E. coli, S. aureus*, or *S. pyogenes*, and changes registered after 35 min of incubation were much lower than for the *Salmonella*. The resistance value was also obtained monitoring the electron transfer between the $Fe(CN_6)^{3-/4-}$ electrolyte solution and the electrode. Similar methodology was used by Jia et al. [88] on the performance of an impedimetric aptasensor for *S. aureus* detection by EIS. In this case, a detection limit of 10 cfu mL^{-1} was obtained and high selectivity over other pathogens was also demonstrated.

Probably, alluding to advantages mentioned previously, the use of aptamers in biosensing by electrochemical techniques will increase in the subsequent years.

4.2.3. Antimicrobial peptides

The use of antimicrobial peptides (AMPs) as biorecognition elements for bacterial detection on impedimetric sensors has progressed in recent years. AMPs are a family of biomolecules that are crucial in the innate immune defense of many organisms that display a broad spectrum of activity against gram-negative and gram-positive bacteria. Basically, the antimicrobial activity has been attributed to their capacity to target and disrupt bacterial membranes [94, 95]. First experiments of AMPs for biosensor applications were conducted by Kulagina et al. [96, 97]. They reported two biosensor assays using magainin I as the recognition molecule in the fluorescent-based detection of *E. coli* and *Salmonella*.

Firsts experiments with electrochemical nonfaradic impedance technique with an APM immobilized on IDEAs were done by Mannoor et al. [98] in 2010. They accentuated the high stability of AMPs in harsh environmental conditions, the durability of AMPs immobilized on sensors under natural ambient environment, and their semiselective binding nature to target cells that allows to bind a variety of pathogens [98]. In this case, the AMPs were immobilized on a gold a microcapacitive electrodes via a C-terminal cysteine residue, and the biosensor was exposed to various bacteria concentrations ranging from 10^3 to 10^7 cfu mL^{-1}. The variation in impedance change at a fixed frequency of 10 Hz was observed directly proportional to the number of bacterial cells bound to the immobilized AMPs and manifested in a logarithmic increase with serially diluted bacterial concentrations. The detection limit of the device to *E. coli* was 10^3 cfu mL^{-1} (1 bacterium/μL). Other bacterial species were tested to investigate the selectivity of AMP-functionalized microelectrodes: gram-negative pathogenic *E. coli* O157:H7, *E. coli* ATCC 35218, *pathogenic S. typhimurium*, and gram-positive pathogenic *L. monocytogenes*. The response of biosensor with magainin I was clear preferential toward pathogenic gram-negative species of *E. coli* and Salmonella, especially toward *E. coli* O157:H7, demonstrating interbacterial strain differentiation and maintaining recognition capabilities toward pathogenic strains of *E. coli* and Salmonella. This research group also demonstrated antimicrobial peptides self-assembling onto a wireless graphene nanosensor integrated on a tooth for remote monitoring of a respiration and bacteria detection in saliva [99].

Similar studies were conducted by Lillehoj et al. [100], who reported a microfluidic chip for the multiplexed detection of bacterial cells using AMPs. Peptide immobilization on the sensors was made via cysteine–gold interactions, revealing robust surface binding. Samples containing *Streptococcus mutans* and *Pseudomonas aeruginosa* were attached to the chip, and both microorganisms were detected at minimum concentrations of 10^5 cfu mL^{-1} in 25 min.

Other works used interdigitated impedimetric arrays for gram-positive bacteria detection with naturally produced AMPs from class IIa bactericins. Etayash et al. [101] used leucocin A, a representative a class IIa bacteriocin, chemically synthesized and immobilized on interdigitated gold microlelectrodes via C-terminal carboxylic acid of the peptide and free amines of a preattached thiolated linker, as antilisterial microbial peptide. In this case, the authors highlighted the narrow activity spectrum of class IIa bacteriocins with high effectiveness with which they act by receptor-mediated mechanism with the target bacterial cells. In this case, leucocin A was used for the real-time detection of *L. monocytogenes*. The detection limit was as low as 10^3 cfu mL^{-1}, which is the equivalent of 1 bacterium/μl. The biosensor also selectively detected *Listeria* in front of other gram-positive strains at 10^3 cfu mL−1. Thus, in this work, high sensitivity and selectivity were obtained.

Finally, Li et al. [102] developed a novel biosensor based on faradic impedance for the detection of *E. coli* O157:H7 using a film formed of ferrocene–peptide conjugate on a gold electrode, and magainin I as antimicrobial peptide. Other bacteria as nonpathogenic *E. coli* K12, *Staphylococcus epidermidis* and *Bacillus subtilis*, were used to evaluate the selectivity of biosensors. Obtained results revealed that *E. coli* O157:H7 was preferentially selected as reported before [98]. For impedance measurements, a redox probe with 5 mM of $K_3[Fe(CN)_6]/K_4[Fe(CN)_6]$ was used to optimized detection and study changes produced in the charge transfer resistance (ΔR_{et}). The detection limit obtained was 10^3 cfu mL^{-1}, similar to previous studies.

Although the use of antimicrobial peptides in biosensing applications offers a robustness and stability compared to other biorecognition elements such as antibodies, the main drawback of these elements is the low or lack of specificity against different species and especially different bacterial strains. We have mentioned some examples of aptamer biosensors with more affinity for certain species than others, but for real biomedical or biosafety applications, where the identification of pathogenic bacteria causing human diseases is really important, AMPs performance remains a challenge and has to be studied more in detail.

4.2.4. Lectins

Another kind of biorecognition element described in literature is lectin, a carbohydrate-binding protein or glycoprotein produced by many organisms (from viruses and microorganisms to plants and mammals) that selectively and reversibly react with mono- and oligosaccharides, widely present on bacterial cell surface [103]. The recognition of these carbohydrates on bacteria surface can be used for the specific detection of bacteria. Carbohydrate–protein interaction is much weaker than protein–protein interaction, but these molecules are more stable and smaller than antibodies, and they can neither be denatured easily nor lose their activity [104]. Moreover, the small size of lectins allows to obtain higher densities of carbohydrate-sensing elements on a sensor surface, leading to higher sensitivity and lower nonspecific adsorption [13].

Gamella et al. [105] reported a lectin-based screen-printed gold electrode for the impedimetric detection of bacteria based on faradic impedimetric measurements. In this case, concanavalin A (ConA), a mannose- and glucose-binding lectin, was used as biorecognition element for interaction with carbohydrate of *E. coli* surface. Biotinylated ConA and *E. coli* formed a complex in solution, and after 1 h at room temperature, the complex was immobilized on the surface. Impedimetric measurements were conducted afterward in a solution of the redox probe of

$K_3[Fe(CN)_6]/K_4[Fe(CN)_6]$. The electron transfer resistance (R_{et}) varied linearly from 5×10^3 to 5×10^7 cfu mL^{-1}. The selectivity was evaluated with different lectins and three different bacteria: *E. coli, S. aureus,* and *Mycobacterium phlei,* and satisfactory conclusions were achieved.

Other studies have been conducted to detect sulfate-reducing bacteria such as *Desulforibrio caledoinensis* by immobilizing ConA using an agglutination assay. Wan et al. [106] immobilized ConA onto a gold electrode using amine coupling on the surface with 11-mercaptoundecanoic acid. A redox probe with $Fe(CN)^{3-/4-}$ was used to obtain faradic impedance spectra, and an electron transfer resistance (R_{et}) was monitored with the increases of bacterial concentration. The system showed high sensitivity with a linear correlation in the concentration range from 1.8×10^0 to 1.8×10^7 cfu mL^{-1}.

Recent studies has been performed by our group [74] developing an impedimetric transducer based on an interdigitated electrode where ConA lectin was utilized as a biorecognition element. Nonfaradic processes were monitored through R_s changes on IDEA surface. ConA was attached on sensor surface by a layer-by-layer method through PEI–ConA interaction. *E. coli* was used as bacterial model, and similar detection limits of 10^4 cfu mL^{-1} were obtained as reported by others [105].

Despite advantages presented by lectins as biorecognition elements for biosensing applications described previously, some drawbacks have to be mentioned. The inherent disadvantage of lectins is that several lectins can bind different carbohydrates as well as different carbohydrates can bind the same lectin [107]. These properties of lectins reduce significantly the specificity between bacterial species and especially between bacterial strains. Therefore, in bacterial detection where bacterial membrane consists of a series of different carbohydrates and lipopolysaccharides (aside from other components), the specific detection of bacterial species can produce false positive in complex samples.

4.2.5. Other biorecognition elements

As a recent alternative, the use of bacteriophages as biorecognition elements has been proposed [108, 109]. Bacteriophages are virus of bacteria that utilize bioreplicative machinery to multiply and bind selectively against outer membrane of the bacterial cell-surface proteins, lipopolysaccharides, pili, and lipoproteins. Therefore, bacteriophages can be used as biorecognition element due to additional properties such as high specificity, low-cost production, long shelf life, and thermostability during handling. Furthermore, metabolic products or intracellular components of bacteria realized by lytic action of phages can be an alternative route for biosensing.

Mejri and co-workers [110] developed a biosensor based on the use of T4 bacteriophage for *E. coli* recognition. In this case, antibodies and phages were compared for *E. coli* biosensing by using EIS. Both biorecognition elements were physisorbed on interdigitated gold microelectrodes. Measurements with phages immobilized on the surface were conducted by monitoring the variations in impedance module (ΔZ) at a fixed frequency of 233 MHz. Results showed an increase of the initial impedance after about 20–25 min, followed by an important decrease in impedance. The initial increase was attributed to the phage–bacteria recognition, while the subsequent decrease presumably happens as consequence of bacterial lysis and release of intracellular components. In case of antibodies, only the initial increase was observed, demonstrating the lytic effect of T4 phage. Linear response was observed for *E. coli* range concentrations from 10^4 to 10^7 cfu mL^{-1}. However, no response was produced in the case of *Lactobacillus* at the same concentrations.

Other recent work conducted by Tlili et al. [111] studies a bacteriophage-impedimetric biosensor for the identification and quantification of *E. coli* with bacteriophage T4. In this case, during the lytic process, the realization of Tuf gene was amplified by a loop-mediated isothermal amplification (LAMP) method and monitored by linear sweep voltammetry (LSV) as a confirmation assay. The phage was attached on a cysteamine-modified gold electrode in the presence of 1,4-phenylene diisothiocyanate, and bacteria adhesion was monitored by changes in electron transfer resistance (ΔR_{et}) in $Fe(CN)^{63-/4-}$ redox pair solution. Electrochemical impedance results reveal a detection limit of 800 cfu mL–1 and a detection time of 15 min, while confirmation assay by LAMP assay and LSV requires 40 min. The reduction of one order of magnitude was obtained by the detection of Tuf gene.

Some drawbacks on the use of bacteriophages in biosensing applications have to be mentioned. During lytic process of bacterium, the signal on a biosensor would be lost or significantly affected due to the components released with bacterial cell disruption [112]. Moreover, some studies suggest that phages bound to the sensor lose their bacterial binding capability upon drying because their tail fibers collapse and are unavailable to bind to the bacterial host [113]. In addition, phages have relatively large sizes, which limit their biosensing applications on particular sensor where detection is limited by distance.

5. Final remarks

Electrochemical biosensors based on impedance detection each year are used more widely due to their high sensitivity and rapid response, which makes this technique extremely useful to detect biological interactions. The detection of pathogenic bacteria using impedance techniques, introduced in this chapter, is an important field that still requires further development.

The detection of bacteria by EIS may be performed in two ways: (1) by the detection of metabolites produced by bacterial growth and involving conductivity changes in the sample and (2) by bacterial detection based on the immobilization of bacteria on electrode surface through biorecognition elements (antibodies, antimicrobial peptides, aptamers, etc.), which is oriented basically on registration of changes in charge transfer resistance (faradic process) and interfacial impedance (nonfaradic process).

The first method is simpler but requires working with low conductivity media and takes longer times. The second method, especially accompanied by some preconcentration technique, may be very fast, selective, and sensitive. Nevertheless, there are still a lot of challenges to be overcome aimed on lower detection limits, shorter detection times, selectivity, and sensitivity. A great help in resolving some of these problems may arrive from using IDEAs. However, analyzing current publications, we may note that there is poor understanding of how biochemical interactions on a sensor surface affect its electrical properties. Without clear knowledge of interfacial chemical processes and their effect on a complex interfacial impedance, it would not be possible to optimize the measurement procedures and sensor geometry, thus improving sensors performance.

Acknowledgements

Authors acknowledge financial support from Spanish Ministry of Economy and Competitiveness (projects CTQ2014-54553-C3-1-R and CTQ2011-29163-C03-02).

Author details

Sergi Brosel-Oliu, Naroa Uria, Natalia Abramova and Andrey Bratov*

*Address all correspondence to: andrei.bratov@imb-cnm.csic.es

BioMEMS Group, Instituto de Microelectrónica de Barcelona, Centro Nacional de Microelectrónica (CSIC), Campus U.A.B., Bellaterra, Barcelona, Spain

References

[1] Demuth A, Aharonowitz Y, Bachmann TT, Blum-Oehler G, Buchrieser C, Covacci A, et al. Pathogenomics: an updated European research agenda. Infection Genetics and Evolution. 2008; 8: 386–393.

[2] Zourob M, Sauna E, Turner APF. Principles of Bacterial Detection: Biosensors, Recognition Receptors and Microsystems. NY: Springer; 2008.

[3] Velusamy V, Arshak K, Korostynska O, Oliwa K, Adley C. An overview of food-borne pathogen detection: in the perspective of biosensors. Biotechnology Advances. 2010; 28: 232–254.

[4] Doyle MP, Buchanan RL. Food Microbiology: Fundamentals and Frontiers. 4th ed. Washington, DC: ASM Press; 2013, 1118.

[5] Radke SA, Alocilja EC. A high density microelectrode array biosensor for detection of E-coli O157:H7. Biosensors and Bioelectronics. 2005; 20: 1662–1667.

[6] Glynn B, Lahiff S, Wernecke M, Barry T, Smith TJ, Maher M. Current and emerging molecular diagnostic technologies applicable to bacterial food safety. International Journal of Dairy Technology. 2006; 59: 126–139.

[7] Lazcka O, Campo FJD, Muñoz FX. Pathogen detection: a perspective of traditional methods and biosensors. Biosensors and Bioelectronics. 2007; 22: 1205–1217.

[8] Skottrup PD, Nicolaisen M, Justesen AF. Towards on-site pathogen detection using antibody-based sensors. Biosensors and Bioelectronics. 2008; 24: 339–348.

[9] Hunt HK, Soteropulos C, Armani AM. Bioconjugation strategies for microtoroidal optical resonators. Sensors. 2010; 10: 9317–9336.

[10] Marusov G, Sweatt A, Pietrosimone K, Benson D, Geary SJ, Silbart LK, et al. A microarray biosensor for multiplexed detection of microbes using grating-coupled surface plasmon resonance imaging. Environmental Science and Technology. 2012; 46: 348–359.

[11] Bratov A, Abramova N. Chemical sensors and biosensors based on impedimetric interdigitated electrode array transducers. In: Suarez DE, editor. Smart Sensor and Sensing Technology. NY: Nova Science Publishers; 2013. p. 155–164.

[12] Varshney M, Li Y. Interdigitated array microelectrodes based impedance biosensors for detection of bacterial cells. Biosensors and Bioelectronics. 2009; 24: 2951–2960.

[13] Wang Y, Ye Z, Ying Y. New trends in impedimetric biosensors for the detection of foodborne pathogenic bacteria. Sensors. 2012; 12: 3449–3471.

[14] Bratov A, Abramova N, Marco MP, Sanchez-Baeza F. Three-dimensional interdigitated electrode array as a tool for surface reactions registration. Electroanalysis. 2012; 24: 69–75.

[15] Bratov A, Ramón-Azcón J, Abramova N, Merlos A, Adrian J, Sánchez-Baeza F, et al. Three-dimensional interdigitated electrode array as a transducer for label-free biosensors. Biosensors and Bioelectronics. 2008; 24: 729–735.

[16] Barsoukov E, Macdonald JR. Impedance Spectroscopy: Theory, Experiment, and Applications. Wiley; 2005.

[17] Lisdat F, Schäfer D. The use of electrochemical impedance spectroscopy for biosensing. Analytical and Bioanalytical Chemistry. 2008; 391: 1555–1567.

[18] Yang L, Bashir R. Electrical/electrochemical impedance for rapid detection of foodborne pathogenic bacteria. Biotechnology Advances. 2008; 26: 135–150.

[19] Ahmed A, Rushworth JV, Hirst NA, Millner PA. Biosensors for whole-cell bacterial detection. Clinical Microbiology Reviews. 2014; 27: 631–646.

[20] Berggren C, Bjarnason B, Johansson G. Capacitive biosensors. Electroanalysis. 2001; 13: 173–180.

[21] Turner AP, Ramsay G, Higgins IJ. Applications of electron transfer between biological systems and electrodes. Biochemical Society Transactions. 1983; 11: 445–448.

[22] Perumal V, Hashim U. Advances in biosensors: principle, architecture and applications. Journal of Applied Biomedicine. 2014; 12: 1–15.

[23] Wilkins J, Young R, Boykin E. Multichannel electrochemical microbial detection unit. Applied and Environmental Microbiology. 1978; 35: 214.

[24] Stewart GN. The charges produced by the growth of bacteria in the molecular concentration and electrical conductivity of culture media. Journal of Experimental Medicine. 1899; 4: 235–243.

[25] Ur A, Brown D. Impedance monitoring of bacterial activity. Journal of medical microbiology. 1975; 8: 19–28.

[26] Cady P. Rapid automated bacterial identification by impedance measurement. London: Wiley & Sons; 1975.

[27] Wheeler T, Goldschmidt MC. Determination of bacterial cell concentrations by electrical measurements. Journal of Clinical Microbiology. 1975; 1: 25–29.

[28] Cady P, Dufour S, Lawless P, Nunke B, Kraeger S. Impedimetric screening for bacteriuria. Journal of Clinical Microbiology. 1978; 7: 273–278.

[29] Cady P, Dufour S, Shaw J, Kraeger S. Electrical impedance measurements: rapid method for detecting and monitoring microorganisms. Journal of Clinical Microbiology. 1978; 7: 265–272.

[30] Richards J, Jason A, Hobbs G, Gibson D, Christie R. Electronic measurement of bacterial growth. Journal of Physics E: Scientific Instruments. 1978; 11: 560.

[31] Harris CM, Kell DB. The estimation of microbial biomass. Biosensors. 1985; 1: 17–84.

[32] Dezenclos T, Ascon-Cabrera M, Ascon D, Lebeault J-M, Pauss A. Optimisation of the indirect impedancemetry technique; a handy technique for microbial growth measurement. Applied Microbiology and Biotechnology. 1994; 42: 232–238.

[33] Palmqvist E, Kriz CB, Khayyami M, Danielsson B, Larsson P-O, Mosbach K, et al. Development of a simple detector for microbial metabolism, based on a polypyrrole dc resistometric device. Biosensors and Bioelectronics. 1994; 9: 551–556.

[34] DeSilva MS, Zhang Y, Hesketh PJ, Maclay GJ, Gendel SM, Stetter JR. Impedance based sensing of the specific binding reaction between Staphylococcus enterotoxin B and its antibody on an ultra-thin platinum film. Biosensors and Bioelectronics. 1995; 10: 675–682.

[35] Ehret R, Baumann W, Brischwein M, Schwinde A, Stegbauer K, Wolf B. Monitoring of cellular behaviour by impedance measurements on interdigitated electrode structures. Biosensors and Bioelectronics. 1997; 12: 29–41.

[36] Dupont J, Ménard D, Hervé C, Chevalier F, Bèliaeff B, Minier B. Rapid estimation of Escherichia coli in live marine bivalve shellfish using automated conductance measurement. Journal of Applied Bacteriology. 1996; 80: 81–90.

[37] Bülte M, Reuter G. Impedance measurement as a rapid method for the determination of the microbial contamination of meat surfaces, testing two different instruments. International Journal of Food Microbiology. 1984; 1: 113–125.

[38] Gibson DM CP, Pimbley DW. Automated conductance method for the detection of *Salmonella* in foods—collaborative study. Journal of AOAC International. 1992; 75: 293–302.

[39] AOAC. *Salmonella* in food, automated conductance methods: AOAC official method 991.38. 16th ed. Gaithersburg, MD: Association of Official Analytical Chemists International; 1996.

[40] Swaminathan B, Feng P. Rapid detection of food-borne pathogenic bacteria. Annual Reviews in Microbiology. 1994; 48: 401–426.

[41] Silley P, Forsythe S. Impedance microbiology—a rapid change for microbiologists. Journal of Applied Bacteriology. 1996; 80: 233–243.

[42] Martins S, Selby M. Evaluation of a rapid method for the quantitative estimation of coliforms in meat by impedimetric procedures. Applied and Environmental Microbiology. 1980; 39: 518–524.

[43] Priego R, Medina L, Jordano R. Bactometer system versus traditional methods for monitoring bacteria populations in Salchichon during its ripening process. Journal of Food Protection. 2011; 74: 145–148.

[44] Edmiston AL, Russell SM. A rapid microbiological method for enumerating Escherichia coli from broiler chicken carcasses. Journal of Food Protection. 1998; 61: 1375–1377.

[45] Dourou D, Ammor MS, Skandamis PN, Nychas G-JE. Growth of *Salmonella enteritidis* and *Salmonella typhimurium* in the presence of quorum sensing signalling compounds produced by spoilage and pathogenic bacteria. Food Microbiology. 2011; 28: 1011–1018.

[46] Jawad GM, Marrow T, Odumeru JA. Assessment of impedance microbiological method for the detection of Escherichia coli in foods. Journal of Rapid Methods and Automation in Microbiology. 1998; 6: 297–305.

[47] Moran L, Kelly C, Cormican M, McGettrick S, Madden RH. Restoring the selectivity of Bolton broth during enrichment for Campylobacter spp. from raw chicken. Letters in Applied Microbiology. 2011; 52: 614–618.

[48] Walker K, Ripandelli N, Flint S. Rapid enumeration of Bifidobacterium lactis in milk powders using impedance. International Dairy Journal. 2005; 15: 183–188.

[49] Glassmoyer KE, Russell SM. Evaluation of a selective broth for detection of *Staphylococcus aureus* using impedance microbiology. Journal of Food Protection. 2001; 64: 44–50.

[50] Salvat G, Rudelle S, Humbert F, Colin P, Lahellec C. A selective medium for the rapid detection by an impedance technique of Pseudomonas spp. associated with poultry meat. Journal of Applied Microbiology. 1997; 83: 456–463.

[51] Felice CJ, Madrid RE, Olivera JM, Rotger VI, Valentinuzzi ME. Impedance microbiology: quantification of bacterial content in milk by means of capacitance growth curves. Journal of Microbiological Methods. 1999; 35: 37–42.

[52] Gómez R, Bashir R, Bhunia AK. Microscale electronic detection of bacterial metabolism. Sensors and Actuators B: Chemical. 2002; 86: 198–208.

[53] Suehiro J, Hamada R, Noutomi D, Shutou M, Hara M. Selective detection of viable bacteria using dielectrophoretic impedance measurement method. Journal of Electrostatics. 2003; 57: 157–168.

[54] Wawerla M, Stolle A, Schalch B, Eisgruber H. Impedance microbiology: applications in food hygiene. Journal of Food Protection. 1999; 62: 1488–1496.

[55] Owens J, Thomas D, Thompson P, Timmerman W. Indirect conductimetry: a novel approach to the conductimetric enumeration of microbial populations. Letters in Applied Microbiology. 1989; 9: 245–249.

[56] Johnson N, Chang Z, Bravo Almeida C, Michel M, Iversen C, Callanan M. Evaluation of indirect impedance for measuring microbial growth in complex food matrices. Food Microbiology. 2014; 42: 8–13.

[57] Schwan HP. Determination of biological impedances. In: Nastuk WL, editor. Electrophysiological Methods: Physical Techniques in Biological Research. NY: Academic Press; 1963. p. 323–407

[58] Ivnitski D, Abdel-Hamid I, Atanasov P, Wilkins E. Biosensors for detection of pathogenic bacteria. Biosensors and Bioelectronics. 1999; 14: 599–624.

[59] Firstenberg-Eden R, Zindulis J. Electrochemical changes in media due to microbial. Journal of Microbiological Methods. 1984; 2: 103–115.

[60] Felice CJ, Valentinuzzi M, Vercellone M, Madrid R. Impedance bacteriometry: medium and interface contributions during bacterial growth. IEEE Transactions on Biomedical Engineering. 1992; 39: 1310–1313.

[61] Hause LL, Komorowski RA, Gayon F. Electrode and electrolyte impedance in the detection of bacterial growth. IEEE Transactions on Biomedical Engineering. 1981; 403–410.

[62] López Rodriguez ML, Madrid RE, Giacomelli CE. The optimization of the culture medium to design Streptomyces sp. M7 based impedimetric biosensors. Sensors and Actuators B: Chemical. 2014; 193: 230–237.

[63] Stulík, K., et al., Microelectrodes. Definitions, characterization, and applications (Technical report). Pure and Applied Chemistry, 2000. 72(8): p. 1483-92.

[64] Varshney M, Li Y. Double interdigitated array microelectrode-based impedance biosensor for detection of viable Escherichia coli O157:H7 in growth medium. Talanta. 2008; 74: 518–525.

[65] Gómez R, Bashir R, Sarikaya A, Ladisch MR, Sturgis J, Robinson JP, et al. Microfluidic biochip for impedance spectroscopy of biological species. Biomedical Microdevices. 2001; 3: 201–209.

[66] Solé M, Rius N, Lorén JG. Rapid extracellular acidification induced by glucose metabolism in non-proliferating cells of Serratia marcescens. International Microbiology. 2010; 3: 39–43.

[67] Uria N, Moral J, Abramova N, Bratov A, Muñoz FX. Fast determination of Escherichia coli in milk samples using impedimetric sensor and a novel calibration method. Unpublished.

[68] Gomez-Sjoberg R, Morisette DT, Bashir R. Impedance microbiology-on-a-chip: microfluidic bioprocessor for rapid detection of bacterial metabolism. Journal of Microelectromechanical Systems. 2005; 14: 829–838.

[69] Guimerà A, Gabriel G, Prats-Alfonso E, Abramova N, Bratov A, Villa R. Effect of surface conductivity on the sensitivity of interdigitated impedimetric sensors and their design considerations. Sensors and Actuators B: Chemical. 2015; 207, Part B: 1010–1018.

[70] Laczka O, Baldrich E, Munoz FX, del Campo FJ. Detection of Escherichia coli and *Salmonella typhimurium* using interdigitated microelectrode capacitive immunosensors: the importance of transducer geometry. Analytical Chemistry. 2008; 80: 7239–7247.

[71] Varshney M, Li Y. Interdigitated array microelectrode based impedance biosensor coupled with magnetic nanoparticle-antibody conjugates for detection of Escherichia coli O157:H7 in food samples. Biosensors and Bioelectronics. 2007; 22: 2408–2414.

[72] Varshney M, Li Y, Srinivasan B, Tung S. A label-free, microfluidics and interdigitated array microelectrode-based impedance biosensor in combination with nanoparticles immunoseparation for detection of Escherichia coli O157:H7 in food samples. Sensors and Actuators B: Chemical. 2007; 128: 99–107.

[73] Kanayeva DA, Wang R, Rhoads D, Erf GF, Slavik MF, Tung S, et al. Efficient separation and sensitive detection of *Listeria monocytogenes* using an impedance immunosensor based on magnetic nanoparticles, a microfluidic chip, and an interdigitated microelectrode. Journal of Food Protection. 2012; 75: 1951–1959.

[74] Brosel-Oliu S, Abramova N, Bratov A, Vigués N, Mas J, Muñoz F-X. Sensitivity and response time of polyethyleneimine modified impedimetric transducer for bacteria detection. Electroanalysis. 2015; 27: 656–662.

[75] Lvov Y, Ariga K, Ichinose I, Kunitake T. Molecular film assembly via layer-by-layer adsorption of oppositely charged macromolecules (linear polymer, protein and clay) and concanavalin A and glycogen. Thin Solid Films. 1996; 284–285: 797–801.

[76] Chambers JP, Arulanandam BP, Matta LL, Weis A, Valdes JJ. Biosensor recognition elements. Current Issues in Molecular Biology. 2008; 10: 1–12.

[77] Guan J-G, Miao Y-Q, Zhang Q-J. Impedimetric biosensors. Journal of Bioscience and Bioengineering. 2004; 97: 219–226.

[78] Sassolas A, Blum LJ, Leca-Bouvier BD. Immobilization strategies to develop enzymatic biosensors. Biotechnology Advances. 2012; 30: 489–511.

[79] Zeng X, Shen Z, Mernaugh R. Recombinant antibodies and their use in biosensors. Analytical and Bioanalytical Chemistry. 2012; 402: 3027–3038.

[80] Ruan CM, Yang LJ, Li YB. Immunobiosensor chips for detection of Escherichia coli O157:H7 using electrochemical impedance spectroscopy. Analytical Chemistry. 2002; 74: 4814–4820.

[81] Yang LJ, Li YB, Erf GF. Interdigitated array microelectrode-based electrochemical impedance immunosensor for detection of Escherichia coli O157:H7. Analytical Chemistry. 2004; 76: 1107–1113.

[82] Mantzila AG, Maipa V, Prodromidis MI. Development of a faradic impedimetric immunosensor for the detection of *Salmonella typhimurium* in milk. Analytical Chemistry. 2008; 80: 1169–1175.

[83] Blaser MJ, Newman LS. A review of human salmonellosis.1. Infective dose. Reviews of Infectious Diseases. 1982; 4: 1096–1106.

[84] Barreiros dos Santos M, Agusil JP, Prieto-Simón B, Sporer C, Teixeira V, Samitier J. Highly sensitive detection of pathogen Escherichia coli O157:H7 by electrochemical impedance spectroscopy. Biosensors and Bioelectronics. 2013; 45: 174–180.

[85] Radke SM, Alocilja EC. Design and fabrication of a microimpedance biosensor for bacterial detection. IEEE Sensors Journal. 2004; 4: 434–440.

[86] Tan F, Leung PHM, Liu Z-b, Zhang Y, Xiao L, Ye W, et al. A PDMS microfluidic impedance immunosensor for *E. coli* O157:H7 and *Staphylococcus aureus* detection via antibody-immobilized nanoporous membrane. Sensors and Actuators B: Chemical. 2011; 159: 328–335.

[87] Dweik M, Stringer RC, Dastider SG, Wu Y, Almasri M, Barizuddin S. Specific and targeted detection of viable Escherichia coli O157:H7 using a sensitive and reusable impedance biosensor with dose and time response studies. Talanta. 2012; 94: 84–89.

[88] Jia F, Duan N, Wu S, Ma X, Xia Y, Wang Z, et al. Impedimetric aptasensor for *Staphylococcus aureus* based on nanocomposite prepared from reduced graphene oxide and gold nanoparticles. Microchimica Acta. 2014; 181: 967–974.

[89] Chen F, Zhou J, Luo F, Mohammed A-B, Zhang X-L. Aptamer from whole-bacterium SELEX as new therapeutic reagent against virulent Mycobacterium tuberculosis. Biochemical and Biophysical Research Communications. 2007; 357: 743–748.

[90] Torres-Chavolla E, Alocilja EC. Aptasensors for detection of microbial and viral pathogens. Biosensors and Bioelectronics. 2009; 24: 3175–3182.

[91] Labib M, Zamay AS, Koloyskaya OS, Reshetneva IT, Zamay GS, Kibbee RJ, et al. Aptamer-based viability impedimetric sensor for bacteria. Analytical Chemistry. 2012; 84: 8966–8969.

[92] Labib M, Zamay AS, Kolovskaya OS, Reshetneva IT, Zamay GS, Kibbee RJ, et al. Aptamer-based impedimetric sensor for bacterial typing. analytical chemistry. 2012; 84: 8114–8117.

[93] Ma X, Jiang Y, Jia F, Yu Y, Chen J, Wang Z. An aptamer-based electrochemical biosensor for the detection of *Salmonella*. Journal of Microbiological Methods. 2014; 98: 94–98.

[94] Zasloff M. Antimicrobial peptides of multicellular organisms. Nature. 2002; 415: 389–395.

[95] Godoy-Gallardo M, Mas-Moruno C, Fernández-Calderón MC, Pérez-Giraldo C, Manero JM, Albericio F, et al. Covalent immobilization of hLf1-11 peptide on a titanium surface reduces bacterial adhesion and biofilm formation. Acta Biomaterialia. 2014; 10: 3522–3534.

[96] Kulagina NV, Lassman ME, Ligler FS, Taitt CR. Antimicrobial peptides for detection of bacteria in biosensor assays. Analytical Chemistry. 2005; 77: 6504–6508.

[97] Kulagina NV, Shaffer KM, Anderson GP, Ligler FS, Taitt CR. Antimicrobial peptide-based array for Escherichia coli and *Salmonella* screening. Analytica Chimica Acta. 2006; 575: 9–15.

[98] Mannoor MS, Zhang S, Link AJ, McAlpine MC. Electrical detection of pathogenic bacteria via immobilized antimicrobial peptides. Proceedings of the National Academy of Sciences of the United States of America. 2010; 107: 19207–19212.

[99] Mannoor MS, Tao H, Clayton JD, Sengupta A, Kaplan DL, Naik RR, et al. Graphene-based wireless bacteria detection on tooth enamel. Nature Communications. 2012; 3: 763.

[100] Lillehoj PB, Kaplan CW, He J, Shi W, Ho C-M. Rapid, Electrical impedance detection of bacterial pathogens using immobilized antimicrobial peptides. Journal of Laboratory Automation. 2014; 19: 42–49.

[101] Etayash H, Jiang K, Thundat T, Kaur K. Impedimetric detection of pathogenic gram-positive bacteria using an antimicrobial peptide from class IIa bacteriocins. Analytical Chemistry. 2014; 86: 1693–1700.

[102] Li Y, Afrasiabi R, Fathi F, Wang N, Xiang C, Love R, et al. Impedance based detection of pathogenic E-coli O157:H7 using a ferrocene-antimicrobial peptide modified biosensor. Biosensors and Bioelectronics. 2014; 58: 193–199.

[103] Ertl P, Mikkelsen SR. Electrochemical biosensor array for the identification of microorganisms based on lectin-lipopolysaccharide recognition. Analytical Chemistry. 2001; 73: 4241–4248.

[104] Zeng X, Andrade CS, Oliveira ML, Sun X-L. Carbohydrate–protein interactions and their biosensing applications. Analytical and Bioanalytical Chemistry. 2012; 402: 3161–3176.

[105] Gamella M, Campuzano S, Parrado C, Reviejo AJ, Pingarron JM. Microorganisms recognition and quantification by lectin adsorptive affinity impedance. Talanta. 2009; 78: 1303–1309.

[106] Wan Y, Zhang D, Hou B. Monitoring microbial populations of sulfate-reducing bacteria using an impedimetric immunosensor based on agglutination assay. Talanta. 2009; 80: 218–223.

[107] Gupta G, Surolia A, Sampathkumar S-G. Lectin microarrays for glycomic analysis. OMICS: A Journal of Integrative Biology. 2010; 14: 419–436.

[108] Shabani A, Zourob M, Allain B, Marquette CA, Lawrence MF, Mandeville R. Bacteriophage-modified microarrays for the direct impedimetric detection of bacteria. Analytical Chemistry. 2008; 80: 9475–9482.

[109] Brigati J, Petrenko V. Thermostability of landscape phage probes. Analytical and Bioanalytical Chemistry. 2005; 382: 1346–1350.

[110] Mejri MB, Baccar H, Baldrich E, Del Campo FJ, Helali S, Ktari T, et al. Impedance biosensing using phages for bacteria detection: generation of dual signals as the clue for in-chip assay confirmation. Biosensors and Bioelectronics. 2010; 26: 1261–1267.

[111] Tlili C, Sokullu E, Safavieh M, Tolba M, Ahmed MU, Zourob M. Bacteria screening, viability, and confirmation assays using bacteriophage-impedimetric/loop-mediated isothermal amplification dual-response biosensors. Analytical Chemistry. 2013; 85: 4893–4901.

[112] Singh A, Poshtiban S, Evoy S. Recent advances in bacteriophage based biosensors for food-borne pathogen detection. Sensors. 2013; 13: 1763–1786.

[113] Singh A, Arya SK, Glass N, Hanifi-Moghaddam P, Naidoo R, Szymanski CM, et al. Bacteriophage tailspike proteins as molecular probes for sensitive and selective bacterial detection. Biosensors and Bioelectronics. 2010; 26: 131–138.

Chapter 10

Bioimpedance Technique for Point-of-Care Devices Relying on Disposable Label-Free Sensors – An Anemia Detection Case

Jaime Punter-Villagrasa, Joan Cid,
Jordi Colomer-Farrarons,
Ivón Rodríguez-Villarreal and Pere Ll. Miribel-Català

Additional information is available at the end of the chapter

http://dx.doi.org/10.5772/60843

Abstract

In this chapter, the development of a point-of-care device for bio-medical applications has been discussed. Our main objective is to research new electronic solutions for the detection, quantification, and monitoring of important biological agents in medical environments. The proposed systems and technologies rely on label-free disposable sensors, with portable electronics for user-friendly, low-cost solutions for medical disease diagnosis, monitoring, and treatment. In this chapter, we will focus on a specific point-of-care device for cellular analysis, applied to the case of anemia detection and monitoring. The methodology used for anemia monitoring is based on hematocrit measurement directly from whole blood samples by means of impedance analysis. The designed device is based on straightforward electronic standards for low power consumption and low-cost disposable sensor for low volume samples, resulting in a robust and low power consumption device for portable monitoring purposes of anemia. The device has been validated through different whole blood samples to prove the response, effectiveness, and robustness to detect anemia.

Keywords: Point of care, electronics, impedance analysis, whole blood, hematocrit

1. Introduction

The integration of medical and electronic technologies allows the development of biomedical devices able to diagnose and/or treat pathologies by detecting, quantifying, and monitoring cellular species in different conductive media. In this chapter, the main attention is focused on the development of instrumentation electronics and sensing systems as well as design protocols for point-of-care (PoC) devices for cellular detection and monitoring, relying on straightforward standards for economic, low power consumption, versatile, safe, and reliable devices.

Several studies have reported that cellular detection, quantification, and monitoring can be accurately monitored by means of its biological electrical impedance (BioZ) [1-10] in different environments, in vivo or ex vivo experiences. One example of such work is the paper presented by Pop et al. (2013) [2], where blood hematocrit (HCT) was continuously in vivo monitored in the human right atrium by a dedicated central venous catheter equipped with an impedance measuring device. Pradhan et al. (2012) [5] studied ex vivo the electrical properties of blood and its constituents using impedance spectroscopy (IS) and three-electrode sensors. Ramaswamy et al. (2013) [6] performed a blood coagulation test based on a custom microfluidic device and the electrical impedance detection of whole blood samples. Li et al. (2012) [8] monitored Legionella serogroups in clinical and environmental samples by means of IS. Dweik et al. (2012) [9] demonstrated the detection of E. coli where bacterial presence was rapidly detected by measuring the antibody-antigen bonding by impedance analysis between 100 Hz and 10 MHz. Also, in the work developed by Grossi et al. (2010) [10], the quantity of bacteria during a culture process was detected by impedance measured at 200 Hz sinusoidal with a 100 mV peak-to-peak signal. Furthermore, other studies have reported the detection of cell-derived microparticles [11] and insulin in blood serum [12] based on an IS technique.

1.1. Biological electrical impedance

Dilute cellular sample suspensions can be described as a network of electrical passive components. The BioZ, identical to the definition from the Ohm's law [13], is defined as the response of applying an electrical stimulus to a biological material through a sensing system and measuring its electrical response. The electrical response is frequency dependant, and depending on the samples under examination and the sensing system, there are different frequency working ranges where an appropriate electric response can be found, so a frequency sweep is needed to evaluate the electrical response of the whole designed system (biological sample along with the sensing system). There are different sensing electrodes topologies to be considered depending on the number of electrodes used. Typical cellular electrical model for dilute cell suspensions can be described as network of electrical components [14] (Figure 1).

Injected current can flow through external media (R_E resistance) or through the cell across the membrane ($R_M \| C_M$) and the intracellular medium (R_I resistance). Considering that R_M resistance is nearly negligible, a simplified model is considered.

$$Z_{CELL} = \frac{R_E\left(1 + j\omega C_M R_1\right)}{1 + j\omega C_M\left(R_1 + R_E\right)}$$

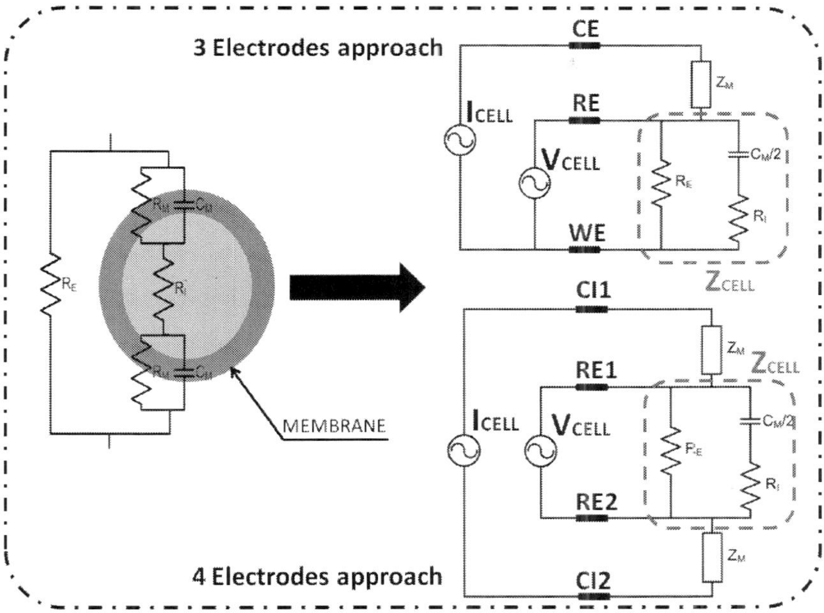

Figure 1. Cellular electrical model. Sensing system based on 3 electrodes and 4 electrodes approach.

A two-electrode topology is defined by the working electrode (WE), where the sample is placed and the electrical signal is applied, in addition to the auxiliary electrode (AE), which tracks the solution potential and supplies the current required for experience. This topology brings some kind of problematic behavior by the AE polarization effects causing a distortion of the applied electrical signal. The three-electrode configuration is defined as follows: the working electrode (WE), where the object is under investigation, the reference electrode (RE), which tracks the electric signal, and the counter or auxiliary electrode (CE), which supplies the required current. This topology avoids the distortion of the applied electrical signal.

Finally, the fourelectrode configuration avoids the measurement distortion due to the WE impedance polarization, as in the threeelectrode topology, the electric signal is directly applied where the single-ended voltage measurement signal is read. The fourelectrode topology is composed of two current-injection electrodes and two voltage-reading electrodes avoiding the electrode polarization distortion in impedance measurement due to a complete differential voltage measurement [15].

BioZ technique allows the use of different sensing topologies and systems, making possible the integration of such electrodes in a great variety of devices and environments, such as biosensors and microfluidic devices for increased functionality and performance. Moreover, the simplicity of the technique implementation makes it possible to accomplish the experience objectives with very simple sensing systems. Hernández et al. (2011) [4] obtained the electric impedance spectrum of human blood using reactive strips of the Bayer's portable glucometer.

The electronics involved in customized PoC devices are implemented as a front-end electronics depending on the sensing topology, and a back-end user interface for data processing and user interface. As a general standard, the whole PoC device can be described in a general outline as it is depicted in Figure 2.

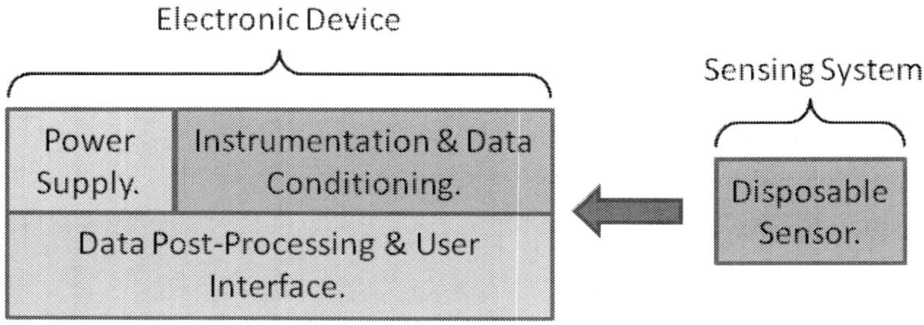

Figure 2. PoC device schematic development view.

2. Point-of-care device electronics design

2.1. Front-end electronics depending on the sensing topology

As it has been stated before, there are different electrodes topologies for the sensing system implementation, depending on the system and applications requirements. Different configurations must be designed for the front-end electronics architecture depending on the sensing topology.

In case of a three-electrode sensing topology, the potentiostat is the key electronic component for this sensing configuration, which is the interface between the biological elements and the instrumentation electronics. The potentiostat can be implemented in different ways, especially in terms of electrodes current, which can be designed as an instrumentation amplifier current readout stage or a transimpedance amplifier readout stage [16].

In figure 3A, a simple potentiostat with an instrumentation amplifier readout stage has been shown. It consists of an operational amplifier (OA), which tracks the input signal (V_{IN}) on to electrodes, and an instrumentation amplifier (IA) that senses the current through the electrodes.

$$I_{CELL} = \frac{V_{IN}}{Z_{CELL}}$$

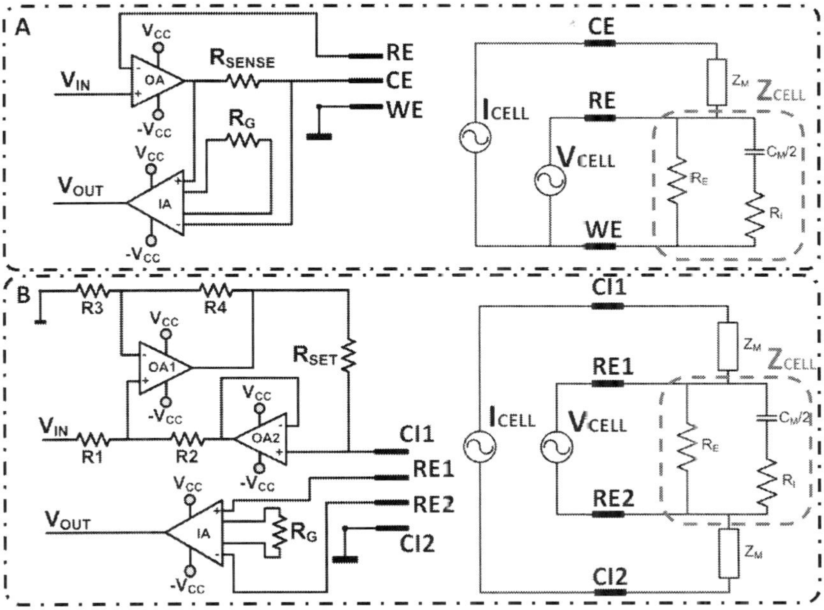

Figure 3. Cellular model, sensing system, and front-end electronics. A: 3-electrode cofiguration. B: 4-electrode configuration.

The current readout electronics consists of the direct conversion of the current through the biological species (I_{CELL}) into a voltage signal by means of a resistor (R_{SENSE}), where an instrumentation amplifier measures the voltage difference. The current through resistor (R_{SENSE}) is equal to the current through the biological species (I_{CELL}), and the voltage between RE and WE is steady due to the direct connection of WE to ground reference. Instrumentation amplifier's gain has been set to 1 to optimize the amplifier's total harmonic distortion.

$$Z_{CELL} = R_{SENSE}\left(\frac{V_{IN}}{V_{OUT}}\right)$$

In case of a four-electrode configuration, a voltage-to-current converter is implemented. In Figure 3B, the voltage-to-current converter is a modified Howland cell based on operational amplifiers (OA1 and OA2), which must guarantee a wide bandwidth and a high slew-rate while maintaining a low spectral noise and a low offset performance. The Howland cell uses R_{SET} and the input signal (V_{IN}) amplitude to define a stable current signal (I_{OUT}) at the output of the circuit regardless of the connected load related to the biological species.

$$I_{CELL} = \left(\frac{1}{R_{SET}}\right) V_{IN}$$

The differential voltage between ER1 and ER2 electrodes is acquired by means of the instrumentation amplifier (IA). The measured voltage (signal V_{IS}) is related to the differential voltage between the reading electrodes (ER1 and ER2), G being the instrumentation amplifier gain.

$$V_{OUT} = G\left(V_{ER1} - V_{ER2}\right) = G.\left(Z_{CELL}.I_{CELL}\right)$$
$$Z_{CELL} = R_{SET}\left(\frac{V_{OUT}}{G.V_{IN}}\right)$$

2.2. Back-end electronics for data post-processing and user interface

The sensing topology and front-end electronics chosen for the implementation of the PoC device outputs a voltage signal related to the biological impedance, which is of complex magnitude. The purpose of the back-end electronics, independent of the front-end electronics and sensing system, is to supply the proper voltage signal to bias the instrumentation electronics, depending on the biological sample, sensor, and experimental setup; process the output data; and present a proper user interface.

Taking into account these considerations, the first approach based on impedance spectroscopy (IS) is a versatile solution in different topics and environments. Two different approaches can be considered for the IS method; the Fast Fourier Transform (FFT) [17] method and the Frequency Response Analyzer (FRA) [18]. In the FFT method, a pulse is applied, ideally a Dirac delta function, to the sample and considering that it contains a wide frequency content, the response provides a full spectrum data of the analyzed sample impedance. The front-end electronics response is analyzed with an FFT algorithm in order to extract the frequency components of the impedance spectra [17]. This method is a simple and fast solution for the IS, but there are several drawbacks in the implementation. It is very difficult to generate a fast step function and a very fast electronic instrumentation capable of driving this step on the electrodes and extracting the resulting signal, producing a distortion in the measurement. Moreover, the important impedance information is contained in a short period of time after the step is applied; so, in addition to a very fast electronic instrumentation, a very fast analog-to-digital converter (ADC) with a high-precision bit resolution is also required, resulting in a high-speed hardware and heavy algorithm implementation device. Considering the PoC characteristics of the device, the FRA approach is a simpler and more efficient solution based on a lock-in amplifier (LIA). This method is much slower, as every frequency component is analyzed separately, to obtain the Bode plot of the measured impedance. The FRA solution is a good solution in terms of the trade-off between speed and complexity, particularly if not too low frequencies need to be measured, as it actually happens on biological cellular samples [3, 5, 19, 20]. Moreover, the implementation of a LIA is useful to reject undesirable harmonics and noise interferences [21-23], which are predominant on biological environments, such as bacteria culture, saline solution buffers, blood plasma, etc.

Complete back-end electronics have been designed to perform the FRA solution based on a real-time mathematical processing digital lock-in amplifier (DLIA) embedded on a real-time platform sbRIO 9632 (National Instruments, Austin, TX, USA), which has been used for fast software prototype development and versatility. To perform the FRA analysis based on a DLIA, the biasing signal of the front-end electronics (V_{IN}) is adopted as a reference signal to analyze the response of the sensing system (V_{OUT}). The sine and cosine signals are derived from V_{IN}, and by means of two multipliers and a filter stage, the real and imaginary components of V_{OUT} are obtained [16] (Figure 4). This measurement must be done for each frequency and front-end electronics output signal to obtain the Bode plot which consists of its magnitude $|V_{OUT}|$ and phase φ_{OUT}, where V_{REAL} and V_{IM} are the real part and imaginary part of the output signal, respectively.

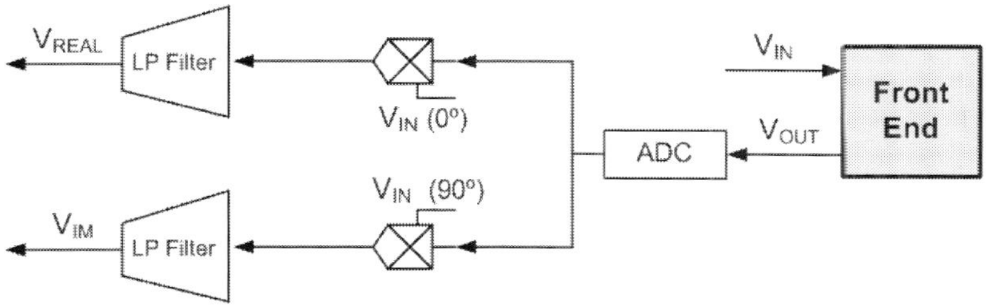

Figure 4. Lock-in software diagram. V_{OUT} is the input signal coming from the front-end electronics, V_{IN} is the reference signal and V_{REAL}, V_{IM}, respectively, are the real and the imaginary parts of the sensing system response.

$$V_{OUT} = \frac{4\sqrt{\left(V_{REAL}^2 + V_{IM}^2\right)}}{V_{IN}}$$

$$\varphi_{OUT} = \arctan\left(\frac{V_{IM}}{V_{REAL}}\right)$$

$$V_{REAL} = \frac{1}{2} V_{OUT}.V_{IN}.\cos\left(\varphi_{OUT}\right)$$

$$V_{IM} = \frac{1}{2} V_{OUT}.V_{IN}.\sin\left(\varphi_{OUT}\right)$$

Figure 5 depicts the back-end electronics architecture for the IA device. It is composed by the sbRIO 9632 real-time platform that allows us to develop different back-end functionalities, such as an oscillator (OSC) (Figure 5) that provides the desired biasing signal (V_{IN}) for the front-end electronics, and the signal conditioning for dual analog-to-digital conversion of both V_{IN} and V_{OUT} signals needed for the FRA approach. The oscillator is based on a signal generator AD9833 (Analog Devices, Norwood, MA, USA) that provides a stable voltage signal with a wide variable frequency range, 0 MHz to 12.5 MHz, which is controlled by a serial peripheral interface (SPI) communication protocol. The signal conditioning consists of a 12-bit dual, low power consuming ADC (Figure 5) known as ADC12D040 (Texas Instruments, Dallas, TX, USA), capable of converting both analog input signals at 40 MSPS simultaneously. The analog inputs are converted from single-ended input to differential output with a differential amplifier (Figure 5) AD8138 (Analog Devices, Norwood, MA, USA).

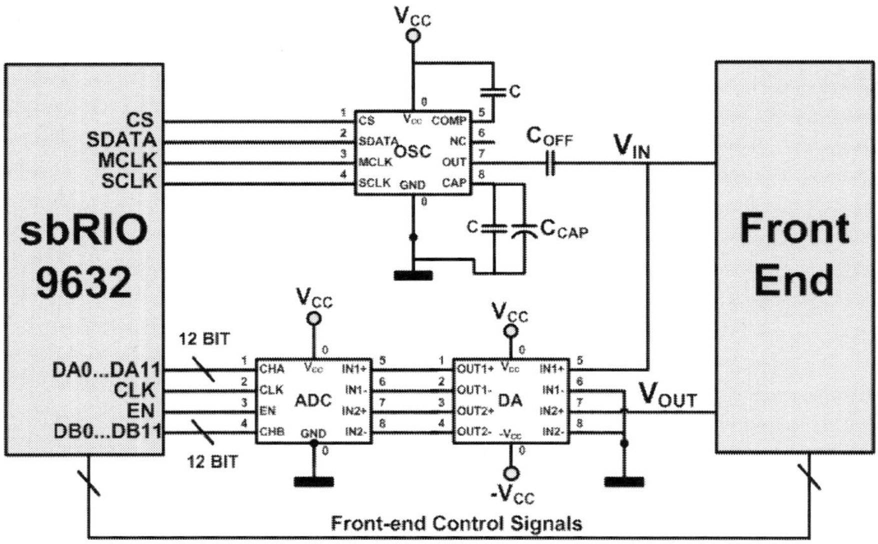

Figure 5. Back-end electronics based on a real-time platform sbRIO 9632.

Finally, a software for system control, data processing, and user interface is embedded on the real-time platform, which offers several functionalities.

First of all, it provides steady clock signals as needed that can be automatically real-time adjusted, allowing complete parallel signal acquisition for all the frequency ranges and more precise control of the oscillator, enabling the development of a signal generator automatic frequency sweep for an automated and complete FRA.

Furthermore, some functionalities on the front-end electronics, like R_{SENSE} or R_{SET} multiplexed auto-scale, are basic features for a precise FRA method and can be implemented by means of additional digital control as it is depicted in Figure 5 (front-end control signals). The real-time platform allows the system configuration and data display, with a user-friendly front-end user panel using Labview (National Instruments, Austin, TX, USA), by means of an external computer connected to the platform with a standard Ethernet connection.

2.3. Back-end electronics test

DLIA is the key component of the whole back-end electronics system, and must be capable to extract information from highly contaminated signals provided by the sensing system, usually electrical signals buried in noise. To estimate the signal-to-noise ratio (SNR) of the DLIA; the parameter that quantifies the ability of the DLIA to extract information of highly contaminated signals; a signal of 10 mV amplitude had been injected at the back-end electronics input V_{OUT}, as the information signal to be processed, for different frequencies ranging from 10 Hz to 100 kHz. Different noise signals representing different noise amplitude values had been added to the information signal for two different frequencies, 50 Hz and 200 kHz.

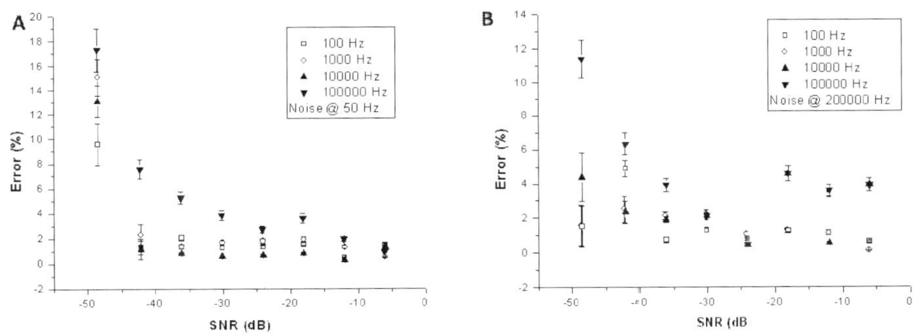

Figure 6. System response to a resistor impedance measurement. A: 50 Hz, low-frequency noise signal. B: 200 kHz, high-frequency noise signal.

Figure 6 shows the error and standard deviation when recovering the information from the different noise-contaminated 10 mV electrical signal. Measurement errors are below 10% for signal-to-noise ratio up to -45 dB, noise levels 200 times higher than the signal amplitude, which means a great environmental noise rejection for both lower (50 Hz) and higher frequencies (200 kHz).

2.4. Combined front-end and back-end electronics test using passive components

The complete PoC device combining both front-end instrumentation for sensing system driving and back-end electronics for data processing and device control, have been tested using passive components (Figure 7). The three-electrode front-end architecture, based on a potentiostat, has been selected for these studies, as three-electrode commercial sensors will be used later. Moreover, the passive components used in the study are in the ranges of typical blood impedance values [4, 5, 8, 12, 24], as the PoC device will be later applied to particular blood analysis. The operational amplifier to bias the sensor is the AD825 (Analog Devices, Norwood, MA, USA), a dual-supply high-speed Junction Gate Field Effect Transistor (JFET) amplifier with low leakage current and low distortion capable of high output driving. The instrumentation amplifier, AD8421 (Analog Devices, Norwood, MA, USA), is a dual-supply high-speed instrumentation amplifier with low noise and ultralow bias current. The amplifier's gain has been set to 1 in order to optimize the amplifier's total harmonic distortion. This stage has been designed with four different multiplexed sensing resistors (R_{SENSE}) taking into account the expected impedance values shown in the literature between 100 Ω and 100 kΩ [4, 5, 8, 12, 24], as was subsequently confirmed in the system validation experiments. These resistors are automatically multiplexed with an auto-scale function controlled by the embedded software.

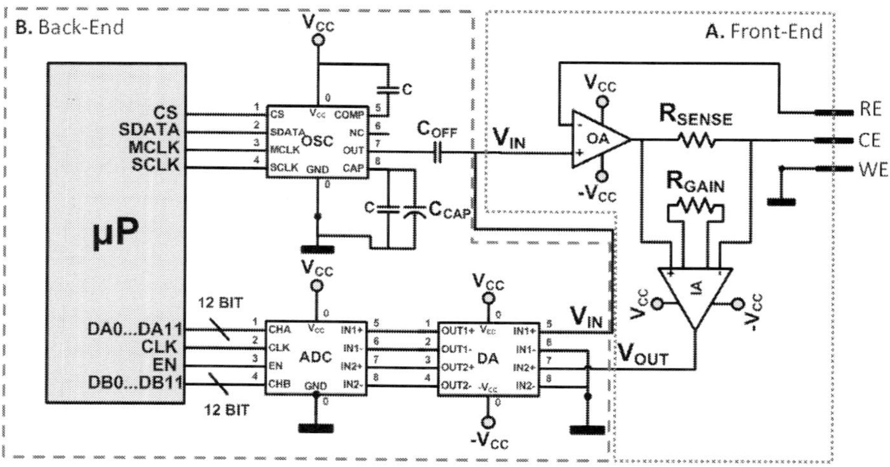

Figure 7. PoC impedance analysis device schematic diagram. A: Front-end electronics. B: Back-end electronics

First of all, we analyze a single resistor of different values as an electrode load in both impedance magnitude (Figure 8A) and phase (Figure 8B). Results demonstrate a great performance and reliability with an impedance magnitude standard deviation of 1%, maximum error of 12.3% in a 100 kHz bandwidth for loads less than 10 kΩ. For loads greater than 10 kΩ, the system performance declines with an impedance magnitude standard deviation of 3% and maximum error of 14% and bandwidth less than 10 kHz. In terms of impedance phase, Figure 8B, the system performance shows a standard deviation of 3.7 degrees in a 10 kHz bandwidth except for higher load values. The potentiostat topology causes a bandwidth limitation, as the instrumentation amplifier current readout system introduces an extra load (R_{SENSE}) on the main amplifier feedback loop, especially when the load increases.

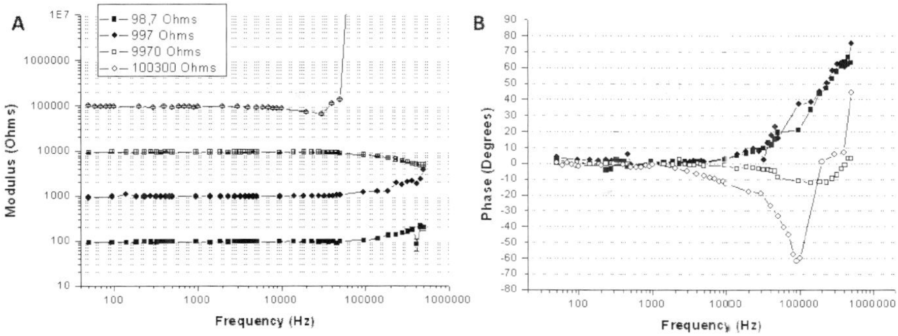

Figure 8. System response for a different resistor values. A: Impedance magnitude. B: Impedance phase.

A transimpedance amplifier topology [16] may solve this problem, as the current readout system and the sensing resistor (R_{SENSE}) are placed outside the feedback loop of the main amplifier. However, the generation of a voltage reference on the electrodes is based on a virtual ground provided by the transimpedance amplifier (Figure 9B), leading to possible errors on electrodes biasing [16].

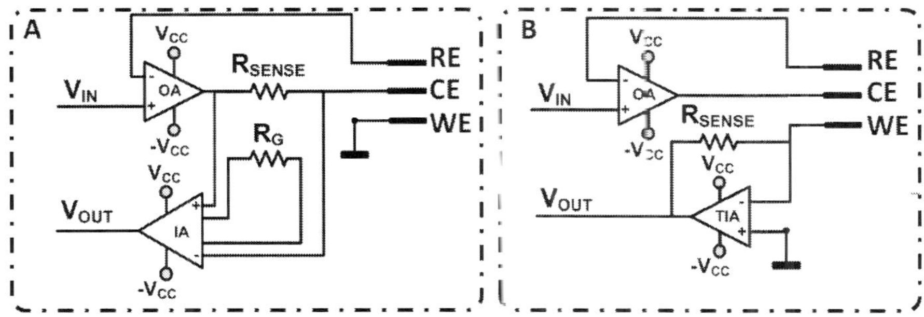

Figure 9. Potentiostat topologies. A: Instrumentation amplifier current readout topology. B: Transimpedance amplifier current readout topology.

Moreover, presence of stray capacitances creates a high-frequency measurement deviation called "Hook Effect." It is caused by leaking currents on the instrumentation [25], producing an impedance measurement error that can be observed above the 100 kHz. This error is frequency and load dependant, as leaking current paths through parasitic capacitances is more conductive at high frequencies and loads. In spite of these limitations, it is not a major drawback to the PoC device application, because the typical impedance and working frequency values for HCT analysis found in literature are below the described limitations [4, 5, 8, 12, 24].

Finally, the whole PoC device has been tested adopting a parallel resistor and capacitor as a load configuration (Table 1). Figure 10 shows the impedance magnitude and phase, in comparison with the theoretical load behavior. Three different resistor and capacitor values have been tested:

	Resistor Value.	Capacitor Value.
Test 1	597Ω	9.98nF
Test 2	994Ω	3.86nF
Test 3	9940Ω	9.98nF

Table 1. Load values.

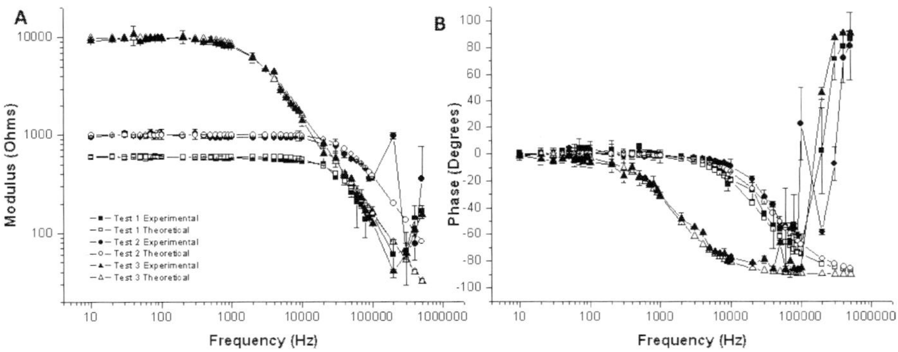

Figure 10. System response for a different parallel resistor and capacitor values. A: Impedance magnitude. B: Impedance phase.

Results in Figure 10 show a great performance with an impedance magnitude standard deviation of 1.5%, a mean error of 3.5% with a defined 100 kHz bandwidth.

2.5. Combined front-end and back-end electronics test using a ferrocyanide/ferricyanide solution

Finally, the whole PoC device has been validated for different sensors topology and compared the device measurements with a commercial equipment SP-150 (BioLogic Science Instruments, Grenoble, France) using a ferrocyanide/ferricyanide solution, a commonly used substance on sensor and equipment characterization [26-28].

Two different sensors have been used: a disposable commercial three screen-printed electrode C223AT (DropSens, Llaneras, Spain); and a standard three-electrode laboratory sensor composed by three different probes. Figures 11A and 11B show the impedance magnitude and phase comparison between the developed device and the commercial equipment SP-150, using the same disposable three screen-printed electrode (C223AT).

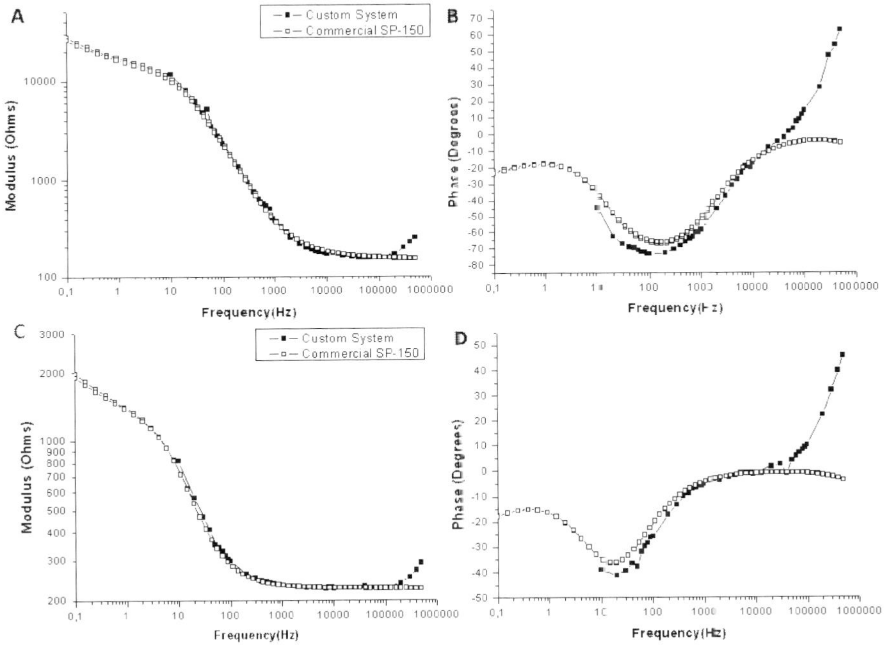

Figure 11. Comparative results with commercial equipment SP-150. A: Commercial sensor C223AT impedance magnitude. B: Commercial sensor C223AT impedance phase. C: Standard laboratory probe impedance magnitude. D: Standard laboratory probe impedance phase.

Figures 11C and 11D show the impedance magnitude and phase comparison between the developed device and the commercial equipment SP-150 using the same standard laboratory three probe electrodes sensor for all the experiments. The system has a proper impedance measurement response in the frequency operating range from 10 to 100 kHz, working within the ranges described earlier, with an impedance magnitude standard deviation of 1% and a maximum error of 1.5%. In terms of phase, maximum error is less than 12.1 degrees and the electrodes electrical pole positions represented are at accurate position A final system characteristics summary is reported in Table 2:

	Conditions (T = 25ºC)	Value
Power Consumption		24 VDC 0.35 A
Applied Voltage Signal Amplitude		10 Vrms
Frequency Working Range	Load < 10 kΩ	< 100 kHz
Frequency Working Range	Load > 10 kΩ	< 10 kHz
Magnitude Maximum Error	In Frequency Working Range	< 12.3%
Phase Maximum Error	In Frequency Working Range	< 12.1º
Lock-In Average SNR Rejected		-43 dB

Table 2. System characteristics summary.

3. Point-of-care device for anemia detection using commercial disposable sensors

3.1. Introduction to anemia

In this section, the previously reported PoC solution is studied for the detection of anemia through whole blood HCT monitoring, using a plug-and-play low-cost disposable commercial sensor.

Anemia is defined by the World Health Organization (WHO) as the stage at which the amount of hemoglobin in blood is below a certain threshold, being considered a world-wide problem [29]. HCT is the proportion of blood volume occupied by red blood cells (RBCs) and is determined by cell number and size, so HCT values below a certain reference range may also indicate anemia or abnormal cell development. 1.62 billion people (24.8% of the global population) are affected by anemia. The highest prevalence of anemia is in less developed countries, such as Africa and south-east Asia (~65%), and around 20% in developed countries, such as the Americas, Europe, and Western Pacific [30]. Severe anemia can lead to major health consequences such as pregnancy disorders, poor physical and cognitive development, and increased risk of morbidity, while less severe cases provoke weakness, fatigue, and dizziness [30, 31]. Nutritional deficiencies, such as severe malnutrition [32], colon cancer [33], or gastrointestinal lesions [34], are the most common cause of anemia, while other anemia originators are hematologic diseases such as sickle cell anemia [35] or thalassemia [36], cancer treatments (chemotherapy and radiation) [37], and indirect causes, such as lower erythropoietin production [38]. Also, frequent blood donations and huge blood draws from hospitalized patients may induce anemia [39]. Non-invasive methods are being studied and developed for anemia screening but have been demonstrated to have lower precision and sensitivity level [40, 41].

These different factors push toward the development of PoC anemia equipment providing an easy-to-use, reliable, and sensitive test with a short response time relying on 50 μL blood sample, which can be collected from capillary by standard medical procedures [42], providing reduced disposition decision time [43] and replacing current venipuncture-based laboratory test and improving patient satisfaction [44]. Moreover, low-volume blood samples avoid inducing anemia or making it worse, as phlebotomy is reported to induce anemia in hospitalized patients [45].

Currently, non-specific huge laboratory devices utilize IS-based applications, involving complex experimental setups and significant expenditure of time. Furthermore, whole blood PoC specific sensing systems also rely on complex microfluidic devices [46] entailing a low environmental integration level to develop autonomous PoC applications [47]. Moreover, IA measurements were performed at 10 mV_{RMS}, to prevent undesired effects like electroporation or irreversible electrical breakdown, which will damage RBCs membranes [48].

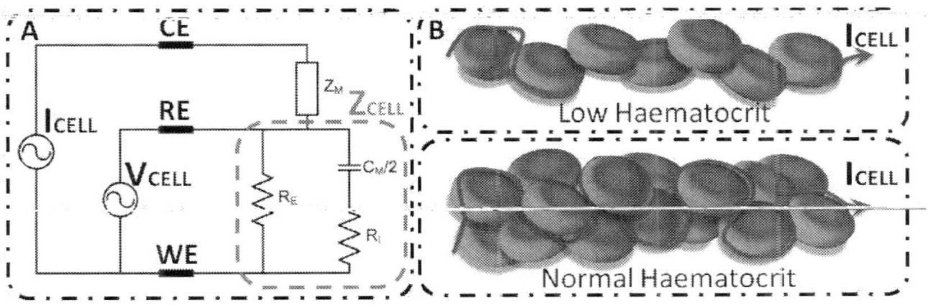

Figure 12. Red Blood Cell (RBC) electrical model. A: Three-electrode model for RBC sample. B: Current flow path through different blood samples with different hematocrit.

The electrical model for dilute cell suspensions in a three-electrode sensor (Figure 12) can be described as an electrical components network, where the close relation between HCT and impedance at low frequencies (up to 100 kHz) has been confirmed [49]. At this frequency range, the current I_{CELL} flows outside the RBCs, across RE impedance. HCT increment on the sample makes the current flow path larger between RE and WE electrodes, becoming an increment on Z_{CELL} impedance. According to literature, this phenomenon occurs at the 10 Hz to 100 kHz frequency range [2, 49].

3.2. Sensor

The sensing system is a low-cost disposable commercial sensor, easy to manipulate by clinical laboratory technicians using standard clinical laboratory tools. It works with 50 μL blood samples, the standard volume for a whole blood drop, easily collected from capillaries [42], and it is made of gold, an acknowledged bio-compatible material. Different commercial sensors have been evaluated, such as AC1 sensor from BVT Technologies (Brno, Czech Republic) or the G-AUG sensor series from Bio-Logic SAS (Claix, France), and the commercial sensor that best meets the defined specifications is the C223AT from Dropsens (Llaneras, Spain).

The device has a plug-and-play sensor system inside a Faraday cage and it is connected to the electronics with a custom-made three-wire coaxial insulated cable in order to provide an easy-to-use setup and reliable performance. Blood samples were put on top of the sensor with an automatic pipette (Labopette Manual 10–100 μL; Hirschmann Laborgeräte, Eberstadt, Germany).

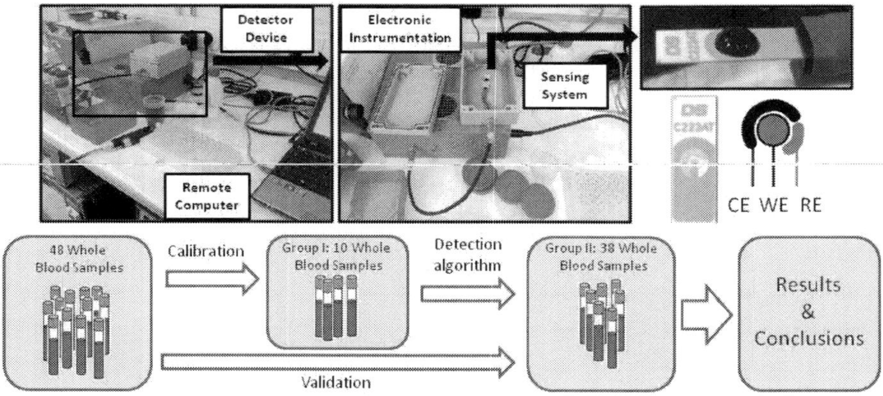

Figure 13. Device prototype electronics: Two custom printed circuit board and a sbRIO 9632 board (National Instruments, Austin, TX, USA) inside a faraday cage. Experimental setup: disposable sensor; electronic instrumentation; external computer with control and data displaying software. Whole blood samples groups and experimental procedure.

3.3. Blood samples

We used 48 whole blood samples obtained from hospitalized patients in Hospital Clínic. However, personal data of the patients was not available to the investigators and samples were randomly selected. Whole blood samples were obtained in 4 mL tubes containing ethylenediaminetetraacetic acid (EDTA 7.2 mg; BD Vacutainer®, Madrid, Spain).

We performed a complete blood count (CBC) of the blood samples with a hematology analyzer (Advia 2120, Siemens AG, Madrid, Spain), which reported the hemoglobin and HCT results as g/dL and percentage (%), respectively. With this methodology, the blood sample stream is divided into two parts, one portion is used for RBC counting and size and other portion is used for hemoglobinometry. RBC counting and size analysis is performed by singly passing the RBCs through a small direct current, where the temporary increase in impedance provides the information about number and volume. HCT is calculated from the measured RBC number and volume [50].

These 48 whole blood samples were distributed in two different groups (Figure 13). The first one (group I), consisted of 10 samples, was used to calibrate the system, whereas the second one (group II), composed by the other randomly collected 38 whole blood samples, was used to validate the whole system performance. The samples in group I have been selected to cover the entire possible HCT range present in human blood.

3.4. Software description

A user-friendly front-end panel on a remote computer has been designed for system configuration and data display. The real-time platform allows us to connect our device to a remote computer by means of a standard Ethernet connection. The user panel, depicted in Figure 14, has two different configurations that can be selected from a menu (A label). Each option allows the system to perform different experiments when the start button (D label) key is pressed. The first one is a complete impedance spectroscopy that provides the Bode diagrams for both impedance magnitude (F label) and phase (G label), where the user could choose the number of points per decade (C label). IA frequency ranges are fixed to the defined bandwidth on section 3.2 and R_{SENSE} values are auto-scaled. The second is an automatic HCT analyzer, where the blood sample HCT is displayed with four significant numbers (B label).

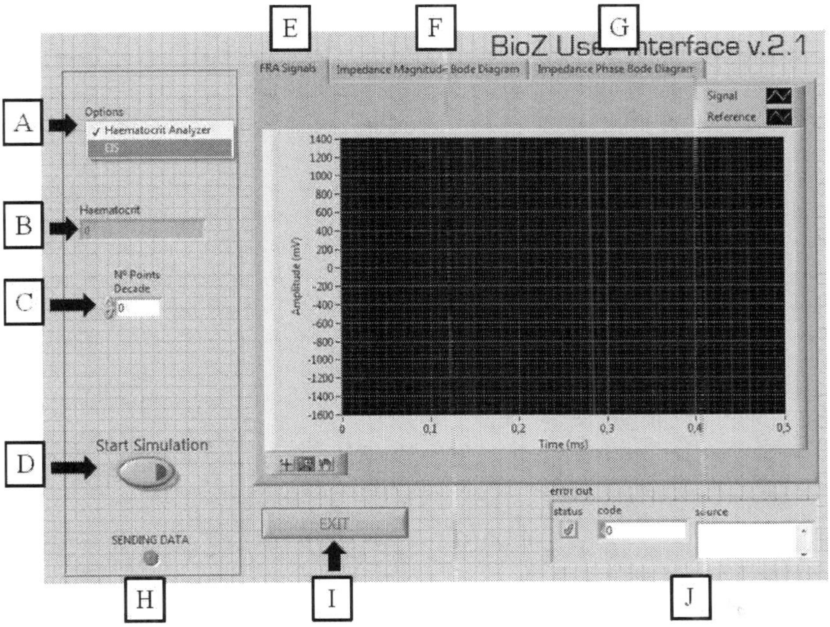

Figure 14. Software front panel for system control and data displaying.

Moreover, the front panel has several displays for user control and error monitoring. The two signals needed for the FRA analysis, reference signal (V_{IN} in Figure 5), and measurement signal (V_{OUT} in Figure 5), are real-time monitored and shown on a graph (E label). Errors in electronics connection with the remote computer are displayed (H label) and generic Labview errors are explained (J label).

3.5. Anemia detection device validation

3.5.1. System calibration: IA spectra analysis

The designed device was validated through IS experiences with the specified sensor described, for each whole blood sample collected, with all the measurements performed at clinical laboratory room temperature. First of all, we measured a complete impedance spectrum for the group I whole blood samples.

The obtained impedance magnitude (Figure 15A) and phase (Figure 15B) Bode diagrams for whole blood samples demonstrate that there exists a difference between measurements for different samples in specific frequency ranges. Furthermore, these differences must be related to whole blood HCT, as the response current I_{CELL} flows outside the RBCs, as discussed in previous section. When HCT is higher, the current flow path becomes larger between the RE and WE electrodes, which represents an increment on measured Z_{CELL} impedance. Therefore, measured impedance differences must be defined as impedance increments related to HCT increments. This phenomenon is present on specific frequency ranges depending on the correlation between impedance measurement and HCT. In terms of impedance magnitude (Figure 15C), the frequency working range is from 10 kHz to 100 kHz and for impedance phase (Figure 15D), it is in the range from 1 kHz to 5 kHz.

The capacity of determining impedance magnitude and phase differences for different HCT samples on a wide frequency working range gives flexibility and data redundancy to the system device, as long as it is not single frequency response dependant, which makes simple statistical data analysis techniques, such as linear regression, feasible, strengthening the robustness and reliability of the device.

Hence, once we have assumed that measured impedance increments, on the defined frequency working range, are related to blood samples HCT increment, we must determine the system's capacity for detection, resolution, and sensitivity. A control software, embedded on the microprocessor from the real-time platform sbRIO 9632, has been created to instantly measure impedance on the previously defined frequency working range, and we have studied the sensitivity, accuracy, and coefficient of variation of the device.

3.5.2. System calibration: Automatic hematocrit detection

To determine the system accuracy, sensitivity, and coefficient of variation, the first approach to the detection system evaluates only the impedance magnitude measurements. Although impedance phase measurements may be related to HCT, as it is shown in Figure 15D, these values, unlike the impedance magnitude, are strongly frequency dependant, so more complex data analysis is needed, resulting in a more complex and slower system. To calibrate the system, the impedance magnitude measurements of the whole blood samples from group I will be compared with the HCT values (%) (Figure 16.). The relation of impedance magnitude

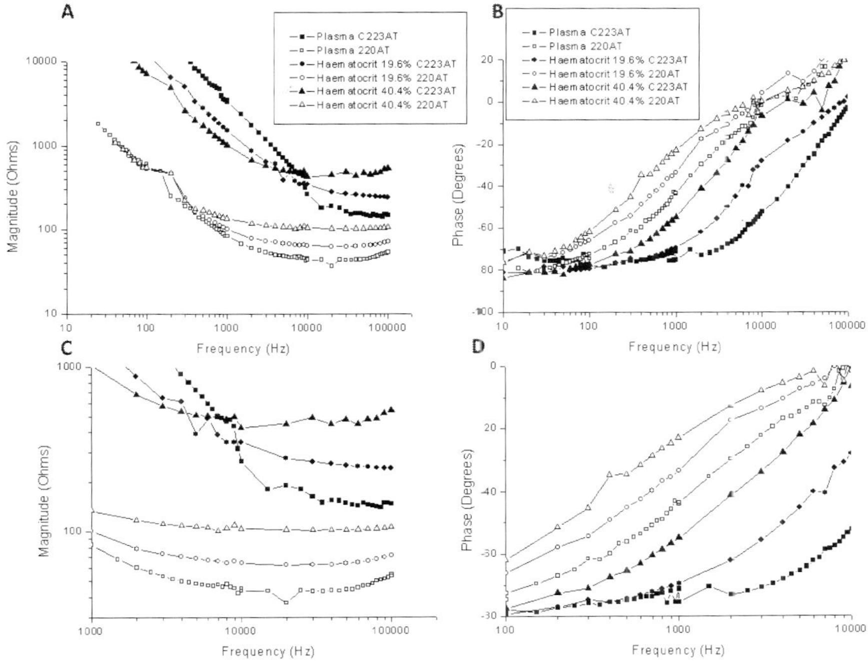

Figure 15. Whole blood impedance magnitude and phase measurement for group I whole blood samples. A: Impedance magnitude over full frequency spectra. B: Impedance phase over full frequency spectra. C: Impedance magnitude over frequency working range. D: Impedance phase over frequency working range.

(|Z|) mean value and whole blood samples HCT had been analyzed using linear regression approach, where the slope (β_Z) defines the sensitivity, in terms of ohms per HCT percentage (Ω/%).

Meanwhile the HCT detection accuracy (%) is the relation between the linear regression standard deviation and β_Z. Precision can be evaluated with the coefficient of variation, that is, the standard deviation divided by the mean value of the five repetitions measurements. In Figure 16, the impedance magnitude for the different HCT values is depicted. The HCT detection system has 10.46 Ω/% sensitivity and a 1.13% accuracy error with a linear correlation of 0.987. The coefficient of variation of the system is acceptable as it is normally below 5%.

3.5.3. Whole blood hematocrit detection validation

Once the whole system has been calibrated, and we have confirmed the HCT relation with impedance measurement, in both magnitude and phase, with the data from these previous studies, we have implemented an automatic real-time anemia detection device that provides instantaneous HCT measurement. An HCT (%) evaluation algorithm, based on impedance

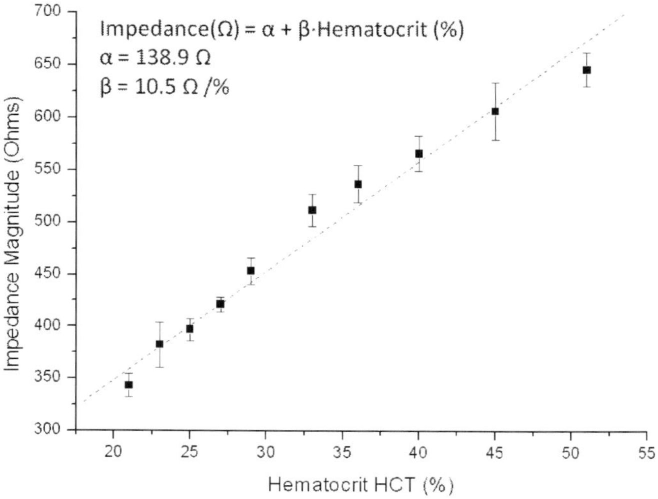

Figure 16. Measured impedance magnitude as a function of whole blood samples HCT value (%). Calibration curve.

magnitude measurement, has been embedded on the microprocessor from the real-time platform. The accuracy, precision, and repeatability of the detection device will be evaluated using 38 whole blood samples from the validation group, described in previous section (group II), which were randomly collected. Five repetitions have been done for each whole blood sample, using different sensors and sub-samples, to evaluate the precision of the device. The predicted HCT from the detection device is the mean value of the five measurements performed with each whole blood sample, and it was compared with the HCT measurement of the CBC performed with the hematology analyzer (Advia 2120, Siemens AG, Madrid, Spain).

In Figure 17, a comparison between detected HCT and HCT calculated with a clinical hematology analyzer is shown. The proposed device presented great accuracy in detecting HCT, with a linearity of 0.93 and an accuracy error of 1.75% with a correlation of 0.98. The coefficient of variation has a mean value of 3.27% for the whole samples, without any particular case with value more than 5%. In quality control procedures in clinical hematology measurements, coefficient of variation of value less than 5% for a test is considered acceptable [50]. As a first approach, which only contemplates impedance magnitude to a HCT analysis algorithm for anemia detection, as previously stated, it would be interesting to develop a more complex algorithm involving both impedance magnitude and phase on a wide frequency working range, for more precision and better performance on detection.

3.5.4. Conclusions

A first approach to a PoC device has been designed, fabricated, and tested for instantaneous anemia detection based on custom instrumentation electronics, electrical IA technique, and

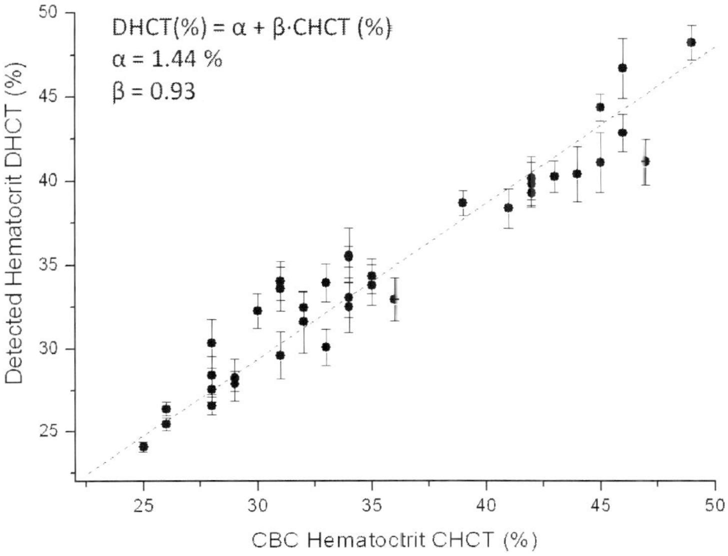

Figure 17. Whole blood HCT measurement from the detection device compared with HCT measurements obtained with a CBC performed by a hematology analyzer. Error bars represent the standard deviations across five repetitions.

disposable sensor. The system performs real-time instrumentation control, data acquisition, and results display by means of an external computer and user-friendly software. The device has been proved to exhibit reliable, robust, and effective results using label-free disposable commercial sensors using 50 μL whole blood samples. Furthermore, unlike actual clinical equipment for blood analysis, whole blood samples are not destroyed in the measurement process.

Considering this device as a first approach algorithm for anemia detection, the development of a more complex algorithm and a more accurate clinical assay with higher testing population will lead to more accurate results to assess the device performance. In addition, as the system is based on straightforward standards on instrumentation electronics and sensing, it represents an economic, portable, safe, and reliable system of anemia detection with a high degree of integration for the clinical environment.

However, further development must be considered for future stages: real-time platform sbRIO 9632 (National Instruments, Austin, TX, USA) has been used for fast prototyping purposes and versatility, but it has a major drawback in terms of power consumption, size, and price. It must be replaced by a low-cost solution for instrumentation control, data acquisition, and post-processing. The future system improvements must push toward the development of a truly autonomous, portable, and versatile device relying on the IA analysis technique.

4. Development of a truly autonomous low-cost point-of-care anemia detection device

4.1. System improvement

In this section, we will develop a truly autonomous PoC anemia detection device, based on the previous approach depicted in Section 3, as a more compact, economic, and portable PoC solution. Once the electronics and sensing system have been calibrated, the electronics involved in the data processing for a full spectrum analysis (DLIA based on FRA), such as a microprocessor, a serial peripheral interface–controlled oscillator, and a specific ADC, are no longer needed. Furthermore, a real-time platform was used which was increasing device power consumption, price and size, a major drawback when aiming for specific low-cost PoC device. The final system must be composed of an economic and reusable electronic device, with the previously reported plug-and-play disposable commercial sensor, based on three screen-printed electrodes for a 50 μL sample.

The main advantages of the proposed device are the facility of use, compactness and small size, and low-cost accessibility. These characteristics are valuable for both patients at risk of anemia and patients with chronic anemia.

Recently, other PoC anemia devices have been introduced, such as color-based diagnostic test for self-screening/self-monitoring of anemia presented by Tybursky et al. [51], which represents one of the more advanced PoC diagnostic test for self-screening/self-monitoring of anemia. This device measures hemoglobin levels, which are visually interpreted by the user using a color scale, and presents a very similar anemia detection performance compared with the device relying on IA measurements. However, there are several drawbacks to be considered; first of all, the readout stage, based on a color scale, relies on the visual interpretation of the user, which could introduce errors on the data interpretation, and reduces considerably the system resolution. Moreover, the principle of operation is based on biochemical reactions, where the blood comes into contact with a reagent solution initiating a redox reaction, which is a slow and destructive procedure.

4.2. System description

The principle of operation of the new device is the same biophysical principle of cellular IS described in Section 3. So, a full custom electronic circuit was specifically designed to carry out the impedance measurements, using a disposable three-electrode C223AT sensor (Dropsens, Llaneras, Spain), and considering the specifications of the previous study. The architecture of the device is divided in three parts: an oscillator that provides the AC voltage signal (V_{CELL}), a sensor driving instrumentation, and an RMS-to-DC converter (Figure 18).

The oscillator is based on a Wien bridge that provides a stable output amplitude signal with low distortion. The operational amplifier (OSC in Figure 18A) called AD8066 from Analog Devices (Norwood, MA, USA) is a low-cost, high-speed JFET amplifier, with low leakage current and distortion, in order to provide a stable AC voltage signal with low offset. The

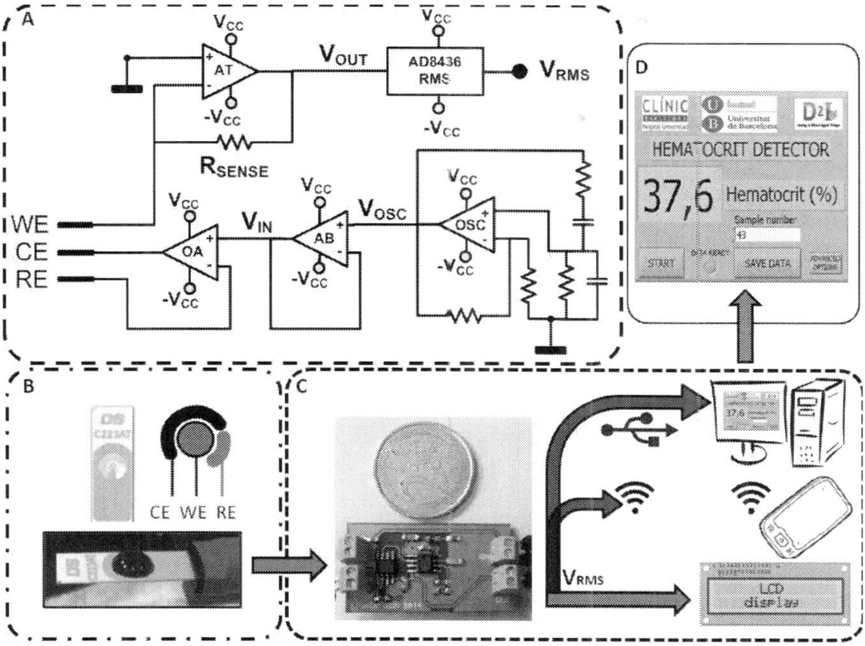

Figure 18. A: Custom electronic instrumentation. B: Commercial C223AT disposable sensor with a 50 μL whole blood drop. C: Device prototype electronics and different suitable and functional user readout interfaces. D: Actual user-friendly front-end user interface develop with Labview.

oscillator has been configured to provide a 33 kHz voltage signal, a well-suited frequency for HCT detection using the C223AT sensor. The AD8066 commercial integrated circuit provides two isolated amplifiers, so the second amplifier has been used as a follower (AB in Figure 18A) due to its high-speed and low-distortion specifications, as an isolator of the oscillator from the instrumentation.

The main electronics involved in the sensor driving is a potentiostat with a transimpedance amplifier current readout stage topology [16]. It is composed of an operational amplifier to bias the sensor and an operational amplifier in transimpedance configuration as a current readout. The operational amplifier in the potentiostat (OA, AT in Figure 18A) is the AD8066 from Analog Devices (Norwood, MA, USA), which is perfectly designed for singly driving the electrodes and track the voltage-biasing signal (V_{IN}) to the electrodes. The JFET high input impedance avoids RE electrode voltage distortion, and the high bandwidth and slew-rate provides stability to the system. The transimpedance amplifier converts the current through the electrodes into a voltage signal (V_{OUT}) by means of a sensing resistor (R_{SENSE} in Figure 18A). The AD8066 have a JFET input to attenuate the current losses through the input impedance, the main drawback of this configuration [16]:

$$Z_{CELL} = \left(\frac{V_{IN}}{I_{CELL}} \right)$$

$$V_{OUT} = -R_{SENSE} . I_{CELL} = -R_{SENSE} \left(\frac{V_{IN}}{Z_{CELL}} \right)$$

The RMS-to-DC converter is the AD8436 from Analog Devices (Norwood, MA, USA) that provides a DC equivalent (V_{RMS}) of the transimpedance amplifier AC signal (V_{OUT}). It is a low power consumption device with a ZFET input buffer for electronic isolation from the instrumentation stage. Considering the biasing voltage (V_{IN}) and the sensing resistor (R_{SENSE}) stable and well known, the variations of DC voltage V_{OUT} are only related to the variations of Z_{CELL}. The device output voltage (V_{RMS}) is inverse compared to the HCT values on the blood sample, so as the HCT increases, V_{RMS} decreases.

A software interface on an external computer, connected by means of a NI USB-6361 data acquisition (National Instruments, Dallas, TX, USA), controls an electric switch to enable the power supply providing the measurement, and presents the resultant data on a user-friendly user interface (Figure 18D).

However, one of the main advantages of the presented device is the readout stage versatility which is capable of addressing different applications (Figure 18C), such as an integrated LCD display for an untrained user self-screening, a remote computer connected by a standardized protocol (USB, ethernet, Bluetooth, etc.) for telemedicine applications. In addition, the presence of an electrical signal directly correlated to HCT, and the high level of integration, allows the device implementation as a controller of other clinical actuators in different environments and situations, increasing functionalities of other devices and applications.

The overall low-cost and low power consumption system composed of optimized straightforward standards for instrumentation electronics results in a reusable, robust, and low power consumption device (<300 mWh), making it completely mobile with a long battery life time.

4.3. Low-cost PoC anemia detection device validation

4.3.1. Blood samples

Twenty four whole blood samples were obtained in 4 mL tubes containing EDTA (7.2 mg; BD Vacutainer, Franklin Lakes, NJ, USA), from four random hospitalized patients in Hospital Clinic. The whole blood samples were centrifuged (Jouan CR412 from DJB Labcare, Newport Pagnell, UK) at 2200 rpm for 15 minutes to separate blood plasma from RBCs. To obtain 24 blood samples, RBCs were diluted in different volumes of blood plasma using an automatic pipette Labopette Manual 10–100 μL (Hirschmann Laborgeräte, Louisville, KY, USA). Blood plasma was used as a reference value. We performed a CBC of the 24 blood samples with an

ABX Micros 60 hematology analyzer (Horiba, Kyoto, Japan) which reported the hemoglobin and HCT results as g/dL and percentage (%), respectively.

4.3.2. PoC device validation

To validate the PoC device for instantaneous detection of anemia, we analyzed 24 consecutive blood samples from patients hospitalized at Hospital Clínic in Barcelona. We tested all samples with the prototype within 2 hours of blood collection and CBC. Every whole blood sample was tested five times consecutively using fresh sensors and fresh sub-samples. Figure 19 depicts the output DC voltage (V_{RMS}) of the device and compares it with the different whole blood samples HCT.

$$V_{RMS}\left(mV\right) = \alpha + \beta \text{HCT}$$

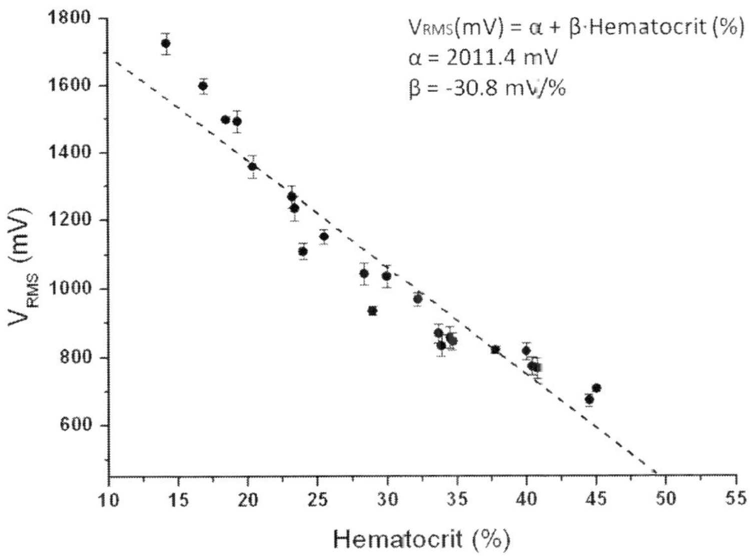

Figure 19. Measured output DC voltage (V_{RMS} (mV)) mean value (n = 5) as a function of blood samples HCT (HCT (%)).

The proposed anemia detector device presented a correlation coefficient of −0.96, an accuracy error of 2.83% HCT, and a coefficient of determination of 92.72%. The mean coefficient of variation is 2.57% without any particular case above 5%. This result represents a great performance and accuracy, according to the acceptable standards in quality control procedures in clinical hematology [50]. The device presents reliable, sensitive, and robust anemia detection compared with other commercial PoC devices, such as AnemiaCheck from Express Diagnostics (Blue Earth, MN, USA), STAT-Site from Stanbio Laboratory (Boerne, TX, USA), or HemoPoint H2 from Alere (Waltham, MA, USA).

5. Conclusions

The PoC device outputs instantaneous reliable results based on electric voltage data directly correlated to HCT. The system has further versatility in terms of applications, compared with other commercial devices, due to the stable DC voltage output. It is an interesting tool as a PoC HCT detector for non-skilled users, a monitoring device for telemedicine patient monitoring, and can be used as a controller of other clinical actuators on more complex lab-on-a-chip devices. Furthermore, one of the most important features of an impedance-based device is that, unlike actual clinical equipment for blood analysis, RBCs are not destroyed in the measurement process, mitigating the adverse effects for patients and samples. Actual commercial devices for PoC anemia detection, based on microfluidic manipulator devices, such as AnemiaCheck from Express Diagnostics (Blue Earth, MN, USA), STAT-Site from Stanbio Laboratory (Boerne, TX, USA), or HemoPoint H2 from Alere (Waltham, MA, USA), rely on slower hemoglobin measurement for a subsequent HCT indirect calculation, with results open to ambiguous readout stages optically interpreted, and decrease functionalities and versatility, with a similar detection performance compared to the proposed PoC device (Table 3).

In summary, this chapter describes the design, development, and test of a PoC anemia detection device with low-cost disposable sensors and specific electronics, highlighting the different advantages and drawbacks of PoC electronic devices for biomedical applications.

Device	Test Time (sec.)	Range (HCT (%) and Hb (g/dL))	Standard Deviation (%)	Coefficient Variation (%)
Presented device.	Instantaneous	HCT: 0%–100%	2.83%	2.57%
STAT-Site (Stanbio Laboratory, Boerne, Texas).	900	HCT: 12%–42%	0.74%	4.10%
HemoPoint H2 System (Alere, Waltham, Massachusetts).	120	Hb: 5.6 g/dL–20.6 g/dL	NA	4.20%
Anemia Check (Express Diagnostics, Blue Earth, Minnesota).	60	Hb: 0 g/dL–25.6 g/dL HCT: 36%–54%	NA	1.5%
Tybursky et al. [52]	60	Hb: < 9 g/dL–"/> 12 g/dL	NA	NA

Table 3. Comparison of Point-of-care devices for HCT and Hb detection.

Author details

Jaime Punter-Villagrasa[1*], Joan Cid[2], Jordi Colomer-Farrarons[1], Ivón Rodríguez-Villarreal[3] and Pere Ll. Miribel-Català[1]

*Address all correspondence to: jpunter@el.ub.es

1 Department of Electronics, University of Barcelona, Barcelona, Spain

2 Department of Hemotherapy and Hemostasis, CDB, IDIBAPS, Hospital Clínic, Villarroel, Barcelona, Spain

3 Centre de Recerca Matemàtica, Campus Bellaterra, UAB, Edifici C, Barcelona, Spain

References

[1] N. Ramirez, A. Regueiro, O. Arias and R. Contreras, "Electrochemical impedance spectroscopy: An effective tool for a fast microbiological diagnosis," Biotecnologia Aplicada, vol. 26, pp. 72–78, 2008.

[2] G.A. Pop, L.L. Bisschops, B. Iliev, et al., "On-line blood viscosity monitoring in vivo with a central venous catheter, using electrical impedance technique," Biosens Bioelectron, vol. 41, pp. 595–601, March 2013.

[3] G.A. Pop, W.J. Hop, M. van der Jagt, et al., "Blood electrical impedance closely matches whole blood viscosity as parameter of hemorheology and inflammation," Appl Rheol, vol. 13 (6), pp. 305–312, 2003.

[4] F. Hernández, C. Guerrero and J. Bernal, "Determinación de las propiedades eléctricas en tejido sangíneo," Ciencia UANL, 8 (4), April 2011.

[5] R. Pradhan, A. Mitra and S. Das, "Impedimetric characterization of human blood using three-electrode based ECIS devices," J Electr Bioimpedance, vol. 3, pp. 12–19, January 2012.

[6] B. Ramaswamy, T.Y. Yin-Ting and Z. Si-Yang, "Microfluidic device and system for point-of-care blood coagulation measurement based on electrical impedance sensing," Sensors Actuators B: Chem, vol. 180, pp. 21–27, April 2013.

[7] N. Ramirez, A. Regueiro, O. Arias and R. Contreras, "Electrochemical impedance spectroscopy: An effective tool for a fast microbiological diagnosis" Biotecnologia Aplicada, vol. 26, pp. 72–78, 2009.

[8] N. Li, A. Brahmendra, A.J. Veloso, et al., "Disposable immunochips for the detection of Legionella pneumophila using electrochemical impedance spectroscopy," Anal Chem, vol. 84, pp. 3485–3488, March 2012.

[9] M. Dweik, R.C. Stringer, S.G. Dastider, et al., "Specific and targeted detection of viable Escherichia coli O157:H7 using a sensitive and reusable impedance biosensor with dose and time response studies,"Talanta, vol. 94, pp. 84–89, 2012.

[10] M. Grossi, M. Lanzoni, A. Pompei, et al., "An embedded portable biosensor system for bacterial concentration detection," Biosens Bioelectron, vol. 26, pp. 983–990, 2010.

[11] V. Lvovich, S. Srikanthan and R.L. Silverstein, "A novel broadband impedance method for detection of cell-derived microparticles," Biosens Bioelectron, vol. 26 (2), pp. 444–451, October 2010.

[12] M. Xu, X. Luo and J.J. Davis, "The label free picomolar detection of insulin in blood serum," Biosens Bioelectron, vol. 39, pp. 21–25, January 2013.

[13] R. Patterson, "Bioelectric Impedance Measurements: The Biomedical Engineering Handbook," J.D. Bronzino (ed.), 2nd edition. Boca Raton: CRC Press, pp. 734–73, 2000.

[14] C. Grosse and H.P. Schwan, "Cellular membrane potentials induced by alternating fields," Biophys J, vol. 63, pp. 1632–1642, 1992.

[15] B. Lenaerts and R. Puers, "Omnidirectional Inductive Powering for Biomedical Implants," Netherlands: Springer, 2009.

[16] J. Punter-Villagrasa, J. Colomer-Farrarons and P. Ll. Miribel, "Bioelectronics for Amperometric Biosensors, State of the Art in Biosensors: General Aspects." Dr. Toonika Rinken (ed.), 1st edition. Rijeka, Croatia: Intech, 2013.

[17] S. Park, J. Yoo, B. Chang and E. Ahn, "Novel instrumentation in electrochemical impedance spectroscopy and a full description of an electrochemical system," J Pure App Chem, vol. 5, pp. 1069–1080, August 2005.

[18] G. Li, M. Zhou, F. He and L. Lin, "A novel algorithm combining oversampling and digital lock-in amplifier of high speed and precision," Rev Sci Instrum, vol. 82, 095106, 2011.

[19] M.S. Cheng, J.S. Ho, S.H. Lau, V.T.K. Chow, and C.S. Toh, "Impedimetric microbial sensor for real-time monitoring of phage infection of Escherichia coli," Biosens Bioelectron, vol. 47, pp. 340–344, 2013.

[20] L. Yang, "Electrical impedance spectroscopy for detection of bacterial cells in suspensions using interdigitated microelectrodes,"Talanta, vol. 74, pp. 1621–1629, 2008.

[21] M. Min, O. Märtens and T. Parve, "Lock-in measurement of bio-impedance variations," J Int Measuret Confeder, vol. 27 (1), pp. 21–28, 2000.

[22] M. Gabal, N. Medrano, B. Calvo, et al., "A single supply analog phase sensitive detection amplifier for embedded applications," Proceedings of 25th Conference on Design of Circuits and Integrated Systems, DCIS2010, 17-19 November 2010, Lanzarote, Spain.

[23] C. Azzolini, A. Magnanini, M. Tonelli, G. Chiorboli and C. Morandi "A CMOS vector lock-in amplifier for sensor applications," Microelectron J, vol. 41 (8), pp. 449–457, 2010.

[24] C. Ribaut, K. Reybier, O. Reynes, et al., "Electrochemical impedance spectroscopy to study physiological changes affecting the red blood cell after invasion by malaria parasites," Biosens Bioelectron, vol. 24 (8), pp. 2721–2725, April 2009.

[25] R. Buendia, F. Seoane and R. Gil-Pita, "A Novel Approach for Removing the Hook Effect Artefact from Electrical Bioimpedance Spectroscopy Measurements," Journal of Physics, Conference on Electrical Bioimpedance, ICEBI2010, 4-8 April 2010, Gainesville, Florida.

[26] V.R. Taliene, T. Ruzgas, V. Razumas and J. Kulys, "Chronoamperometric and cyclic voltammetric study of carbon paste electrodes using ferricyanide and ferrocenemonocarboxylic acid," J Electroanal Chem, vol. 372, pp. 85–89, 1994.

[27] C. Beriet and D. Pletcher, "A microelectrode study of the mechanism and kinetics of the ferro/ferricyanide couple in aqueous media: The influence of the electrolyte and its concentration," J Electroanal Chem, vol. 361, pp. 93–101, 1993.

[28] S. Petrovic, "Cyclic voltammetry of hexachloroiridate(IV): An alternative to the electrochemical study of the ferricyanide ion," J Chem Educ, vol. 5 (5), pp. 231–235, 2000.

[29] C.A. Northrop-Clewes and D.I. Thurnham, "Biomarkers for the differentiation of anemia and their clinical usefulness," J Blood Med, vol. 4, pp. 11–22. March 2013.

[30] B. de Benoist, E. McLean, I. Egli and M.E. Cogswell, "Worldwide Prevalence of Anaemia 1993–2005: WHO Global Database on Anaemia" Geneva, Switzerland, 2008.

[31] World Health Organization, "Iron Deficiency Anaemia: Assessment, Prevention and Control: A Guide for Programme Managers," Geneva, Switzerland, 2001.

[32] N. Thakur, J. Chandra, H. Pemde and V. Singh, "Anemia in severe acute malnutrition," Nutrition. vol. 30, pp. 440-442, 2014.

[33] M. Rottoli, T. Sabharwal, A.M. Schizas and M.L. George, "Bleeding pseudoaneurysm of the internal iliac artery after extended resection for advanced rectal cancer: Report of two cases," Int J Colorectal Dis, 2014. DOI: 10.1007/s00384014195З4.

[34] B. Ong and D.C. Rockey, "The syndrome of a large drop in hematocrit in hospitalized patients: Clinical features and gastrointestinal bleeding outcomes," J Investig Med, 2014. DOI: 10.1097/JIM.0000000000000109.

[35] National Heart, Lung, and Blood Institute. "What Is Sickle Cell Anemia?" Available at http://www.nhlbi.nih.gov/health/health-topics/topics/sca/ (accessed November 10, 2014).

[36] Medline Plus, "Thalassemia," Available at http://www.nlm.nih.gov/medlineplus/ency/article/000587.htm (accessed November 10, 2014).

[37] American Cancer Society, "Anemia in People With Cancer," Available at http://www.cancer.org/treatment/treatmentsandsideeffects/physicalsideeffects/anemia/anemia-in-people-with-cancer (accessed November 10, 2014).

[38] J. Portolés, J.L. Gorriz, E. Rubio, et al., "The development of anemia is associated to poor prognosis in NKF/KDOQI stage 3 chronic kidney disease," BMC Nephrol, 2013, 14, doi:10.1186/1471-2369-14-2.

[39] M. Pinto, M.L. Barjas-Castro, S. Nascimento, et al., "The new noninvasive occlusion spectroscopy hemoglobin measurement method: A reliable and easy anemia screening test for blood donors," Transfusion, vol. 53, pp. 766–769, April 2013.

[40] A. Belardinelli, M. Benni, P.L. Tazzari and P. Pagliari, "Noninvasive methods for haemoglobin screening in prospective blood donrs," Vox Sang, vol.105, pp. 116–120, April 2013.

[41] R.B. Thompson, "A Short Textbook of Haematology," fourth edition. Kent, England: Pitman Medical Publishing Co, 1975.

[42] D.J. Ernst, L.O. Ballance, R.R. Calam, et al., "Procedures and Devices for the Collection of Diagnostic Capillary Blood Specimens: Approved Standard," sixth edition. Wayne, Pennsylvania: CLSI, 2008.

[43] S.D. Asha, A.C.F. Chan, E. Walter, et al., "Impact from point-of-care devices on emergency department patient processing times compared with central laboratory testing of blood samples: a randomised controlled trial and cost-effectiveness analysis," Emerg Med J, vol. 31,pp. 714–719, September 2014.

[44] C.D.H. Jones, J. Howick, N.W. Roberts, et al., "Primary care clinicians' attitudes towards point-of-care blood testing: A systematic review of qualitative studies," BMC Fam Pract, vol. 14, pp. 117, August 2013.

[45] P. Thavendiranathan, A. Bagai, A. Ebidia, A.S. Detsky and N.K. Choudhry, "Do blood tests cause anemia in hospitalized patients? The effect of diagnostic phlebotomy on hemoglobin and hematocrit levels," J Gen Intern Med, vol. 20, pp. 520–524, June 2005.

[46] Y. Zheng, E. Shojaei-Baghini, A. Azad, C. Wang and Y. Sun, "High-throughput biophysical measurement of human red blood cells," Lab Chip, vol. 12, pp. 2560–2567, July 2012.

[47] J.P. Esquivel, J. Colomer-Farrarons, M. Castellarnau, at al., "Fuel cell-powered microfluidic platform for lab-on-a-chip applications: Integration into an autonomous amperometric sensing device," Lab Chip, vol. 12, pp. 4232–4235, November 2012.

[48] R. Patterson, "Bioelectric Impedance Measurements" in The Biomedical Engineering Handbook, J.D. Bronzino (ed.), 2nd edition. Boca Raton: CRC Press, 2000, pp. 734–773.

[49] D.W. Hill and F.D. Thompson, "The effect of haematocrit on the resistivity of human blood at 37ºC and 100 kHz," J Med Bioll Eng, vol. 13, pp. 182–186, March 1975.

[50] L. Corash, "Laboratory Hematology: Methods for the Analysis of Blood," in Blood, Principles and Practice of Hematology, R.I. Handin, S.E. Lux and T.P. Stossel, (eds.), 1st edition. Philadelphia: JB Lippincott & Co, 1995, chap. 2, pp. 23–61.

[51] E.A. Tyburski, S.E. Gillespie, W.A. Stoy, et al., "Disposable platform provides visual and color-based point-of-care anemia self-testing," J Clin Invest, vol. 124, pp. 4387–4394, 2014.

Chapter 11

Electrochemical and Optical Biosensors in Medical Applications

Jadwiga Sołoducho and Joanna Cabaj

Additional information is available at the end of the chapter

http://dx.doi.org/10.5772/60967

Abstract

Analysis of many biochemical processes is of great significance for clinical, biological, food, environmental as well as bioterror applications. But, exchanging of the biochemical information to kind of electronic signal is a defiance due to connecting an electronic tool directly to a biological surrounding. Electrochemical detection instrument due to its advantageous to analyze the subject of a biological sample has a great potential in conversion of a biochemical occurrence to an electronic signal.

In this chapter we presented the advancements made in the field of electrochemical biosensors targeted at medical field as well as optical elements used in modern biosensors. This material encompasses the technology, performance and commercialization of this brood of sensors. We have focused on the future development perspectives of this class of sensors, based on the experience gathered by different groups researching this field.

Keywords: biosensors, electrochemistry, optical biosensors, medical applications, nucleic acids, lactate, optical fiber

1. Introduction

Electrochemical detection instrument due to its advantageous to analyze the subject of a biological sample has a great potential in conversion of a biochemical occurrence to an electronic signal [1]. Due to its advances reported earlier [2] they are great alternative for other analytical detection methods used elsewhere, Fig. 1. The connection among quick, sensitive, selective, precise, miniaturizable and inexpensive electrochemistry-based biosensing [3–5] and areas as genetics, proteomics, biochemistry, and molecular biology, conduct to the development of electrochemical sensing tools of unique selectivity and sensitivity for a mass of the samples being tested on this truck.

As was reported earlier by Almeida et al. [6], since Clark glucose sensor [7], this system was improved highly [6-9], and novel sensing systems for detection i.e. urea [10], creatinine [11], cholesterol [12], insulin [13], cancer cells [14] were developed. Electrochemical biosensors were also found for determination of microorganisms as *Escherichia coli* [15] or *Salmonella typhimurium* [16].

Furthermore, the optical fiber as biosensor part is suitable for rough and unsafe surroundings, because of their notably potent, flexible and persistent arrangements. This is non-electrical; consequently, it may be utilized in varied disrepair electric current adoptions. Optical fibers are usually applied since upstanding quality and the base price for sensing appliances. Especially, the major impressive features of optical fibers may allow transmission of multiple signals synchronously. Due to the fact, it may receive variety capacities for detecting of analyte [17].

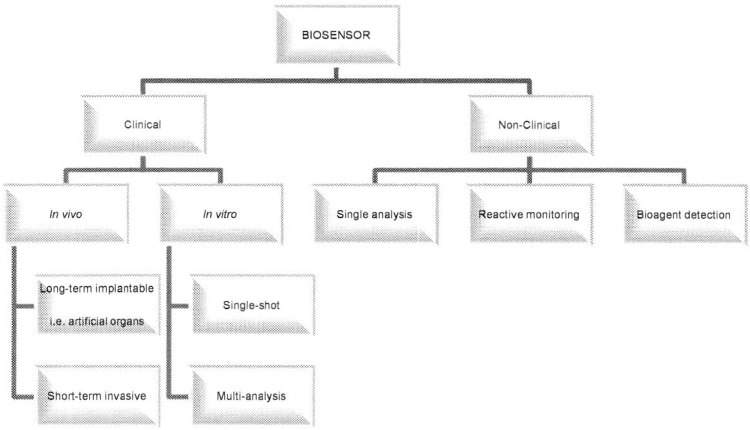

Figure 1. Various applications of biosensors

2. Electrochemical detection in clinical analysis-overview

Electrochemical parameters do not present direct correlations with biological data. However, there are several factors that influence the action mechanisms of drugs, such as stereochemistry, diffusion, solubility, membrane permeability, bioavailability and others, and these parameters can be related to electrochemical reactions. Electron transfer plays an important role in describing the action mechanisms of many drugs, such as alkylating agents. Other metabolic pathways that involve electron transfer and redox processes include the generation of reactive oxygen species and free radicals and the metabolism of xenobiotics [6,18]. After a brief search of the literature, several biological studies that include electrochemical mechanisms were found. Moreover, it is important to consider the type of the appropriate electrochemical technique that has to be developed to achieve the measurements and the associated parameters.

Among the four common electrochemical techniques, amperometry, pulsed chronoamperometry, differential normal pulse amperometry and differential normal pulse voltamperometry, only the sampled and/or differential techniques permit to determine the analytical signals independent of background and capacitive components [19]. For example, the differential normal pulse amperometry guarantees a stability to the sensor system since this further is polarised at the measurement potential of NO.

The determination of nitric oxide is valid for direct examination of its regulatory roles in biochemical systems. NO assists in the control of crucial cell activities as proliferation or apoptosis. Its unsuitable metabolism can lead to the varied pathologies, as cardiovascular dysfunctions.

One of the NO detection systems was developed by Brunet et al. [19]. The system concerns an appropriate analysis of interfering compounds (hydrogen peroxide, $ONOO^-$, ascorbate) that should be done to evaluate the specificity of the electrochemical sensing of NO in solution. Measuring setup is designed of a nickel porphyrin and Nafion©-coated carbon microfiber. Ames and Kovacic [20] suggested that anti-ulcer agent, as omeprazole could inhibit the H^+/K^+ ATPase from an electron transfer process at the binding site. The electrochemical techniques used in the experiments were cyclic voltammetry and normal pulse polarography.

Liang et al. proposed an electrochemical sensor fabricated by imprinting D- and L-tyrosine on polypyrrole at the surface of Ni electrode [21]. The enantioselectivity of the designed system was determined by coulometry applying positive potential to prompt adsorption of the objective analyte. The L-tyrosine imprinted polypyrrole layer rendered good selectivity for its own template and presented only a minimal relation for the tyrosine derivatives, norepinephrine, epinephrine and dopamine.

Winkler et al. designed catechol-modified chitosan redox cycling system to amplify and detect the electrochemical clozapine signal in undiluted human serum [22]. Clozapine is the antipsychotic ratified by the FDA for treatment-resistant schizophrenia [23]. The authors exemplify the critical function of the reducing mediator in the redox cycling approach, where accurate attention has to be paid to the choice of concentration, contributing up to 1.75-fold to signal reinforcement. The study of electrical potential adoption displays electrophoretic transfer of clozapine playing a similarly significant role in signal amplification, especially in diffusion-limited surroundings, accounting for a factor of up to 2.47 [22]. Recently, Keeley et al. has utilized cyclic voltammetry as well as differential normal pulse voltamperometry in electrochemical biosensing tool for determination of paracetamol in human serum [24]. The electrochemical active layer was designed as pyrolytic carbon films created using non-catalytic chemical vapour depositionon Si/SiO_2 wafers. The proposed sensor was selective, stable and reproducible, with a low detection limit (μM), high sensitivity and broad linear range. Its accuracy has been verified using human serum and commercially available pharmaceutical products. The electroanalytical properties of pyrolytic carbon were found to be competitive with the best available carbon electrodes [24].

Electrochemistry in biosensing/clinical context, as was mentioned earlier [25], has many profits, which make it an appealing choice for clinical analysis. Most electrochemical methods have excellent limits of detection and a wide dynamic range. Electroanalytical techniques require only very small sample volumes, often in the microliter range, coupled with the low detection limits allowing analysis on even subpicogram amounts of analyte [25]. Moreover, the selectivity of electrochemical detection in complex samples is excellent because fewer electroactive interferents are often encountered than spectroscopic interferents.

2.1. Nucleic acid electrochemical biosensors

The rapid detection of nucleic acid sequences (DNA o rRNA) represents a challenging research of bioanalytical chemistry applied to biosensors. Over the last few years, many works have been focused on this aim, with applications ranging from environmental monitoring to food and clinical analysis, developing platforms based on different transduction principles, i.e. piezoelectric, optical and electrochemical [26].

Present clinical diagnostics and monitoring demands rapid and accurate analyses. Vital factors complicating these procedures are often cost of analytical proceedings. Moreover, if it is considered that a qualified personnel is needed for clinical analyses, the natural necessity for alternative analytical technologies is present. For this purpose, biosensors can serve as an option for solving problems mentioned before, or become a helpful in a personalized medicine [27].

Electrochemical biosensors are used to the moment for detection several pathogens (Table 1). These types of detection are the ones that test with complex samples such as tissue, cell culture (without the aid of PCR), or food. It should be noted that many studies utilize synthetic nucleotides [28,29]. Some of the more striking innovations in DNA detection have occurred in the field of virus detection, especially with respect to probe design. Researchers who examined sensing of viruses investigate various capture probes that are tagged with redox species [30].

In one approach, polythiolated DNA probe tagged with ferrocene was evaluated [30]. When the capture probe was not hybridized, the ferrocene molecule on the flexible ssDNA is more mobile and migrates to the electrode surface. When the target DNA hybridizes with the probe, a stiff double helix is formed, which significantly decreases the DPV signal due to relatively immobile ferrocene. Use of peptide nucleic acids (PNA) as capture probes has also been investigated, as it provides for neutral DNA analogs, which lowers electrostatic repulsion, and form "triplex" with dsDNA [31]. In the study reported by Aguilar and Fritsch, an electrochemical approach targeted 121-mer mRNA of the hsp70 heat shock protein in *Cryptosporidium parvum* [28]. A sample of 2.6× 06 oocysts/ml was heat shocked for 10 min to induce transcription of hsp70. A capture DNA probe was immobilized onto an aminated Au/SiO_2 wafer. Then, 500 μl of a 50-μg/ml solution of heat-shocked oocysts was incubated with the functionalized wafer for 1 h for hybridization to be completed. Subsequently, incubation with a reporter probe consisting of 42-base ssDNA conjugated to alkaline phosphatase was used. The modified wafer was rinsed and transferred to a solution containing the substrate for the enzyme, p-aminophenyl phosphate (PAPP). Over a 12-h period, alkaline phosphatase generated the electroactive species, p-aminophenyl, which was measured by cyclic voltammetry. The authors established that there was very limited cross-reactivity with several common pathogens such as *Cryptosporidium parvum*, *Listeria monocytogenes*, *Campylobacter lari*, *E. coli* O157:H7, and *Salmonella typhi*.

Detection method	Target	Sample matrix	Analysis time (hours)	Detection limit	Ref.
DPV	*Escherichia coli* ssDNA	50 μL sample of separated SSDNA	7	5 cfu/mL	[34]
SWV	*Escherichia coli* ssDNA	1 μL buffer	~1	0.75 amol nucleotide in 1 μL sample	[35]
Potentiometric sensor	*Escherichia coli* rRNA	4 μL buffer	1	10 cfu (0.2 amol 16S rRNA) in 4 μL sample	[32]
Chronoamperometry	*Escherichia coli*	"meat juice" lysate	~7	1 cfu/mL	[36]
Cyclic voltammetry	*Cryptosporidium parvum*	500 μL lysate	16	2 μg mRNA (37 oocysts)/1 mL	[28]
Impedance spectroscopy	*Salmonella choleraesuis* amplicon	PCR mix	~3	1 nM	[37]
DPV	*Staphylococcus saprophyticus* DNA	cell lysate	0.5	1 cfu/μL	[38]
DPV	HBV ssDNA	10 μL buffer	3	0.300 pM	[39]

Table 1. Electrochemical detection of nucleic acids of pathogens

The limit of detection for *Cryptosporidium* is poor in comparison to the femtomol and attomol values in the *E. coli* experiments. The 2 μg/ml limit of detection was established using a synthetic target as in all the work summarized in Table 1, with the exception of Wu et al. [32] where they used bacterial lysate with no PCR. The limits of detection given in Table 1 would be more relevant to food safety and environmental monitoring if they were extracted from experiments that used actual samples of the pathogens as in Wu et al. [32]. Electrochemical sensors, in combination with PCR techniques, are a powerful approach to detecting nucleic acid sequences in batch assays. The cyclic voltammetry and impedimetric assays also provide characterization of adsorbed monolayers. An alternative sensing approach is to use electromechanical sensors to detect nucleic acids with flow-through sample analysis, where the sample volume is larger than the 1- to 50 μl volumes reported for the electrochemical sensors [33].

2.2. Electrochemical lactate biosensors

L-Lactate plays a vital role in various biochemical processes. Physiological lactate levels are related to the status of anaerobic metabolism associated with muscle contraction. Under normal conditions, healthy persons have lactate concentrations between 0.6 and 2 mM, but during physical activity the value of this parameter can rise up to 20 or 30 mM. That is the reason, L-lactate quantification is vital in i.e. sports medicine [40].

Formation of lactate is followed in the wake of growth in H^+ concentration within the cells; when the rapidity of lactate production is high enough, the H^+ buffering ability may be oversteped, the cellular pH reduces, and the fact may influence on cell acidosis which results on performance of muscles [41].

Even though glucose is usually assumed to be the main energy source for living tissues, there are some evidences that it is lactate, and not glucose, that is preferentially metabolized by neurons in the brain of several mammalian species (i.e. humans) [42], and its monitoring provides a crucial signal in brain stroke and head trauma [43,44]. There was also also revealed that lactate is an indicator for sleep periods [45].

The significance of lactate monitoring in medical diagnostics is indicated by the certainty that blood lactate levels have long been a vital marker of sport as well as clinical states. The facts display the considerable regard in the development of efficient lactate sensors for diagnostics.

2.2.1. Amperometric lactate biosensors

Amperometric transduction is based on the survey of the current appearing from the electrochemical oxidation or reduction of an electroactive species. The resulting current is immediately correlated to the bulk connection of the electroactive species or its production or consumption rate within the adjacent biocatalytic layer. Because of the fact, amperometric sensors alter the concentration of the analyte in their closest vicinity, that is within the diffusion layer [46]. The simplest amperometric biosensors in common usage involve the Clark oxygen electrode.

The enzymes commonly utilized in the development of amperometric biosensors for L-lactate detection are lactate oxidase as well as L-lactate dehydrogenase. In the case of lactate oxidase, oxygen consumption or hydrogen peroxide formation is verified – H_2O_2 provides a current proportional to amount of L-lactate. In dehydrogenase-based sensor, the protein catalyzes the oxidation of lactate to pyruvate in the presence of nicotinamide NAD^+ and the reduced NADH is amperometrically determined. As was reviewed by Rassaei et al. [47] cytochrome b2 [48,49] and lactate monooxygenase [50] have been used in the structure of lactate sensors rarely.

One of the first amperometric lactate biosensor was presented by Faridnia et al. [51]. An amperometric hydrogen peroxide-based biosensor has been developed for non-invasive determination of L-lactate. The biosensor utilizes lactate oxidase immobilized between a polycarbonate membrane and a polytetrafluoroethylene blocking membrane to effectively eliminate electrochemical interferences. The response times were found as 2 min and 10 s, respectively. The biosensor was applied to the analysis of sweat L-lactate content of healthy subjects during physical exercise [51].

Romero et al. has developed [40] biosensor requires only 0.2 U of lactate oxidase, which is immobilized in a mucin/albumin hydrogel matrix. By protecting the platinum surface with a Nafion membrane, typical interference related to negatively charged species such as ascorbic acid has been minimized to practically undetectable levels. A detection limit was found as 0.8 μM. The lactate biosensor presents remarkable operational stability and sensitivity (0.537 ± 0.007) mA M^{-1}. In this regard, the sensor keeps practically the same sensitivity for 5 months.

Because of the fact, the sensors based on L-lactate oxidase involve the electrochemical detection of enzymatically generated H_2O_2. The lactate oxidase based sensors invariably suffers from the interference due to other electroactive compounds such as ascorbate, as the detection of H_2O_2 requires high overpotential. Moreover, the sensitivity of the sensor depends on the oxygen concentration [52]. The method to avoid the oxygen depletion [53] is replacement of lactate oxidase on lactate dehydrogenase, which needs to operate NADH or NADPH.

As was reviewed earlier [47], the character of NAD+/NADH is mediating to shuttle the electrons between the enzyme and the electrode. Nevertheless, hardship in introducing the coenzymes into biosensing device as well as its regeneration, which is connected with high oxidation potential, may frequently cause the need of optimization of collateral parameters. That is the main reason the L-lactate oxidase is mostly used in design and production of amperometric L-lactate sensing devices.

But there are still several models of sensing tools based on L-lactate dehydrogenase as biosensor proposed by Pereira et al. [54]. The amperometric biosensor is based on lactate dehydrogenase and Meldola's Blue. The measured response was based on the electrocatalytical properties of Meldola's Blue to oxidize NADH, which was generated in the enzymatic reaction of lactate with NAD^+ under catalytic activity of lactate dehydrogenase. The amperometric response for lactate using this biosensor showed excellent sensitivity (3.46 μA cm^{-2} mmol L^{-1}), operational stability (around 96.5%) and wide linear response range (0.10–10 mmol L^{-1}). Another example of lactate dehydrogenase-based biosensor was reported by Jena and Raj [52]. The setup was developed using L-lactate dehydrogenase and hydroxylamine enlarged gold nanoparticles (GNPs). Lactate dehydrogenase and GNPs have been integrated with the sol–gel 3-D silicate network derived from 3-(mercaptopropyl)trimethoxysilane. The biosensing of L-lactate is based on the electrocatalytic determination of enzymatically generated NADH by GNPs of the integrated assembly.

As was reported earlier by Rassaei et al. [47] in the design and construction of amperometric lactate biosensor the most attention has centered on electrode, technologies of protein immobilization, coenzymes, mediators, the lifetime and biocatalyst stability.

According to IUPAC definition, the sensor sensitivity and selectivity may be improved by using of mediators. Fulz and Durst [55], as well as Nikolaus and Strehlitz [46] reported that the irreversible electrochemical characteristic of several biological species is due to slow heterogeneous electron transport at electrode. This keeps to a sever electrode fouling by adsorption of the bioelement on the electrode or insulation of the active protein center by the surrounding proteins. To avoid these problems the mediators (electroactive agents which operate as an electron shuttle) are utilized. Mediators may also extend the slow rate of electron transfer and reduce the overpotential [46]. In L-lactate detection is suggested to use the polymeric FAD as environment-friendly mediator [56].

As was mentioned earlier by Rassaei et al. [47], enzymes used in lactate biosensor technology may be stabilized on the electrode by a number of different strategies, such as cross-linking [57], entrapment in conductive or nonconductive polymers [58], confinement in a sol–gel [59, 60] or hydrogel [61] matrix, immobilization with another biological component, such as bovine serum albumin [62], or covalent attachment of the enzyme to the electrode [63]. Since the biosensor technologies have emerged, the researches permanently work on the stability, activity as well as efficiency of used enzymes [64].

2.3. Electrochemical biosensors in pharmaceutical assay

Taking into account the fact, that the pharmacological methods are often time-consuming and not cheap, biosensing tools have appeared an entertaining option. Many investigations have been conducted in consideration of the optimization of immobilization procedures as well as the obtaining of the analytical rate. Modern biomimetic identifying setups have been investigated [65]. Moreover, redox properties of drugs can provide insight into metabolic fate, into their *in vivo* redox processes, and their pharmacological effect.

The analytical policy adapted in the expansion of biosensing tools extend from the identification of a signal created by the straight interaction amongst the analyte and the recognizing factor, or indirectly, engaging mediating elements and coupling processes.

The magnitude of biosensors depend on biocatalytic reactions, therewith the essence of the species in the biochemical process, as the molecular features of the analyte, establish the finest group of transducer to be applied [66].

Transport of electron between the enzyme and the electrode surface can arrive by three pathways, qualifying the kind of amperometric tools used, which may be sorted into three classes:

1. The 1st class of electrodes are dependent on the redox proceeding of cofactors (*ß*-NAD(P)$^{+}$).
2. The 2nd class has included the employment of electron mediators due to reduce potential and rationalize electron transport and sensitivity.
3. The 3rd class of electrodes - the electron transport takes place fairly amongst the active center of protein and electrode [67].

Towards drug specifics inclosing hydrolysable and unstable clusters and/or when proton release is observed, the employment of ion-selective electrodes or potentiometric biosensing tools is the most routinely used variant [68].

Amongst the favorable points of view affected the future of biosensors is the employment of biomimetic systems, which link updated chemical stability with almost identical activity of the particular enzyme [66, 69].

2.3.1. *Antiviral drugs*

Drug analysis has a large influence on social health. Many analytical techniques have been investigated for the quantitative determination of drugs in both pharmaceutical and biological samples. Even though e.g. chromatographic methods are sensitive and selective, they are tedious and time consuming due to the requisite pretreatment of samples and optimization of chromatogram conditions. Electrochemical methods have been utilized for the determination of wide range of drug compounds, repeatedly without derivatization [70].

Substituted purines represent an important category of compounds actively studied as potential therapeutics against viral infection. Acyclovir is the most commonly utilized nucleoside analog antiviral drug. It is mainly used for the treatment herpes simplex as well as herpes zoster infections. But also may be used for treatment of primary genital herpes or herpetic encephalitis, it is effective in preventing HSV (human herpes virus) infections. This drug provides significant therapeutic benefits in the treatment of viral diseases, and no serious side effects associated with its use have been reported [71]. Based on above description the quantitative determination of ACV become very important and has been widely studied. Several analytical methods have been proposed for individual measurement of acyclovir by fluorometric, high-performance liquid chromatography (HPLC), radioimmunoassay, spectrophotometric methods, GCV by HPLC and PCV by fluorometric [72].

Recently, Heli et al. [70] has reported the electrochemical assay for acyclovir determination by electrocatalytic oxidation of acyclovir on copper nanopartciles-modified carbon paste electrode. Due to investigate simple and time-saving method for the analysis of acyclovir in pure form as well as pharmaceutical formulations, amperometry method was used. Also electrochemical technique coupled with carbon nanotube (CNT) modified electrodes, as was presented by Shah et al. [72] as comfortable and cost-effective. Another example of acyclovir electrochemical analysis was shown by Shetti et al. [73]. The oxidation of acyclovir was investigated at fullerene-C_{60}-modified glassy carbon by cyclic and differential pulse voltammetry. The authors have shown that no reduction peak was observed in the reverse scan, suggesting that the electrochemical reaction was a totally irreversible process. They have assumed also that the oxidation process is located on the guanine moiety in the molecule. It can be concluded that the electrochemical oxidation of acyclovir involves two electron and two proton transfer process to the formation of 8-oxoacyclovir, which is analogous to the initial oxidation product of guanine [73].

2.3.2. Neurotransmitters drugs

The useful in electrochemistry procedures are particularly advantageous in the survey of neurotransmitters and neuronal medicines, not only according to a quality verification, but also for physiological aims. Voltammetry paired to microelectrodes altered with medicine or neurotransmitter receptors are a subservient tool for rating the physiopathological process, clarifying their pharmacological effect, as well as for the extension of novel drugs [74].

Catecholamines

As was reported earlier by de Souza Gil and Rodrigues de Melo, catecholamines apply a great deal of duties at the central as well as autonomous nervous systems [66]. They possess an oxidized catecholic grouping which can be catalyzed by a numerous of oxidases.

The utilization of vegetal material as an origin of oxidases for biosensor destinations images a favorable choice for the quantification of this group of compounds [66, 75-77].

The enzymatic protein - polyphenoloxidase, inherent in coconut fiber material [78], apple tissue [79], avocado pulp [76], zucchini [66], and a lot of sorts of palms [75] have been utilized as source of biocatalysts for biological sensor design concentrating on the assay of particles including catecholic and phenolic groups. Biosensors including palm fruit (*Livistona chinensis*) have been utilized for the test of epinephrine in pharmaceutical formulations when concentration is ranging between 0.05 - 0.35mM. In parallel to classical procedures, the effects rendered significant notability (3.1%) and repeatability [80]. Another palm material-based biosensing tools for catecholamine detection contain the coconut biosensor [81] and the guariroba (*Syagrus oleracea)* biosensor [75].

A biosensor enclosing 25% (p/p) polyphenoloxidase isolated from banana (*Annona muricata L.*), 30% (p/p) graphite, 30% (p/p) silicone oil, and 15% (p/p) of 7,7,8,8-tetracyanoquinodimethane (TCNQ) linked to flow injection systems (FIA) was efficiently employed for the analysis of dopamine in various pharmaceutical species [81].

Benzodiazepines

Benzodiazepines are central nervous system inhibitors, usually utilized for anxiety curing. Since the curing triumph of this group of medicines depends on the support of accurate serum levels, a lot of occurrence can take place regarding to the intake of benzodiazepines. Due to typical antipsychotics, they are not net dopaminergic antagonists but possess other receiver antagonistic activities (e.g. acting on serotonin and/or adrenergic receivers); these specific advantages make them convenient in a wide confine of psychotic diseases (mania, schizophrenia and delirium) with a decreased risk of inducing extrapyramidal effects and acting on negative symptoms of schizophrenia (alogia, anhedonia and avolition) [82].

Thereby, the invention of procedures for benzodiazepines determination is significant in a lot of areas, i.e. quality control, clinical diagnostics and forensic investigations [66].

As was reviewed before [66], reports representing the electrochemical character of temazepam, oxazepam and diazepam on carbon paste electrodes were developed by cyclic and pulse differential voltammetry. For determination of olanzapine (2-methyl-4-(4-methyl-1-piperazinyl)-10H-thieno[2,3-b][1,5]benzodiazepine) Brunton et al. used modified carbon glass electrode with multiwalled nanotubes and layer of TiO_2. The optimal pH for temazepam and oxazepam was found ranging from 3.0 to 5.0, in case of olanzapine it was found as 5.0 as well as no influences of various foreign species on the determination of olanzapine were investigated by square wave voltammetry [82]. On the other hand, Liu et al. reported the combination of electrochemistry and mass spectrometry (nanospray desorption electrospray ionization) as a powerful analytical tool for identification of benzodiazepines [83]. The advantage of the system was found as high sensitivity and the capability of analyzing small volumes of electrolyzed solution. This capability is crucial in analyzing trace amounts of biological samples.

3. Optical sensors

Recently, the optical biosensing devices had been developed rapidly and have been utilized in a numerous of significant fields enclosing food safety, security, environmental monitoring and clinical analysis [47,84]. In clinical analysis, the optical sensor has been used in both regular clinical diagnostic and for clinical research adoptions [85]. The word *optrode* is a fusion of the *optical* and *electrode*. It is used to qualify optical tools [86]. This technique of transduction has been adapted in different kinds of biosensor regarding to the varied classes of spectroscopy [87]. These transduction techniques are efficient to measure another features of the target/analyte. The optical based biosensor is efficient to supply tab free, real time and parallel determination [88]. The fluorescence or surface plasmon resonance which is united with optical fiber is the most exoteric procedure accessible for optical based biosensors [87]. It occurs that sensing tools based on the optical fiber concept acquire research concern in the area of biosensor design [47].

3.1. Optical fiber sensor

Optical fiber (occasionally named *optrodes*, Fig.2) has acquired substantial concern for investigation as a biosensing element notably in lower detection limit (LOC) sensing surveys. *Optrodes* based sensors are built of few principal constituents: a source of light, a biorecognition factor (immobilized), an optical fiber utilized in light transmission and useful as the base, and a detector (i.e., spectrophotometer) where the output light indicates the operation [89]. As a consequence, when the objective analyte communicates with the biological recognition agents at the surface of the fiber, a biocatalytical process occurs generating the alterations in the optical behavior. These alterations may be compared to the analyte concentration. The biorecognition occurs on the fiber optic, when the light source is transmitted through. The identical or another fiber is applied to direct the output light to the detecting element. The frequent directions of biosensing utilization of optical fiber are acquiring scientist concern due to its principal benefits as a miniaturized fine efficiency sensor, small fiber optical sensors, high sensitivity, rapid result, fine selectivity, and low detection limits [90]. Optical fiber sensors as well possess a number of profits over electrochemical and another biosensing due to their advanced sensitivity, security, independence of electromagnetic involvement and their ability for real time control [86]. Moreover, this kind of transduction is free of reagent and adaptable, consequently relevant for distant sensing as well as and singular molecule determination [86]. Nevertheless, there are respective disadvantages as the deficient durability of biorecognition agents and sensitivity to surrounding light [86]. Majority of the causes notified in the literature on optical fiber biological sensor appliances have adopted the optic fiber with alternative optical procedures [91]. Moreover, these techniques have been utilized for the determination of numerous biological species as antigen, nucleic acids [92]. Optical fiber biosensing is comprehensive and being commonly utilized for microorganisms identification [91] Leung et al. [93] have described label free determination of DNA hybridization for pathogen testing by applying a fiber optic biosensor.

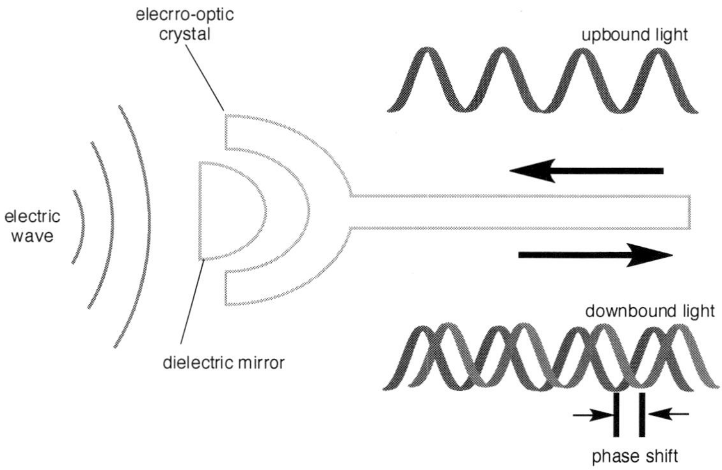

Figure 2. Scheme of optical fiber sensor

3.2. Plasmonic sensors

Surface plasmon resonance (SPR) refers to the collective oscillations of the conduction electrons in metallic nanostructures [94]. Both the intensity and the position of SPR strongly depend on the size, shape and composition of the nanostructures, as well as the dielectric properties of the ambient environment [95,96]. This sundries of responsive variables allows for optical sensors to be created using plasmonic metallic nanostructures. Thereby plasmon-enhanced optical sensors are finding increasing application in detection of analytes in clinical diagnosis, security, food safety and environmental control [97,98]. Herein the expression *plasmonic sensor* refers to sensors that immediately utilize shifts in spectral behavior of the plasmon to operate as the transducer of the sensing signal. Plasmonic sensors are constructed either with 2D chips that support surface plasmon polariton (SPP) mode or with nanoparticles that support localized surface plasmon resonance (LSPR).

3.2.1. Chip-based plasmonic sensors

Planar plasmonic base support expanding SPP course or blended SPP/LSPR course which can be adapted for plasmonic sensing along variations in the refractive index of the ambient environment. SPP may merely be excited utilizing a prism or a grating to provide the extra speed essential to adapt to free-space light. The typical arrangement in chip-based SPP sensing tools is the Kretschmann layout (Fig. 3), [99,100] which allows time- and angle determined reflectivity surveys of a noble metal layer's SPP course through a glass prism. In this collocation, the metal layer is extremely reflective excepting at a concrete angle when the SPP is excited, related to as the SPR angle [101,102]. When target species hinder to ligands immobilized on the plasmonic metal layer, the SPP strand red-shifts regarding to the upper value of refractive index of the species than the water solution, acting as a sensor [98]. The wavelength of the SPR signal differs linearly with the refractive index of the ambient environment due to the Drude pattern [103]. Thereby, the refractive index sensitivity (S=$\Delta\lambda_p/\Delta n$) is signified in entities of nm per RIU, where λ_p is the plasmon frequency of the metal and n is the refractive index of the ambient environment. Figure of merit (FOM=S/FWHM) can be qualified to account the sensing projection of a plasmonic sensor, where FWHM places for the full spread at half-maximum and S is the refractive index sensitivity.

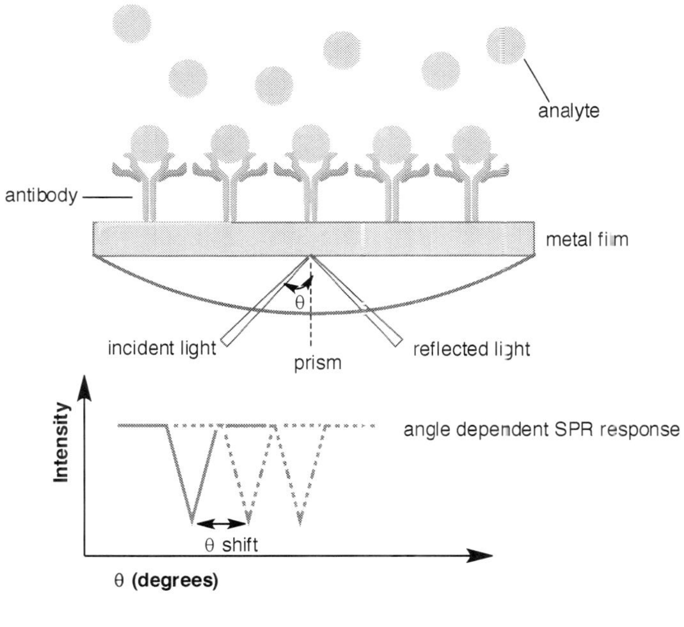

Figure 3. Outline of plasmonic sensing system based on the Kretschmann layout The incident light is reflected by the metal layer along a prism, and the reflected beam presents a dark line regarding to the SPR absorption. This plasmonic sensing layout can determine time and angle-resolved SPR results upon the binding of analytes.

The method to accomplish the refractive index sensitivity of a plasmonic tool is to decline the SPR signal of the metal layer with the absorption band of the chromophore that ties to the plasmonic metal [104]. Otherwise, the refractive index sensitivity may be increased by exchanging the planar metal layer with a broad-space cyclic nano-array model as a nano-disc array, nanosphere array, or nano-triangle array which assists both a heavier regional EM field and upper sensing plate than planar layers. Conjugated plasmonic nanosized particles may be either combined to a gold layer in a sandwich manner as an analyte linker is prevalent, subsequent enlarging the shift of the SPR signal related to the gold layer alone [105]. Plasmonic chip-based tools possess three clear benefits [99]: (a) label-free determination facilitates the structure and action of the sensor, and excludes the utilization of multiple antibodies as adapted in ELISA, the Au pattern of analytical resolution [106]; (b) real time study of process kinetics as the plasmonic chip is combined with a flow-cell, which supplies a mighty implement for survey the binding incidents [107]; (c) long-range SPP (LRSPP) modes may be created to demonstrate arrangements of size less damping than classic SPP [108]. The smaller damping in LRSPP permits tighter band spreads and higher FOM, as well the penetration deepness of more than 1 mm [109]. It seems to be perfect for study of living organisms [110]. LRSPP modes can be useful when a fine metal layer is settled among two dielectrics with almost identical refractive indexes [111].

3.2.2. *Plasmon-enhanced fluorescence (PEF) sensors*

The first PEF sensing tool was described in 1991 [112]. As was mentioned before [99], PEF is related either to a surface plasmon-enhanced fluorescence spectroscopy (SPFS) [113], surface-enhanced fluorescence spectroscopy (SEFS), or metal-enhanced fluorescence (MEF). Till now a typical naming convenant for this type of tool does not subsist. The PEF studies were not observable before 2007, but recently, several nanosized elements have been examined to plasmonically enhance fluorescence, compatible with the growth of navigable nanoparticle synthesis and lithography-based formation of nanosized chips. The plasmon-induced fluorescence enhancement agent commonly decreases in the range of 10–100 but may be upper in optimized plasmonic nanosized structures, even rising up to 1340 [114]. The sudden growth of PEF sensors is easy to imagine observed the last tendencies in progress in the idea of PEF [99].

3.3. Optical lactate sensors

A numerous literature reports of optical sensing systems for the determination of lactate reveals that sensor designs largely utilization either fluorometry or (electro)chemiluminescence techniques [115-118]. Also an optrode is capable to classify by the kind of enzymatic protein and the pathway it is combined to the transducer. First time, the optical sensing implement for the determination of lactate in blood was described by Broder and Weil [119] in 1964. The detection tool was relied on a photospectrometric technique, and the relation light/concentration was calculated due to the Lambert-Beer law [119].

Current tendency to miniaturize the analytical tools have led scientists to design optical sensing elements on a chip dually with identification pieces. Wu et al. [120,121], in example, combined optical fibers with a microfluidic tool, which was consisted of an SU-8/polydimethylsiloxane waveguide to monitor online the lactate (Fig. 4). The determination of lactate relied on the absorbance of phenol red (max. 540 nm) through the oxidative reaction of chromogen precursors in the presence of H_2O_2 (produced by the lactate oxidase catalyzed process).

As was reported by Rassaei et al. [47] chemiluminescent biosensing tools are mainly based on incorporation the leading biocatalyzed reaction with a luminescent substrate. Therefore, lactate sensors are based on the chemiluminescent reaction among H_2O_2 formed in enzymatic process and luminol. Still, this reaction demands a basic surroundings (pH 8.5–9) as the light emission intensity decreases severely with pH downfall to inert worth essential for maximal enzyme efficiency. Marquette and Blum [122] invented a fiber-optic flow injection biosensing tool for the monitoring of glucose and lactate concentration employing luminol. The authors immobilized glucose oxidase as well as lactate oxidase on membranes built of polyamide and collagen. Lately, Zheng et al. [123] developed a lactate sensing tool based on an exceptional nanosized optofiber with high spatial resolution to detect the extracellular lactate of a single cancer cell. Their great achievement was to indicating the capability of the biosensor to differentiate higher extracellular lactate planes for cancer cell lines.

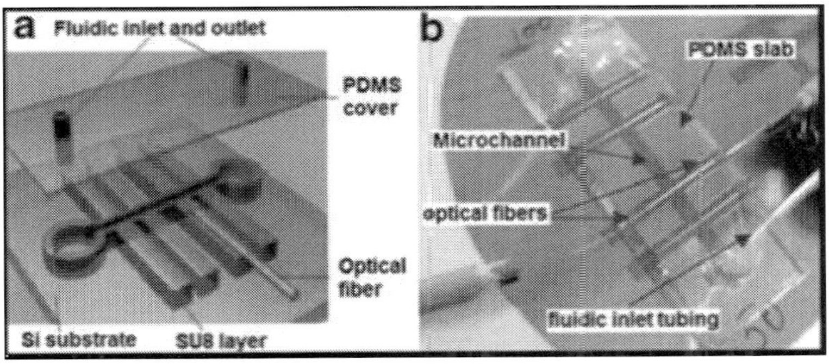

Figure 4. A microfluidic tool built of SU-8/polydimethylsiloxane (PDMS) (a) with incorporated optical fiber as detecting element, and (b) a microfabricated instrument, adopted from [120]

4. Outlook and conclusions

Recently, there has been noticed the significant developments in the area of biosensor. This potent instrument has been employed in a several territories of life. Nevertheless, literature indicates that there are a lot of researches that have been assumed by middle survey with straight clean buffer solutions in place of immediate measurement for real-sample monitoring which is much less crucial. Technological benefits have supplied with the instruments and matters required to fabricate a biosensor which can be combined with a microfluidic arrangement, sound, sampler, detector, amplifier and logic circuitry. This type of implement may be perspective candidate for label free, reagentless, real time controlling, miniaturization and reasonable adoption cost. For clinical application, this benefit would permit the growth of exceptionally low cost, available biosensors that could be utilized for domestic medical diagnostics without the exigency of transporting analytical samples.

Acknowledgements

The authors gratefully acknowledge to the financial support from NCN Grants No. 2012/05/B/ST5/00749 and support of Wrocław University of Technology.

Author details

Jadwiga Sołoducho* and Joanna Cabaj

*Address all correspondence to: jadwiga.soloducho@pwr.wroc.pl

Wroclaw University of Technology, Faculty of Chemistry, Wrocław, Poland

References

[1] Stradiotto N.R., Yamanaka H., Valnice M., Zanoni B. Electrochemical sensors: A powerful tool in analytical chemistry, J. Braz. Chem. Soc. 2003, 14, 159-173

[2] Lee T.M.-H., Hsing I.-M. DNA-based bioanalytical microsystems for handheld device applications, Anal. Chim. Acta 2006, 556, 26-37

[3] Drummond T.G., Hill M.G., Barton J.K. Electrochemical DNA sensors, Nat. Biotechnol. 2003, 21, 1192-1199

[4] Mikkelsen S.R. Electrochecmical biosensors for DNA sequence detection, Electroanalysis 1996, 8, 15-19

[5] Palecek E., Fojta E. Detecting DNA Hybridization and Damage, Anal. Chem. 2001, 3, 73A-83A

[6] Almeida M.O., Maltarollo V.G., de Toledo R.A., Shim H., Santos M.C. and Honorio K.M. Medicinal Electrochemistry: Integration of Electrochemistry, Medicinal Chemistry and Computational Chemistry, Cur. Med. Chem. 2014, 21, 2266-2275

[7] Clark Jr, L.C., Lyons C. Electrode system for continuous monitoring in cardiovascular surgery, Ann. N.Y. Acad. Sci., 1962, 102, 29-45

[8] Thandavan K., Gandhi S., Sethuraman S., Rayappan J.B.B., Krishnan U.M. Development of electrochemical biosensor with nano interface for xanthine sensing - A novel approach for fish freshness estimation. Food Chem., 2013, 139, 963-969

[9] Wang J. Glucose biosensors: 40 years of advances and challenges, Electroanal. 2001, 13, 983-988

[10] Barhoumi H., Maaref A., Rammah M., Martelet C., Jaffrezic N., Mousty C., Vial S., Forano C. Urea biosensor based on Zn_3Al–Urease layered double hydroxides nanohybrid coated on insulated silicon structures, Mater. Sci. Eng., C, 2006, 26, 328-333

[11] Serafín V., Hernández P., Agüí L., Yáñez-Sedeño P., Pingarrón J. Electrochemical biosensor for creatinine based on the immobilization of creatininase, creatinase and sarcosine oxidase onto a ferrocene/horseradish peroxidase/gold nanoparticles/multi-walled carbon nanotubes/Teflon composite electrode, Electrochim. Acta, 2013, 97, 175-183

[12] Ahmad R., Tripathy N., BongHahn Y. High-performance cholesterol sensor based on the solution-gated field effect transistor fabricated with ZnO nanorods, Biosens. Bioelectron. 2013, 45, 281-286

[13] Xu M., Luo X., Davis J.J. The label free picomolar detection of insulin in blood serum, Biosens. Bioelectron. 2013, 39, 21-25

[14] Hu Y., Zuo P., Ye B.C. Label-free electrochemical impedance spectroscopy biosensor for direct detection of cancer cells based on the interaction between carbohydrate and lectin, Biosens. Bioelectron. 2013, 43, 79-83

[15] Pyun J.C., Beutel H., Meyer J.U., Ruf H.H. Development of a biosensor for *E. coli* based on a flexural plate wave (FPW) transducer, Biosens. Bioelectron. 1998, 13, 839-845

[16] Prusak-Sochaczewski E., Luong J.H.T. Development of a piezoelectric immunosensor for the detection of *Salmonella typhimurium*, Enzyme Microb. Technol. 1990, 12, 173-177

[17] Biran I., Walt D.R. Optrode- Based Fiber Optic Biosensors (Bio-Optrode). In: Ligler F.S, Taitt C.A.R, Ed. Optical Biosensors Present and Future, Elsevier, Amsterdam, 2004

[18] Abreu F.C.D., Ferraz P.A.D.L., Goulart M.O.F. Some Applications of Electrochemistry in Biomedical Chemistry. Emphasis on the Correlation of Electrochemical and Bioactive Properties, J. Braz. Chem. Soc. 2002. 13, 19-35

[19] Brunet A., Pailleret A., Devynck M.A., Devynck J., Bedioui F. Electrochemical sensing of nitric oxide for biological systems: methodological approach and new insights in examining interfering compounds, Talanta, 2003, 61, 53-59

[20] Ames J.R., Kovacic P. Electrochemistry of omeprazole, active metabolites and a bound enzyme model. Possible involvement of electron transfer in anti-ulcer action, Bioelectrochem. Bioenerg. 1992, 28, 443-450

[21] Liang H.-J., Ling T.-R., Rick J.F., Chou T.-C., Molecularly imprinted electrochemical sensor able to enantroselectivly recognize d and l-tyrosine, Anal. Chim. Acta, 2005, 542, 83–89

[22] Winkler T.E., Ben-Yoav H., Chocron S.E., Kim E., Kelly D.L., Payne G.F., Ghodssi R., Electrochemical Study of the Catechol-Modified Chitosan System for Clozapine Treatment Monitoring, *Langmuir*, 2014, 30 (48),14686–14693

[23] Conley R.R., Tamminga C.A., Kelly D.L., Richardson C.M., Treatment-Resistant Schizophrenic Patients Respond to Clozapine after Olanzapine Non-Response Biol. Psychiatry 1999, 46, 73–77

[24] Keeley G.P., McEvoy N., Nolan H., Kumar S., Rezvani E., Holzinger M., Cosnier S. and Duesberg G.S., Simultaneous electrochemical determination of dopamine and paracetamol based on thin pyrolytic carbon films, Anal. Methods, 2012, 4, 2048-2053

[25] Lunte C.E., Osbourn D. Electrochemical detection for pharmaceutical analysis, in: Swarbrick, J. Encyclopedia of Pharmaceutical Technology, 3rd.; CRC Press: New York, 2006

[26] Ermini M. L., Mariani S., Scarano S., Minunni M. Bioanalytical approaches for the detection of single nucleotide polymorphisms by Surface Plasmon Resonance biosensors, Biosen. Bioelectron. 2014, 61, 28–37

[27] Monosık R., Stredansky M., and Sturdık E., Application of Electrochemical Biosensors in Clinical Diagnosis, J. Clinic. Lab. Anal. 2012, 26, 22–34

[28] Aguilar Z.P., Fritsch I. Immobilized enzyme-linked DNA-hybridization assay with electrochemical detection for *Cryptosporidium parvum* hsp70 mRNA, Anal. Chem. 2003, 75, 3890–3897

[29] Sun W., Qi X., Zhang Y., Yang H., Gao H., Chen Y., Sun Z. Electrochemical DNA biosensor for the detection of *Listeria monocytogenes* with dendritic nanogold and electrochemical reduced graphene modified carbon ionic liquid electrode, Electrochim. Acta, 2012, 85, 145–151

[30] Lereau M., Fournier-Wirth C., Mayen J., Farre C., Meyer A., Dugas V., Cantaloube J.-F., Chaix C., Vasseur J.-J., Morvan F. Development of innovative and versatile polythiol probes for use on ELOSA or electrochemical biosensors: application in hepatitis C virus genotyping, Anal. Chem. 2013, 85, 9204–9212

[31] Ahour F., Pournaghi-Azar M.H., Alipour E., Hejazi M.S. Detection and discrimination of recombinant plasmid encoding hepatitis C virus core/E1 gene based on PNA and double-stranded DNA hybridization, Biosens. Bioelectron. 2013, 45, 287–291

[32] Wu J., Chumbimuni-Torres K. Y., Galik M., Thammakhet C., Haake D.A., Wang J. Potentiometric detection of DNA hybridization using enzyme-induced metallization and a silver ion selective electrode, Anal. Chem. 2009, 81, 10007–10012

[33] Rosario R., Mutharasan R. Nucleic acid electrochemical and electromechanical biosensors: a review of techniques and developments, Rev. Anal. Chem. 2014; 33(4), 213–230

[34] Anderson M.J., Miller H.R., Alocilja E.C. PCR-less DNA co-polymerization detection of Shiga like toxin 1 (stx1) in Escherichia coli O157:H7. Biosens. Bioelectron. 2013, 42, 581–585

[35] Liao W.-C., Ho J.-A.A. Attomole DNA Electrochemical Sensor for the Detection of *Escherichia coli* O157. Anal. Chem. 2009, 81, 2470–2476

[36] Heidenreich B., Poehlmann C., Sprinzl M., Gareis M. Detection of *Escherichia coli* in Meat with an Electrochemical Biochip. J. Food Prot. 2010, 73, 2025–2033

[37] Berdat D., Rodriguez A.C.M., Herrera F., Gijs M.A.M. Label-free detection of DNA with interdigitated micro-electrodes in a fluidic cell. Lab Chip, 2008, 8, 302–308

[38] Lam B., Fang Z., Sargent E.H., Kelley S.O. Polymerase chain reaction-free, sample-to-answer bacterial detection in 30 minutes with integrated cell lysis. Anal. Chem. 2012, 84, 21–25

[39] Zheng J., Chen C., Wang X., Zhang F., He P. A sequence-specific DNA sensor for *Hepatitis B* virus diagnostics based on the host–guest recognition. Sens. Actuat. B, 2014, 199,168–174

[40] Romero M.R., Ahumada F., Garay F., Baruzzi A.M., Amperometric biosensor for direct blood lactate detection, Anal. Chem., 2010, 82, 5568–5572

[41] Kemp G. Lactate accumulation, proton buffering, and pH change in ischemically exercising muscle, Am J Physiol Regul Integr Comp Physiol. 2005, 289(3), R895–R901

[42] Wyss M.T., Jolivet R., Buck A., Magistretti P.J., Weber B. In vivo evidence for lactate as a neuronal energy source, J. Neurosci. 2011, 31(20), 7477–7485

[43] Cureton E.L., Kwan R.O., Dozier K.C., Sadjadi J., Pal J.D., Victorino G.P. A different view of lactate in trauma patients: protecting the injured brain, J. Surg. Res. 2010, 159(1), 468–473

[44] Goodman J.C., Valadka A.B., Gopinath S.P., Uzura M., Robertson C.S. Extracellular lactate and glucose alterations in the brain after head injury measured by microdialysis, Crit. Care Med. 1999 27(9), 1965–1973

[45] Naylor E., Aillon D.V., Barrett B.S., Wilson G.S., Johnson D.A., Harmon H.P., Gabbert S., Petillo P.A. Lactate as a biomarker for sleep, Sleep, 2012, 35(9),1209–1222

[46] Nikolaus N., Strehlitz B., Amperometric lactate biosensors and their application in (sports) medicine, for life quality and wellbeing, Microchim. Acta, 2008, 160, 15-55

[47] Rassaei L., Olthuis W., Tsujimura S., Sudholter E.J.R., van den Berg A. Lactate biosensors: current status and outlook, Anal. Bioanal. Chem. 2014, 406,123–137

[48] Shkil H., Stoica L., Dmytruk K., Smutok O., Gonchar M., Sibirny A., Schuhmann W. Bioelectrochemical detection of L lactate respiration using genetically modified *Hansenula polymorpha* yeast cells overexpressing flavocytochrome b2, Bioelectrochemistry, 2009, 76(1–2), 175–179

[49] Smutok O., Dmytruk K., Gonchar M., Sibirny A., Schuhmann W. Permeabilized cells of flavocytochrome b2 over-producing recombinant yeast *Hansenula polymorpha* as biological recognition element in amperometric lactate biosensors, Biosens Bioelectron. 2007, 23(5), 599–605

[50] Romero M.R., Ahumada F., Garay F., Baruzzi A.M. Amperometric biosensor for direct blood lactate detection, Anal. Chem. 2010, 82(13), 5568–5572

[51] Faridnia M.H., Palleschi G., Lubrano G.J., Amperometric biosensor for determination of lactate in sweat, Anal. Chim. Acta, 1993, 278, 35-40

[52] Jena B.K., Raj C.R., Amperometric L-Lactate Biosensor Based on Gold Nanoparticles. Electroanal. 19, 2007, 7-8, 816–822

[53] Makovos E.B., Liu C.C. Measurements of lactate concentration using lactate oxidase and an electrochemical oxygen sensor, Biotechnol. Bioeng. 1985, 27(2), 167–170

[54] Pereira A.C., Aguiar M.R., Kisner A., Macedo D.V., Kubota L.T., Amperometric biosensor for lactate based on lactate dehydrogenase and Meldola Blue coimmobilized on multi-wall carbon-nanotube, Sensor. Actuat. B, 2007, 124, 269-276

[55] Fultz M.L., Durst R.A., Mediator compounds for the electrochemical study of biological redox systems: a compilation. Anal. Chim. Acta. 1982, 140, 1-18

[56] Leonida M.D., Starczynowski D.T., Waldman R., Aurian-Blajeni B., Polymeric FAD used as enzyme-friendly mediator in lactate detection, Anal. Bioanal. Chem. 2003, 376, 832-837

[57] Romero M.R., Garay F., Baruzzi A.M. Design and optimization of a lactate amperometric biosensor based on lactate oxidase cross-linked with polymeric matrixes, Sens Actuat. B, 2008, 131(2):590–595

[58] Rahman M.M., Shiddiky M.J.A., Rahman M.A., Shim Y.B. A lactate biosensor based on lactate dehydrogenase/nictotinamide adenine dinucleotide (oxidized form) immobilized on a conducting polymer/multiwall carbon nanotube composite film, Anal. Biochem. 2009, 384(1), 159–165

[59] Yashina E.I., Borisova A.V., Karyakina E.E., Shchegolikhina O.I., Vagin M.Y., Sakharov D.A., Tonevitsky A.G., Karyakin A.A. Sol–gel immobilization of lactate oxidase

from organic solvent: toward the advanced lactate biosensor, Anal. Chem. 2010, 82(5), 1601–1604

[60] Tsuchiya M., Matsuhisa H., Hasebe Y. Selective amperometric response to hydrogen peroxide at a protein-incorporated sol–gel hybrid film-modified platinum electrode, Bunseki Kagaku, 2012, 61(5), 425–428

[61] Zanini V.P., de Mishima B.L., Solis V. An amperometric biosensor based on lactate oxidase immobilized in Laponite-chitosan hydrogel on a glassy carbon electrode. Application to the analysis of L-lactate in food samples, Sens. Actuat. B, 2011, 155(1), 75–80

[62] Palmisano F., Rizzi R., Centonze D., Zambonin P.G. Simultaneous monitoring of glucose and lactate by an interference and cross-talk free dual electrode amperometric biosensor based on electropolymerized thin films, Biosens. Bioelectron. 2000, 15(9–10), 531–539

[63] Palmisano F., DeBenedetto G.E., Zambonin C.G. Lactate amperometric biosensor based on an electrosynthesized bilayer film with covalently immobilized enzyme, Analyst, 1997, 122(4), 365–369

[64] Campas M., Prieto-Simon B., Marty J.L. A review of the use of genetically engineered enzymes in electrochemical biosensors, Semin. Cell. Dev. Biol. 2009, 20(1), 3–9

[65] Fatibello-Filho O., Lupetti K.O., Leite O.D., Vieira I.C. Electrochemical biosensors based on vegetable tissues and crude extracts for environmental, food and pharmaceutical analysis, Compr. Anal. Chem., 2007, 49, 357-377

[66] de Souza Gil E., Rodrigues de Melo G. Electrochemical biosensors in pharmaceutical analysis, Braz. J. Pharmac. Sci. 2010, 46, 375-391

[67] Oliveira-Neto G., Rover-Junior L., Kubota L.T. Electrochemical biosensors for salicylate and its derivatives. Electroanalysis, 1999, 11, 527-533

[68] Ozoemena K.I., Stefan R.-I., Van-Staden J.F., Aboul-Enein H.Y. Enantioanalysis of S-perindopril using different cyclodextrin-based potentiometric sensors, Sens. Act. B, 2005, 105, 425-429

[69] Kindschy L.M., Alocilja E.C. Development of a molecularly imprinted biomimetic electrode, Sensors, 2007, 7, 1630-1642

[70] Heli H., Zarghan M., Jabbari A., Parsaei A., Moosavi-Movahedi A.A., Electrocatalytic oxidation of the antivirial drug acyclovir on copper nanoparticles-modified carbon paste electrode, J. Solid State Electrochem. 2010, 14, 787-795

[71] Demirtas C., Yilmaz S., Saglikoglu G., Sadikoglu M., Electrochemical Determination of Phenazopyridine Hydrocloride using Poly(*p*-Aminobenzene Sulfonic Acid) Film Modified Glassy Carbon Electrode. Int. J. Electrochem. Sci. 2015, 10, 1883-1892

[72] Shah B., Lafleur T., Chen A., Carbon nanotube based electrochemical sensor for the sensitive detection of valacyclovir, Faraday Discuss., 2013, 164, 135-146

[73] Shetti N.P., Malode S.J., Nandibewoor S.T., Electrochemical behavior of an antiviral drug acyclovir at fullerene-C_{60}-modified glassy carbon electrode, Bioelectrochem. 2012, 88, 76-83

[74] Crespi F. In vivo voltammetry and concomitant electrophysiology at a single micro-biosensor to analyse ischaemia, depression and drug dependence, J. Neurosci. Meth. 2002, 119, 173-184

[75] Gil E.S., Moraes F.C., Oliveira A., Santana J.C., De-Siqueira E.Q., Machado S.A.S. Development of a modified biosensor by a crude extract of Guariroba (*Syagrus oleracea*) for the analysis of phenolic compounds in environmental samples. In: Spring Meeting of International Society of Electrochemistry, 2008, 6., 148, Book of Abstracts. Lausanne: International Society of Electrochemistry

[76] Fatibello-Filho O., Lupetti K.O., Vieira I.C. Chronoamperometric determination of paracetamol using an avocado tissue (*Persea americana*) biosensor, Talanta, 2001, 55, 685-692

[77] Vieira I.C., Lupetti K.O., Fatibello-Filho O. Determination of paracetamol in pharmaceutical products using a carbon paste biosensor modified with crude extract of zucchini (*Cucúrbita pepo*). Quim. Nova, 2003, 26, 39-43

[78] Lima A.W.O., Nascimento V.B., Pedrotti J.J., Angnes L. Coconut-based plant tissue reactor for biosensing of cathecol in flow injection analysis, Anal. Chim. Acta, 1997, 354, 325-331

[79] Cummings E.A., Mailley P., Linquette-Mailley S., Eggins B.R., Mcadams E.T., McFadden S. Amperometric carbon paste biosensor based on plant tissue for the determination of total flavanol content in beers, Analyst, 1998, 123, 1975-1980

[80] Felix F.S., Yamashita M., Angnes L. Epinephrine quantification in pharmaceutical formulations utilizing plant tissue biosensors, Bios. Bioelec. 2006, 21, 2283-2289

[81] Bezerra V.S., De-Lima-Filho J.L., Montenegro M.C.B.S.M., Araujo A.N., Da-Silva V.L. Flow-injection amperometric determination of dopamine in pharmaceuticals using a polyphenol oxidase biosensor obtained from soursop pulp, J. Pharm. Biomed. Anal. 2003, 33, 1025-1031

[82] Brunton L.L., Blumenthal D.K., Murri N., Dandan R.H., Knollmann B.C. (Eds.), Goodman & Gilman's The Pharmacological Basis of Therapeutics, McGraw-Hill 2011

[83] Liu P., Lanekoff I.T., Laskin J., Dewald H.D., Chen H., Study of Electrochemical Reactions Using Nanospray Desorption Electrospray Ionization Mass Spectrometry, Anal. Chem. 2012, 84, 5737-5743

[84] Tsoka S., Gill A., Brookman J.L., Hoare M. Rapid monitoring of virus-like particles using an optical biosensor: a feasibility study, J. Biotechnol. 1998, 63, 147–153

[85] Caygill R.L., Blair G.E., Millner P.A. A review on viral biosensors to detect human pathogens, Anal. Chim. Acta, 2010, 681, 8–15

[86] Biran I., Yu X., Walt D.R. Optrode-based fiber optic biosensors (bio-optrode). In: Ligler, F., Taitt, C. (Eds.), Optical Biosensors: Today and Tomorrow. Elsevier, Amsterdam, 2008

[87] Abdulhalim I., Zourob M., Lakhtakia A. Overview of optical biosensing techniques. In: Marks R.S., Lowe C.R., Cullen D.C., Weetall H.H., Karube I. (Eds.), Handbook of Biosensors and Biochips. Wiley, Weinheim, 2007

[88] Francia G.D., Ferrara V.L., Manzo S., Chiavarini S. Towards a label-free optical porous silicon DNA sensor, Biosens. Bioelectron. 2005, 21, 661–665

[89] Perumal V., Hashim U. Advances in biosensors: Principle, architecture and applications, J. Appl. Biomed. 2014, 12, 1–15

[90] Lee B., Roh S., Park J. Current status of micro- and nanostructured optical fiber sensors, Opt. Fiber Technol. 2009, 15, 209–221

[91] Atias D., Liebes Y., Chalifa-Caspi V., Bremand L., Lobel L., Marks R.S., Dussart P. Chemiluminescent optical fiber immunosensor for the detection of IgM antibody to dengue virus in humans, Sens. Actuat. B, 2009, 140, 206–215

[92] Brogan K.L., Walt D.R. Optical fiber-based sensors: application to chemical biology, Curr. Opin. Chem. Biol. 2005, 9, 494–500

[93] Leung A., Shankar P.M., Mutharasan R. Label-free detection of DNA hybridization using gold-coated tapered fiber optic biosensors (TFOBS) in a flow cell at 1310 nm and 1550 nm. Sens. Actuat. B, 2008, 131, 640–645

[94] Jones M.R., Osberg K.D., Macfarlane R.J., Langille M.R. and Mirkin C.A. Templated techniques for the synthesis and assembly of plasmonic nanostructures, Chem. Rev., 2011, 111, 3736–3827

[95] Ringe E., Zhang J., Langille M.R., Sohn K., Cobley C., Au L., Xia Y., Mirkin C.A., Huang J., Marks L.D. and Van Duyne R.P. Effect of Size, Shape, Composition, and Support Film on Localized Surface Plasmon Resonance Frequency: A Single Particle Approach Applied to Silver Bipyramids and Gold Nanocubes, Mater. Res. Soc. Symp. Proc. 2010, 1208, O10-02

[96] Ringe E., Langille M.R., Sohn K., Zhang J., Huang J., Mirkin C.A , Van Duyne R.P. and Marks L.D., Plasmon Length: A Universal Parameter to Describe Size Effects in Gold Nanoparticles, J. Phys. Chem. Lett., 2012, 3, 1479–1483.

[97] Sendroiu I.E., Warner and Corn R.M. Fabrication of Silica-Coated Gold Nanorods Functionalized with DNA for Enhanced Surface Plasmon Resonance Imaging Biosensing Applications, Langmuir, 2009, 25, 11282–11284

[98] Mayer K.M. and Hafner J.H. Localized surface plasmon resonance sensors, Chem. Rev., 2011, 111, 3828–3857

[99] Li M., Cushing S.K. and Wu N., Plasmon-enhanced optical sensors: a review, Analyst, 2015, 140, 386-406

[100] Brolo A.G. Plasmonics for Future Biosensors, Nat. Photonics, 2012, 6, 709–713

[101] Liu Y. and Cheng Q. Detection of membrane-binding proteins by surface plasmon resonance with an all-aqueous amplification scheme, Anal. Chem. 2012, 84, 3179–3186

[102] Springer T., Piliarik M. and Homola J. Surface plasmon resonance sensor with dispersionless microfluidics for direct detection of nucleic acids at the low femtomole level, Sens. Actuat. B, 2010, 145, 588-591

[103] Oates T.W.H., Wormeester H. and Arwin H. Characterization of plasmonic effects in thin films and metamaterials using spectroscopic ellipsometry, Prog. Surf. Sci., 2011, 86, 328–376

[104] Zhao J., Jensen L., Sung J., Zou S., Schatz G.C. and Van Duyne R.P. Interaction of plasmon and molecular resonances for rhodamine 6G adsorbed on silver nanoparticles, J. Am. Chem. Soc., 2007, 129, 7647–7656

[105] Law W.C., Yong K.T., Baev A. and Prasad, P.N. Sensitivity improved surface plasmon resonance biosensor for cancer biomarker detection based on plasmonic enhancement, ACS Nano, 2011, 5, 4858–4864

[106] Lahav M., Vaskevich A. and Rubinstein I. Biological sensing using transmission surface plasmon resonance spectroscopy, Langmuir, 2004, 20, 7365–7367

[107] Malinsky M.D., Kelly K.L., Schatz and Van Duyne R.P. Chain Length Dependence and Sensing Capabilities of the Localized Surface Plasmon Resonance of Silver Nanoparticles Chemically Modified with Alkanethiol Self-Assembled Monolayers, J. Am. Chem. Soc., 2001, 123, 1471–1482

[108] Liu Z.D., Li Y.F., Ling J. and Huang C.Z. A localized surface plasmon resonance light-scattering assay of mercury (Ii) on the basis of Hg(2+)-DNA complex induced aggregation of gold nanoparticles, Environ. Sci. Technol. 2009, 43, 5022–5027

[109] Mejard R., Dostalek J., Huang C.-J., Griesser H. and Thierry B. Tunable and robust long range surface plasmon resonance for biosensor applications, Opt. Mater. 2013, 35, 2507–2513

[110] Vala M., Etheridge S., Roach J.A. and Homola J. Long-range surface plasmons for sensitive detection of bacterial analytes, Sens. Actuat. B, 2009, 139, 59–63

[111] Sarid D. Long-range surface-plasma waves on very thin metal films, Phys. Rev. Lett., 1981, 47, 1927–1930

[112] Attridge J.W., Daniels P.B., Deacon J.K., Robinson G.A. and Davidson G.P. Sensitivity enhancement of optical immunosensors by the use of a surface plasmon resonance fluoroimmunoassay, Biosens. Bioelectron. 1991, 6, 201-214

[113] Vasdekis A.E. and Laporte G.P.J. Enhancing Single Molecule Imaging in Optofluidics and Microfluidics, Int. J. Mol. Sci., 2011, 12, 5135–5156

[114] Kinkhabwala A., Yu Z., Fan S., Avlasevich Y., Mullen K. and Moerner W.E. Large single-molecule fluorescence enhancements produced by a bowtie nanoantenna, Nat. Photonics, 2009, 3, 654–657

[115] Galban J., de Marcos S., Castillo J.R., Fluorometric-enzymatic lactate determination based on enzyme cytochrome b2 flurescence, Anal. Chem. 1993, 65, 3076-3080

[116] Zhu A., Romero R., Howard P.R. a sensitive fluorimetric assay for npyruvate, Anal. Biochem. 2010, 396, 146-151

[117] Haghighi B., Bozorgzadeh S., Fabrication of a highly sensitive electrochemiluminescence lactate biosensor using ZnO nanoparticles decorated multiwalled carbon nanotubes, Talanta, 2011, 85, 2189-2193

[118] Han S., Li H., Hu L. electrochemiluminescence of tris(2,2′-bipyridyl)ruthenium(II)/pyruvate system in the absence of cerium(III), Anal. Methods, 2010, 2, 479-483

[119] Broder G,Weil MH (1964) Excess lactate - index of reversibility of shock in human patients. Science 143 (3613):1457–1459

[120] Wu M.H., Cai H.Y., Xu X., Urban J.P.G., Cui Z.F., Cui Z. A SU-8/PDMS hybrid microfluidic device with integrated optical fibers for online monitoring of lactate, Biomed. Microdevices, 2005, 7(4), 323–329

[121] Wu M.H,, Wang J.B., Taha T., Cui Z.F., Urban J.P.G., Cui Z. Study of on-line monitoring of lactate based on optical fiber sensor and inchannel mixing mechanism, Biomed. Microdevices, 2007, 9(2), 167–174

[122] Marquette C.A., Blum L.J. Luminol electrochemiluminescence based fibre optic biosensors for flow injection analysis of glucose and lactate in natural samples, Anal. Chim. Acta, 1999, 381(1), 1–10

[123] Zheng X.T., Yang H.B., Li C.M. Optical detection of single cell lactate release for cancer metabolic analysis, Anal. Chem. 2010, 82(12), 5082–5087

Chapter 12

Nanomaterials for Advancing the Health Immunosensor

Blanca A.G. Rodriguez, Erika K.G. Trindade,
Diego G.A. Cabral, Erika C.L. Soares,
Cayo E.L. Menezes, Danielle C.M. Ferreira,
Renata K. Mendes and Rosa F. Dutra

Additional information is available at the end of the chapter

http://dx.doi.org/10.5772/61149

Abstract

Nanotechnology has exerted a significant impact in the development of biosensors allowing more sensible analytical methods. In health applications, the main challenge of the immunoassay is to reach the suitable limit of detection, recognizing different analytes in complex samples like whole blood, serum, urine, and other biological fluids. Different nanomaterials, including metallic, silica and magnetic nanoparticles, quantum dots, carbon nanotubes, and graphene, have been applied, mainly to improve charge electron transfer, catalytic activity, amount of immobilized biomolecules, low-background current, signal-to-noise ratio that consequently increase the sensitivity of immunosensors. Given the great impact of nanotechnology, this chapter intends to discuss new aspects of nanomaterials relating to immunosensor advancement.

Keywords: Immunosensors, immunoassay, nanosensor, nanomaterial

1. Introduction

A major challenge faced by health programs is the maintenance and availability of diagnostic tests that are required not only in inpatient or outpatient hospital but also for an improved epidemiological survey. In many cases, the absence of laboratory testing or delay of diagnosis generates negative economic impacts, resulting in unnecessary hospitalization, intercurrence, and in some cases implications on the global life quality of patients and underreporting of surveillance. In this context, the development of practical, fast, and reliable analytical methods is imperative.

Biosensors have been considered one of the more attractive analytical methods. They are biodevices capable of transforming an interaction with specific analytes into an electrical signal by a transducer, including a biorecognition element. [1] Pharmaceutical industries and users of rapid tests from the United States and Europe are unanimous in stating that biosensors, mainly those based on point-of-care testing (POCT), or bedside testings are a practical technology, regarded as a short-, medium-, and long-term trend. Among several advantages, POCT can provide immediate responses (results in few minutes or in real time), samples do not need to be transported to the analytical phase (in situ monitoring) and require generally small volumes of samples, and the users can be skilled or unskilled and present better cost-effective analyses compared with conventional technologies used in clinical diagnostic (user-friendly technology). One of the most widely useful POCT is the glucometer, which measures glucose levels with accuracy by requiring a single drop of blood. The rapid glucose measurement is very important in trials to avoid serious adverse effects stemming from diabetes, including seizures, coma, or even death. Worldwide, some diabetic outpatients have been benefited by POCTs.

Although there is a great promising market dedicated to health for the detection of diseases and therapeutic monitoring, biosensors are not yet entirely broadcast, especially those devoted to nonenzymatic reactions, i.e., biosensors based on the affinity between antigen–antibodies, DNA–RNA, DNA–DNA, etc. So far, there are focused studies to develop affinity biosensors for a wide number of applications. Some of these include environmental, agriculture, veterinary, safety food analysis, and health diagnostic in attempting to detect pesticides in water, in monitoring of environmental pollutants in soil, and in determining contaminants and pathogens in food and many others.[2]

Regarding the health diagnostic, affinity biosensors devoted to immunoserological diagnosis have demonstrated to be more accurate, feasible, practical, and advantageous for POCTs than nucleic acid biosensors. First, the levels of antibodies or antigens circulating in whole blood, serum, or other biological fluids are in higher amounts compared to RNA or DNA sequences. Second, blood samples of immunosensors do not need cell lysis before measurements to release the analytes. Third, antigen or antibody samples do not need pretreatment before measurements as amplification by polymerase chain reaction (PCR) or transcriptase reverse polymerase chain reaction (TR-PCR). Fourth, antibodies are more chemically stable than RNA or DNA sequences that are easily contaminated by attacking the RNases or DNases enzymes present in digital samples.

Due especially to nanotechnology, biosensors dedicated to immunoserological diagnosis have emerged, in the last decade, with the possibility of very promising point-of-care diagnosis. The contribution of nanomaterials has made possible the development of new immobilization matrices with improved features, increased sensor surface area, greater amount of biomolecules per area/volume, and major electrical conductivity, making it possible to achieve a lower limit of detection compared to existing bulk biosensors. Currently, several studies have highlighted the following nanomaterials: metallic, silica and magnetic nanoparticles, quantum dots (semiconductor nanoparticle), carbon nanotubes, graphene, and nanostructured surface.

Selective monoclonal antibodies, recombinant antigens, fragments, and aptamers associated with the nanomaterial advancements to mediate the antigen–antibody responses have also allowed several nanostructured devices with optical, piezoelectric, and electrochemical improved transductions, besides the integration of microfluidics and portable approaches.

Given the great impact of nanotechnology, this chapter intends to discuss new aspects of nanomaterials concerning to the development of immunosensors that resulted in more accurate, reliable, and practical analytical methods for health.

2. Important aspects for immunosensor development

Immunosensor technology has shown an exponential growth in the number of publications over the last decade (Figure 1(a)). Although, there were significant advances in all the areas mainly in the food analyses, immunosensors devoted to health still have huge challenges to overcome in order to yield commercial uses (Figure 1(b)). Some difficulties can be attributed to the biomolecules specificity, immobilization matrix stability, transduction mode employed and pretreatment of complex samples like whole blood, serum, or other fluidic biologics for a reliable detection.

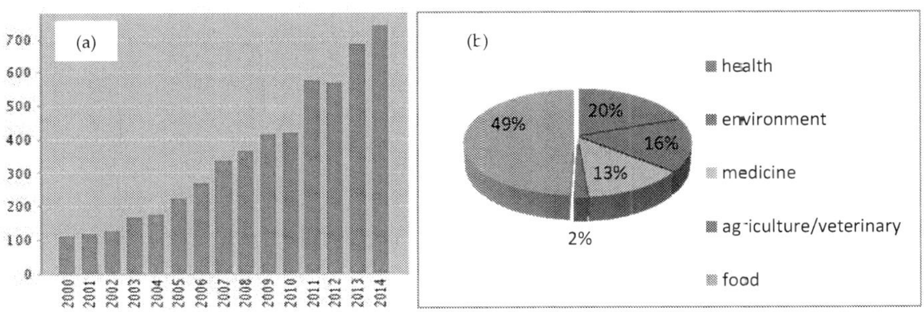

Figure 1. Immunosensors research published over the last decade (a) and main application areas (b). (Extracted from ISI of Knowledge base)

Three aspects are considered crucial in the development of an ideal immunosensor: (a) the bioreceptor, i.e., biomolecule used to recognize the antigen or antibodies in the sample; (b) the matrix assembled for immune compound immobilization; and (c) the transducer type employed.

The choosing of bioreceptors for analyte recognition is a fundamental aspect to ensure an optimal selectivity of immunosensor. Different immunomolecules have been used to detect antigens or antibodies in different samples, besides monoclonal or polyclonal antibodies, and antigens, recently recombinant antibodies, [3] aptamers, [4] and antibody fragments [5] have been also assayed. Immunoglobulin classes IgG and IgM are the most commonly employed in immunosensors. IgG is a Y-shaped structure with two binding sites for antigens recognizing (two paratopes), with approximately 150 kDa. Meanwhile, IgM immunoglobulins are pentamers comprising of ten antigen sites, called natural antibodies. However, due to IgG being more prevalent and most abundant in the circulation (73%), this immunglobulin is more frequently used in all immunoassays. IgM immunoglobulins are detected in specific assays when is important to identify diseases in their acute Phase, Kidwai et al. [6] developed a rapid immunochromatographic (ICT) assay detection for IgM and IgG detection in serum.

Immunosensor performance is directly dependent on the immobilization matrix used and orientation and density of affinity biomolecules (antibodies and antigens) on the electrode surfaces. There are different strategies used to immobilize the recognition element, either directly on the electrode surface or on other solid supports. [7] Conventionally, there are noncovalent and covalent techniques employed to immobilize antibodies, which are based on adsorption, encapsulation, and entrapment in polymers, covalent binding, and cross-linking of antibodies aggregates (Figure 2). Developments in these techniques have great interest and potential application in other areas of biotechnology, including purification of proteins, [8] medicine and drug delivery, [9] regenerative medicine, tissue engineering, and many other applications. [10]

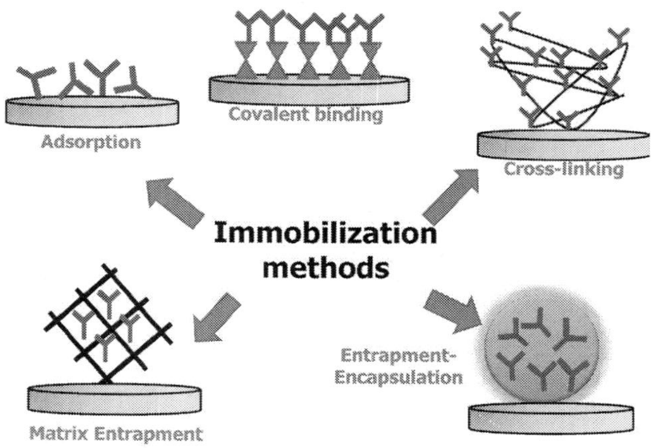

Figure 2. Illustration of different methods of antibodies immobilization.

Although sensor surfaces prepared with antibodies immobilized in a random manner yield satisfactory results, the site-directed immobilization of the sensing molecules significantly improves the immunosensor sensitivity. [11] In this sense, antibodies should be immobilized with optimal capability to recognize the antigens, while fully maintaining their preserved structures. The Fab region needs to be sufficiently free in order to be exposed to the medium, i.e., epitopes of antigens. The best approach is to immobilize antibodies by their Fc regions. [12] This configuration has been achieved using different strategies, including by the use of protein A [13] or protein G, [14] via covalent immobilization through the oxidized sugar chains of the antibody, [15] and others. Besides orientation, it's important to consider antibody density on the electrode surface. A higher density increases the sensor response, however is likely to increase the steric hindrance on planar substrates causing a low immobilization efficiency and low assay sensitivity. To solve these problems, researchers are focused on modifying the substrates for forming the 3D network, which ensures high percentage availability of antibody binding sites. Nanomaterials contribute to increase the amount of protein immobilization because of their capability to form 3D nanostructured surfaces with innumerous cavities and valleys.

Choosing the transducer is another important and fundamental aspect to achieve the sensitivity and response time desired. Bioaffinity sensors (immunosensor) have been explored by using different transduction modes: optical, acoustic and electrochemical by using different approaches. Surface plasmon resonance, [16] localized surface plasmon resonance [17] and fluorescence resonance energy transfer (FRET) [18] are examples of optical transducers. Quartz crystal microbalance (QCM), also entitled mass-sensitive, is the most explored as acoustic transducer. Electrochemical transducers are comprised of different ways to generate an electrical signal, for instance, by amperometric, impedimetric, potentiometric and capacitive changes. [19] Regarding the response time, two classes of immunosensor operation mode are distinguished by, a) Label-free or nonlabeled immunosensors that readily convert the species interaction response with the complementary species into an electrical signal, denominated as direct transduction, and b) Labeled immunosensors that need a second antibody or antigen conjugated to chemical species to generate the analytical response, such as enzymes, fluorescent labels, etc. [20] Although labeled immunosensors are more time consuming than label-free immunosensors, they provide more specificity due to the second antibody which minimize the nonspecific binding negative effects.

The design of label-free affinity biosensors has been extensively studied in academy and industry. One source of stimulation is the demand of POCTs for health, which requires rapid response, lower cost-effective analyses and simplicity for potential analysis. The main technologies of label-free immunosensors currently in use or under development are: surface plasmon resonance (SPR) devices, mass-sensitive, field effect transistor (FET), and electrochemical sensors, including impedimetric and capacitive. Recently, due to advances of nanomaterial based-immunosensors, new categories of label-free amperometric sensors using screen printed electrodes have been successfully developed. [21, 22, 23, 24]

3. Immunosensor based on nanomaterials

Nanomaterial is composed of unique functional materials that display incomparable characteristics related to their shape, structure and size (in the order of 1 to 100 nm). Nanostructured materials are interesting because they can bridge the gap between the bulk and molecular levels and lead to entirely new avenues for applications, especially in electronics, optoelectronics and biology. The contribution of nanomaterials has allowed powerful immunosensor assemblies, creating platforms with increasing detection limit. [25]

In recent years, various nanomaterials with different physical and chemical properties have been applied to achieve the immobilization of immunocompounds. They can modify the sensing surface, improving the immobilization of procedures and transduction properties of immunosensors. A great number of electrochemical advantages have been mentioned, such as possessing low-background current, high signal-to-noise ratio, and fast electron transfer, including an increased amount of immobilized biomolecules, with consequent increase on the sensitivity of sensors. Nanomaterials with zero-dimensional space (metallic, silica, and magnetic nanoparticles and quantum dots or semiconductor nanoparticles), one-dimensional space (carbon nanotubes), and two-dimensional space (graphene) have been show as potential for different transducers in many immunosensor applications.

3.1. Metallic, silica, and magnetic nanoparticles

Nanoparticles (NPs) obtained from commercial sources or properly produced in laboratories have attracted much attention in biological studies due to their low toxicity, biocompatibility, and unique optical properties. Nanoparticles and nanospheres can be divided into magnetic, metallic, semiconducting, or insulating nanoparticles based on their conductivity.

NPs have high surface areas and unique physical–chemical properties that can be easily tuned, making them ideal candidates for developing immunosensors devices. The basic function of nanoparticles in an immunosensor can be summarized as follows: immobilizing the biomolecules on the electrode surface, catalyzing electrochemical reaction, enhancing electron transfer charge, and acting as a reactant or labeling biomolecules for further experiments, among others. [26]

Biological tests measuring the presence or activity of selected analytes become quicker, more sensitive, and flexible when nanoscale particles are combined, with numerous advantages over more traditional procedures. In recent years, various nanomaterials with distinct physical and chemical properties have been applied to improve the immobilization of immunocompounds. [27] These have many electrochemical advantages, such as possessing low-background current, high signal-to-noise ratio, and fast electron transfer, besides increased amount of immobilized biomolecules, with consequent increase on the sensitivity of sensors. [28]

Surface modification using nanoparticles composites have shown an increase of sensitivity and help adsorb a large amount of antibodies on electrode surface. Lu et al. [29] constructed an immunosensor based on a nanocomposite formed with CeO_2 and gold nanoparticles on the Au electrode via cysteine to detect a cardiac marker, the myeloperoxidase.[29] Thereat, the nanoparticles enhanced the active surface area available for antibody binding. The high stability of this sensor was attributed to the good biocompatibility of the composite. Another study shows an increase in immunosensor response. Fe_2O_3 nanoparticles were used in the construction of an electrochemical device to detect cancer biomarker prostate antigen (PSA) via horseradish peroxidase (HRP) signal. The high amount of nitrodopamine (film coated on nanomaterial to immobilize the anti-PSA) anchored onto Fe_3O_4 increased the loading of biomolecules onto the surface, which increased the electrochemical immunosensor sensitivity. [30]

The carcinoembryonic antigen (CEA) is a protein used as a tumor marker and has been frequently investigated in immunoreactions. An elevated CEA level in serum may be an early indication of lung cancer, ovarian carcinoma, colon cancer, breast cancer, and cystadenocarcinoma. Recently, an interesting work was reported involving this protein investigation using an immunosensor constructed by Pt hollow nanospheres modified with anti-CEA as label for a 3D Au-TiO_2 hybrid platform. [31] The immunoassay exhibited a high sensitivity and a low detection limit compared with conventional label methods. Another way to detect CEA antigen was developed by Gao et al. [32] using a label-free voltammetric sensor with chitosan and gold nanoparticles (AuNPs) to immobilize anti-CEA on carbon surface. The detection is based on the variation of current responses before and after the immunoreaction. When the immobilized antibodies have bounded with antigens, the antigen–antibody complex formed on the surface inhibited the electron transfer. Then a decrease of the electrochemical signal was verified as the concentration of antigen on surface increased. Another method also using a composite of chitosan and AuNPs for CEA determination, but with multiwalled carbon nanotubes, was described by Huang et al. [33] The nanocomposite film exhibited high current response intensity, good biocompatibility, and high stability. Similar CEA detection was also performed using a gold nanoparticle–thionine-reduced graphene oxide composite that possesses as advantage fast electron transfer kinetics and large specific surface area. [34] Another work for CEA analysis described the use of silver nanoparticles on SiO_2 surfaces. [35] The high stability of the immunosensor was assigned to the stable nanocomposite produced.

The sensitivity of electrochemical immunosensors can also be improved by using the association of AuNPs and dendrimers that are three-dimensional macromolecules, with hundreds of functional groups at the periphery, for surface modification. This architecture was employed by An et al. [36] to detect α-synuclein, a very important neuron protein. The dendrimer (PAMAM)-encapsulated AuNPs were covalently bound on the poly-o-aminobenzoic acid (ABA) electropolymerized on a glassy carbon electrode surface to achieve abundant carboxyl groups, which allowed a highly dense antigen immobilization and facilitated the improvement of electrochemical responses as well. Subsequently, the enhanced gold nanoparticle labels were fabricated by immobilizing a horseradish peroxidase secondary antibody (HRP-Ab_2) on the AuNPs surface. After an immunoassay process, the labels were introduced onto the electrode surface to produce an electrocatalytic response with hydrogen peroxide. The presence of dendrimer Au not only increased the covalent coupling of more protein but also accelerated electron transfer when compared to immunosensor without dual signal amplification strategy.

The picogram detection limit of estradiol was achieved using an immunosensor constructed with AuNPs and protein G scaffold to modify a gold electrode. [37] Coupled with the amperometric determination of the hormone in a flow system, the device exhibited superior linear range, sensitivity, and stability in blood serum samples spiked with estradiol.

Other applications for metallic nanoparticle have included optical transduction. Krishnan constructed an optical immunosensor in a quartz glass surface for the detection of *Escherichia coli*, using core-shell nanoparticles (silver-silica) anchoring labeled antibodies. The results show that changes in photoluminescent standards are consistent with the immobilization of various species. Thus, the optical immunosensor demonstrated improved sensitivity and specificity in comparison to the usual methods, detecting as low as 5 CFU/mL.

Using a great number of luminescence molecules as stabilizers coated on the surface of the AuNPs, Shen et al. [40] developed an electrochemiluminescence immunosensor to detect human cardiac troponin, an important acute myocardial infarction biomarker. First, the sensor was constructed by using streptavidin-coated gold nanoparticles as the immobilization matrix for biotinylated antibody. Meanwhile, the three-dimensional nanostructures increased the surface-to-volume ratio, allowing more biomolecules to be immobilized. The sandwich-type immunosensor was fabricated by reacting with antigen and AuNPs modified with luminescence molecules labeled with the secondary antibody, forming a nanoprobe. The enhanced sensitivity of the proposed apparatus mainly derives from the novel nanoprobe, which achieves a large amount of luminescence molecules loading toward each sandwich immunological reaction event.

Another strategy in immunocomplex detection involves the use of magnetic nanoparticles as solid support for biomolecule immobilization. The magnetic particles offer the convenience of magnetic separation. These particles respond to a magnetic field but demagnetize completely when the field is removed. Thus, the nanoparticles can easily be separated from the liquid phase with a small magnet but can be dispersed again immediately after the magnet is removed. The use of magnetic nanoparticles as solid phase for the immunosensor development improves the bioreaction performance due to surface area increase and has better immunoassay kinetics because the particles are in suspension and the target species does not need to diffuse very far. [41]

An interesting work was described by Shen et al., [42] who developed a device to detect *E. coli*, an intestinal pathogenic bacterium, using a quartz crystal microbalance (QCM) immunosensor based on beacon magnetic nanoparticles. A polyclonal antibody was immobilized on iron nanoparticles with subsequent addition of *E. coli*. AuNPs were inserted in the system to amplify the signal. Weakly bound biomolecules were removed with a magnetic plate. Finally, the crystal was modified with protein A and monoclonal antibody. The frequency shift of the QCM immunosensor is amplified using *E. coli* immobilized on to magnetic particles and enlarged gold particles for the bacterium detection. The signal was amplified three times, and the crystal was regenerated without difficulty and could be used at least 10 times. In a recent work, the use of magnetic nanoparticles as an amplification means for QCM signal for avian influenza H5N1 virus detection has been reported. [43] Polyclonal antibodies against the virus were immobilized on the gold surface of the crystal through self-assembled monolayer (SAM). Target H5N1 viruses were then captured by the immobilized antibodies, resulting in a change in the frequency. Magnetic nanoparticles coated with anti-H5 antibodies were used for further amplification of the binding reaction between antibody and antigen (virus).

AuNPs have a remarkably high extinction coefficient and strong distance-dependent optical properties. Different aggregation states of AuNPs correspond to distinctive color, which can be appreciably discerned with the naked eye and be used in immunoassay. Based on this, Yuan et al. [44] developed a label-free colorimetric immunoanalysis for the simple detection of neurogenin3, a marker for pancreatic endocrine precursor cells, using glutathione functionalized gold nanoparticles. The antibody-conjugated AuNPs were formed through electrostatic interaction upon the addition of the antibody to the modified AuNPs solution. The antigen positively charged to the negatively charged AuNP antibody will minimize the electrostatic repulsion between nanoparticles by neutralizing the surface charge and then agglomeration is induced by an increasing NaCl salt concentration, noticeably revealed by the color change of the solution from red to purple or blue. The concentration of neurogenin3 can be conveniently accessed by the optical absorption spectra. Another important property of the AuNPs is that they could catalyze silver reduction and act as the nuclei for silver precipitation. [45] In this interesting work, the core mechanism of the method to quantify cardiac troponin is that the catalytic capability of the AuNPs was inhibited by immunocompounds covering their surface. This covering is influenced by the amount of reduced silver of the reaction, resulting in a color difference.

In state-of-the-art improved sensor devices for health applications, the possibility of assembling nanoparticles and biomolecules in different ways by using different sizes, formats, and compound types allow more sensitive, simple, robust, and especially faster analysis.

3.2. Quantum dots

Quantum dots (QDs) represent one class of nanostructured materials. They are spherical nanocrystals of semiconductor, 1–10 nm in diameters, made of elements of the IIB–VIA or IIIA–VA groups. The use of QD properties requires sufficient control during their synthesis because their intrinsic properties are determined by different factors, such as size, shape, defect, impurities, and crystallinity. [46, 47] Two of the most widely used commercial QDs come with a core of CdSe or CdTe and a shell of ZnS and emissions from 405 to 805 nm [48, 49] The shell stabilizes the structure, helping to overcome quenching compared to a QD made only from a core and provides a large surface area available for further modification.

Analogous dimensions of QD and biological materials, such as enzymes, antigens/antibodies, protein receptors, or nucleic acids, show great promise as photonic labels for bioanalytical applications and suggest that electronic communication between the QD and the specific recognition site or biocatalytic processes of the biomaterials can exist. These electronic interactions may lead to the optical or photoelectrochemical transduction of the biological events. [47, 50]

Generally, monodispersed QDs are developed by introducing organic molecules that adsorb on the surface and act as capping agents. The efficacy of QD in a biological application is critically dependent on coating properties. The liabilities of these initial methods require the continued development of QD coatings. Important criteria for an ideal QD coating include high-affinity for the QD surface, long-term colloidal stability across a broad range of pH and ionic strengths capacity for bioconjugation, minimization of hydrodynamic size, and biocompatibility with nonspecific binding. [51] However, the selection of organic ligands that bond with surface atoms of the QD is a very delicate issue. In general, phosphenes or mercaptans (-SH) are the most widely used ligands. [52]

In order to make QD suitable for biological imaging and use in a biological environment, they also have to be rendered water soluble. This is done by capping the shell with a polymer layer that contains a hydrophobic segment facing inward, the shell, and a hydrophilic segment facing outward. The hydrophilic layer can be modified to include functional groups such as –COOH and $-NH_2$ groups for further conjugation to proteins and antibodies or oligonulceotides. [49, 53] The single-step synthesis of thiolated cyclodextrin-modified CdSe/CdS core-shell QD resulted in a water-soluble QD, keeping the luminescence properties of the QD in aqueous systems. This is an important aspect since biorecognition events require aqueous environments for reaction. [54]

Thiol ligands and amphiphilic polymers are the most common types of QD coating available. They allow two essential design elements: a moiety that anchors to the QD surface and a hydrophilic functionality for aqueous dispersion. The selection of these groups determines the degree to which a QD coating can approach the ligand/amphiphilic polymer structures. [51] For example, small molecules with thiol groups can bind to the quantum dot surface, and distal carboxylated group provides aqueous colloidal stability. [55] Another strategy for QD coating that provides aqueous dispersion, improves the biocompatibility, and minimizes nonspecific binding was developed by Mattoussi et al. (2000). They combined dihydrolipoic acid, a dithiol ligand that binds the QD more closely, which is attached to a poly(ethylene glycol) oligomer.

Biomolecule conjugation on to the QD is achieved by different ways like electrostatic binding, noncovalent biotin–streptavidin bonding, or covalent bonding. The most widely used conjugation technique of all is the covalent bond formation between the QD surface and the biomolecules. Surface modifications on QD allow easier covalent bond formation. In one of the most widely used methods, amine-terminated QDs are used for conjugating antibodies. The amine-terminated QDs are activated with maleimide containing a cross-linker molecule, which can then be conjugated to a fragment or whole antibody molecule. Some of the most commonly employed QD conjugation methods are based on cross-linking reactions between amine and sulfhydryl groups, carboxylic acid, and amine and aldehyde and hydrazide groups. The carboxylic-amine bond has one advantage over all other methods, seeing as this method does not require any antibody modification before QD conjugation. In the case of amine and sulfhydryl bond formation, the antibody should be reduced to expose their interchain -SH bonds. In relation to aldehyde and hydrazide bonds, carbohydrate groups on the antibody Fc portion are oxidized. These modifications on antibodies may affect their performance to a certain extent. [56]

Functionalized semiconductor quantum dots have been used as fluorescence labels in numerous biorecognition events. For example, Liu et al. (2004) developed an immunosensor with simultaneous measurements of four proteins based on antibodies linked to the inorganic nanocrystal. Stripping voltammetric immunoassay was used to observe the response of a mixture containing microglobulin, IgG, bovine serum albumin, and C-reactive protein connected with ZnS-, CdS-, PbS-, and CuS-labeled antibodies, respectively. The system was obtained by using carbamate linkage for conjugating the hydroxyl-terminated nanocrystal with the secondary antibody. [57]

Li et al. (2011) [58] used a novel strategy to modify the surface of graphene quantum dots composites. A layer-by-layer assembling process was employed via electrostatic interactions between negatively charged thioglycolic acid modified CdSe QD and positively charged graphene, which was noncovalently functionalized with poly (diallyldimethylammonium chloride) (PDDA) via an exfoliation in situ reduction of graphene oxide in the presence of PDDA. This process allowed excellent conductivity, extraordinary electron transport properties, and large specific surface area, which resulted in high electroluminescence (ECL) intensity and excellent film-forming ability and made it a promising candidate for the development of ECL immunosensors.

Luminescent quantum dots are viable optical markers and have been used in a direct assay for IgG. Protein A was labeled with CdSe/Zn QD (λ_{max} of 655 nm) and then was immobilized at the tip of an optical fiber. Once the immunoreaction with IgG occurs, a decrease in fluorescence intensity is observed as a result of the fluorescence resonance energy transfer from the QD to the bound protein. [59] Lingerfelt et al. [60] reported the preparation of QD-biotin conjugates and their use in immunochromatographic assays. The detection of immunoglobulin G was carried out on a glass chip through a sandwich assay approach using a secondary antibody conjugated to the QD. [61] A sandwich immunoassay for the detection of staphylococcal enterotoxin B was run using polyclonal sheep anti-staphylococcal enterotoxin B antibody conjugated with QD and microtiter plates coated with monoclonal staphylococcal enterotoxin B antibody. [62]

Kerman et al. (2007) applied conjugated QD streptavidin in a model immunoassay system for the detection of a total prostate-specific antigen cancer marker from the spiked and undiluted serum samples. Immunorecognition was carried out on a carbon substrate using a sandwich assay approach. After the recognition event, the substrate was exposed to the biotinylated secondary antibodies and, subsequently, fluorescence imaging of the substrate surface illuminated the QD. [63]

QDs based on narrow photoemission spectra, with high resistance to photobleaching and broad excitation spectra, are widely used as tags in immunoassay. A carcinoembryonic antigen immunosensor was fabricated using biofunctionalized QD probes. This immunosensor array was designed to detect a wide range of analytes using the inherent characteristics of QD and the flexibility of engineered elastin-like polypeptides. [64]

There are some studies based on thioalkyl-functionalized QD, which are pH sensitive, [65] suggesting many different biological applications. In this context, mercaptoacetic acid-CdSe/ZnSe/ZnS QDs have been used as an intracellular pH sensor by observing a quenching of fluorescent QDs in acidic pH. [66]

Another approach was based on the direct conjugation of CdSe/ZnS QD–IgG complexes using a genetically engineered tripartite fusion protein. This fusion protein was made up of a histidine tag for QD conjugation, an elastin-like peptide for stimuli-responsive purification and the protein L (cell-wall component of *Peptostreptococcus magnus*) that has high affinity to IgG. The functionality of this sensitive immunofluorescent probe was demonstrated in the detection of a representative tumor antigen. [64]

Despite recent progress, more work still needs to be done to achieve reproducible and robust surface functionalization and develop flexible bioconjugation techniques. The potential of QD in biology has just begun.

3.3. Carbon nanotubes

Carbon nanotubes (CNTs) can be highlighted as the most important nanomaterials for biosensors. Their excellent optical and mechanical conductivity, high surface-to-volume ratio, good chemical stability, biocompatibility, and easy functionality have revolutionized the biosensors area for the last decade.

Since their discovery by Ijima in 1991, CNTs are being used in large volumes for different purposes in many industrial areas, i.e., in nanocomposites for sporting materials, as a battery in supercapacitors, transparent films, and liquid crystal displays. Other limited-volume carbon applications include their use as components in wind turbine blades, scanning probe tips, membrane filters and sorbents, flat panel displays, memory devices, transistors, drug delivery systems, and other medical and analytical chemistry applications. [67]

Carbon nanotubes can be described as hollow cylindrical tubes of graphene sheets with high aspect ratios (length/diameter). [68] The structure of graphene is a planar atomic sheet consisting of covalently bonded carbon atoms. The atoms in graphene are sp^2 carbon units, forming a two-dimensional (2D) network with a hexagonal lattice. A graphene layer wrapped as a cylinder forms a single-walled carbon nanotube (SWNT). A multiwalled carbon nanotube (MWNT) is nothing more than multiple SWNTs packed in a tight concentric frame. All the carbon nanotubes have several nanometers in diameter and many microns in length. SWNTs have the smallest diameter (0.8–5 nm), whereas MWNTs have a larger diameter (~3 to >100 nm), both variable in length (from millimeters to tens of nanometers). The proper architecture is reflected in the highly anisotropic properties. Most of the extraordinary electrical, thermal, and mechanical characteristics are localized specifically along the axial direction. The strong sp^2 bonding between the carbon atoms in CNTs yields remarkable mechanical strength, making them one of the most resilient materials. Moreover, it is known that an SWNT presents metallic and semiconducting properties where such electronic features depend on its chirality. [69] They have three different structures: armchair, zigzag, and chiral. [70]

The applications of CNTs in biosensors have been hindered for a long time due to the drawback of insolubility. CNTs present a high molecular weight, an ability to entangle (tendency to individually interact with each other through van der Waals forces), aggregating into bundles and ropes. However, these bundles can be quite large that they become insoluble in any solvent; thereby, it can be difficult to disperse them in either aqueous or nonaqueous medium. [71]

Ultrasonication is an effective method to disperse CNTs in liquids that have low viscosity, such as water, acetone, and ethanol. However, most polymers are either in a solid or viscous liquid state, which require the polymer to be dissolved or diluted using a solvent to reduce the viscosity before dispersion of CNTs. [72] The simplest stable dispersions have been achieved by using a solvent able to efficiently interact with CNTs, such as phenylethyl alcohol, *N*-methylpyrrolidone (NMP), *N,N*-dimethylformamide (DMF), and *N,N*-diethylacetamide (DEA). An additional strategy to favor dispersion in organic solvents is to coat CNTs with a molecule characterized by a high affinity toward nanotube sidewalls and at the same time soluble in the selected solvent. Both small molecules and polymers formed by repetitive units of alkyl chains and aromatic compounds have been used as dispersants. Thus, the adsorption of different polycyclic aromatic hydrocarbons such as pyrene, anthracene, tetracene, and phenanthrene on SWCNTs has been extensively investigated. In order to favor the dispersion of CNTs in water, the widely and most used approach is the adsorption of surfactants. These small molecules typically have a hydrophobic tail and a hydrophilic head group—the former is intended to favor adsorption onto the hydrophobic carbon nanotube and the latter to promote affinity with the aqueous solvent. Over the years, stable aqueous CNT dispersions were obtained with differently charged and nonionic surfactants such as sodium dodecylbenzene sulfonate (SDBS), cetyltrimethylammonium p-toluenesulfonate (CTAT), cetyltrimethylammonium bromide (CTAB), and sodium cholate (SC) enhanced by sonication. Additionally, polymers have been employed for CNT dispersion in water. The majority of polymers and block copolymers have been used to wrap CNT by exposing their polar domains toward the aqueous environments while favoring the contact of their hydrophobic domains with the nanotube surface. [73]

Strategies for the immobilization of biomolecules on CNTs have been widely explored aiming to improve sensitivity on an immunosensor. The high aspect-ratio of CNT allows a great amount of anchored biomolecules by noncovalent and covalent functionalization for different types of transduction (Figure 3).

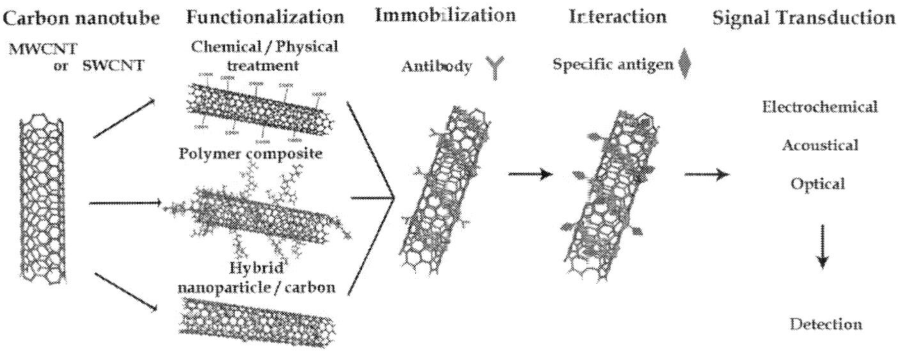

Figure 3. The schematic diagram steps and transducer types of immunosensors based on carbon nanotubes.

Noncovalent functionalization enables reversible adsorption of biomolecules on the CNT surface. For this purpose, CNTs are added to a dispersant solution, and the mixture is agitated in an ultrasound bath. The CNTs are mechanically debundled and then stabilized by dispersant molecules through noncovalent interactions. This does not cause changes in the chemical structures, electronic, and mechanical properties of the carbon nanotubes, and therefore it is a very attractive method. Surfactants, biomolecules, and polymers are widely used as dispersants and noncovalent modifiers. Among them, the polymers are quite efficient dispersants because of their long chain structure that can wrap themselves around CNTs by disrupting the van der Waals interactions between the walls of nanotubes. In biosensors, polymers are particularly interesting and have been widely employed to prepare CNT composites for electrochemical detection, especially for conductive polymers due to their native electron trans-mediation, high conductivity, good environmental stability, and specific organic groups. Furthermore, they can be overoxidized to create an electrically insulating layer. Many reports have demonstrated that CNTs coated with polymers, including polypyrrole, poly(methylene blue), poly(neutral red), poly(acrylic acid), and poly(3-methylthiophene) have become a popular strategy [74]

The covalent chemical functionalization arises mainly from organic molecules reacting with carboxyl groups of CNTs treated by oxidation, which depends on the hydrophilicity/hydrophobicity of the species attached, which can make carbon nanotubes soluble in water or organic solvents. The modification of carbon nanotube surfaces by covalent attachments of soluble groups usually alters intrinsic properties such as conductivity, mechanical strength, and optical properties. [73, 75] Nevertheless, the functionalization involving the introduction of carboxyl, amine, thiol, and other reactive groups are attractive strategies because antibodies or antigens can be covalently immobilized, improving the stability and, in some cases, the sensitivity and selectivity of the immunosensors. Figure 4 exhibits different covalent and noncovalent methods of functionalization of the carbon nanotubes.

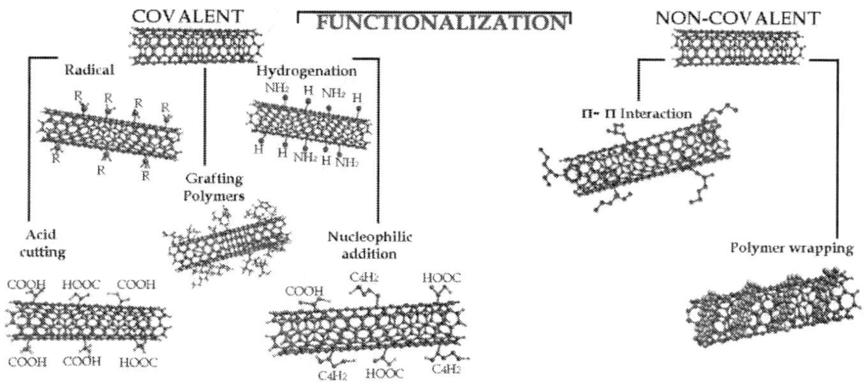

Figure 4. Illustration of different carbon nanotube functionalization methods.

In some practical applications, Sánchez and coworkers [76] have constructed immunosensors where the biomolecules are immobilized on an MWCNT–polysulfone composite film. The layer was applied onto screen printed working electrodes to provide a suitable immunosensor for the rapid determination of human chorionic gonadotropin hormone. The detection limit was 14.6 mIU/mL with a linear range up to 600 mIU/mL. Viswanathan et al. [77] developed another disposable electrochemical immunosensor based on CNTs for the detection of carcinoembryonic antigen with a detection limit of 1 pg/mL in saliva and serum.

For better detecting performance toward interleukin-6, in cases of oral cancer, Malhotra et al. [78] made an ultrasensitive immunosensor sandwich assay on an electrically conductive and high surface area platform, featuring densely packed and upright SWCNTs with capture antibodies attached to their ends. This biosensor had the highest sensitivity at 19.3 nA/mL (pg IL-6)$^{-1}$ cm^{-2} and the best detection limit (DL) of 0.5 pg/mL (25 fM) for IL-6 in 10 μL of calf serum. Similarly, Munge et al. [79] have presented a novel electrochemical sensor using a sandwich immunoassay for the detection of metalloproteinase-3, a cancer biomarker, based on vertically aligned SWCNT arrays. The multilabeled polymeric bead amplification method demonstrated a detection limit of 0.4 ng/mL in 10 μL of calf serum. This showed great potential for these elements in future cancer diagnostics.

The self-assembly of oxidative SWCNTs on gold was attempted for the detection of bovine serum albumin, BSA, by cyclic voltammetry. This sensor has shown excellent sensitivity and dynamic linear response at the range of 0.1 to 1.2 μM. [80] A conductive multilayer composed of Nafion-coated MWCNTs, thionine (Thi), and AuNPs was prepared using an innovative self-assembly strategy to form an immunosensor for α-1-fetoprotein. This reagentless amperometric sensor presented broader linear response in two ranges between 0.5–20 ng/mL and 20–200 ng/mL with a detection limit of 0.26 ng/mL. [81]

There are many studies demonstrating that CNTs can provide high electrocatalytic activity to the electrochemical devices and minimize surface fouling effect. Their unique properties enable them to promote a fast electron transfer, play the role of a biomolecular immobilization platform, and be compatible with different materials for construction of different electrodes. The sensitivity of electrochemical sensors has been greatly enhanced due to these materials, which promotes high active surface area and conductivity CNTs play an important role in recent trends for immunosensor fabrication. They can function as transducers, act as carriers and labels of immunoassay due to the transfer of large amounts of electroactive species for amplifying electrochemical signals, and also offer an easy way to protect and stabilize these bioactive species. [82] In this section, different strategies were described like the easy adsorption of CNT on the electrode surface, biomolecule immobilization by simple adsorptions and covalent binding, and preparation of screen printed electrode.

Based on a simple amino-functionalization method for MWCNT, Dutra's group developed an electrochemical immunosensor for the detection of human cardiac troponin T (cTnT), an important marker for acute myocardial infarction. It showed a broad linear range (0.02 to 0.32 ng/mL) and a low limit of detection, 0.016 ng/mL. [83] Another sandwich-type immunosensor for the detection of cTnT based on carbon nanotubes supported by a conductive polyethyleneimine film has achieved a low limit of the detection of 0.033 ng/mL and a linear range between 0.1 and 10 ng/mL. [24] Amperometric response is generated by peroxidase reaction with substrate in chronoamperometry detection. The high electronic transfer and catalytic response helped by the CNT was essentially important to dispense the mediator in order to generate the analytical responses. Due to the high conductivity achieved by incorporation of CNTs in screen printed electrodes, a label-free amperometric immunosensor was fabricated, presenting new strategies based on differential pulse amperometry. The immunosensing device for cTnT, with amine-functionalized carbon nanotubes incorporated in screen-printed electrode ink, reached a lower detection limit of 0.0035 ng/Ml, better than any previously described immunologic sensors. [21]

3.4. Graphene

Graphene is a two-dimensional material, formed by carbon atoms that are densely packed in a regular sp^2-bonded atomic scale as a hexagonal pattern, [84] which was produced in laboratory for the first time in 2004. [85] This is the base construction block for other carbon allotropes such as fullerene, carbon nanotubes, graphite, nanoribbons, and others. [86] It is a transparent (optical transmittance of ~97.7%), very thin sheet with large theoretical surface area (2630 $m^2\ g^{-1}$), one atom thick, stronger than steel (mechanical stiffness of 1TPa). In addition, it is a good heat conductor (thermal conductivity of 500 $W\ m^{-1}\ K^{-1}$), chemically inert, and a semimetal with high electron transfer (charge-carrier mobility of 250 000 $cm^2\ V^{-1}\ s^{-1}$ at room temperature). [87, 88] These properties make them attractive for many applications. [84] There are a variety of synthesis methods for obtaining graphene such as chemical vapor deposition, chemical vapor deposition by plasma, the graphite intercalation of metal sheets, mechanical or thermal exfoliation of graphite oxide, intercalation, and exfoliation of graphite, among other variants of these. Despite all these syntheses methods, the mass production is still difficult, making it hard to develop some applications. [88]

Graphene oxide (GO) synthesis has been an alternative to graphene mass production. It is produced from the oxidation of graphite and has polar oxygen functional groups. GO is rich in carboxylic acids at its edges, and epoxy and hydroxyl groups at basal planes, which grants many functionalization routes and good dispersion in water. [89] Furthermore, the functional groups are responsible for the exfoliation of graphite, seeing as they increase the interplanar distance due to the formation of hydrogen bonds between the graphite sheets. The hydrogen bonds are weak and can be easily broken by ultrasound bath, resulting in monolayer or a few sheets of carbon, known as GO. This is an excellent material for biological applications attributed to the functional groups that readily interact with nucleic acids, proteins, cells, and other organic molecules. However, GO is not a good electrical conductor because of the disruption of its sp^2 bonding as functional groups increase, which can narrow its nanobiotechnology applications. To overcome these difficulties, the reduced form of GO has been chosen as an alternative.

Reduced graphene oxide (RGO) has more commonly been used to form nanocomposites with nanoparticles or polymers to develop biomedical applications such as biosensors, controlled drug delivery, therapeutic modalities for cancer treatment, substrates for antibacterial effects, scaffolds for mammalian cell culture, and gene delivery among others. [90]

In the RGO synthesis, functional groups are removed, and the conductivity is increased again. This removal can be done in different ways such as electrochemical, optical, hydrothermal, microwave, or heating procedures. These methods for removing the functional groups form different shapes and therefore the conductivity recovery is variable. Also, they form different functional groups, becoming favorable in a wide number of applications. [91]

The most common method for obtaining RGO is chemical reduction, which is used in colloidal dispersing of GO. Hydrazine monohydrate is the most used reduction agent, seeing as it does not react with water, which makes it attractive for aqueous dispersions. The reduction process mediated by hydrazine normally occurs through the addition of H_2 groups and removal of N_2, and it is gentle enough not to affect the cyano and nitro groups. The second most used reducing agent is sodium borohydride ($NaBH_4$), which is more effective than hydrazine and easily hydrolyzed in water. The hydrolysis process should be slow enough so it does not affect the reduction process. The $NaBH_4$ reduces C=O species and has a low effect in epoxy and carboxylic acid groups. Other reduction agents such as hydroquinone, alkaline solutions, and gaseous hydrogen are also being described as mediators. [92]

Another low-cost mean of producing RGO is by thermally reducing GO, heating it in a furnace at 1050 °C, which creates thermodynamically stable oxide carbon species. Electrochemistry can also be used in the reduction process of GO, removing oxygen functionalities. Thermal and electrochemical reduction techniques have the advantage of avoiding dangerous reducers and the problem with their disposal, but they are still less used than chemical reduction. The reduction processes frequently provide RGO with functional groups, but in some cases, its functionalization is still necessary prior to use. Covalent and noncovalent methods for functionalization of RGO have been studied, whereas noncovalent bonds are the most common used, for instance, the physical adsorption of both polymers and small molecules via van der Waals interactions onto the basal planes of RGO sheets. [92]

An initial and successful approach using RGO to create biosensors was its combination with nanoparticles. An example is the work of Shan et al., [93] who used Au nanoparticles associated with RGO and chitosan as a nanocomposite film onto a gold electrode for developing an electrochemical glucose sensor obtaining a linear response range from 2 to 10 mM. Copper nanoparticles were also used to modify RGO sheets to create an electrochemical sensor for glucose obtaining a detection limit of 0.5 μM. [94]

Afterward, RGO was applied to the production of immunosensors, with and without nanoparticles. An example is the work of Mao et al., [95] who reported the use of RGO sheets coated with Au nanoparticles, which were initially functionalized with human immunoglobulin G (IgG) to create conjugates. These conjugates were immobilized onto a field effect transistors (FETs) biosensor platform for the detection of human proteins.

A developing area for immunosensors is the detection of cancer markers. It is a recent and very attractive field, with growing publication numbers, including the use of RGO for these. An example is the work of Zhong et al., [96] who used a gold nanoparticle enwrapped graphene nanocomposite on a glassy carbon electrode in a sandwich-type immunoassay format. The detection limit obtained for this assay was 10.0 pg mL^{-1}. Another CEA immunosensor was developed by Huang et al. [97] using Ag/Au nanoparticles coated with RGO in a clinical immunoassay for the detection of carcinoembryonic antigen (CEA). The nanoparticles were used as means for amplification of the signal and the method showed a detection limit of 8.0pg mL^{-1} in human serum.

Different cancer markers were the focus of other works, such as the one developed by Tang et al., [98] which aimed to create an electrochemical immunosensor for the simultaneous detection of alpha fetoprotein (AFP) and carcinoembryonic antigen (CEA), using biofunctionalized magnetic RGO nanosheets (MGO) coated with iron oxide nanoparticles as immunosensing probes, obtaining detection limits of 1 pg mL^{-1} for CEA and 1 pg mL^{-1} for AFP. Also, Teixeira et al. [99] created a chemically modified epitaxial graphene diagnostic sensor for the detection of human chorionic gonadotropin, which is a main marker for pregnancy and can also indicate some types of tumors. They obtained a detection limit of 0.62 ng/mL.

For optical transductors in cancer marker detection, RGO was used by Xu et al. [100] in a modified glassy carbon electrode using luminol to create a electrogenerated chemiluminescence (ECL) immunosensor for prostate specific antigen, using two antibodies in a sandwich immunoassay, which achieved a detection limit 8.0pg mL^{-1}. RGO was also used in the development of an ECL immunosensor using CdTe quantum dots (semiconductor nanocrystals) along with Au nanoparticles for signal amplification in the detection of human IgG with detection limit of 0.005 pg/mL. [101]

4. Conclusions

The different concepts of nanomaterials applied to immunosensors have been discussed. Nanomaterials can be utilized for a wide variety of immobilization matrices intending to improve the immunosensor sensitivity, allowing lower limit of detection. The potential of nanomaterials on immunosensors has resulted in a positive impact on the clinical outcome of various diseases, including cancer, cardiac injuries, parasitic infections, and viruses, among others. It is well known that carbon nanotubes and grapheme nanostructures are more favorable to amperometric transducers due to their electrochemical proprieties, which increase the electronic transfer charge and electrocatalytic activity. Metallic and magnetic nanoparticles have successfully been applied to different transducers, especially electrochemical, by enlarging the electroactive surface area. Quantum dots, a semiconductor nanoparticle, present a promising potential for many transducers mainly due to their photostability and luminescence characteristics. Nevertheless, more challenging studies involving nanomaterial sciences, biochemistry, electronic, and molecular engineering should be done in attempting to achieve faster, more practical, and more reliable biosensors. More specifically, biomolecules and a deeper knowledge in nanomaterial science associated to new electronic designs represent a promising field in the development of portable and integrated point of care devices for health applications and other areas of diagnostic.

Acknowledgements

The authors thank the financial support of the National Council for Scientific and Technological Development (CNPq), CAPES (Coordination for the Improvement of Higher Level Personnel), and the Foundation for Research of the State of Pernambuco (FACEPE).

Author details

Blanca A.G. Rodriguez[1], Erika K.G. Trindade[1], Diego G.A. Cabral[1], Erika C.L. Soares[1], Cayo E.L. Menezes[1], Danielle C.M. Ferreira[2], Renata K. Mendes[3] and Rosa F. Dutra[1*]

*Address all correspondence to: rosa.dutra@ufpe.br

1 Laboratory of Biomedical Engineering, Federal University of Pernambuco, Recife, Brazil

2 National Center for Research in Energy and Materials (CNPEM), Campinas, Brazil

3 Pontifical Catholic University of Campinas, São Paulo, Brazil

References

[1] Morgan, C.; Newman, D.; Price, C., Immunosensors: technology and opportunities in laboratory medicine. *Clinical Chemistry* 1996, *42* (2), 193–209.

[2] Holford, T. R. J.; Davis, F.; Higson, S. P. J., Recent trends in antibody based sensors. *Biosensors and Bioelectronics* 2012, *34* (1), 12–24.

[3] Zeng, X. Q.; Shen, Z. H.; Mernaugh, R., Recombinant antibodies and their use in biosensors. *Analytical and Bioanalytical Chemistry* 2012, *402* (10), 3027–3038.

[4] Lim, Y. C.; Kouzani, A. Z.; Duan, W., Aptasensors: a review *Journal of Biomedical Nanotechnology* 2010, *6* (2), 93–105.

[5] Rousseaux, J.; Rousseaux-Prevost, R.; Bazin, H., Optimal conditions for the preparation of Fab and F(ab)2 fragments from monoclonal IgG of different rat IgG subclasses. *Journal of Immunological Methods* 1983, 64, 141–146.

[6] Kidwai, A. A.; Jamal, Q.; Mehrunnisa. S.; Farooqi, F. U. R.; Saleem, U., Serodiagnosis of dengue infection using rapid immunochromatography test in patients with probable dengue infection. *Journal of the Pakistan Medical Association* 2010, *60* (11), 936–939.

[7] Pei, X.; Zhang, B.; Tang, J.; Liu, B.; Lai, W.; Tang, D., Sandwich-type immunosensors and immunoassays exploiting nanostructure labels: a review. *Analytica Chimica Acta* 2013, *758*, 1–18.

[8] Pfaunmiller, E.; Moser, A. C.; Hage, D. S., Biointeraction analysis of immobilized antibodies and related agents by high-performance immunoaffinity chromatography. *Methods* 2012, *56* (2), 130–135.

[9] Sanchez-Moreno, P.; Ortega-Vinuesa, J. L.; Boulaiz, H.; Marchal, J. A.; Peula-Garcia, J. M., Synthesis and characterization of lipid immuno-nanocapsules for directed drug

delivery: selective antitumor activity against HER2 positive breast-cancer cells. *Biomacromolecules* 2013, *14* (12), 4248–4259.

[10] Shalumon, K. T.; Chen, J. P., Scaffold-based drug delivery for cartilage tissue regeneration. *Current Pharmaceutical Design* 2015.

[11] Makaraviciute, A.; Ramanaviciene, A., Site-directed antibody immobilization techniques for immunosensors. *Biosensors and Bioelectronics* 2013, *50*, 460–471.

[12] Bergstrom, G.; Mandenius, C.-F., Orientation and capturing of antibody affinity ligands: applications to surface plasmon resonance biochips. *Sensors and Actuators B: Chemical* 2011, *158* (1), 265–270.

[13] Cavalcanti, I. T.; Silva, B. V. M.; Peres, N. G.; Moura, P.; Sotomayor, M. D. P. T.; Guedes, M. I. F.; Dutra, R. F., A disposable chitosan-modified carbon fiber electrode for dengue virus envelope protein detection. *Talanta* 2012, *91*, 41–46.

[14] Lee, J.; Park, H.; Jung, Y.; Kim, J.; Jung, S.; Chung, B., Direct immobilization of protein G variants with various numbers of cysteine residues on a gold surface. *Analytical Chemistry* 2007, *79* (7), 2680–2687.

[15] Batalla, P.; Fuentes, M.; Grazu, V.; Mateo, C.; Fernandez-Lafuente, R.; Guisan, J. M., Oriented covalent immobilization of antibodies on physically inert and hydrophilic support surfaces through their glycosidic chains. *Biomacromolecules* 2008, *9* (2), 719–723.

[16] Dutra, R. F.; Kubota, L. T., An SPR immunosensor for human cardiac troponin T using specific binding avidin to biotin at carboxymethyldextran-modified gold chip. *Clin Chim Acta* 2007, *376* (1–2), 114–120.

[17] Lee, J. H.; Kim, B. C.; Oh, B. K.; Choi, J. W., Highly sensitive localized surface plasmon resonance immunosensor for label-free detection of HIV-1. *Nanomedicine* 2013, *9* (7), 1018–1026.

[18] Grant, S. A.; Pierce, M. E.; Lichlyter, D. J.; Grant, D. A., Effects of immobilization on a FRET immunosensor for the detection of myocardial infarction. *Analytical and Bioanalytical Chemistry* 2005, *381* (5), 1012–1028.

[19] Mehrvar, M.; Abdi, M., Recent developments, characteristics, and potential applications of electrochemical biosensors. *Analytical Sciences* 2004, *20* (8), 1113–1126.

[20] Mistry, K. K.; Layek, K.; Mahapatra, A.; RoyChaudhuri, C.; Saha, H., A review on amperometric-type immunosensors based on screen-printed electrodes. *Analyst* 2014, *139* (10), 2289–2311.

[21] Silva, B.; Cavalcanti, I.; Silva, M.; Dutra, R., A carbon nanotube screen-printed electrode for label-free detection of the human cardiac troponin T. *Talanta* 2013, *117*, 431–437.

[22] Qi, T.; Liao, J.; Li, Y.; Peng, J.; Li, W.; Chu, B.; Li, H.; Wei, Y.; Qian, Z., Label-free alpha fetoprotein immunosensor established by the facile synthesis of a palladium-graphene nanocomposite. *Biosens Bioelectron* 2014, *61*, 245–250.

[23] Pandiaraj, M.; Sethy, N. K.; Bhargava, K.; Kameswararao, V.; Karunakaran, C., Designing label-free electrochemical immunosensors for cytochrome *c* using nanocomposites functionalized screen printed electrodes. *Biosensors and Bioelectronics* 2014, *54*, 115–121.

[24] Gomes-Filho, S. L. R.; Dias, A. C. M. S.; Silva, M. M. S.; Silva, B. V. M.; Dutra, R. F., A carbon nanotube-based electrochemical immunosensor for cardiac troponin T. *Microchemical Journal* 2013, *109*, 10–15.

[25] Dolatabadi, J.; de la Guardia, M., Nanomaterial-based electrochemical immunosensors as advanced diagnostic tools. *Analytical Methods* 2014, *6* (12), 3891–3900.

[26] Luo, X.; Morrin, A.; Killard, A. J.; Smyth, M. R., Application of Nanoparticles in Electrochemical Sensors and Biosensors. 2006; *319*–326.

[27] Vidotti, M.; Carvalhal, R. F.; Mendes, R. K.; Ferreira, D. C. M.; Kubota, L. T., Biosensors based on gold nanostructures. *Journal of the Brazilian Chemical Society* 2011, *22* (1), 3–20.

[28] Campbell, F.; Compton, R., The use of nanoparticles in electroanalysis: an updated review. *Analytical and Bioanalytical Chemistry* 2010, *396* (1), 241–259.

[29] Lu, L.; Liu, B.; Li, S.; Zhang, W.; Xie, G., Improved electrochemical immunosensor for myeloperoxidase in human serum based on nanogold/cerium dioxide-BMIMPF6/L-Cysteine composite film. *Colloids and Surfaces B: Biointerfaces* 2011, *86* (2), 339–344.

[30] Li, H.; Wei, Q.; Wang, G.; Yang, M.; Qu, F.; Qian, Z., Sensitive electrochemical immunosensor for cancer biomarker with signal enhancement based on nitrodopamine-functionalized iron oxide nanoparticles. *Biosensors and Bioelectronics* 2015, *26* (6), 3044–3049.

[31] Yang, H.; Yuan, R.; Chai, Y.; Mao, L.; Su, H.; Jiang, W.; Liang, M., Electrochemical immunosensor for detecting carcinoembryonic antigen using hollow Pt nanospheres-labeled multiple enzyme-linked antibodies as labels for signal amplification. *Biochemical Engineering Journal* 2011, *56* (3), 116–124.

[32] Gao, X.; Zhang, Y.; Wu, Q.; Chen, H.; Chen, Z.; Lin, X., One step electrochemically deposited nanocomposite film of chitosan-carbon nanotubes-gold nanoparticles for carcinoembryonic antigen immunosensor application. *Talanta* 2011, *85* (4), 1980–1985.

[33] Huang, K.; Niu, D.; Xie, W.; Wang, W., A disposable electrochemical immunosensor for carcinoembryonic antigen based on nano-Au/multi-walled carbon nanotubes-chitosans nanocomposite film modified glassy carbon electrode. *Analytica Chimica Acta* 2010, *659* (1–2), 102–108.

[34] Kong, F.; Xu, M.; Xu, J.; Chen, H., A novel label-free electrochemical immunosensor for carcinoembryonic antigen based on gold nanoparticles-thionine-reduced graphene oxide nanocomposite film modified glassy carbon electrode. *Talanta* 2011, *85* (5), 2620–2625.

[35] Yuan, Y.; Yuan, R.; Chai, Y.; Zhuo, Y.; Mao, L.; Yuan, S., A novel label-free electrochemical immunosensor for carcinoembryonic antigen detection based on the [Ag-Ag2O]/SiO2 nanocomposite material as a redox probe. *Journal of Electroanalytical Chemistry* 2010, *643* (1–2), 15–19.

[36] An, Y.; Jiang, X.; Bi, W.; Chen, H.; Jin, L.; Zhang, S.; Wang, C.; Zhang, W., Sensitive electrochemical immunosensor for alpha-synuclein based on dual signal amplification using PAMAM dendrimer-encapsulated Au and enhanced gold nanoparticle labels. *Biosensors and Bioelectronics* 2012, *32* (1), 224–230.

[37] Liu, X.; Wong, D., Picogram-detection of estradiol at an electrochemical immunosensor with a gold nanoparticle vertical bar protein G-(LC-SPDP)-scaffold. *Talanta* 2009, *77* (4), 1437–1443.

[38] Krishnan, S.; Chinnasamy, T.; Veerappan, S.; Senthilkumar, K.; Kannaiyan, D., Dual labeled Ag@SiO2 core-shell nanoparticle based optical immunosensor for sensitive detection of *E. coli*. *Materials Science and Engineering C: Materials for Biological Applications* 2014, *45*, 337–342.

[39] Zhang, S.; Ma, H.; Yan, L.; Cao, W.; Yan, T.; Wei, Q.; Du, B., Copper-doped titanium dioxide nanoparticles as dual-functional labels for fabrication of electrochemical immunosensors. *Biosensors and Bioelectronics* 2014, *59*, 335–341.

[40] Shen, W.; Tian, D.; Cui, H.; Yang, D.; Bian, Z., Nanoparticle-based electrochemiluminescence immunosensor with enhanced sensitivity for cardiac troponin I using *N*-(aminobutyl)-*N*-(ethylisoluminol)-functionalized gold nanoparticles as labels. *Biosensors and Bioelectronics* 2011, *27* (1), 18–24.

[41] Mendes, R.; Laschi, S.; Stach-Machado, D.; Kubota, L.; Marrazza, G., A disposable voltammetric immunosensor based on magnetic beads for early diagnosis of soybean rust. *Sensors and Actuators B-Chemical* 2012, *166*, 135–140.

[42] Shen, Z.; Wang, J.; Qiu, Z.; Jin, M.; Wang, X.; Chen, Z.; Li, J.; Cao, F., QCM immunosensor detection of *Escherichia coli* O157:H7 based on beacon immunomagnetic nanoparticles and catalytic growth of colloidal gold. *Biosensors and Bioelectronics* 2011, *26* (7), 3376–3381.

[43] Li, D.; Wang, J.; Wang, R.; Li, Y.; Abi-Ghanem, D.; Berghman, L.; Hargis, B.; Lu, H., A nanobeads amplified QCM immunosensor for the detection of avian influenza virus H5N1. *Biosensors and Bioelectronics* 2011, *26* (10), 4146–4154.

[44] Yuan, Y.; Zhang, J.; Zhang, H.; Yang, X., Label-free colorimetric immunoassay for the simple and sensitive detection of neurogenin3 using gold nanoparticles. *Biosensors and Bioelectronics* 2011, *26* (10), 4245–4248.

[45] Wu, W.; Bian, Z.; Wang, W.; Zhu, J., PDMS gold nanoparticle composite film-based silver enhanced colorimetric detection of cardiac troponin I. *Sensors and Actuators B: Chemical* 2010, *147* (1), 298–303.

[46] Bera, D.; Qian, L.; Tseng, T.; Holloway, P., Quantum dots and their multimodal applications: a review. *Materials* 2010, *3* (4), 2260–2345.

[47] Gill, R.; Zayats, M.; Willner, I., Semiconductor quantum dots for bioanalysis. *Angewandte Chemie-International Edition* 2008, *47* (40), 7602–7625.

[48] Medintz, I.; Uyeda, H.; Goldman, E.; Mattoussi, H., Quantum dot bioconjugates for imaging, labelling and sensing. *Nature Materials* 2005, *4* (6), 435–446.

[49] Chan, W.; Nie, S., Quantum dot bioconjugates for ultrasensitive nonisotopic detection. *Science* 1998, *281* (5385), 2016–2018

[50] Somers, R.; Bawendi, M.; Nocera, D., CdSe nanocrystal based chem-/bio-sensors. *Chemical Society Reviews* 2007, *36* (4), 579–591.

[51] Algar, W.; Susumu, K.; Delehanty, J.; Medintz, I., Semiconductor quantum dots in bioanalysis: crossing the valley of death. *Analytical Chemistry* 2011, *83* (23), 8826–8837.

[52] Colvin, V.; Goldstein, A.; Alivisatos, A., Semiconductor nanocrystals covalently bound to metal-surfaces with self-assembled monolayers. *Journal of the American Chemical Society* 1992, *114* (13), 5221–5230.

[53] Lidke, D.; Nagy, P.; Heintzmann, R.; Arndt-Jovin, D.; Post, J.; Grecco, H.; Jares-Erijman, E.; Jovin, T., Quantum dot ligands provide new insights into erbB/HER receptor-mediated signal transduction. *Nature Biotechnology* 2004, *22* (2), 198–203.

[54] Palaniappan, K.; Xue, C.; Arumugam, G.; Hackney, S.; Liu, J., Water-soluble, cyclodextrin-modified CdSe-CdS core-shell structured quantum dots. *Chemistry of Materials* 2006, *18* (5), 1275–1280.

[55] Mattoussi, H.; Mauro, J.; Goldman, E.; Anderson, G.; Sundar, V.; Mikulec, F.; Bawendi, M., Self-assembly of CdSe-ZnS quantum dot bioconjugates using an engineered recombinant protein. *Journal of the American Chemical Society* 2000, *122* (49), 12142–12150.

[56] Xing, Y.; Chaudry, Q.; Shen, C.; Kong, K.; Zhau, H.; Chung, L.; Petros, J.; O'Regan, R.; Yezhelyev, M.; Simons, J.; Wang, M.; Nie, S., Bioconjugated quantum dots for multiplexed and quantitative immunohistochemistry. *Nature Protocols* 2007, *2* (5), 1152–1165.

[57] Liu, G.; Wang, J.; Kim, J.; Jan, M.; Collins, G., Electrochemical coding for multiplexed immunoassays of proteins. *Analytical Chemistry* 2004, *76* (23), 7126–7130.

[58] Li, L.; Liu, K.; Yang, G.; Wang, C.; Zhang, J.; Zhu, J., Fabrication of graphene-quantum dots composites for sensitive electrogenerated chemiluminescence immunosensing. *Advanced Functional Materials* 2011, *21* (5), 869–878.

[59] Aoyagi, S.; Kudo, M., Observation of fluorescence-labeled protein A on a biosensor surface by means of TOF-SIMS. *Sensors and Actuators B-Chemical* 2005, *108* (1–2), 708–712.

[60] Lingerfelt, B.; Mattoussi, H.; Goldman, E.; Mauro, J.; Anderson, G., Preparation of quantum dot-biotin conjugates and their use in immunochromatography assays. *Analytical Chemistry* 2003, *75* (16), 4043–4049.

[61] Sun, B.; Xie, W.; Yi, G.; Chen, D.; Zhou, Y.; Cheng, J., Microminiaturized immunoassays using quantum dots as fluorescent label by laser confocal scanning fluorescence detection. *Journal of Immunological Methods* 2001, *249* (1–2), 85–89.

[62] Goldman, E.; Balighian, E.; Mattoussi, H.; Kuno, M.; Mauro, J.; Tran, P.; Anderson, G., Avidin: a natural bridge for quantum dot-antibody conjugates. *Journal of the American Chemical Society* 2002, *124* (22), 6378–6382.

[63] Kerman, K.; Endo, T.; Tsukamoto, M.; Chikae, M.; Takamura, Y.; Tamiya, E., Quantum dot-based immunosensor for the detection of prostate-specific antigen using fluorescence microscopy. *Talanta* 2007, *71* (4), 1494–1499.

[64] Lao, U.; Mulchandani, A.; Chen, W., Simple conjugation and purification of quantum dot-antibody complexes using a thermally responsive elastin-protein L scaffold as immunofluorescent agents. *Journal of the American Chemical Society* 2006, *128* (46), 14756–14757.

[65] Susha, A.; Javier, A.; Parak, W.; Rogach, A., Luminescent CdTe nanocrystals as ion probes and pH sensors in aqueous solutions. *Colloids and Surfaces A: Physicochemical and Engineering Aspects* 2006, *281* (1–3), 40–43.

[66] Liu, Y.; Sun, Y.; Vernier, P.; Liang, C.; Chong, S.; Gundersen, M., pH-sensitive photoluminescence of CdSe/ZnSe/ZnS quantum dots in human ovarian cancer cells. *Journal of Physical Chemistry C* 2007, *111* (7), 2872–2878.

[67] Herrero-Latorre, C.; Alvarez-Mendez, J.; Barciela-Garcia, J.; Garcia-Martin, S.; Pena-Crecente, R. M., Characterization of carbon nanotubes and analytical methods for their determination in environmental and biological samples: a review. *Analytica Chimica Acta* 2015, *853*, 77–94.

[68] Vashist, S.; Zheng, D.; Al-Rubeaan, K.; Luong, J.; Sheu, F., Advances in carbon nanotube based electrochemical sensors for bioanalytical applications. *Biotechnology Advances* 2011, *29* (2), 169–188.

[69] Li, J.; Pandey, G. P., Advanced physical chemistry of carbon nanotubes. *Annual Review of Physical Chemistry* 2015.

[70] Yang, N.; Chen, X. P.; Ren, T. L.; Zhang, P.; Yang, D. G., Carbon nanotube based biosensors. *Sensors and Actuators B: Chemical* 2015, *207*, 690–715.

[71] Girifalco, L.; Hodak, M.; Lee, R., Carbon nanotubes, buckyballs, ropes, and a universal graphitic potential. *Physical Review B* 2000, *62* (19), 13104–13110.

[72] Peng-Cheng Maa, N. A. S., Gad Maromb, Jang-Kyo Kima, Dispersion and functionalization of carbon nanotubes for polymer-based nanocomposites a review. 2010, *41* (10), 1345–1367.

[73] Di Crescenzo, A.; Ettorre, V.; Fontana, A., Non-covalent and reversible functionalization of carbon nanotubes. *Beilstein Journal of Nanotechnology* 2014, *5*, 1675–1690.

[74] Gao, C.; Guo, Z.; Liu, J.; Huang, X., The new age of carbon nanotubes: an updated review of functionalized carbon nanotubes in electrochemical sensors. *Nanoscale* 2012, *4* (6), 1948–1963.

[75] Tuncel, D., Non-covalent interactions between carbon nanotubes and conjugated polymers. *Nanoscale* 2011, *3* (9), 3545–3554.

[76] Sanchez, S.; Roldan, M.; Perez, S.; Fabregas, E., Toward a fast, easy, and versatile immobilization of biomolecules into carbon nanotube/polysulfone-based biosensors for the detection of hCG hormone. *Analytical Chemistry* 2008, *80* (17), 6508–6514.

[77] Viswanathan, S.; Rani, C.; Anand, A.; Ho, J., Disposable electrochemical immunosensor for carcinoembryonic antigen using ferrocene liposomes and MWCNT screen-printed electrode. *Biosensors and Bioelectronics* 2009, *24* (7), 1984–1989.

[78] Malhotra, R.; Patel, V.; Vaque, J.; Gutkind, J.; Rusling, J., Ultrasensitive electrochemical immunosensor for oral cancer biomarker IL-6 using carbon nanotube forest electrodes and multilabel amplification. *Analytical Chemistry* 2010, *82* (8), 3118–3123.

[79] Munge, B.; Fisher, J.; Millord, L.; Krause, C.; Dowd, R.; Rusling, J., Sensitive electrochemical immunosensor for matrix metalloproteinase-3 based on single-wall carbon nanotubes. *Analyst* 2010, *135* (6), 1345–1350.

[80] Chen, Y.; Huang, J., Electrochemical sensing of bovine serum albumin at self-assembled SWCNTs on gold. *Diamond and Related Materials* 2009, *18* (2–3), 516–519.

[81] Su, H.; Yuan, R.; Chai, Y.; Zhuo, Y.; Hong, C.; Liu, Z.; Yang, X., Multilayer structured amperometric immunosensor built by self-assembly of a redox multi-wall carbon nanotube composite. *Electrochimica Acta* 2009, *54* (17), 4149–4154.

[82] Jacobs, C.; Peairs, M.; Venton, B., Review: carbon nanotube based electrochemical sensors for biomolecules. *Analytica Chimica Acta* 2010, *662* (2), 105–127.

[83] Freitas, T.; Mattos, A.; Silva, B.; Dutra, R., Amino-functionalization of carbon nanotubes by using a factorial design: human cardiac troponin T immunosensing application. *Biomed Research International* 2014.

[84] Yao, J.; Sun, Y.; Yang, M.; Duan, Y. X., Chemistry, physics and biology of graphene-based nanomaterials: new horizons for sensing, imaging and medicine. *Journal of Materials Chemistry* 2012, *22* (29), 14313–14329.

[85] Novoselov, K. S.; Geim, A. K.; Morozov, S. V.; Jiang, D.; Zhang, Y.; Dubonos, S. V.; Grigorieva, I. V.; Firsov, A. A., Electric field effect in atomically thin carbon films. *Science* 2004, *306* (5696), 666–669.

[86] Zhang, L.; Zhou, R.; Zhao, X., Graphene-based materials as supercapacitor electrodes. *Journal of Materials Chemistry* 2010, *20* (29), 5983–5992.

[87] Lawal, A. T., Synthesis and utilisation of graphene for fabrication of electrochemical sensors. *Talanta* 2015, *131*, 424–443.

[88] Zhu, Y.; Murali, S.; Cai, W.; Li, X.; Suk, J.; Potts, J.; Ruoff, R., Graphene and graphene oxide: synthesis, properties, and applications. *Advanced Materials* 2010, *22* (46), 5226–5226.

[89] Zhu, Y.; Murali, S.; Cai, W.; Li, X.; Suk, J.; Potts, J.; Ruoff, R., Graphene and graphene oxide: synthesis, properties, and applications. *Advanced Materials* 2010, *22* (35), 3906–3924.

[90] Chung, C.; Kim, Y.; Shin, D.; Ryoo, S.; Hong, B.; Min, D., Biomedical applications of graphene and graphene oxide. *Accounts of Chemical Research* 2013, *46* (10), 2211–2224.

[91] Chen, D.; Feng, H. B.; Li, J. H., Graphene oxide: preparation, functionalization, and electrochemical applications. *Chemical Reviews* 2012, *112* (11), 6027–6053.

[92] Dreyer, D.; Park, S.; Bielawski, C.; Ruoff, R., The chemistry of graphene oxide. *Chemical Society Reviews* 2010, *39* (1), 228–240.

[93] Shan, C.; Yang, H.; Han, D.; Zhang, Q.; Ivaska, A.; Niu, L., Graphene/AuNPs/chitosan nanocomposites film for glucose biosensing. *Biosensors and Bioelectronics* 2010, *25* (5), 1070–1074.

[94] Luo, J.; Jiang, S. S.; Zhang, H. Y.; Jiang, J. Q.; Liu, X. Y., A novel non-enzymatic glucose sensor based on Cu nanoparticle modified graphene sheets electrode. *Analytica Chimica Acta* 2012, *709*, 47–53.

[95] Mao, S.; Lu, G.; Yu, K.; Bo, Z.; Chen, J., Specific protein detection using thermally reduced graphene oxide sheet decorated with gold nanoparticle-antibody conjugates. *Advanced Materials* 2010, *22* (32), 3521–+.

[96] Zhong, Z.; Wu, W.; Wang, D.; Wang, D.; Shan, J.; Qing, Y.; Zhang, Z., Nanogold-enwrapped graphene nanocomposites as trace labels for sensitivity enhancement of electrochemical immunosensors in clinical immunoassays: carcinoembryonic antigen as a model. *Biosensors and Bioelectronics* 2010, *25* (10), 2379–2383.

[97] Huang, J. L.; Tian, J. N.; Zhao, Y. C.; Zhao, S. L., Ag/Au nanoparticles coated graphene electrochemical sensor for ultrasensitive analysis of carcinoembryonic antigen in clinical immunoassay. *Sensors and Actuators B: Chemical* 2015, *206*, 570–576.

[98] Tang, J.; Tang, D.; Niessner, R.; Chen, G.; Knopp, D., Magneto-controlled graphene immunosensing platform for simultaneous multiplexed electrochemical immunoassay using distinguishable signal tags. *Analytical Chemistry* 2011, *83* (13), 5407–5414.

[99] Teixeira, S.; Burwell, G.; Castaing, A.; Gonzalez, D.; Conlan, R. S.; Guy, O. J., Epitaxial graphene immunosensor for human chorionic gonadotropin. *Sensors and Actuators B-Chemical* 2014, *190*, 723–729.

[100] Xu, S.; Liu, Y.; Wang, T.; Li, J., Positive potential operation of a cathodic electrogenerated chemiluminescence immunosensor based on luminol and graphene for cancer biomarker detection. *Analytical Chemistry* 2011, *83* (10), 3817–3823.

[101] Li, L. L.; Liu, K. P.; Yang, G. H.; Wang, C. M.; Zhang, J. R.; Zhu, J. J., Fabrication of graphene-quantum dots composites for sensitive electrogenerated chemiluminescence immunosensing. *Advanced Functional Materials* 2011, *21* (5), 869–878.

Chapter 13

Real-Time Detection of Nitric Oxide Release in Live Cells Utilizing Fluorinated Xerogel-Derived Nitric Oxide Sensor

Gi-Ja Lee, Sung Wook Kang, Bochan Seo, Jae Ho Shin and Hun-Kuk Park

Additional information is available at the end of the chapter

http://dx.doi.org/10.5772/60614

Abstract

Nitric oxide (NO) is an important signaling molecule that regulates a diverse range of physiological and cellular processes in many tissues. Therefore, the accurate detection of physiological NO concentration is crucial to the understanding of NO signaling and its biological role. There has been growing interest in the development of electrochemical sensors for direct and real-time monitoring of NO. As the direct electrooxidation of NO requires a relatively high working potential, further surface modification with permselective membranes is required to achieve the desired selectivity for NO via size exclusion or electrostatic repulsion. Here we reported a planar-type NO sensor with a fluorinated xerogel-derived gas permeable membrane for real-time detection of NO release in live cells. First, we evaluated the biocompatibility of xerogel-derived NO permeable membranes modified with fluorinated functional groups by growing RAW 264.7 macrophages on them. And we performed the AFM measurements to examine the morphology of RAW 264.7 macrophages on xerogel membrane. Finally, we successfully detected NO release in RAW 264.7 macrophages, using a planar-type xerogel-derived NO sensor. As a result, fluorinated xerogel-derived membrane could be utilized as both NO permeable and cell-adhesive membranes. Besides, planar-type xerogel-based NO sensors can be easily applied to the cellular sensing system, with a simple coating procedure.

Keywords: Nitric oxide, Live cells, Xerogel, Real-time detection, Cellular sensing

1. Introduction

Nitric oxide (NO) is an important mediator that regulates a diverse range of physiological and cellular processes in many tissues [1,2]. In particular, it plays an important role in a variety of biological processes including neurotransmission, immune defense, regulation of cell death (apoptosis), and cell motility [3–5]. The accurate detection of physiological NO concentration is crucial to the understanding of NO signaling and its biological role. However, it is difficult to accurately measure NO levels in vivo because NO is present at nanomolar concentrations in the body and has a half-life of 2–6 s [6]. Additionally, it is highly reactive with numerous endogenous species including free radicals, peroxides, and oxygen [7,8]. Consequently, most methods for NO detection are indirect including spectroscopic approaches such as the Griess assay for nitrite, detection of nitrate and nitrite with reductase enzymes, and detection of methemoglobin after NO reaction with oxyhemoglobin [9–11]. Unfortunately, these methods often fail to accurately reflect the dynamics of NO in vivo and in real time [12]. Direct measurement strategies are therefore necessary for examining biological process and diseases related to NO in biological conditions, in particular the action of endogenously produced NO.

There has been growing interest in the development of electrochemical sensors for direct and real-time monitoring of NO. Electrochemical methods provide simplicity, fast response times, good sensitivity, high selectivity, and long-term calibration stability [8,12], and thus have significant advantages over other techniques such as electron paramagnetic resonance (EPR) [13–15], chemiluminescence [16–18], and fluorescence [19,20]. As the direct electrooxidation of NO requires a relatively high working potential, further surface modification with permselective membranes is required to achieve the desired selectivity for NO via size exclusion or electrostatic repulsion. Many polymeric materials have been reported as NO-permeable and permselective membranes, including polycarbazole [21], collodion [22], Nafion [23], cellulose acetate [24], poly(tetrafluoroethylene) [25], polydimethylsiloxane (silicone rubber) [26], and multilayer hybrids of these polymers. However, the utility of these polymeric membranes for improving sensor performance *in vivo* or in cells remains limited [8,12].

The sol-gel process is an inorganic polymerization process for the fabrication of materials under mild conditions [27] through the hydrolysis and condensation of suitable alkoxysilane precursors [28]. These sol-gel derived materials (xerogels) have emerged as a class of materials suitable for a wide range of sensing applications [8]. Organically modified xerogels are particularly advantageous for gas sensing applications [29]. The porous inorganic network provides an open and rigid structure for rapid diffusion of gaseous molecules, while the organic groups impart hydrophobicity to the membrane for preventing leakage of the internal electrolyte [30,31]. A previous study reported that sol-gel derived electrochemical sensors showed good sensitivity and selectivity for NO detection in vitro [21,30]. These studies were mainly focused on improving sensor performance such as sensitivity and selectivity, in particular optimization of permselective membrane in vitro.

An amperometric NO sensor for cellular biosensing is significantly affected by the distance between the sensor and the NO-producing cells because NO is rapidly diluted by diffusion under cell-culture conditions [32,33]. However, it is difficult to adjust the exact distance between the needle or disk type sensors and cells without special apparatus [34]. Therefore, it would be better to grow the cells directly on the electrode surface with a NO permeable membrane. Our previous study reported good sensitivity and selectivity of electrochemical NO sensors with a xerogel membrane in standard solution, but it was mainly focused on membrane optimization including composition, permeability, and contact angle to improve the sensor performance [8,12].

In this study, we developed a planar-type NO sensor with a fluorinated xerogel-derived gas permeable membrane for real-time detection of NO release in live cells including RAW 264.7 macrophages. First, we evaluated the cell viability of RAW 264.7 macrophages on xerogel-derived NO permeable membranes. And we performed the AFM measurement to examine the morphology of RAW 264.7 macrophages on xerogel membrane. Finally, we performed the real-time electrochemical detection of NO release in RAW 264.7 macrophages by lipopolysaccharide (LPS) stimulation utilizing a planar-type NO sensor.

2. Methods used

2.1. Materials

Methyltrimethoxysilane (MTMOS) was purchased from Fluka (Buchs, Switzerland). (Heptadecafluoro-1,1,2,2-tetrahydrodecyl) trimethoxysilane (17FTMS) was purchased from Gelest (Tullytown, PA, USA). Silver and dielectric insulator pastes were purchased from Acheson (Tokyo, Japan) and Jujo (Tokyo, Japan), respectively. Chloroplatinic acid hexahydrate, and lead(II) acetate trihydrate were purchased from Sigma (St. Louis, MO, USA). Nitric oxide (NO; 99.5%), carbon monoxide (CO; 99.5%), and nitrogen (N_2) gases were obtained from Dong-A Scientific (Seoul, South Korea). Other solvents and chemicals were analytical-reagent grade and were used as received. All aqueous solutions were prepared using distilled water of 18.3 MΩ cm resistivity.

2.2. Preparation and biocompatibility test of fluorinated xerogel membrane

A silane solution was prepared by dissolving 18 μL of MTMOS and 4.5 μL of 17FTMS in 727.5 μL of ethanol, as previously reported [12]. The solution was mixed with 160 μL of water, followed by 10μL of HCl (0.5 M) for 1 h. The ensuing solution was deposited on a culture dish and cured for 1 day under ambient conditions.

RAW 264.7 macrophages were cultured in Dulbecco's modified Eagle's medium (DMEM) media with 10% fetal bovine serum (FBS), 1% penicillin, and 1% streptomycin. The cells were then maintained in a humidified atmosphere that contained 5% CO_2 at 37 °C and the medium was changed every second day. Cells were detached by washing and scraping. After washing, cells were resuspended in fresh medium and used for subsequent experiments.

To evaluate the biocompatibility of xerogel membrane, we performed by 3-(4,5-dimethylthiazol- 2-yl)-2,5-diphenyltetrazolium bromide (MTT) assay (Sigma, St.Louis, MO, USA) using RAW 264.7 macrophages. The MTT assay, which is one of the most versatile and popular methods to measure cell viability, involves the conversion of water soluble MTT to an insoluble formazan. The amount of formazan in the MTT assay reflects the total mitochondrial enzymatic activity of live cells. Cells (1 × 10^4 cells per well) were seeded in a xerogel membrane coated 96-well plate and incubated for 24, 48, and 72 h. After MTT was added to each group, the cells were incubated for 4 h. The viability was then measured with a microplate reader (Bio-Tek, Winooski, VT, USA) at a test wavelength of 540 nm. The optical density (OD) was calculated as the difference between the reference wavelength and the test wavelength.

2.3. AFM measurement of cells on fluorinated xerogel membrane

RAW 264.7 macrophages on xerogel membrane were characterized using atomic force microscopy (AFM). RAW 264.7 macrophages were washed twice with filtered phosphate buffered saline (PBS, 0.1 M, pH 7.4) and fixed for 20 min in 2.5% glutaraldehyde in PBS at room temperature and 5 ml PBS was added to xerogel-modified culture dish containing fixed cells. Contact mode AFM images were obtained using a NANOS N8 NEOS (Bruker, Herzogenrath, Germany), equipped with a 42.5 μm XY/4 μm Z scanner and a Zeiss optical microscope (Epiplan 500x). The AFM was placed on an active vibration isolation table (TS-150, S.I.S., Herzogenrath, Germany), inside a passive vibration isolation table (Pucotech, Seoul, Korea) to eliminate external noise. Data acquisition and image processing were performed with SPIP (Scanning Probe Image Processor Version 4.1, Image Metrology, Denmark).

Silicon cantilevers with the reflective side coated with gold (Budget Sensor, Bulgaria) were used for the measurements under liquid conditions. The AFM probe tips were stabilized with PBS for at least 10 min prior to scanning. Fixed RAW macrophages were scanned in PBS solution at a resolution of 512 × 512 pixels, at a scan speed of 0.5 line/s.

2.4. Fabrication of planar-type NO sensor

The planar-type NO sensor with three-configuration (i.e., working, reference, and counter electrodes) was prepared using a screen-printing technique, as shown in Figure 1(A). Briefly, a polyester substrate (500 μm thick; 3 M, St. Paul, MN, USA) was thermally treated at 130 °C for 3 h before screen-printing. The silver (for reference electrode and electrical connections), carbon (for working and counter electrodes), and dielectric insulator patterns were sequentially screen-printed onto the thermally pretreated polyester substrate using a LS-150 semi-automated screen printer (Newlong Seimitsu Kogyo; Tokyo, Japan). Each pattern was cured at 130 °C for 13 min. The carbon working electrode (7.1 mm^2) was platinized in 3% chloroplatinic acid hexahydrate and 0.03% lead(II) acetate trihydrate (w/w in water) by cycling the potential from +0.6 to −0.35 V (vs. Ag/AgCl) at a scan rate of 20 mV/s using a CH Instruments 760B bipotentiostat (Austin, TX, USA). The Ag/AgCl reference electrode was prepared by oxidative treatment using 0.3 M $FeCl_3$ solution on the dried silver paste for 15 minutes. A silane solution for xerogel membrane was deposited onto platinized carbon working electrode and allowed to cure for 24 h under ambient conditions.

2.5. Evaluation of planar-type NO sensors

To evaluate the analytical performance of the planar-type NO sensor, amperometric measurements were performed using a CH Instruments 760B bipotentiostat or a portable MicroStat

(Dropsens, Spain). The sensor was prepolarized for 100 min and tested in DMEM solution under ambient conditions with constant stirring. Currents were recorded at an applied potential of +0.8 V (vs. Ag/AgCl). For *in vitro* calibration, a NO standard solution (1.9 mM) was prepared by purging DMEM solution with N_2 gas for 30 min to remove any oxygen, followed by purging with NO (99.5%) for 30 min. The CO solution (0.9 mM) was similarly prepared by consecutively purging DMEM with N_2 for 30 min and CO (99.5%) for another 30 min. (Caution! *The NO and CO purging process must be carried out in a fume hood since NO and CO gases are toxic!*) The solutions of NO and interfering species (nitrite, ascorbic acid, uric acid, acetaminophen, dopamine, hydrogen peroxide, and carbon monoxide) were freshly prepared every second day and stored at 4 °C.

The selectivity of the xerogel-modified sensors for NO in the presence of interfering species was also evaluated using the separate solution method [12], for which the amperometric selectivity coefficients ($\log K_{NO,j}^{amp}$) were calculated using the following equation:

$$\log K_{NO,j}^{amp} = \log\left(\frac{\Delta I_j / c_j}{\Delta I_{NO} / c_{NO}}\right) \tag{1}$$

where ΔI_{NO} and ΔI_j are the measured current values for the target analyte (NO) and interfering species (j = nitrite, ascorbic acid, uric acid, acetaminophen, dopamine, hydrogen peroxide, and carbon monoxide), respectively. The concentration of each interfering substance (c_j) was selected to be 1 μM, the same concentration of NO (c_{NO}).

2.6. Real-time detection of NO release in live cells

RAW 264.7 macrophages (1×10⁴ cells/electrode) were grown on top of the electrode surface which was covered with sol–gel-derived membrane. Stimulation of RAW macrophages with LPS (1 μg/ml) was performed after stabilizing the background current for 2~3 h at 37 °C. The experimental scheme is shown in Figure 1(B).

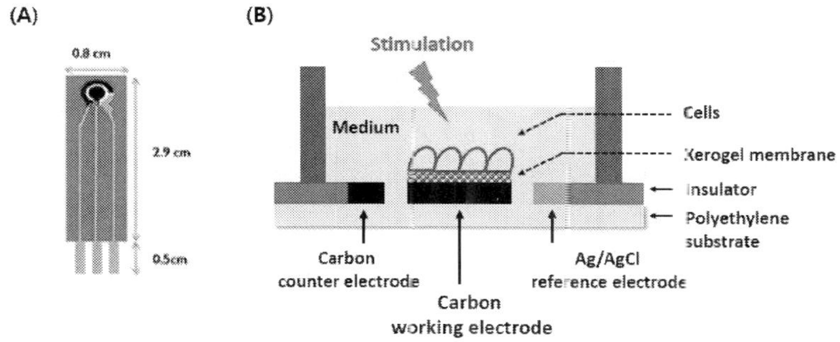

Figure 1. Schematic of (A) screen-printed electrode and (B) experimental design for real-time detection of nitric oxide release in live cells utilizing fluorinated xerogel-derived nitric oxide sensor.

3. Results and discussion

3.1. Cytotoxicity assessment of fluorinated xerogel membrane

In general, amorphous, hydrophobic fluoropolymers are highly permeable to many gaseous species including oxygen, nitric oxide, carbon dioxide, ammonia, and volatile organic compounds [12,35]. Indeed, a microporous poly(tetrafluoroethylene) (PTFE) film has been extensively employed to improve a Clark-type oxygen sensor as a gas-permeable membrane [36,37]. The PTFE membrane effectively prevented the biofouling at the interface between the sensor surface and the biological samples (e.g., blood), while enabling high permeation and rapid diffusion of oxygen [36]. However, the fabrication of such sensors remains complicated. The PTFE film must be stretched across the electrode and fixed to the sensor. This process is often irreproducible, impacting both the thickness of the gas-permeable membrane and the sensor's response to oxygen [29,36].

In this study, we employed a fluorinated xerogel gas-permeable membrane to take advantage of both the versatility of sol–gel chemistry and high and fast gas permselectivity like PTFE. Our previous study reported good sensitivity and selectivity of electrochemical NO sensors with a xerogel membrane in standard solution [8,12]. In addition, we had preliminarily evaluated the cell viability, adhesion, and growth of cells on xerogel-derived NO permeable membrane, comparing with that of other cell-adhesive matrices such as collagen and poly L-lysine [34].

Here, we performed an MTT assay to reconfirm the cytotoxicity of fluorinated xerogel-derived NO permeable membrane, prior to real-time detection of NO release in live cells. RAW 264.7 macrophages were incubated on fluorinated xerogel-derived NO permeable membrane and collagen for 24, 48, and 72 h. Similar to preliminary result, the viability of RAW 264.7 cells exposed to xerogel membrane for 24 h was 102.64 ± 3.52%. This result showed that fluorinated xerogel-derived NO permeable membrane was not cytotoxic to RAW 264.7 macrophages. Moreover, the proliferation of RAW 264.7 macrophages was not influenced during 48 and 72 h of growth on the xerogel membrane.

3.2. Morphology of RAW macrophages on fluorinated xerogel membrane

Macrophages, which was derived from circulating blood monocytes and reside in most tissues, play an important role in host defense and tissue homeostasis as they ingest dead tissue and fight invading pathogens [38,39]. Therefore, the activation of macrophages is a general characteristic for the early stages of pathogens infection. LPS, a major outer membrane component of Gram-negative bacteria, plays an important role in the pathogenesis of Gram-negative bacterial infection [40] and has been served as an important active component for pathogen-induced macrophage inflammation studies [41]. In general, LPS stimulation caused some morphological changes in human macrophages such as the formation of lamellipodia and the extensive cell spreading [42]. But these LPS induced morphological changes in RAW 264.7 macrophages by conventional microscopy could not provide the ultrastructural changes in RAW 264.7 macrophages by LPS stimulation.

http://dx.doi.org/10.5772/60614

AFM has emerged as an imaging technique that provides 3D topographic information and structural details of biological samples including microorganisms, cell membranes, biopolymers, and more. Therefore, AFM is now a common tool for nanotechnology, not only for imaging but also to measure the interaction forces in biological systems [43]. We performed AFM measurement to observe the morphological changes in RAW macrophages on xerogel membrane following LPS treatment. Figure 2 shows representative AFM 3D topography images taken from LPS-stimulated RAW cells for 0, 0.5, 1, 1.5, 6, and 24 h. As shown in Figure 2(A), control RAW macrophage showed a typical spherical shapes with ruffled membranes and a few lamellipodia surround them. But the morphology of RAW cells after 1 h LPS stimulation exhibited the extension of the cytoplasm toward its outside, as shown in Figure 2(C). After the LPS exposure for 1.5 h, they appeared spindle and long-stretched shapes (Figure 2(D)). The number of microspikes was increased from 1.5 h after stimulation of LPS. As a result, RAW macrophages for 24 h LPS stimulation showed strong lamellipodia with a lot of dendritic pseudopod. In addition, LPS stimulated RAW macrophages for 24 h showed some holes on their surface. Saxena et al. [44] reported that LPS induced differentiation of RAW 264.7 macrophage cells into dendritic like cells. And Pi et al. [39] reported the LPS activated RAW 264.7 macrophages changed to be much bigger than control cells with some holes emerged on cell surface. From AFM results, we suggested that RAW macrophages on fluorinated xerogel-derived NO permeable membrane were normally activated by LPS stimulation.

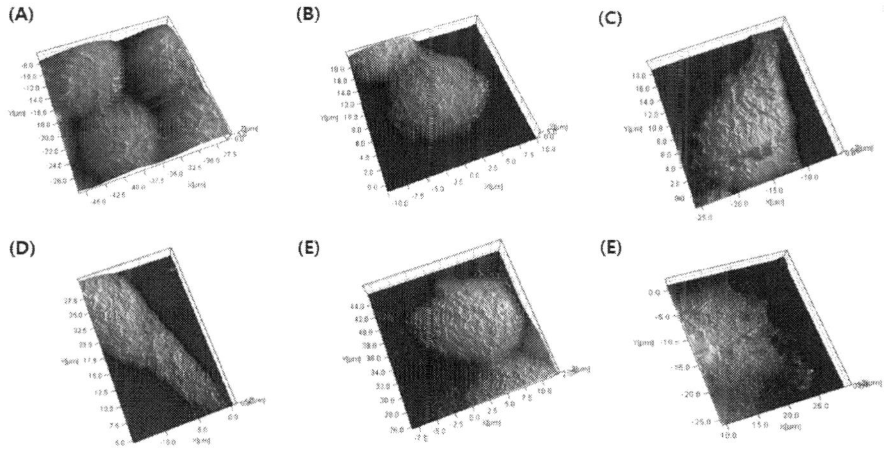

Figure 2. Representative AFM 3D topography images of RAW 264.7 macrophages on fluorinated xerogel-membrane after LPS (1 μg/ml) stimulation for (A) 0, (B) 0.5, (C) 1, (D) 1.5, (E) 6, and (F) 24 h.

3.3. Amperometric response of xerogel-derived planar-type NO sensor

The sensor performance of the resulting NO microelectrode was investigated in DMEM solution. The calibration and dynamic response curves for NO and interfering species (i.e., nitrite) are shown in Figure 3. A sensor coated with fluorinated xerogel membrane did not exhibit serious interference from nitrite concentrations of up to 20 μM, which is a concentration much greater than the highest level present in physiological samples. Indeed, the fluorinated xerogel-modified sensors exhibited excellent selectivity ($\log K_{NO,j}^{amp}$) of < –5, <–5, <–5, <–5, <–5, –4.38, and –3.85 for j = nitrite, ascorbic acid, uric acid, acetaminophen, dopamine, hydrogen peroxide, and carbon monoxide, respectively. In addition, the xerogel-derived sensor responded to NO concentrations of up to 1,000 nM with respect to sensitivity (0.15 ± 0.012 nA/nM), linearity (R^2 = 0.9981, 20~1,000 nM NO range), and response time ($t_{90\%}$ < 10 s).

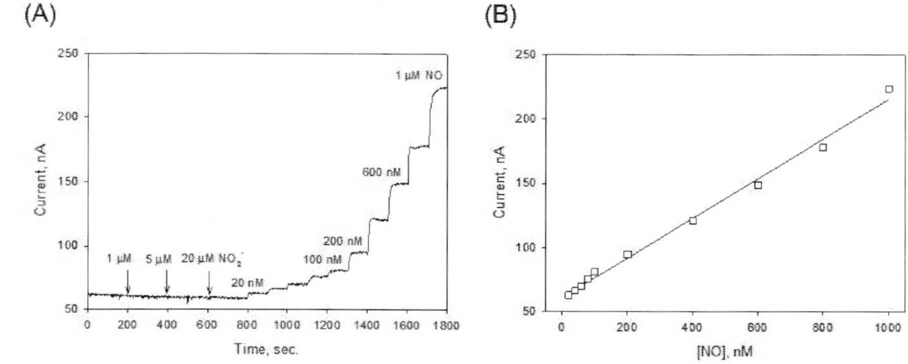

Figure 3. Dynamic response (A) to nitric oxide and an interfering species (nitrite) and calibration curve (B) of the nitric oxide sensor modified with fluorinated xerogel-derived membrane.

3.4. Real-time detection of NO release in live cells utilizing planar-type NO sensor

NO plays an important role in a variety of cellular functions, including blood vessel relaxation, neurotransmission, and immune response [34]. It is produced through the conversion of L-arginine into L-citrulline by diverse cell types such as macrophages, neutrophils, neuronal cells, and endothelial cells as both an extra and intracellular signaling molecule [45]. The mechanisms of NO formation by three kinds of nitric oxide synthase (NOS), including endothelial (eNOS), neuronal (nNOS), and inducible NOS (iNOS), are well established [46]. Among them, the expression of iNOS is induced by stimulation of cells with LPS and/or proinflammatory cytokines [46,47]. Macrophage iNOS is responsible for producing large quantities of NO, which are synthesized over the period of several hours after cells are stimulated [34]. NO plays an important role as part of a nonspecific defense mechanism, as a signaling molecule between macrophages and other cells and as an intracellular messenger within the macrophages themselves [45]. Besides, NO produced by endothelium cells plays a critical role in regulation of vascular tone and activity. The continuous detection of NO release from endothelium could provide new and important insights into the modulatory role of NO in cardiovascular system [48].

However, most methods for NO quantification are indirect including spectroscopic approaches such as the Griess assay which determines the major and stable products of NO degradation—nitrite and nitrate, as well as the detection of methemoglobin after NO reaction with oxyhemoglobin [9–11]. Unfortunately, these methods often fail to accurately reflect the dynamics of NO *in vivo, in vitro*, and in real time [12]. In contrast to other electrochemical NO sensor designs, xerogel-based NO sensors are fabricated using a simple coating procedure, with high sensitivity, selectivity, and reproducibility [8].

Figure 4 represents real-time detection monitoring of NO release by a planar-type xerogel-derived NO sensor in RAW 264.7 macrophages stimulated by LPS. In control RAW macrophages, there was no increase in current throughout the whole experimental period. But the current resulting from NO production started to rise in 5,500 s after LPS stimulation. After reaching the maximum current in 6,000 s, a slow decline in NO production lasted for the rest of the experiment. The maximum change in NO concentration was about 60 nM, based on *in vitro* calibration results.

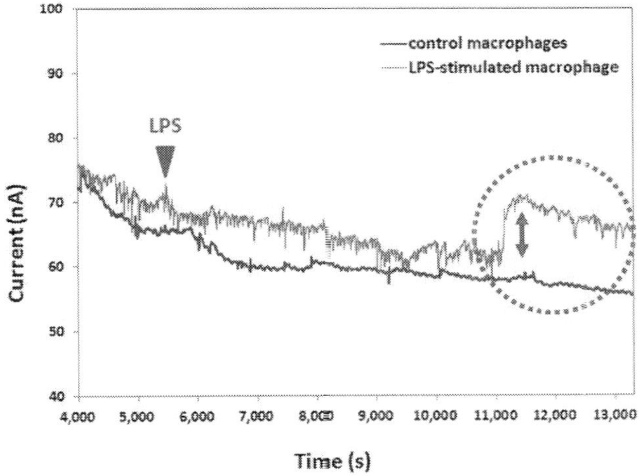

Figure 4. Representative real-time measurement of NO release in RAW 264.7 macrophages stimulated by lipopolysaccharide.

4. Conclusions

In this study, we evaluated the biocompatibility of sol-gel derived NO permeable membranes modified with fluorinated functional groups with RAW 264.7 macrophages. Our results showed that xerogel-derived NO permeable membrane was nontoxic and could provide good cell adhesion. In addition, the morphology of these adhering RAW 264.7 macrophages did not show any differences compared to those grown on other biocompatible membranes. Therefore, we expected that fluorinated xerogel-derived membrane could be utilized as both NO gas permeable and cell-adhesive membranes on which cells are directly grown for a cellular NO sensor. Besides, in the planar-type sensor with NO permeable membrane, there is no need to adjust the distance between sensor and cells. Overall, in contrast to other electrochemical NO sensor designs, xerogel-based NO sensors can be easily applied to the cellular sensing system, with a simple coating procedure.

In summary, a planar-type NO sensor with fluorinated xerogel membrane has good sensitivity and specificity for detecting biologically released NO in live cells. And we successfully measured the NO release in RAW 264.7 macrophages stimulated by LPS, utilizing our planar-type xerogel-derived NO sensor. Therefore, we expect that our sensor can be applied for real-time monitoring of NO dynamics in various cells, tissues, and organs.

Acknowledgements

This study was supported by the National Research Foundation of Korea (NRF) funded by the Ministry of Science, ICT & Future Planning (2010-0019917), and the Ministry of Education (2013R1A1A2065149).

Author details

Gi-Ja Lee[1,2], Sung Wook Kang[1], Bochan Seo[3], Jae Ho Shin[3] and Hun-Kuk Park[1,2*]

*Address all correspondence to: sigmoidus@khu.ac.kr

1 Department of Biomedical Engineering & Healthcare Industry Research Institute, College of Medicine, Kyung Hee University, Seoul, Korea

2 Department of Medical Engineering, Graduate School, Kyung Hee University, Seoul, Korea

3 Department of Chemistry, College of Natural Science, Kwangwoon University, Seoul, Korea

References

[1] Brown FO, Finnerty NJ, Lowry JP. Nitric oxide monitoring in brain extracellular fluid: Characterisation of Nafion-modified Pt electrodes *in vitro* and *in vivo*. Analyst. 2009;134:2012–2020. DOI: 10.1039/B909005C

[2] Deng XD, Wang F, Chen Z. A novel electrochemical sensor based on nano-structured film electrode for monitoring nitric oxide in living tissues. Talanta. 2010;82:1218–1224. DOI: 10.1016/j.talanta.2010.06.035

[3] Hibbs JB, Taintor RR, Vavrin Z, Rachelin EM. Nitric oxide: A cytotoxic activated macrophage effector molecule. Biochem. Biophys. Res. Commun. 1988;157:87–94. DOI: 10.1016/S0006-291X(88)80015-9

[4] Palmer PMJ, Ferrige AG, Moncada S. Nitric oxide release accounts for the biological activity of endothelium-derived relaxing factor. 1987;327:524–526. DOI: 10.1038/327524a0

[5] Szabo C. Physiological and pathophysiological roles of nitric oxide in the central nervous system. Brain Res. Bull. 1996;44:131–141. DOI: 10.1016/0361-9230(96)00159-1

[6] Lee GJ, Chae SJ, Kim SK, Lee S, Shin JH, Kim W, Park HK. Real time dynamics of nitric oxide during cardiac ischemia-reperfusion of the rat. Sens. Actuators B Chem. 2012;161:480–485. DOI: 10.1016/j.snb.2011.10.064

[7] Bedioui F, Trevin S, Devynck J. Chemically modified microelectrodes designed for the electrochemical determination of nitric oxide in biological systems. Electroanalysis. 1996;8:1085–1091. DOI: 10.1002/elan.1140081202

[8] Shin JH, Weinman SW, Schoenfisch MH. Sol-gel derived amperometric nitric oxide microsensor. Anal. Chem. 2005;77:3494–3501. DOI: 10.1021/ac048153i

[9] Archer S. Measurement of nitric oxide in biological models. FASEB J. 1993;7:349–360. DOI: 0892-6638/93/0007-0349

[10] Taha TH. Nitric oxide measurements in biological samples. Talanta. 2003;61:3–10. DOI: 10.1016/S0039-9140(03)00354-0

[11] Wang Y, Hu S. Nitric oxide sensor based on poly (p-phenylenevinylene) derivative modified electrode and its application in rat heart. Bioelectrochemistry. 2009;74:301–305. DOI: 10.1016/j.bioelechem.2008.11.002

[12] Shin JH, Privett BJ, Kita JM, Wightman RM, Schoenfisch MH. Fluorinated xerogel-derived microelectrodes for amperometric nitric oxide sensing. Anal. Chem. 2008;80:6850–6859. DOI: 10.1021/ac800185x

[13] Fujii S, Yoshimura T. A new trend in iron–dithiocarbamate complexes: as an endogenous NO trapping agent. Coord. Chem. Rev. 2000;198:89–99. DOI: 10.1016/S0010-8545(99)00196-4

[14] Han TH, Hyduke DR, Vaughn MW, Fukuto JM, Liao JC. Nitric oxide reaction with red blood cells and hemoglobin under heterogeneous conditions. PNAS. 2002;99:7763–7768. DOI: 10.1073/pnas.122118299

[15] Takahashi S, Omori Y, Miyazaki H, Yoshino F, Shoji H, Lee MC, Todoki K, Kamibayashi M, Murakami E. Real-time monitoring of nitric oxide in ischemic myocardium using an NO-selective electrode calibrated by electron spin resonance. Life Sci. 2003;74:75–85. DOI: 10.1016/j.lfs.2003.06.027

[16] Kikuchi K, Nagano T, Hayakawa H, Hirata Y, Hirobe M. Detection of nitric oxide production from a perfused organ by a luminol-hydrogen peroxide system. Anal. Chem. 1993;65:1794–1799. DOI: 10.1021/ac00061a025

[17] Kikuchi K, Nagano T, Hayakawa H, Hirata Y, Hirobe M. Real time measurement of nitric oxide produced ex vivo by luminol-H_2O_2 chemiluminescence method. J. Biol. Chem. 1993;268:23106–23110 / 8226828.

[18] Robinson JK, Bollinger MJ, Birks JW. Luminol/H_2O_2 chemiluminescence detector for the analysis of nitric oxide in exhaled breath. Anal. Chem. 1999;71:5131–5136. DOI: 10.1021/ac990646d

[19] Gabe Y, Urano Y, Kikuchi K, Kojima H, Nagano T. Highly sensitive fluorescence probes for nitric oxide based on boron dipyrromethene chromophore-rational design of potentially useful bioimaging fluorescence probe. J. Am. Chem. Soc. 2004;126:3357–3367. DOI: 10.1021/ja037944j

[20] Kojima H, Nakatsubo N, Kikuchi K, Kawahara S, Kirino Y, Nagoshi H, Hirata Y, Nagano T. Detection and imaging of nitric oxide with novel fluorescent indicators: Diaminofluoresceins. Anal Chem. 1998;70:2446–2453. DOI: 10.1021/ac9801723

[21] Prakash R. Srivastava RC, Seth PK. Polycarbazole modified electrode; nitric oxide sensor. Polym. Bull. 2001;46:487–490. DOI: 10.1007/s002890170036

[22] Kitamura Y, Uzawa T, Oka K, Komai Y, Takizawa N, Kobayashi H, Tanishita K. Operation of a miniature redox hydrogel-based pyruvate sensor in undiluted deoxygenated calf serum. Anal. Chem. 2000;72:2957–2968. DOI: 10.1021/ac991021i

[23] Zhang X, Cardosa L, Broderick M, Fein H, Lin J. An integrated nitric oxide sensor based on carbon fiber coated with selective membranes. Electroanalysis. 2000;12:1113–1117. DOI: 10.1002/1521-4109(200010)

[24] Cserey A, Gratzl M. Stationary-state oxidized platinum microsensor for selective and on-line monitoring of nitric oxide in biological preparations. Anal. Chem. 2001;73:3965–3974. DOI: 10.1021/ac010123h

[25] Lee Y, Oh BK, Meyerhoff ME. Improved planar amperometric nitric oxide sensor based on platinized platinum anode. 1. Experimental results and theory when applied for monitoring NO release from diazeniumdiolate-doped polymeric films. Anal. Chem. 2004;76:536–544. DOI: 10.1021/ac035064h

[26] Mizutani F, Hirata Y, Yabuki S, Iijima S. Amperometric measurement of nitric oxide (NO) using an electrode coated with polydimethylsiloxane. Chem. Lett. 2000;29:802–803. DOI: 10.1246/cl.2000.802

[27] Coradin T, Boissiere M, Livage J. Sol-gel chemistry in medicinal science. Curr. Med. Chem. 2006;13:99–108. DOI: 10.2174/092986706789803044

[28] Islamnezhad A, Valizadeh M. Taguchi OA25 orthogonal array design for the optimization of La^{3+}-selective electrode prepared by sol-gel technique. Anal. Bioanal. Electrochem. 2012;4:386–398.

[29] Lee GJ, Kim SK, Kang SW, Kim OK, Chae SJ, Choi S, Shin JH, Park HK, Chung JH. Real time measurement of myocardial oxygen dynamics during cardiac ischemia-reperfusion of rats. Analyst. 2012;137:5312–5319. DOI: 10.1039/c2an35208g

[30] Tsionsky M, Lev O. Electrochemical composite carbon-ceramic gas sensors: introduction and oxygen sensing. Anal. Chem. 1995;67:2409–2414. DOI: 10.1021/ac00110a014

[31] Rabinovich L, Lev O, Tsirlina GA. Electrochemical characterisation of Pd modified ceramic / carbon electrodes: Partially flooded versus wetted channel hydrophobic gas electrodes. J. Electroanal. Chem. 1999;466:45–59. DOI: 10.1016/S0022-0728(99)00118-7

[32] Asakawa H, Mochitate K, Haruyama T. Seamless signal transduction from live cells to an NO Sensor via a cell-adhesive sensing matrix. Anal. Chem. 2008;80:1505–1511. DOI: 10.1021/ac702001u

[33] Isik S, Etienne M, Oni J, Blochl A, Reiter S, Schuhmann W. Dual microelectrodes for distance control and detection of nitric oxide from endothelial cells by means of scanning electrochemical microscope. Anal. Chem. 2004;76:6389–6394. DOI: 10.1021/ac049182w

[34] Kang SW, Seo B, Kim JH, Kim OK, Shin JH, Lee GJ, Park HK. Cell viability, adhesion and function of RAW 264.7 macrophages on fluorinated xerogel-derived nitric oxide permeable membrane for the application of cellular sensing. J. Nanosci. Nanotechnol. 2014;14:8398–8404. DOI: 10.1166/jnn.2014.9928

[35] Moskvin LN, Nikitina TG. Membrane methods of substance separation in analytical chemistry. J. Anal. Chem. 2004;59:2–16. DOI: 10.1023/B:JANC.0000011661.47796.b2

[36] Park SS, Hong M, Song CK, Jhon GJ, Lee Y, Suh M. Real-time in vivo simultaneous measurements of nitric oxide and oxygen using an amperometric dual microsensor. Anal. Chem. 2010;82:7618–7624. DOI: 10.1021/ac1013496

[37] Wang Z, Lin P, Baker GA, Stetter J, Zeng X. Ionic liquids as electrolytes for the development of a robust amperometric oxygen sensor. Anal. Chem. 2011;83:7066–7073. DOI: 10.1021/ac201235w

[38] Murray PJ, Wynn TA. Protective and pathogenic functions of macrophage subsets. Nat. Rev. Immunol. 2011;11:723–737. DOI: 10.1038/nri3073

[39] Pi J, Li T, Liu J, Su X, Wang R, Yang F, Bai H, Jin H, Cai J. Detection of lipopolysaccharide induced inflammatory responses in RAW 264.7 macrophages using atomic force microscope. Micron. 2014;65:1–9. DOI: 10.1016/j.micron.2014.03.012

[40] Khan SA, Everest P, Servos S, Foxwell N, Zahringer U, Brade H, Rietschel ET, Dougan G, Charles IG, Maskell DJ. A lethal role for lipid A in Salmonella infections. Mol. Microbiol. 1998;29:571–579. DOI: 10.1046/j.1365-2958.1998.00952.x

[41] Smith PD, Smythies LE, Mosteller-Barnum M, Sibley DA, Russell MW, Merger M, Sellers MT, Orenstein JM, Shimada T, Graham MF, Kuba-gawa H. Intestinal macrophages lack CD14 and CD89 and consequently are down-regulated for LPS- and IgA-mediated activities. J. Immunol. 2001;167:2651–2656. DOI: 10.4049/jimmunol.167.5.2651

[42] Williams LM, Ridley AJ. Lipopolysaccharide induces actin reorganization and tyrosine phosphorylation of Pyk2 and paxillin in monocytes and macrophages. J. Immunol. 2000;164:2028–2036. DOI: 10.4049/jimmunol.164.4.2028

[43] Choi S, Jung GB, Kim KS, Lee GJ, Park HK. Medical applications of atomic force microscopy and raman spectroscopy. J. Nanosci. Nanotechnol. 2014;14:71–97. DOI: 10.1166/jnn.2014.9112

[44] Saxena RK, Vallyathan V, Lewis DM. Evidence for lipopolysaccharide-induced differentiation of RAW 264.7 murine macrophage cell line into dendriticlike cells. J. Biosci. 2003;28:129–134. DOI: 10.1007/BF02970143

[45] Pekarova M, Kralova J, Kubala L, Ciz M, Lojek A, Gregor C, Hrbac J. Continuous electrochemical monitoring of nitric oxide production in murine macrophage cell line RAW 264.7. Anal. Bioanal. Chem. 2009;394:1497–1504. DOI: 10.1007/s00216-009-2813-x

[46] Mayer B, Hemmens B. Biosynthesis and action of nitric oxide in mammalian cells. Trends Biochem. Sci. 1997;22:477–481. DOI: 10.1016/S0968-0004(97)01147-X

[47] Yao D, Vlessidis AG, Evmiridis NP. Determination of nitric oxide in biological samples. Microchim. Acta. 2004;147:1–20. DOI: 10.1007/s00604-004-0212-8

[48] Du F, Huang W, Shi Y, Wang Z, Cheng J. Real-time monitoring of NO release from single cells using carbon fiber microdisk electrodes modified with single-walled carbon nanotubes. Biosens. Bioelectron. 2008;24:415–421. DOI: 10.1016/j.bios.2008.04.020

Chapter 14

Evaluation of the Structure-Activity Relationship of Hemoproteins through Physicochemical Studies: Hemoglobins as a Prototype of Biosensor

Leonardo M. Moreira, Juliana P. Lyon,
Vanessa J. S. V. Santos and Fabio V. Santos

Additional information is available at the end of the chapter

http://dx.doi.org/10.5772/60576

Abstract

In the present work, we have studied a group of prerequisites in terms of "structure-function relationship" of hemoproteins, especially hemoglobins, emphasizing the role of the heme and its chemical environment in the biochemical and physicochemical properties of the biomolecule. We have discussed the ferrous center and its properties as coordination center; the macrocyclic ligands, especially the porphyrins; the esterochemical and electronic properties of the iron-porphyrins (heme groups); and the interaction between heme groups and globins, which is related to several redox and oligomeric properties of hemoprotein systems and its potential applications with respect to novel materials. One of the main uses of hemoglobins in new materials is also discussed, which is its employment as a biosensor. Therefore, we have discussed the development of novel biosensors based on hemoglobins and their physico-chemical properties as well as on the main molecules of biological relevance that have been detected by these biosensors, such as hydrogen peroxide (H_2O_2), nitric oxide (NO), and cholesterol, among others. Indeed, several important biomolecules and biological processes can be detected and/or evaluated by devices that present hemoglobins as leading chemical components. Different apparatus are covered with respect to distinct characteristics, such as chemical stability, sensitivity, selectivity, reproducibility, durability, optimum conditions of measurements, etc. and their respective characteristics are analyzed.

Keywords: iron, hemoprotein structure-function relationship, macrocyclic ligands, porphyrin flexibility, biological molecules

1. Introduction

1.1. Hemoproteins

Hemoproteins constitute a very relevant group of proteins which present the heme as a prosthetic group. A prosthetic group is a chemical compound inserted in a polypeptide chain by chemical binding, which is not amino acid residue or peptide binding. The heme group is a coordination compound (metallic complex) that presents a ferric (Fe(III)) or ferrous (Fe(II)) ion as coordination center, which is strongly ligated to a macrocyclic ligand (Lewis base with more than three electron-donor sites), which, in the case of the heme, is a porphyrin. Therefore, the prosthetic group of hemoproteins is an iron-porphyrin compound. It is important to emphasize that the heme group is the main active site of the hemoproteins in which the iron ion plays an important role.

There are several types of hemoprotein in nature. One of the main differences that involve the several kinds of hemoproteins is associated to the number of polypeptide chains that can occur in each hemoprotein. Indeed, we can find hemoproteins with only one polypeptide chain, i.e., hemoproteins that do not have a quaternary protein structure (obviously, these hemoproteins present lower molecular mass), such as, for example, the cytochrome *c* [1]. On the other hand, there are giant hemoproteins with a great number of subunits (therefore, presenting a quaternary protein structure), such as the giant extracellular hemoglobin of *Lumbricus terrestris* (HbLt), which presents around 144 globin chains (subunits with heme group) and approximately 36 *Linker* chains (subunits with structural function that do not have any heme group). This kind of giant hemoprotein presents an extraordinarily high supramolecular mass, which, in the case of *HbLt*, is approximately of 3.6 MDa, presenting a very complex and well-organized tridimensional arrangement [2-4].

A second important difference between the various types of hemoproteins is related to the isolation level of the heme pocket. In fact, in most hemoproteins, there is a hydrophobic isolation around the heme (that is variable in intensity, depending on each specific hemoprotein), which is constituted by the lateral chains of the amino acid residues that are encountered in the neighborhood of the iron-porphyrin system. This isolation is relevant to preclude a significant grade of water molecule penetration in the heme pocket, which can provoke several chemical reactions involving the metallic center of this prosthetic group and, consequently, the loss of its biological function. Furthermore, heme isolation avoids dimerization between heme groups (mainly in the hemoproteins with a great number of globin subunits), which also causes function loss to the several hemoproteins.

Hemoproteins are very interesting chemical systems that can be applied with distinct objectives. In this context, the elaboration of the called "novel materials" have received special emphasis in order to optimize the employment and the durability of the hemoproteins with respect to its application as biosensor, considering the respective chemical and biochemical functions. Alternative employments, such as biosensors, catalysts, blood substitutes, surface modifiers, among others, have been considered and evaluated by several research groups. The results of these initial studies are very promising.

1.2. Understanding hemoproteins and its potential to application as biosensor

Bioelectronic sensing devices, which are well known as biosensors, represent a growing research area at the interface of chemistry, biochemistry, physics, and nanotechnology, in which the interest focused on applications of redox-active biomolecules (e.g. enzymes and carrier proteins) are often used for the development of this type of analytical devices [5]. In this context, the present study is focused on the structure-activity relationship of hemoproteins through physicochemical evaluations, which includes biochemical behavior and inferences obtained from instrumental analytical methods, with an especial emphasis on spectroscopic and electrochemical analyses. These data sources furnish relevant information regarding several aspects of the chemical reactivity of these extraordinary biological macromolecules. Indeed, the so-called "structure-function relationship" of hemoproteins has different types of implications in terms of biochemical processes, such as heme biological synthesis, physiological disturbances, ligand changes, polypeptide unfolding generated by pH changes and surfactant interaction, thermal denaturation, autoxidation mechanisms, and other processes. These studies constitute relevant contributions to understand the structure of the heme, the globin, and their mutual interactions. This basic research is an important prerequisite to optimize several types of applications of hemproteins. Indeed, these applications require the elaboration of novel materials, which includes, in several cases, an especial role to the hemoproteins, considering its chemical properties. Shumyantseva and coworkers, for example, reported significant advancements in biosensor development based on nanocomposite materials, with hemoproteins as fundamental components [6]. The previous knowledge of the principal chemical agent of the system will favor the interpretation of the first results and the resolution of the eventual initial limitations.

Considering the great potential of employment of hemoproteins in the constitution of new materials, it is relevant to emphasize some important areas in which hemoproteins have been applied, such as biosensors, blood substitutes, and agents of surface modification.

We have studied the great potential, in terms of chemical properties, of hemoglobins regarding their application as biosensors and the main molecules that have been evaluated in the literature through the biosensor based on hemoglobins and/or other hemoproteins.

2. The iron element and its action as metallic coordination center

Iron constitutes 4.7% of the Earth's crust, and it is the second most abundant metal (the first most abundant metal is the aluminum). This remarkable presence is associated to its extraordinary mechanical properties, mainly in the form of steel, making it a very important element with respect to its technological application potential [7]. Due to its electronic configuration, iron (Z=26) presents two more stable oxidation states: iron(III), which is the most stable oxidation state, known as the ferric ion or met-form (this denomination is mainly used by biologists and biochemists when they discuss biological iron), and iron(II), which is known as the ferrous form. Depending on the chemical environment, iron can present other less usual oxidation states, such as iron(I) and iron(IV), which are much less chemically stable than the ferric and ferrous forms.

Iron as metallic center, mainly when its more stable oxidation states are considered, 2+ and 3+, tends to occur, in terms of coordination chemistry, in hexacoordinated and octahedral complexes, since its electronic configurations are the well-known d^6 and d^5 electronic distributions, respectively. Considering that iron is an element of the first series of transition metals (first period of the transition metals in the periodic table), its electronic configuration can be observed in the high- and low-spin states, since the 3*d* orbital is relatively small, precluding a more intense *d* orbital splitting, when in contact with weak-field ligand, such as F^- and Cl^-. Therefore, the complexes with Iron(II) and Iron(III) as metallic center can or not to obey Hund's rule.

3. The bioinorganic chemistry of iron

Small quantities of iron are essential for animals and plants. However, similarly to copper and selenium, iron is a very toxic element in higher concentrations in the biological medium [8]. Biologically, iron is the most important transition metal, being associated to several relevant biochemical processes:

a. Oxygen transport in the blood of mammalians, fishes, and other animals: hemoglobin;

b. Storage of Oxygen (O_2) in the muscular tissue: mioglobin;

c. Electron transport in plants, animals, and bacteria (cytochromes and, to plants and bacteria, ferrodoxines);

d. Storage and release of iron in animals (ferritin and transferrin);

e. Nitrogenase components (enzyme of the bacteria that makes fixation of Nitrogen (N_2));

f. Relevant role in several other enzymes, such as aldehyde-oxidase (oxidation of aldehydes), catalases, and peroxidases (decomposition of hydrogen peroxide (H_2O_2)), and succinate dehydrogenase (aerobic oxidation of carbohydrates) [Lee, 1996].

4. The macrocyclic ligands

Macrocyclic ligands contain at least three donor atoms and the macrocyclic ring should consist of a minimum of nine atoms [9]. Therefore, all macrocyclic ligands are chelants; however, the chelant ligands are not necessarily macrocyclic ligands, since the chelant compounds include ligands that make only two ligations to the center of coordination of the metallic complex. Among the more studied macrocyclic compounds nowadays are the cryptates of lanthanides due to their ability to coordinate with antigens and antibodies [10]. Indeed, Jean-Marie Lehn, who was awarded the Nobel Prize in Chemistry in 1987, obtained a macromolecular system with the presence of donor groups in a tridimensional arrangement of flexible form, which allowed a cation to be complexed and, moreover, encapsulated. The first species that presented this capability was denominated "criptantings", and its metallic complexes, "cryptates" [10].

It is important to note that there are excellent reasons for nature choosing macrocyclic derivatives for the relevant bioinorganic complexes, such as enhanced kinetic, thermodynamic, and thermal stabilities [9].

5. Porphyrinoid macrocylic ligands

Considering the structural complexity presented by macrocyclic systems, such as natural photosynthetic ones, significant scientific effort has been devoted toward the preparation and study of structurally simpler systems. These works are focused on the reproduction of some of the fundamental steps that occur in nature with macrocyclic ligands. For instance, we can cite natural photosynthesis (one of the most important being photoinduced charge separation (CS)) [Bottari, 2010]. Among the chromophores that have been used as molecular components in artificial photosynthetic systems, the large family of porphyrinoid systems, which constitute the ubiquitous molecular building blocks employed by nature in photosynthesis, have been the preferred and obvious choice, due to their intense optical absorption and rich redox chemistry [11].

Tetrapyrroles, which includes porphyrins, are conjugated heteromacrocyclic molecules consisting of four pyrrole rings joined by bridging carbon atoms. They are excellent ligands for many metal ions and, as such, are widely employed in biological systems as prosthetic groups involved in central biochemical processes such as oxygen transport, photosynthesis, respiration, etc. The insertion of metals into various tetrapyrroles is catalyzed by a group of enzymes named chelatases, e.g., the nickel, cobalt, and magnesium chelatases and ferrochelatases, with ferrochelatases being the best-studied enzyme from this group [12].

6. The porphyrin as biological molecule: a versatile macrocycle

Natural porphyrin derivatives, including hemes, chlorophylls, and bacteriochlorophylls, are integrated into protein scaffolds that are essential for their biological activities. For example, the protein matrices of hemoproteins provide active centers with a special hydrophobic pocket having a specific steric hindrance. In the case of hemoglobin and myoglobin, the embedded active iron centers are protected sterically and hydrophobically against irreversible oxidation, thereby enabling reversible O_2-binding [13]. In the light harvesting complexes of purple bacteria, wheel-like supramolecular assemblies of many bacteriochlorophyll units are also performed with the assistance of proteins; they play an important role in the efficient capture of energy from sunlight and transference of this energy to the photosynthetic reaction center. Because of their morphological similarities to proteins, three-dimensional dendrimer architectures can act as attractive scaffolds for the site-specific positioning of porphyrin functionalities in nanoscale-size regimes [13].

The order of stability of porphyrin complexes with metallic ions in 2+ oxidation state is Ni^{2+}>Cu^{+2}>Co^{+2}>Fe^{2+}>Zn^{+2}. The kinetics of formation of these metalloporphyrins has also been measured. It has been found to be in the order Cu^{2+}>Co^{2+}>Fe^{2+}>Ni^{2+} [14]. Considering the higher abundance of iron porphyrins, interesting questions are inferred regarding the origin and evolution of biological systems if the natural abundance of iron was not over a thousandfold greater than those of cobalt and copper [15].

Cobalt, for example, is the 30th most abundant element in the Earth's crust. However, the only known cobalt-containing biological system is the coenzyme B_{12}, which contains cobalt bonded to five nitrogen atoms and a carbon atom of an adenosine ligand. The carbon-cobalt bond in this molecule qualifies coenzyme B_{12} as the first example of a biological organometallic compound, as the closely related vitamin B_{12} contains a cyano ligand in place of the adenosine ligand and other cobalt-bound organic substituents are encountered (the denomination "cobalamin" is used for this class of B_{12} derivatives) [16].

It is important to note that porphyrins and related molecules are of importance not only in biological systems but also in several applications, such as organic semiconductors and nonlinear optical materials, due to extended π-electron systems. Recently, many studies have been focused on the use of artificial photosynthesis to develop light-energy conversion systems [17].

7. The heme group (iron porphyrins)

Many, though not all, naturally occurring porphyrins contain iron, and a significant portion of porphyrin-containing proteins possess heme as a prosthetic group. The most abundant heme, heme *b*, is found in the hemoproteins myoglobin and hemoglobin and contains two propionate, two vinyl, and four methyl side chains [18]. Oxidation of a methyl side chain to a formyl group and substitution of a vinyl side chain with a 17-carbon isoprenoid side chain converts heme *b* to heme *a*, the prosthetic group of the mitochondrial enzyme cytochrome *c* oxidase. C-type hemoproteins, such as cytochrome *c* and the *bc1* complex, contain heme *c*, in which the two vinyl side chains of heme *b* are covalently attached to the protein [18].

The great number of studies that employ porphyrin-like compounds in different chemical contexts denotes the extraordinary interdisciplinary and multidisciplinary characters of these macrocyclic compounds. The application of porphyrin-like compounds, metallated or not, in PDT [19], catalysis, electrochemical studies, biomimetic studies, and others is a definitive fingerprint of the great biochemical and physicochemical relevance of this area.

8. Porphyrins and metalloporphyrins: very mechanically flexible chemical systems

Studies focused on the non-planar distortions affecting the structures of sterically crowded porphyrins have shown that these macrocycles are considerably more flexible than originally suspected. This conformational flexibility may play an important role in controlling a wide range of physicochemical properties of the heme-cofactors present in heme proteins [20]. In fact, porphyrins and other tetrapyrroles exhibit a considerable flexibility of the macrocycle conformation. Examples of non-planar tetrapyrrole conformations have been observed in a variety of protein complexes, such as chlorophyll-reaction center and antenna systems, heme proteins, and cytochromes. Different macrocycle conformations are believed to be responsible for diverse functions of the same chromophore in different protein environments. This allows an explanation of the spectral properties of antenna complexes and the unidirectionality of electron transfer in reaction centers [21].

The types of distortion are often conserved for proteins with the same function but isolated from different species [22]. It suggests that the kind of interaction between a prosthetic group and a polypeptide chain is, at least partially, related to the spatial conformation presented by the prosthetic group. In this way, since non-planar distortion is energetically unfavorable for iron-porphyrins, for example, the conservation of the heme conformation strongly suggests that the biological function of hemoproteins might be modulated by protein control over the conformation of the heme prosthetic group [22]. Indeed, in the absence of external forces and steric crowding of the porphyrin substituents the macrocycle exhibits *D4h* symmetry. This is because the energy of the π electronic system is lowest for a planar conformation [23].

Therefore, porphyrins and metalloporphyrins are conformationally flexible chemical structures, and a great number of examples of distorted macrocycle conformations have been observed crystallographically. In biological systems, distortions form planarity have been observed in photosynthetic reaction centers, light-harvesting complexes, hemoproteins, and methyl coenzyme M reductase [24]. The conformation of the heme group of each of the more than 25 peroxidases and their mutants is predominantly saddled, while the heme groups of cytochrome *c* tend to be ruffled [25]. Metalloporphyrins undergo remarkable non-planar distortions of the macrocycle that disturb the chemical and photochemical properties of these important protein cofactors. In fact, mechanical (different substituents or polypeptide interactions) and/or electronic factors can significantly alter the chemical reactivity of the ferrous center, affecting the ligand affinity and several consequent aspects. Furthermore, the tertiary structure of the surrounding protein can manipulate these distortions as a way of regulating biological functions. Cytochromes *c*, for example, presents an energetically unfavorable conserved non-planar distortion of the heme, which probably is associated to its role in its electron transfer function [26].

Indeed, redox potentials, axial ligand binding affinities and dynamics, catalytic selectivity, and basicity, as well as both the biological and non-biological insertion of metal ions are among the many functional properties shown to be sensitive to deformations of the macrocycle, mainly with respect to out-of-plane distortion [27].

9. Hemoproteins as biosensors

The great ability to bind relevant biological molecules, which is an important property of the heme ferrous center, makes that the hemoglobins can be considered prototypes of biosensors.

Biologically relevant molecules, such as NO, CN^-, CO, F^-, Cl^-, and others, are typical potential ligands to bind ferrous ion. It is important to mention, for example, the work of Michael K. Chang [28], that employed heme to sense molecular oxygen (O_2); carbon monoxide (CO); and nitric oxide (NO), which identified heme-driven conformational changes can happen of a very peculiar form. Considering the group formed by NO, CN^-, CO, F^-, and Cl^-, we have to mention that some of these ligands can also bind ferric ion, which can be generated depending of the medium conditions. Indeed, the met-hemoglobin (ferric hemoglobin) presents a very stable ferric oxidation state (Fe(III)), which represents an oxidized state to the hemoglobin (which presents biological function when its coordination center is the ferrous center), therefore, without native biological function (other hemoproteins present functional character in its ferric state). Thus, considering the mechanical stability of the hemoglobin, together with its significant redox stability (two possible oxidation states to the coordination center of the heme (prosthetic group), the hemoglobins, as well as other hemoproteins, could be applied as biosensors.

Considering the relevant biomolecules mentioned above, we can emphasize, for example, the work of Palaniappan and co-workers [Palaniappan, 2008], which attempted to develop a sensor to detect nitric oxide (NO) in gaseous state, employing the NO-specific hemoprotein known as guanylyl cyclase (sGC) entrapped in a mesoporous silica network [29].

In this context, it is important to note that the level of spatial arrangement and redox stabilities is decisive to the efficiency and durability of the hemoproteins in its application as a biosensor. Normally, hemoproteins with higher supramolecular mass and/or higher number of polypeptide chains (protein subunits) are more stable considering the spatial configuration (quaternary structure) and redox state. In this way, a priori, hemoproteins with high mass, number of polypeptide chains, and redox stability are potential prototypes of biosensors. This is the case of several groups of hemoproteins as, for example, the giant extracellular hemoglobins [30-32]. The giant extracellular hemoglobin of *Glossoscolex paulistus* (HbGp), for example, has a mass of approximately 3.6 MDa, 180 subunits (144 globins with heme, and 36 *Linker* chains, without prosthetic group) and great stability against autoxidation and other chemical processes [33-36]. Hemoproteins with these properties can be prepared to act as different types of biosensors, depending on previous knowledge regarding its structure-activity relationship.

On the other hand, there is an intrinsic difficulty to elaborate the novel materials to form the biosensor based on hemoglobins, mainly the ones with higher supramolecular mass. Indeed, the electron transfer reactivity of hemoglobin on conventional electrode surfaces is physiologically limited as the function of the normal electro-active center, which is the iron atom in the heme, is deeply buried in its electrochemically "insulated" peptide backbone [37]. In this way, the hydrophobic isolation of the ferrous ion in the heme pocket, which is favorable to avoid the oxidation of the heme and the dimerization between heme groups through μ-oxo bridges (chemical bounds between the metallic centers (iron center) of two heme groups (iron porphyrins), is also associated to a significant difficulty, in terms of construction of the biosensor, which is the significant distance between the ferrous metallic center and the chemical neighborhood, limiting the electron transfer associated to the metallic center as well as the chemical contact among relevant biological molecules and the metallic center. In fact, several studies based on hemoproteins adsorbed or immobilized on different electrodes demonstrated that high supramolecular mass is not amenable to direct electrical communication with the electrode [38].

Thus, the redox and structural stabilities are significant advantages in order to apply the hemoprotein as biosensor, but the hydrophobic isolation of the heme precludes a high intense level of contact with the electrode surface (difficulting the "electric contact", i.e., the electric current transfer between the hemoglobin the the electrode) in the elaboration of modified electrodes. However, the redox and structural stabilities as well as the hydrophobic isolation of heme can be altered, depending on the chemical medium conditions. Indeed, these modifications can favor or disfavor the biosensor performance, which makes it necessary for several chemical conditions to be tested, aiming at the optimization of the best possible conditions of each hemoprotein in each biosensor prototype. In fact, depending on the chemical properties of the hemoprotein employed as biosensor and its respective matrix in which the protein is immobilized (chemically or physically adsorbed), the electrochemical and/or electroanalytical performance can be drastically changed. Chemical factors as pH, ionic force, contaminant presence, and level of chemical interaction with the matrix, among others, provoke polypeptide unfolding of the hemoprotein, increasing, frequently, the permeability of water and other relevant molecules into heme pockets, which, in its native state, presents high hydrophobic character. In this way, the loss of wild polypeptide arrangement changes the total spatial arrangement of the protein, changing the state of the polypeptide arrangement and the spatial conformations of all globins. Nevertheless, each hemoprotein has chemical and structural peculiarities that can sense the modifications in different ways, depending on all characteristics. Consequently, the evaluation of chemical environment conditions that are suitable to each hemoprotein and each type of biosensor can vary significantly depending on the molecular or supramolecular mass, number of polypeptide chains (protein subunits), type of amino acid residues around the heme pocket, matrix chemical constitution etc. in addition to the characteristics of the own electrode, which are also important factors to the hemoglobins as principal component of the electrodes [39]. In fact, in order to present high sensitivity, reproducibility, and long stability, a simple biomolecule/solid (electrode) interface is required with features of good biocompatibility, depending mainly of the type of applications and sensitive electrochemical activity, being that the electrodes used, i.e. the matrix constitution of the electrode, in these employments play key roles [39]. All these above-mentioned factors can affect the hemoprotein sensibility of the biochemical molecules that must be determined as final task of these type of biosensor. Therefore, each biosensor should be submitted to several tests, in a preliminary way, in various different chemical (or biochemical) conditions, in order to obtain the best standard reference that is possible in a specific context, i.e., the optimum conditions in terms of sensibility, selectivity or specificity, durability, and reproducibility, among other relevant parameters that determine the quality of the biosensor.

Paola Ringhieri, for instance, in her PhD thesis, developed an interesting strategy to overcome this limitation, which is the exploitation of artificial low-molecular mass proteins, to accentuate the electric contact [38]. Ringhieri evaluated a new class of heme-peptide conjugate that are known as "mimochromes", aiming to understand the effects of the peptide chain composition and conformation in modulating the redox properties of the heme and, consequently, the novel biosensor [38].

This above-mentioned limitation can also be solved through adjustments in the medium conditions of the chemical environment around the hemoprotein. In fact, variations of pH, ionic force, polypeptide arrangement, and humidity, among others can decrease the compaction of the polypeptide chains and the consequent heme hydrophobic isolation, favoring the electron transfer as well as the chemical interaction with several relevant molecules.

In this context, many efforts have been made to enhance the electron transfer rate of hemoglobins through mediators, promoters, and several immobilization materials, such as polymer films, surfactants, and nanomaterials [37].

Moreover, it is necessary to optimize an apparatus that can maintain the hemoglobin supported on this surface. The stability of this surface modification is a prerequisite to the precision grade of the methodology. It is important to note that the acid or basic conditions of the surface are decisive in determining the adsorption of the globin, mainly if the fixation is caused by ionic contacts. In this way, it must be considered the isoelectric point (pI) of the hemoglobin, since the predominance of positive or negative charges on the protein surface depends on the acid/basic conditions of the medium in relation to the pI of the hemoglobin.

The detection of hydrogen peroxide (H_2O_2) and nitric oxide (NO) has received special attention from several groups that apply hemoglobins as biosensors, as hemoglobins have been considered a new class of sensitive, stable, and auspicious electrochemical biosensors [40]. Indeed, electrodes elaborated with hemoglobins as relevant components have generated direct electron transfer of hemoglobins [Dang, 2013]. Dang and co-workers, for example, evaluating immobilized hemoglobin, encountered good response in terms of biocatalytic action in nitric oxide (NO) and hydrogen peroxide (H_2O_2) [41]. Gu and co-workers, for example, aiming to determine amperometrically nitric oxide with immobilized hemoglobin, found electrocatalytic responses that were proportional to the nitric oxide concentration, without any interference of several relevant biomolecules [42].

In this context, the detection of hydrogen peroxide has generated a great number of excellent works due to the long-term stability, good reproducibility, and high selectivity of biosensors produced to this kind of determination [43]. Furthermore, the determination of hydrogen peroxide is highly relevant as a function of its applications as mediator in food, pharmaceutical, clinical, industrial, and environmental analyses [44]. For example, a recent study describes the immobilization of hemoglobin on a Clark electrode surface to develop a novel electrochemical biosensor for the detection of hydrogen peroxide. The principle of the measurements was based on the electrocatalytic activity of the immobilized hemoglobin to the reduction of hydrogen peroxide [45]. A highly sensitive and selective amperometric hydrogen peroxide (H_2O_2) biosensor based on immobilization of hemoglobin (Hb) at multiwalled carbon nanotube–zinc oxide (MWCNT/ZnO) composite modified glassy carbon electrode (GCE) is also reported [46]. Zeolites are also employed as matrix to immobilization of hemoglobin, so this biomolecule can act reducing H_2O_2 [47]. An interesting study demonstrates that titanium dioxide (TiO_2) nanoparticles, through the photovoltaic effect generated by ultraviolet radiation, can significantly improve the catalytic action of hemoglobin as a peroxidase, denoting that the elaboration of photocontrolled protein-based biosensors is very promising [48].

It is important to point out that a recent important work presented a cholesterol amperometric biosensor containing hemoglobin, being that this hemoprotein acts as an efficient electron mediator, without any interference from phospholipids [Souza, 2013]. The authors claim that the cholesterol could also be detected in real samples from chicken egg yolk, without any influences from phospholipids [49].

It is important to register that the immobilization of hemoglobins has generated highly stable new materials. Batra and co-workers [50], for instance, developed a methodology to detect acrylamide, which was based on covalent immobilization of hemoglobin on carboxylated multi-walled carbon nanotube, with the respective electrode being employed 120 times over a period of 100 days (stored at 4°C) [50]. On the other hand, Saleh Ahammad [51] claims that little attention has been paid to the basic research related to the elucidation of the mechanisms associated to the electrochemical performance of electrodes based on hemoproteins, such as horseradish peroxidase and hemoglobin and that, considering the cost of production of these novel biosensors, higher efforts are required to control the influences of temperature, pH, humidity, and toxic chemicals [51].

10. Conclusions

The perspectives for the applications of hemoproteins, mainly hemoglobins, as biosensors are very promising. However, a more consistent previous study, involving basic research, is a fundamental prerequisite in order to optimize the efficiency of these respective biosensors. The very complex chemistry associated to hemoproteins requires a continuous study focused on its structure-activity relationship to propitiate a higher understanding of the substrate-matrix interaction, surface modification (chemical and/or physical adsorption), and instrumental results. In any case, there has been great progress in this area in the last few years, which has generated very positive perspectives to the future applications of hemoglobins as biosensors.

Author details

Leonardo M. Moreira[1*], Juliana P. Lyon[2], Vanessa J. S. V. Santos[3] and Fabio V. Santos[3]

*Address all correspondence to: leonardomarmo@gmail.com

1 Departamento de Zootecnia (DEZOO), Universidade Federal de São João del-Rei (UFSJ), São João del-Rei, Minas Gerais, Brazil

2 Departamento de Ciências Naturais (DCNAT), Universidade Federal de São João del-Rei (UFSJ), São João del-Rei, Minas Gerais, Brazil

3 Campus Centro-Oeste "Dona Lindu" (CCO), Universidade Federal de São João del-Rei (UFSJ), Divinópolis, Minas Gerais, Brazil

References

[1] Ow YP, Green DR, Hao Z, Mak TW. Cytochrome *c*: functions beyond respiration. Nature Reviews Molecular Cell Biology. 2008; 9:532-542.

[2] Ownby DW, Zhu H, Schneider K, Beavis RC, Chait BT, Riggs AF. The Extracellular Hemoglobin of the Earthworm, *Lumbricus terretris* – Determination of Subunit Stoichiometry. The Journal of Biological Chemistry. 1993; 268: 13539-13547.

[3] Zhu H, Ownby DW, Riggs CK, Nolasco NJ, Stoops JK, Riggs AF. Assembly of the Gigantic Hemoglobin of the Earthworm *Lumbricus terrestris* – Roles of Subunit Equilibria, Non-globin Linker Chains, and Valence of the Heme Iron. The Journal of Biological Chemistry. 1996; 271: 30007-30021.

[4] Kao W, Qin J, Fushitani K, Smith SS, Gorr TA, Riggs CK, Knapp JE, Chait BT, Riggs AF. Linker Chains of the Gigantic Hemoglobin of the Earthworm *Lumbricus terrestris*: Primary Structures of Linkers L2, L3, and L4 and Analysis of the Connectivity of the Disulfide Bonds in Linker L1. Proteins: Structure, Function, and Bioinformatics. 2006; 63: 174-187.

[5] Vitale R, Balland V, Lista L, Maglio O, Nastri F, Pavone V, Lombardi A. Artificial heme-enzymes with enhanced peroxidase-like activity. Journal of Biological Inorganic Chemistry. 2014; 19:1.

[6] Shumyantseva VV, Suprun EV, Bulko TV, Dobrynina OV, Archakov AI. Sensor Systems for medical application based on hemoproteins and nanocomposite materials. Biochemistry (Moscow) Supplement Series B: Biomedical Chemistry. 2010; 4:25-36.

[7] Mahan B, Myers RJ. University Chemistry. The Benjamin/Cummings Publishing Company, Inc. 1987; 4.ed.

[8] Lee JD. Concise Inorganic Chemistry. Chapman & Hall. 1996.

[9] Lindoy LF. The Chemistry of Macrocyclic Ligand Complexes. Cambridge University Press. 1989.

[10] Vila Nova SP, Pereira GAL, de Sá GF, Alves Júnior S, Bazin H, Autiero H, Mathis G. Synthesis, Characterization and Spectroscopic Properties of $[Lnc(bipy)_2py(CO_2Et)_2]^{3+}$ Lanthanide Cryptates. Química Nova. 2004; 27: 709-714.

[11] Bottari G, de la Torre G., Guldi DM, Torres T. Covalent and Noncovalent Phthalocyanine-Carbon Nanostructure Systems: Synthesis, Photoinduced Electron Transfer, and Application to Molecular Photovoltaics. Chemical Reviews. 2010; 110: 6768-6816.

[12] Shipovskov S, Karlberg T, Fodje M, Hansson MD, Ferreira GC, Hansson M, Reimann CT, Al-Karadaghi S. Metallation of the Transition-state Inhibitor N-methyMesoporphyrin by Ferrochelatase: Implications for the Catalytic Reaction Mechanism. Journal of Molecular Biology. 2005; 352: 1081-1090.

[13] Li W, Aida T. Dendrimer Porphyrins and Phthalocyanines. Chemical Reviews. 2009; 109:6047-6076.

[14] Bishop DG, Reed ML. Photochemistry and Photobiology Reviews. 1976; 1.

[15] Huheey JE, Keiter EA, Keiter RL. Inorganic Chemistry: Principles of Structure and Reactivity. Harper Collins College Publishers. 1993; 4.ed.

[16] Shriver DF, Atkins PW. Inorganic Chemistry, Chapter 19,Third Edition, Oxford University Press, 1999.

[17] Park SH, Kim SJ, Kim JD, Park S, Huh DS, Shim YK, Choe SJ. The Studies on Molecular Geometries and Electronic Structures of Substituted *meso*-Catecholic Porphyrins: DFT Methods and NSD. Bulletin of the Korean Chemical Society. 2008; 29: 1141-1148.

[18] Severance S, Hamza I. Trafficking of Heme and Porphyrins in Metazoa. Chemical Reviews. 2009; 109: 4596-4616.

[19] Moreira LM, Santos FV, Lyon JP, Maftoum-Costa M, Pacheco-Soares C, da Silva NS. Photodynamic Therapy: Porphyrins and Phthalocyanines as Photosensitizers. Australian Journal of Chemistry. 2008; 61:741-754.

[20] Weiss R, Fischer J, Bulach V, Schünemann V, Gerdan M, Trautwein AX, Shelnutt JA, Gros CP, Tabard A, Guilard R. Structure and Mixed Spin State of the Chloroiron(III) Complex of 2,3,7,8,12,13,17,18-Octaphenyl-5,10,15,20 Tetraphenylporphyrin, Fe(dpp)Cl. Inorganica Chimica Acta. 2002; 337: 223-232.

[21] Senge MO. New Trends in photobiology: The Conformational Flexibility of Tetrapyrroles — Current Model Studies and Photobiological Relevance. Journal of Photochemistry and Photobiology B. 1992; 16: 3-36.

[22] Ma J, Zhang J, Franco R, Jia S, Moura I, Moura JJG, Kroneck PMH, Shelnutt JA. The Structural Origin of Nonplanar Heme Distortions in Tetraheme Ferricytochromes *c*3. Biochemistry. 1998; 37: 12431-12442.

[23] Unger E, Dreybrodt W, Schweitzer-Stenner R. Conformational Properties of Nickel(II) *meso*-Tetraphenylporphyrin in Solution. Raman Dispersion Spectroscopy Reveals the Symmetry of Distortions for a Nonplanar Conformer. The Journal of Physical Chemistry A. 1997; 101: 5997-6007.

[24] Ogura H, Yatsunyk L, Medforth CJ, Smith KM, Barkigia KM, Renner MW, Melamed D, Walker FA. Molecular Structures and Magnetic Resonance Spectroscopic Investigations of Highly Distorted Six-Coordinate Low-Spin Iron(III) Porphyrinate Complexes. Journal of the American Chemical Society. 2001; 123: 6564-6578.

[25] Neal TJ, Cheng B, Ma J, Shelnutt JA, Schulz CE, Scheidt WR. Conformational diversity in (octaethylporphinato)(trichloroacetato)iron(III) derivatives Inorganica Chimica Acta. 1999; 291: 49-59.

[26] Jentzen W, Simpson MC, Hobbs JD, Song X, Ema T, Nelson NY, Medforth CJ, Smith KM, Veyrat M, Mazzanti M, Ramasseul R, Marchon JC, Takeuchi T, Goddard III WA,

Shelnutt JA. Ruffling in a Series of Nickel(II) *meso*-Tetrasubstituted Porphyrins as a Model for the Conserved Ruffling of the Heme of Cytochromes *c*. Journal of the American Chemical Society. 1995; 117: 11085-11097.

[27] Tsai H, Simpson MC. Isolated Impact of Ruffling on the Vibrational Spectrum of Ni Porphyrins. Diagnosing Out-of-Plane Distortions. The Journal of Physical Chemistry A. 2004; 108: 1224-1232.

[28] Chang MK. Recent Advances in Heme-Proteins Sensors. Current Opinion in Chemical Biology. 2001; 5: 216-222.

[29] Palaniappan AI, Moochhala S, Tay FEH, Phua NCL, Su X. Selective and Enhanced Nitric Oxide Detection using Hemoprotein/Silica hybrids. Sensors and Actuators, B: Chemical. 2008; 133: 241-243.

[30] Lamy JN, Green BN, Toulmond A, Wall JS, Weber RE, Vinogradov SN. Giant Hexagonal Bilayer Hemoglobins. Chemical Reviews. 1996; 96:3113-3124. Doi: 10.1021/cr9600058

[31] Weber RE, Vinogradov SN. Nonvertebrate Hemoglobins: Functions and Molecular Adaptations. Physiological Reviews. 2001; 81:569-628.

[32] Vinogradov SN. The Structure of Invertebrate Extracellular Hemoglobins (erythrocruorins and chlorocruorins). Comparative Biochemistry and Physiology Part B: Comparative Biochemistry. 1985; 82: 1-15. Doi: 10.1016/0305-0491(85)90120-8

[33] Poli AL, Moreira LM, Imasato H. Autoxidation of Giant Extracellular Hemoglobin of *Glossoscolex paulistus*: Molecular Mechanism and Oligomeric Implications. Spectrochimica Acta Part A: Molecular and Biomolecular Spectroscopy. 2011; 82:306-315. Doi: 10.1016/j.saa.2011.07.053

[34] Poli AL, Moreira LM, Tabak M, Imasato H. SDS (sodium dodecyl sulfate) Effect on the Autoxidation of the *Glossoscolex paulistus* Giant Extracellular Hemoglobin: Kinetic studies at pH 7.0 and 9.0. Colloids and Surfaces B: Biointerfaces 2006: 52:96-104.

[35] Moreira LM, Santiago PS, de Almeida EV, Tabak M. Interaction of Giant Extracellular *Glossoscolex paulistus* Hemoglobin (HbGp) with Zwitterionic Surfactant *N*-hexadecyl-*N,N*-dimethyl-3-ammonio-1-propanesulfonate (HPS): Effects of Oligomeric Dissociation. Colloids and Surfaces B: Biointerfaces 2008; 61:153-163.

[36] Moreira LM, Poli AL, Lyon JP, Aimbire F, Toledo JC, Costa-Filho AJ, Imasato H. Ligand Changes in Ferric Species of the Giant Extracellular Hemoglobin of *Glossoscolex paulistus* as Function of pH: Correlations between Redox, Spectroscopic and Oligomeric Properties and General Implications with Different Hemoproteins. Journal of Porphyrins and Phthlacyanines. 2010; 14:199-218. Doi: 10.1142/S108842461000201

[37] Tan X, Zhang J, Tan S, Zhao D, Huang Z, Mi Y, Huang Z. Amperometric Hydrogen Peroxide Biosensor Based on Immobilization of Hemoglobin on a Glassy Carbon

Electrode Modified with Fe_3O_4/Chitosan Core-Shell Microspheres. Sensors. 2009; 9: 6185-6199.

[38] Ringhieri P. Synthetic Heme-Proteins in Biosensors Development. PhD Thesis. Università Degli Studi Di Napoli Federico II. 2010: 167.

[39] Liu M, Zhao G, Tang Y, Shi H, Yang N. Direct Electrochemistry of Hemoglobin on Vertically Aligned Carbon Hybrid TiO_2 Nanotubes and Its Highly Sensitive Biosensor Performance. Chinese Journal of Chemistry. 2013; 31: 215-220. Doi: 10.1002/cjoc.201200883

[40] Wang F, Yang C, Duan C, Xiao D, Tang Y, Zhu J. An Organ-Like Titanium Carbide Material (MXene) with Multilayer Structure Encapsulating Hemoglobin for a Mediator-Free Biosensor. Journal of the Electrochemical Society. 2015; 162:816-821. Doi: 10.1149/2.0371501jes

[41] Dang X, Zheng J, Hu C, Wang S, Hu S. Hemoglobin Biosensor Based on Reduced Graphite Oxide Modified Gold Electrode Array Printed on Paper. Chemical Sensors. 2013; 3:17.

[42] Gu H, Yu A, Yuan S, Chen H. Amperometric nitric oxide biosensor based on the immobilization of hemoglobina on a nanometer-sized gold colloid modified Au electrode. Analytical Letters. 2002; 35:647-661.

[43] Wu JP, Yin, F. Novel Hydrogen Peroxide Biosensor Based on Hemoglobin Combined with Electrospinning Composite Nanofibers. Analytical Letters. 2013; 46:818-830. Doi: 10.1080/00032719.2012.733900

[44] Norouzi P, Larijan B, Faridbod F, Ganjali MR. Hydrogen Peroxide Biosensor Based on Hemoglobin Immobilization on Gold Nanoparticle in FFT Continuous Cyclic Voltammetry Flow Injection System. International Journal of Electrochemical Science. 2010; 5:1550-1562.

[45] Sezgintürk MK, Dinçkaya E, H_2O_2 Determination by a Biosensor Based on Hemoglobin. Preparative Biochemistry and Biotechnology. 2009; 39: 1-10.

[46] Palanisamy S, Cheemalapati S, Chen S. Highly Sensitive and Selective Hydrogen Peroxide Biosensor Based on Hemoglobin Immobilized at Multiwalled Carbon Nanotubes–Zinc oxide Composite Electrode. Analytical Chemistry. 2012; 429: 108-115.

[47] L. Ren, J. Dong, X. Cheng, J. Chu, P. Hu, Hydrogen Peroxide Biosensor Based on Direct Electrochemistry of Hemoglobin Immobilized on Gold Nanoparticles in a Hierarchically Porous Zeolite. Microchimica Acta. 2013; 180, 1333-1340.

[48] Zhou H, Gan X, Wang J, Zhu X, Li G. Hemoglobin-Based Hydrogen Peroxide Biosensor Tuned by the Photovoltaic Effect of Nano Titanium Dioxide. Analytical Chemistry. 2005; 77:6102-6104. Doi; 10.1021/ac050924a.

[49] Souza TTL, Moraes ML, Ferreira M. Use of Hemoglobina as Alternative to Peroxidases in Cholesterol Amperometric Biosensors. Sensors and Actuators B: Chemical. 2013; 178:101-106.

[50] Batra B, Lata S, Sharma M, Pundir CS. An Acrylamide Biosensor Based on Immobilization of Hemoglobin onto Multiwalled Carbon Nanotube/Copper Nanoparticles/Polyaniline Hybrid Film. Analytical Biochemistry. 2013; 433:210-217.

[51] Saleh Ahammad AJ. Hydrogen Peroxide Biosensors Based on Horseradish Peroxidase and Hemoglobin. Biosensors & Bioelectronics. 2013; S9:1-11. Doi: 10.4172/2155-6210.S9-001.

Chapter 15

Graphene-Polyaniline Biosensor for Carbamate Pesticide Determination in Fruit Samples

Luleka Luzi-Thafeni, B. Silwana, E. Iwuoha and V. Somerset

Additional information is available at the end of the chapter

http://dx.doi.org/10.5772/61220

Abstract

In this study, a simple, sensitive, and low cost electrochemical biosensor for the quantitative determination of carbamate pesticides has been constructed. A composite consisting of polyaniline (PANI) and graphene oxide was electrochemically synthesised on a platinum electrode. This sensor platform was then used in the biosensor construction by electrostatic attachment of the enzyme, horseradish peroxidase (HRP) onto the surface of the Pt/GO-PANI electrode. Voltammetric results concluded that HRP immobilised on the Pt/GO-PANI composite retained its bioelectrocatalytic activity towards the reduction of H_2O_2 and was not changed during its immobilisation. The Pt/GO-PANI/HRP biosensor was then applied to successfully detect standard carbamate pesticides in a 0.1 M phosphate buffer (PB; pH = 6.8) solution. Various performance and stability parameters were evaluated for the Pt/GO-PANI/HRP biosensor, which included the optimal enzyme loading, effect of pH and long-term stability of the biosensor on its amperometric behaviour. The Pt/GO-PANI/HRP biosensor was finally applied to the detection of three carbamate pesticides of carbaryl, carbofuran, and methomyl using the enzyme inhibition method. Carbaryl, carbofuran, and methomyl analyses were amperometrically determined using spiked real samples of orange, pear, and grapes, within a concentration range of 0.01–0.3 mg/L. These results indicated that the biosensor is sensitive enough to detect carbamate pesticides in real fruit matrices. The detection limit for carbaryl, carbofuran, and methomyl in real fruit samples by amperometric method was determined to be 0.136 mg/L, 0.145 mg/L, and 0.203 mg/L, respectively. The application of the Pt/GO-PANI/HRP biosensor has demonstrated that the biosensor is sensitive enough for amperometric detection and could be a useful tool in the screening of these pesticides at low concentrations.

Keywords: Poly(2,5-dimethyaniline), Graphene, Organic phase biosensor, Carbamate pesticides, Horseradish peroxidase

1. Introduction

Pesticides are known to be very toxic compounds that pose a threat to aquatic ecosystems and human health. One such group of pesticides that have been applied in agriculture to control insects are the organophosphates and carbamates. These pesticides are of particular concern for human health since they inhibit the activity and functioning of the enzyme called acetylcholinesterase (AChE) in insects and mammals. When inhibition of AChE occurs, it may lead to respiratory paralysis and consequently death. It is therefore important for food safety and environmental protection to employ fast and effective detection technologies [1-3].

Many standard methods are available for the detection of pesticide compounds in various matrices and include gas chromatography (GC), high performance liquid chromatography (HPLC), liquid chromatography (LC) coupled to sensitive and specific detectors that includes nitrogen-phosphorous detectors (NPD), flame ionisation detectors (FID), ultraviolet detectors (UVD), diode array detector (DAD), or mass spectrometry (MS) [1, 4-5].

However, the preceding methods are known to be expensive, time-consuming, and requiring highly trained personnel to operate this equipment. Furthermore, this equipment is not always suitable for *in situ* and real-time detection of carbamate pesticides [1, 4-5].

Alternative methods that have been utilised for pesticide detection involve the use of electrochemical sensors, especially biosensors constructed with AChE in enzyme-modified electrodes. Biosensors based on the inhibition of AChE by carbamates have been successfully implemented for detection, since they provide advantages such as rapid detection, simplicity, and low cost [1, 6-7].

The development of biosensors involves one most critical step such as immobilisation, whereby the biological recognition element is associated with a physicochemical transducer [8]. Biosensor performance can be negatively affected by the immobilisation process. Therefore, intensive efforts are needed for the development of effective immobilisation methods, allowing for improvements in operational and stability storage, response time, linear range, and sensitivity, while preserving the enzyme affinity for the substrates and inhibitors [9-10].

This study involved the utilisation of graphene, modified with a conducting polymer (e.g., polyaniline) on a suitable transducer surface, for immobilisation of the enzyme during biosensor construction. Graphene is known to consist of a single layer of carbon atoms in a closely packed honeycombed two-dimensional lattice. Graphene has unique electronic, mechanical, and thermal properties that has seen this compound been extensively applied in fields such as batteries, field-effect transistors, ultrasensitive sensors, and electrochemical resonators [11-13].

However, graphene has some limitations that include poor solubility and the synthesis of graphene oxide was proposed to overcome this limitation. Graphene oxide (GO) sheets are known to be hydrophilic and offer the potential of preparing graphene film that is more processable. Secondly, the properties of GO are similar to that of graphene, which include a single atomic plane of graphite structure into which target ions, molecules, and other macromolecules can be adsorbed [11, 13].

Functionalising the GO with a conducting polymer such as polyaniline (PANI) increased the success of the sensor platform devised for this study. It is known that PANI, one of the mostly used conducting polymers in biosensor construction, has some unique properties. Some of these properties include ease of synthesis, high capacitive characteristics, low cost, supportive conducting platform for enzyme entrapment, etc. [9-10, 12]. Therefore, this study has seen the synthesis of a graphene-PANI composite film that was synthesised from graphene oxide as the starting material to tap into both the properties of graphene and PANI, but also to overcome the limitations of each of the conducting films [12].

The primary goal of the present study was therefore the construction and application of the Pt/GO-PANI/HRP biosensor that were constructed to determine the amount of carbamate pesticide compounds such as carbaryl, methomyl, and carbofuran in deciduous and citrus fruit (grapes, pears, and oranges) samples. The enzyme horseradish peroxidase (HRP) replaced the AChE usually employed in carbamate biosensor studies and was found to perform well in the enzyme inhibition studies. The results obtained for the determination of the carbamates using voltammetric (e.g., differential pulse voltammetry) analysis are discussed in this chapter.

2. Materials and methods

2.1. Chemical and reagents

Three carbamate pesticides (e.g., carbaryl, carbofuran, and methomyl) were selected for this work. Stock solutions were prepared from the 1000 mg/L certified reference materials (CRMs; Sigma-Aldrich, South Africa) of carbamate pesticides using acetonitrile organic solvent. Working standard such as 0.01, 0.1, 0.15, 0.2, and 0.3 ppm were also prepared from the stock solution and used for the spiking of real fruit samples. The enzyme horseradish peroxidase (EC 1.11.1.7 type IV from horseradish, 250–330 units/mg) was also purchased from Sigma-Aldrich (Germany). The hydrogen peroxide (30%), sulphuric acid (99 %), potassium permanganate ($KMnO_4$), potassium chloride (KCl), and sodium nitrate ($NaNO_3$) (84.99%) were all purchased from Merck, South Africa. The aniline (99.5 %), graphite (fine powder synthetic), and the following organic solvents such as n-hexane (96% HPLC grade), acetonitrile (HPLC grade), methanol (HPLC grade), and iso-octane (95%) were bought from Sigma-Aldrich (South Africa). All solutions were prepared with Millipore deionised water and experiments were performed at room temperature.

2.2. Apparatus

2.2.1. Voltammetric measurements

Electrochemical measurements were conducted in a three electrode electrochemical cell under controlled temperature (25°C). The working electrode was Pt (diameter = 1.6 mm). The reference electrode was comprised of an Ag/AgCl in 3 M NaCl system, and a Pt wire was used as the auxiliary electrode. All the electrochemical measurements were performed with an Epsilon electrochemical analyser (BASi Instruments, 2701 Kent Ave, Westr Lafayette, IN 47906, USA), utilising cyclic voltammetry (CV) and differential pulse voltammetry (DPV) modes. Electrocatalytic responses of the Pt/GO-PANI/HRP biosensor to H_2O_2 substrate were investigated by amperometric mode in the presence and absence of carbaryl, carbamate, and methomyl standards in the test solutions [14-15].

2.3. Preparation of graphene oxide and polyaniline mixture

Approximately 2.01 g of graphite powder, 1.03 g of $NaNO_3$, and 4.02 g of $KMnO_4$ were weighed and dissolved in 100 mL of concentrated sulphuric acid (H_2SO_4) solution and the resulting mixture was stirred vigorously for 7 hours at room temperature. Subsequently, 250 mL of 5% H_2SO_4 aqueous solution was added and the solution was kept at 98°C for 2 hours. Temperature was reduced to room temperature and the 10 mL of 30% H_2O_2 was slowly added and the reaction was further stirred for 2 hours. A light brown graphene oxide (GO) precipitate was obtained by washing it with 0.1 M H_2SO_4 solution and then distilled water until the pH of the supernatant was neutral [16].

The precipitate obtained was transferred to a glass vial to prepare the 1 g GO precipitate for mixing with the aniline monomer, before electrosynthesis of the graphene oxide-aniline mixture was performed. A 10 mL aqueous solution of 0.5 M aniline in 1 M H_2SO_4 solution was prepared separately and then transferred to the glass vial containing the GO precipitate. The mixture was allowed to mix for approximately 5–10 minutes using a sonicator bath. Electrosynthesis of the graphene oxide-polyaniline (GO-PANI) film was obtained by scanning the potential repeatedly between -0.2 V and + 1.1 V (vs. Ag/AgCl) for 10 cycles at a scan rate of 40 mV/s. The obtained GO-PANI polymer film was then characterised using CV, DPV measurements, including UV-Vis and FTIR spectrometry (results not reported here). The modified constructed electrode was referred to as Pt/GO-PANI and stored in 0.1 M phosphate buffer (pH 7.2) solution at 4 °C.

2.4. HRP biosensor construction

The Pt/GO-PANI electrode was stored in phosphate buffer (PB) solution at 4°C, when not in use. This was followed by enzyme incorporation to obtain the horseradish peroxidase (HRP) biosensor. The biosensor was constructed, using the prepared Pt/GO-PANI electrode that was transferred to a batch cell, containing a 1 mL solution of argon degassed 0.1 M phosphate buffer (pH 6.8). The GO-PANI film was next reduced at a potential of – 500 mV (vs. Ag/AgCl) until a steady current was achieved (approximately 30 minutes). Afterwards, the electrode was transferred to a second batch cell containing 50 mL of 2 mg/L of HRP in a 0.1 M phosphate buffer (pH 6.8) solution. This solution was also argon degassed, before enzyme immobilisation through covalent binding to the GO-PANI film was performed. This was achieved by oxidation of the PANI film in the presence of HRP at a potential of + 700 mV (vs. Ag/AgCl) until a steady current was achieved (approximately 40 minutes) [10]. The achievement of a steady state current was an indication that bulk electrolysis was complete and the polymer film was charged and ready for enzyme attachment.

Following enzyme attachment, the constructed Pt/GO-PANI/HRP biosensor was thoroughly rinsed with double-distilled water, in order to remove any unbound enzyme followed by storage in the working 0.1 M phosphate buffer (pH 6.8) solution at 4°C.

2.5. Stability and reproducibility studies

The Pt/GO-PANI/HRP biosensor was further evaluated at different 7 day intervals, using the same specific constructed biosensor. A test solution consisting 1 ml of 0.1 M phosphate buffer, 0.1 M KCl (pH = 6.8) solution was degassed before any H_2O_2 as substrate was added. The Pt/GO-PANI/HRP biosensor was then evaluated using a 1 ml test solution to which small aliquots of H_2O_2 as substrate was added and peak current collected. This procedure was repeated for one month on 7 day intervals, on the 7th, 14th, 21st, and 28th day after the specific sensor was constructed. The peak current collected were then compared to determine the stability of the biosensor constructed [9, 17].

The reproducibility of the Pt/GO-PANI/HRP biosensor was also evaluated by the construction of 5 HRP biosensors, for which the sensing of H_2O_2 as substrate was evaluated.

3. Electrochemical detection

3.1. Determination of the limit of detection

The limit of detection of the Pt/GO-PANI/HRP biosensor was evaluated by performing 10 replicate measurements in a 0.1 M phosphate buffer, 0.1 M KCl (pH = 6.8) solution, or on any one of the analyte (standard pesticide) solutions at the lowest working concentration. This was followed by the construction of a calibration graph of current (*Ip*) versus saline phosphate buffer (PBS) or analyte concentration for which the slope and the linear range was then determined. The limit of detection (LOD) was then calculated with the following equation:

$$LOD = \frac{3 \cdot s}{m}$$

In which s is the standard deviation of the 10 replicate measurements in PBS, or on any one of the analyte (standard pesticide) solutions at the lowest working concentration? The variable m represents the slope of the calibration graph in the linear range, which was also used to estimate the sensitivity of the measurements performed [9, 18].

3.2. Amperometric detection of carbamate pesticides

The detection of the carbamate pesticides (e.g., carbaryl, carbofuran, and methomyl) was performed using the percentage inhibition method. During this procedure, the Pt/GO-PANI/HRP biosensor was placed in a stirred 1 mL of 0.1 M phosphate (0.1 M KCl, pH 6.8) solution (anaerobic conditions), followed by multiple additions of a standard hydrogen peroxide (H_2O_2) substrate solution, until no relative change in current response was obtained. No increase in peak current during the catalytic monitoring of the HRP biosensor to substrate addition indicated that substrate saturation was reached.

The biosensor was thereafter rinsed with double-distilled water and incubated in a carbaryl solution of specific concentration and the HRP enzyme was exposed for 20 minutes to the chosen pesticide concentration in a 0.1 M phosphate (0.1 M KCl, pH 6.8) solution solution. After pesticide exposure, the HRP biosensor was thoroughly rinsed with deionised water and placed into a freshly stirred 1 mL of 0.1 M phosphate (0.1 M KCl, pH 6.8) solution.

Thereafter, the Pt/GO-PANI/HRP biosensor was again exposed to sequential addition of H_2O_2 substrate solution in a 0.1 M phosphate (0.1 M KCl, pH 6.8) solution. This represented the biosensor current response after enzyme inhibition was performed, which was evident in the reduced biosensor current responses obtained.

This procedure was repeated for several different concentrations of carbaryl (followed by carbofuran and methomyl), to obtain a standard calibration curve. The percentage inhibition ($I\%$) for each concentration of carbaryl inhibition of the enzyme was determined using the formula in Eqn 1:

$$I\% = \frac{(I_1 - I_2)}{I_1} \times 100 \qquad (1)$$

where I% is the degree of inhibition, I_1 is the steady-state current obtained in buffer solution, I_2 is the steady-state current obtained in buffer solution after the biosensor was exposed to carbaryl standard solution [10].

The anodic difference differential pulse voltammogram (DPV) responses were collected in an oxidation direction also, by applying a linear potential scan between – 0.40 V and – 1.0 V (vs. Ag/AgCl). The DPV responses were performed at a scan rate of 10 mV/s and a pulse amplitude of 0.50 V. The sample width, pulse width, and pulse period were 17 ms, 50 ms, and 200 ms, respectively [9, 18].

4. Results and discussion

4.1. Optimisation of solution pH for Pt/GO-PANI/HRP biosensor

After successful construction of the Pt/GO-PANI/HRP biosensor, evaluation of the biosensor was performed in the pH range from 5.0 to 7.2, to evaluate and determine the optimum current response for the constructed biosensor. A fresh biosensor was constructed and evaluated at each of the pH values evaluated from 5.0 to 7.2. The results obtained are displayed in Figure 1.

The optimum pH of the Pt/GO-PANI/HRP biosensor was determined as follows. A series of 0.1 M phosphate buffer, 0.1 M KCl solutions were prepared at different pH values of 5.0, 5.5,

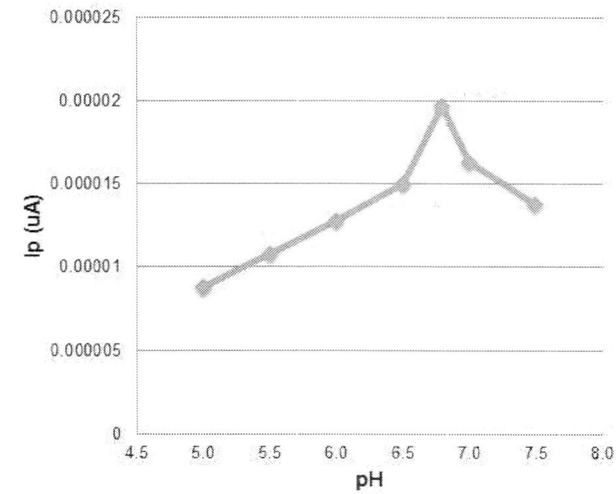

Figure 1. Results obtained for the optimisation of the pH of the Pt/GO-PANI/HRP biosensor evaluated in 0.1 M phosphate buffer, 0.1 KCl solution with the addition of H_2O_2 as substrate.

6.0, 6.5, 6.8, 7.0, and 7.5. An aliquot of 1 ml test solution containing 0.1 M phosphate buffer, 0.1 M KCl solution was degassed with argon before any substrate was added [9].

The Pt/GO-PANI/HRP biosensor was then evaluated in the 1 ml test solution by sequential addition of the 1 μM H_2O_2 substrate to the test solution. After degassing, the maximum current response of the biosensor was then obtained at the different pH values after a total of 0.8–1.2 μM of the H_2O_2 substrate was added. In Figure 1, the results obtained have shown that the optimum peak current response was obtained at pH = 6.8. This pH was then used in all subsequent Pt/GO-PANI/HRP biosensor investigations.

4.2. Voltammetric characterisation of Pt/GO-PANI/HRP biosensor

The differential pulse voltammetric (DPV) responses of the Pt/GO-PANI/HRP biosensor for the analysis of standard carbamates pesticide samples, incubated in acetonitrile-saline phosphate buffer (pH = 6.8; 0.1 M KCl) solution, were recorded for each of the three different pesticides investigated.

In Figure 2 the optimum DPV responses for the Pt/GO-PANI/HRP biosensor to sequential hydrogen peroxide (H_2O_2) substrate addition up to 1.2 μM is shown. The results shown are for the substrate addition (before pesticide exposure) and the respective responses obtained after incubation in different carbaryl pesticide concentrations.

Figure 2 shows a decrease in the maximum cathodic current obtained after incubation of the biosensor in a 0.01 mg/L carbaryl standard solution, when the Pt/GO-PANI/HRP biosensor was subjected to successive additions of H_2O_2 as substrate. A similar decrease was shown when

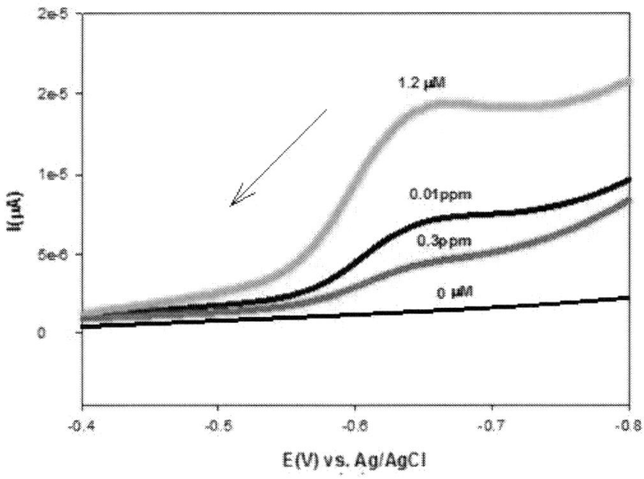

Figure 2. Results for the Pt/GO-PANI/HRP biosensor responses to an optimum 1.2 μM H_2O_2 substrate addition before incubation, and responses obtained after incubation in different carbaryl pesticide concentrations of 0.01 ppm (n = 3; RSD = 4.67%) and 0.3 ppm (n = 3; RSD = 6.92%), respectively.

the Pt/GO-PANI/HRP biosensor was incubated in a 0.3 mg/L carbaryl standard solution. It was observed that the maximum cathodic current obtained decreased as the carbaryl pesticide concentration was increased.

The results obtained for the stability of the Pt/GO-PANI/HRP biosensor have shown that a gradual and steady decrease in peak current (not shown here) was observed over time. The decrease in peak current was more apparent after 28 days. The relative standard deviations (RSD) were found to be 10.3%.

Similarly, the results obtained for the reproducibility of the Pt/GO-PANI/HRP biosensor have shown that the HRP enzyme immobilised in the GO-PANI matrix was relatively stable. The five sensors evaluated have shown a relative standard deviation (RSD) of 6.5% for the evaluation of H_2O_2 as substrate.

Figure 3 shows the calibration curves of peak current versus substrate concentration for the successive addition of H_2O_2 as substrate in the absence of carbaryl pesticide, followed by measured responses after exposure to 0.01 and 0.3 mg/L carbaryl pesticide solutions.

The results obtained for the responses of the Pt/GO-PANI/HRP biosensor to carbaryl pesticide inhibition have shown clear trends. It was firstly observed that the highest peak current responses were obtained in the absence of pesticide exposure (Figure 3). This was followed by a decrease in peak current responses observed as the carbaryl pesticide concentration was increased from 0.01 to 0.3 mg/L, with the lowest peak current responses observed for the highest pesticide concentration of 0.3 mg/L evaluated in Figure 3.

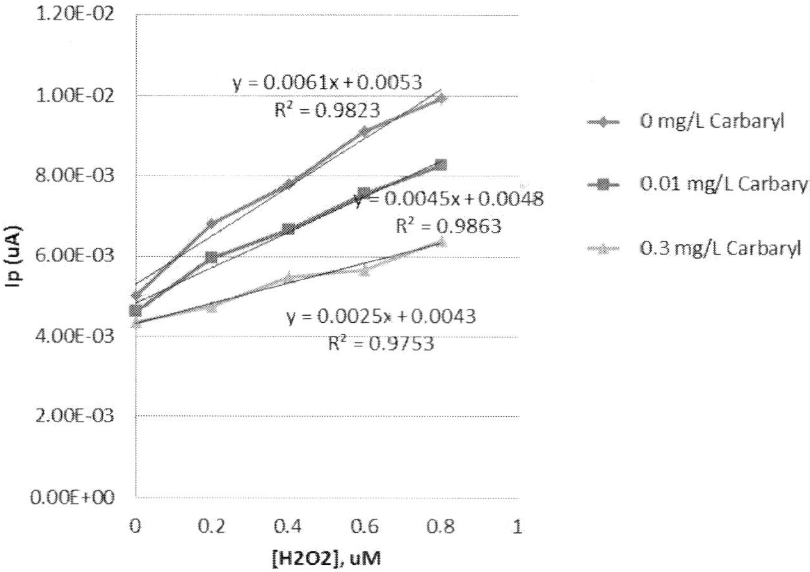

Figure 3. Results for the combined calibration plots of the Pt/GO-PANI/HRP biosensor responses to H_2O_2 substrate addition in the absence of carbaryl, followed by the addition of different carbaryl concentrations of 0.01 mg/L (*n* = 3; RSD = 10.4%) and 0.3 mg/L (*n* = 3; RSD = 17.3%), respectively.

The responses of the Pt/GO-PANI/HRP biosensor to 1.2 μM of the substrate H_2O_2 added before incubation, followed by the respective responses obtained after incubation in different carbofuran pesticide concentrations, are shown in Figure 4.

The first result indicated that a shift in cathodic peak potential was obtained after 1.2 μM of the H_2O_2 substrate was added to the Pt/GO-PANI/HRP biosensor. Carbofuran pesticide standard solutions had a similar behaviour as for carbaryl as pesticide when the HRP biosensor was incubated and evaluated. However, the results obtained have shown that the cathodic peak potentials were obtained at different potentials after substrate addition to the biosensor, compared to the carbaryl investigation.

The results in Figure 4 further showed that not only was a peak potential shift observed for carbofuran pesticide evaluation, but the carbofuran behaviour in the Pt/GO-PANI/HRP biosensor further showed a smaller decrease in cathodic peak current obtained in comparison to the results obtained for carbaryl.

The combined Lineweaver-Burk plots for the Pt/GO-PANI/HRP biosensor responses to 1.2 μM of the substrate H_2O_2 added before incubation, followed by the respective responses obtained after incubation in different carbofuran pesticide concentrations are shown in Figure 5.

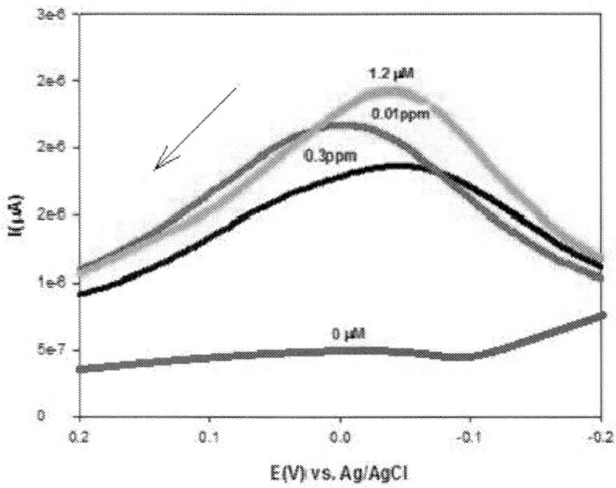

Figure 4. Results for the Pt/GO-PANI/HRP biosensor responses to an optimum 1.2 μM H_2O_2 substrate addition before incubation, and responses obtained after incubation in different carbofuran pesticide concentrations of 0.01 ppm (n= 3; RSD = 8.35%) and 0.3 ppm (n= 3; RSD = 10.4%), respectively.

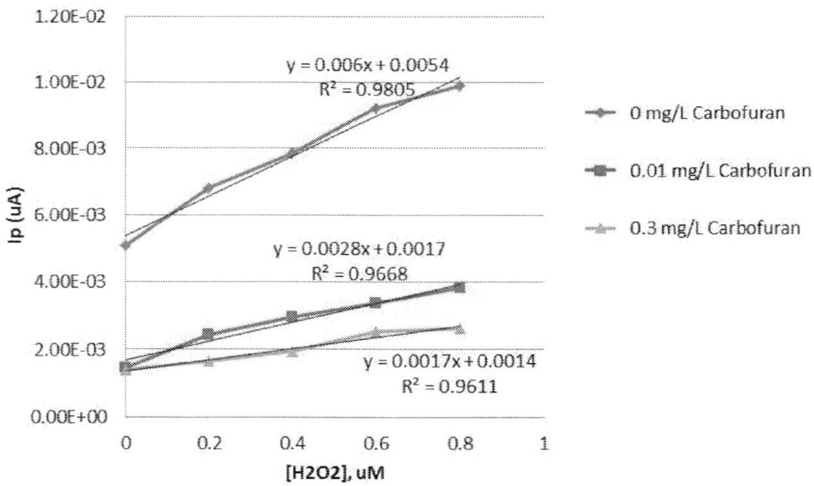

Figure 5. Results for the combined calibration plots of the Pt/GO-PANI/HRP biosensor responses to H_2O_2 substrate addition in the absence of carbofuran, followed by the addition of different carbofuran concentrations of 0.01 mg/L (n = 3; RSD = 31.5%) and 0.3 mg/L (n = 3; RSD = 38.9%), respectively.

The results in Figure 5 show the calibration curves of current versus substrate concentration for the successive addition of H_2O_2, clearly showing a decrease in the cathodic current after incubation of the biosensor in a 0.01 mg/L and 0.3 mg/L carbofuran standard solutions, respectively.

The Pt/GO-PANI/HRP biosensor results for the 0.01 mg/L and 0.3 mg/L carbofuran standard solutions showed that a relatively small difference in the peak current responses of the biosensor was observed. This was a very important observation made, especially since the difference in pesticide concentration evaluated was rather tenfold bigger.

Next, the Pt/GO-PANI/HRP biosensor was evaluated against methomyl as pesticide (Figure 6). The results obtained for methomyl also showed the same behaviour as observed for carbaryl and carbofuran pesticides standard solutions. Figure 6 represents the DPV responses for the Pt/GO-PANI/HRP biosensor to 1.0 μM substrate (H_2O_2) before incubation and the respective responses obtained after incubation in.01 mg/L and 0.3 mg/L methomyl pesticide concentrations.

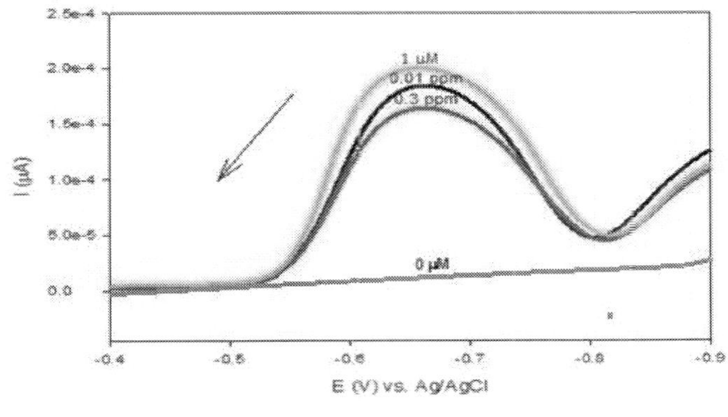

Figure 6. Results for the Pt/GO-PANI/HRP biosensor responses to an optimum 1.0 μM H_2O_2 substrate addition before incubation, and responses obtained after incubation in different methomyl pesticide concentrations of 0.01 ppm (n = 3; RSD = 3.86%) and 0.3 ppm (n = 3; RSD = 9.25%), respectively.

Analysis of the results in Figure 6 has shown a relatively small decrease in the cathodic current after incubation of the biosensor in a 0.01 mg/L methomyl standard solution, compared to the peak current results obtained for the biosensor response in H_2O_2 substrate solution.

Figure 7 confirmed the results obtained in Figure 6. The Pt/GO-PANI/HRP biosensor evaluation has confirmed the methomyl pesticide results to be similar to that of carbaryl and carbofuran, indicating that the lowest peak current responses were observed for the highest pesticide concentration evaluated in Figure 7.

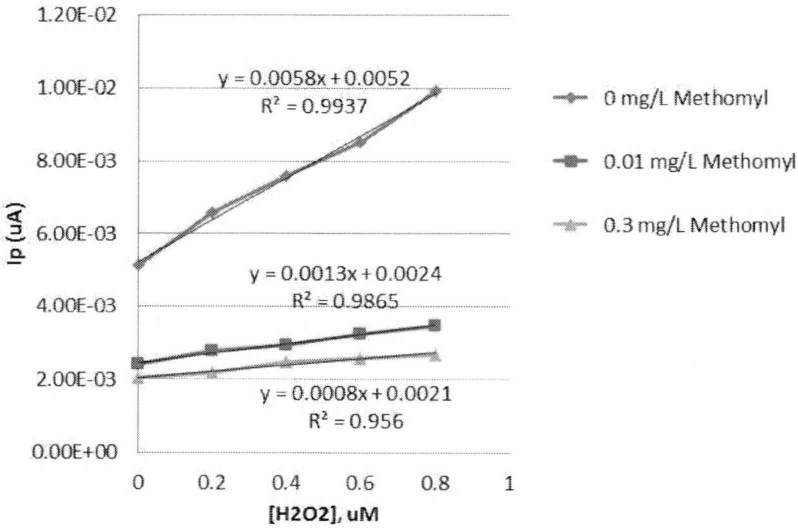

Figure 7. Results for the combined calibration plots of the Pt/GO-PANI/HRP biosensor responses to H_2O_2 substrate addition in the absence of methomyl, followed by the addition of different methomyl concentrations of 0.01 mg/L (n = 3; RSD = 32.5%) and 0.3 mg/L (n = 3; RSD = 39.8%), respectively.

4.3. Inhibition studies of standard carbamate pesticide samples

The Pt/GO-PANI/HRP biosensor was further evaluated using the percentage inhibition method described in Section 2.5.

Analyses of the results in Figure 8 have shown that the HRP biosensor obtained different results for each of the carbamate pesticides evaluated. A steady increase in the percentage inhibition was observed as the individual concentrations for each of the respective carbamate pesticides was increased (see each individual graph). The highest percentage inhibition was observed for carbaryl (58%), followed by methomyl (52%), and the least for carbofuran (35%) [9, 18].

Analysis of the percentage inhibition results in Figure 8 have also shown that for carbaryl, the highest percentage inhibitions were obtained ranging from 39% to 58% over pesticide concentrations ranging from 0.01 to 0.3 ppm (or mg/L). In the case of methomyl, the percentage inhibitions ranged between 28% to 52%, while in the case of carbofuran it ranged between 16% to 35% over the same concentration range.

The Pt/GO-PANI/HRP biosensor evaluation has further shown that although HRP can replace acetylcholinesterase (AChE) as enzyme in the evaluation of carbamate pesticides, reduced percentage results were obtained for HRP as enzyme [9, 18].

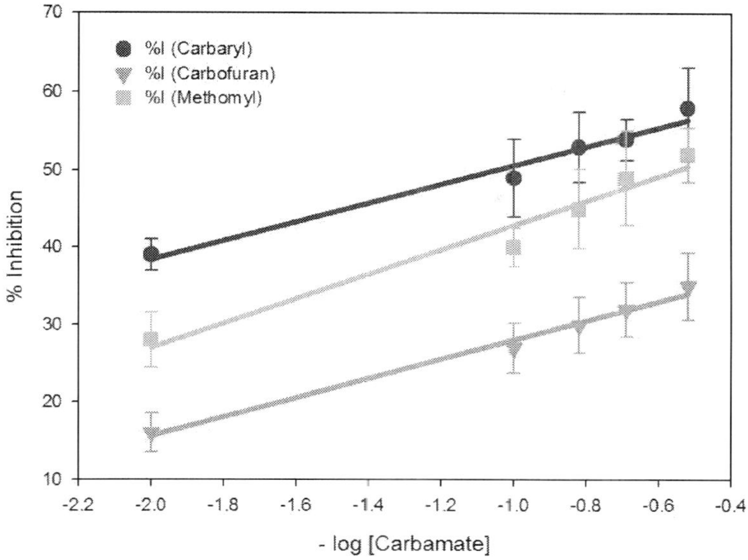

Figure 8. Graphs for the percentage inhibition vs. log [carbamate pesticide] for all three different carbamate pesticides investigated are shown. Inhibition results were collected with the application of the Pt/PDMA-PVS/HRP biosensor.

4.4. Carbaryl, carbofuran, and methomyl pesticide analysis in real samples

The applicability of the HRP biosensor for carbamate pesticide evaluation was further demonstrated by evaluating three different fruit samples such as pear, grapes, and oranges for the determination of carbaryl, carbamate, and methomyl concentrations in spiked samples.

In this work, the method applied for the sample preparation was a liquid–liquid extraction of the pear, grape, and orange samples. The method of standard addition was used to determine the concentrations of carbaryl, carbofuran, and methomyl in these fruit samples.

In this regard, the Pt/GO-PANI/HRP biosensor was immersed in an electrochemical cell containing a fresh 1 ml of 0.1 M phosphate buffer (0.1 M KCl, pH = 6.8) solution and a 1000 μL aliquot of fruit sample was added to the PBS solution, after which the amperometric responses of the biosensor was measured.

The same procedure was then used to evaluate the fruit samples spiked with carbaryl, carbofuran, and methomyl standard solutions, respectively. A known concentration of analyte was added to the extracted fruit sample solution in order to account for any impurities in the calibration of extracted samples. The concentrations of carbaryl, carbofuran, and methomyl in the samples were then determined by extrapolation.

Figure 9 shows the calibration curves of peak current vs. carbamate concentration for the determination of carbaryl, carbofuran, and methomyl in real fruit samples.

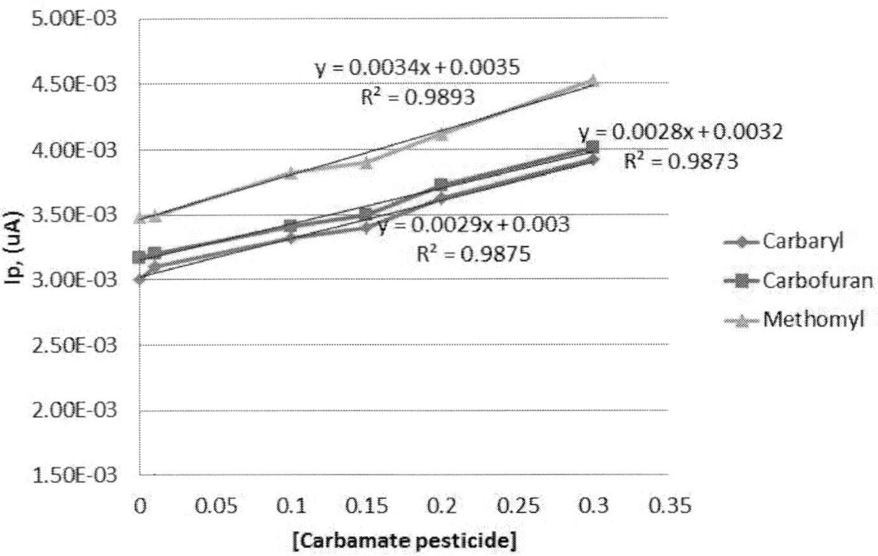

Figure 9. Analytical calibration curves and results obtained for the determination of carbaryl, carbofuran, and methomyl concentrations in real fruit samples by the method of standard addition (n = 4).

The calibration curve for carbaryl (Figure 9) determination in pear fruit samples was found to be relatively linear within a concentration range of 0–0.3 mg/L and with a very good correlation coefficient ($r^2 = 0.9875$; $n = 4$). The carbaryl detection limits (LOD) in pear fruit sample achieved by this method was 0.136 mg/L. The biosensor reproducibility for successive measurements was good with a R.S.D. value of 3.5%. The concentration of carbaryl in replicate measurements (n = 4) of pear samples was found to be 0.08 mg/L.

The calibration curve for carbofuran (Figure 9) was also found to be linear within a concentration range of 0–0.3 mg/L and with a very good correlation coefficient ($r^2 = 0.9873$; $n = 4$). Carbofuran was analysed in orange fruit samples and the detection limit (LOD) achieved by this method was found to be 0.145 mg/L, which is slightly less than for carbaryl. The concentration of carbofuran in replicate measurements (n = 4) of orange samples was found to be 0.05 mg/L.

The calibration curve and results for methomyl determination in grape samples is also displayed in Figure 9. This curve was again found to be linear within a concentration range of 0–0.3 mg/L and with a very good correlation coefficient ($r^2 = 0.9893$; $n = 4$). The methomyl detection limits (LOD) in grape fruit sample achieved by this method was 0.203 mg/L. The concentration of methomyl in replicate measurements (n = 4) of grape samples was found to be 0.05 mg/L.

The results reported for the LOD values in this study compared relatively well to previous studies done by this research group for the evaluation of carbamate pesticides [9, 18-19].

A summary of the results obtained for the application of the Pt/GO-PANI/HRP biosensor to real fruit samples analysis is shown in Table 1.

Pesticide Evaluated	Matrix	Added (mg/L)	Detected (mg/L)	%Recovery
Carbofuran	Pears	0.01	0.0053	53.0
		0.10	0.0599	59.9
	Grapes	0.01	0.0048	48.0
		0.10	0.0601	60.1
	Oranges	0.01	0.0051	51.0
		0.10	0.0608	60.8
Carbaryl	Pears	0.01	0.0080	80.2
		0.10	0.0801	80.1
	Grapes	0.01	0.0080	80.6
		0.10	0.0799	79.9
	Oranges	0.01	0.0070	70.3
		0.10	0.0699	69.9
Methomyl	Pears	0.01	0.0050	50.4
		0.10	0.0601	60.1
	Grapes	0.01	0.0051	51.7
		0.10	0.0608	60.8
	Oranges	0.01	0.0050	50.3
		0.10	0.0599	59.9

Table 1. The results obtained for the application of the Pt/GO-PANI/HRP biosensor in the evaluation and analysis of carbofuran, carbaryl, and methomyl concentrations in pear, grape, and orange samples, respectively.

Analysis of the results in Table 1 has shown very interesting trends for the application of the Pt/GO-PANI/HRP biosensor to the determination of carbamate pesticides in fruit samples.

Analysis of the percentage recoveries for the spiked samples (Table 1) have shown variability between the different fruits and pesticides evaluated. In the case of carbofuran results, it was found that for the lower spiked concentration of 0.01 mg/L, the highest percentage recovery was obtained in the grapes, while the lowest percentage recovery was observed in the pears. Evaluation of the high carbofuran concentration of 0.10 mg/L has shown that the lowest percentage recovery was obtained in the pears, with the slightly higher recovery obtained in the oranges.

Evaluation of the carbaryl results has shown overall higher percentage recoveries, compared to the results of carbofuran and methomyl. For the low 0.01 mg/L concentration, the percentage recoveries ranged between 70.3% (oranges) to 80.6% (grapes). In the case of the high 0.10 mg/L concentration, the recoveries ranged between 69.9% (oranges) and 80.1% (pears).

The percentage recovery results for methomyl showed the lowest percentage recoveries compared to the other carbamates evaluated. For the low 0.01 mg/L concentration, the percentage recoveries ranged between 50.3% (oranges) and 51.7% (grapes). In the case of the high 0.10 mg/L concentration, the recoveries ranged between 59.9% (oranges) and 60.8% (grapes). Future work will investigate and compare the results obtained for a Pt/GO-PANI/AChE biosensor.

The results obtained for the analysis of real fruit samples (Table 1), using the Pt/GO-PANI/AChE biosensor, have further highlighted some difficulties obtained with the recoveries. In fact, the relatively lower recoveries in real samples obtained for carbofuran (50–60%), carbaryl (70–80%), and methomyl (50–60%) indicate that some matrix effects may also have hampered better results. It further highlights the difficulty of organic phase biosensor determination of carbamate in fruit samples, indicating that some matrix effects should be further investigated in future biosensor construction and investigations.

5. Conclusion

The results for work done in this study have shown that the Pt/GO-PANI/HRP biosensor was successfully developed for the detection and quantification of carbamates such as carbaryl, carbofuran, and methomyl in real fruit samples. The principle of the detection was based on the reduced biosensor response measurements, which occurred as a result of the catalytic activity of HRP, immobilised on a GO-PANI matrix immobilised on a platinum electrode. The effect of the inhibition was found to be increasing with increasing concentrations of the carbamates utilised in the biosensor inhibition studies. An incubation period of 20 min was applied and the response of the HRP biosensor was measured before and after incubation in each of the three carbamates investigated. The percentage inhibition results showed that the Pt/GO-PANI/HRP biosensor was more inhibited by the carbofuran pesticide exposure, and the least inhibited in the carbaryl pesticide. Due to the sensitivity of HRP biosensor for these carbamates, the development of this method is a way forward for the analysis of carbaryl, carbofuran, and methomyl at residue levels such as those occurring in the environment. The Pt/GO-PANI/HRP biosensors further demonstrated a detection limit of 0.136 mg/L for carbaryl determination, followed by 0.145 mg/L for carbofuran and 0.203 mg/L for methomyl determination in real fruit samples. The Pt/GO-PANI/HRP biosensor evaluation has further shown that although HRP can replace acetylcholinesterase (AChE) as enzyme in the evaluation of carbamate pesticides, reduced percentage inhibition results were obtained for HRP as enzyme. This result may also have affected the percentage recovery results obtained for the determination of carbaryl, carbofuran, and methomyl in spiked pear, orange, and grape samples. It further highlights the difficulty of organic phase biosensor determination of carbamate in fruit

http://dx.doi.org/10.5772/61220

samples, indicating that some matrix effects should be further investigated in future biosensor construction and investigations.

Acknowledgements

The authors wish to express their gratitude to the National Research Foundation (NRF), South Africa, for financial and student support to perform this study. The assistance of the researchers in the SensorLab, Chemistry Department, University of the Western Cape, are also greatly acknowledged.

Author details

Luleka Luzi-Thafeni[1,2], B. Silwana[1,2], E. Iwuoha[2] and V. Somerset[1*]

*Address all correspondence to: vsomerset@csir.co.za; vsomerset@gmail.com

1 NRE, CSIR, Stellenbosch, South Africa, South Africa

2 SensorLab, Department of Chemistry, University of the Western Cape, Bellville, South Africa

References

[1] Raghu, P., Reddy, T.M., Kumara Swamy, B.E., Chandrashekar, B.N., Reddaiah, K. and Sreedhar, M. (2012). Development of AChE biosensor for the determination of methyl parathion and monocrotophos in water and fruit samples: A cyclic voltammetric study. *Journal of Electroanalytical Chemistry*, 665: 76-82.

[2] Liu, Y. and Wei, M. (2014). Development of acetylcholinesterase biosensor based on platinum-carbon aerogels composite for determination of organophosphorus pesticides. *Food Control*, 36: 49-54.

[3] Zhai, C., Guo, Y., Sun, X., Zheng, Y. and Wang, X. (2014). An acetylcholinesterase biosensor based on graphene–goldnanocomposite and calcined layered double hydroxide. *Enzyme and Microbial Technology*, 58-59: 8-13.

[4] Andreescu, S., Noguer, T., Magearu, V. and Marty, J-L. (2002). Screen-printed electrode based on AChE for the detection of pesticides in presence of organic solvents. *Talanta*, 57: 169-176.

[5] Uygun, Z.O. and Dilgin, Y. (2013). A novel impedimetric sensor based on molecularly imprinted polypyrrole modified pencil graphite electrode for trace level determination of chlorpyrifos. *Sensors and Actuators B*, 188: 78-84.

[6] Zhao, Q., Chai, Y., Yuan, R. and Luo, J. (2013). Square wave anodic stripping voltammetry determination of lead based on the Hg(II) immobilized graphene oxide composite film as an enhanced sensing platform. *Sensors and Actuators B*, 178: 379-384.

[7] Zhang, J., Luo, A, Liu, P., Wei, S., Wang, G. and Wei, S. (2009). Detection of organophosphorus pesticides using potentiometric enzymatic membrane biosensor based on methylcellulose immobilization. *Analytical Sciences*, 25 (4): 511-515.

[8] Zhang, W.L. and Hyoung, J.C. (2011). Dynamic response of a graphene oxide-polyaniline composite suspension under an applied electric field. *Journal of the Korean Physical Science*, 61(9): 1422-1425.

[9] Somerset, V., Baker, P. and Iwuoha, E. (2013). Mercaptobenzothiazole-on-Gold Organic Phase Biosensor Systems: 2. Enhanced Carbamate Pesticide Determination. *International Journal of Chemical Research*, 5(1): 138-146.

[10] Silwana, B., Van der Horst, C., Iwuoha, E. and Somerset, V. (2014). Amperometric determination of cadmium, lead, and mercury metal ions using a novel polymer immobilised horseradish peroxidase biosensor system. *Journal of Environmental Science and Health*, A49: 1501-1511.

[11] Shan, C., Wang, L., Han, D., Li, F., Zhang, Q., Zhang, X. and Niu, L. (2013). Polyethyleneimine-functionalized graphene and its layer-by-layer assembly with Prussian blue. *Thin Solid Films*, 534: 572-576.

[12] Dong, Y-P., Zhang, J., Ding, Y., Chu, X-F. and Chen, J. (2013). Electrogenerated chemiluminescence of luminol at a polyaniline/graphene modified electrode in neutral solution. *Electrochimica Acta*, 91: 240-245.

[13] Iwuoha, E. I., Saenz de Villaverde, D., Garcia, N. P., Smyth, M. R. and Pingarront, J. M. (1997). Reactivities of organic phase biosensors. 2. The amperometric behaviour of horseradish peroxidase immobilized on a platinum electrode modified with an electro synthetic polyaniline film. *Biosensor and Bioelectronics,* 12: 749-761.

[14] Somerset, V.S., Klink, M.J., Sekota, M.M.C., Baker, P.G.L. and Iwuoha, E.I. (2006). Polyaniline-mercaptobenzothiazole biosensor for organophosphate and carbamate pesticides. *Analytical Letters,* 39 (8): 1683-1698.

[15] Somerset, V., Baker, P. and Iwuoha, E. (2010). Chapter 9: Mercaptobenzothiazole-on-gold organic phase biosensor systems: 3. Thick-film biosensors for organophosphate and carbamate pesticide determination. In V.S. Somerset (Ed.), *Intelligent and Biosensors*. (pp. 185-204). INTECH, Croatia. ISBN: 978-953-7619-58-9.

[16] Basavaraja, C., Kim, W.J., Kim, Y.D. and Huh, D.S. (2011). Synthesis of polyaniline-gold/graphene oxide composite and microwave absorption characteristics of the composite films. *Materials Letters*, 65(19-20): 3120-3123.

[17] Luzi-Thafeni, L. (2014). Graphenated Polyanilino-Horse Radish Peroxidate Nanobiosensor for the determination of carbamate pesticides in fruits. Unpublished Masters Thesis. University of the Western Cape, Bellville, South Africa.

[18] Somerset, V., Baker, P. and Iwuoha, E. (2009). Mercaptobenzothiazole-on-gold organic phase biosensor systems: 1. Enhanced organosphosphate pesticide determination. *Journal of Environmental Science and Health B*, 44: 164-178.

[19] Somerset, V.S., Klink, M.J., Priscilla, F.G.L. and Iwuoha, E.I (2007). Acetylcholinesterase – polyaniline biosensor investigation of organophosphate pesticides in selected organic solvents. *Journal of Environmental Sci*ence and Health, B42: 297-304.

Chapter 16

Biosensors for the Detection of Antibiotic Residues in Milk

Kairi Kivirand, Margarita Kagan and Toonika Rinken

Additional information is available at the end of the chapter

http://dx.doi.org/10.5772/60464

Abstract

Milk and dairy products are important nutrients for all age groups. However, the use of antibiotics for the treatment of food-producing animals generates the risk to human health, as these compounds and their metabolites can be transferred into milk. Rapid testing of the presence of antibiotics in raw milk to grant its quality has become a major task for farmers and dairy industry. The conventional analytical methods are either too slow or do not enable quantitative detection of antibiotic residues, so alternative methods that are rapid, cost effective, and easy to perform should be considered. The present chapter gives an overview of the recent developments and issues of the construction of different biosensors for the detection of antibiotic residues in milk.

Keywords: Biosensor, antibiotic residues, milk, detection limit

1. Introduction

Milk and dairy products, generally considered to be healthy and nutritionally balanced natural food, comprise essential nutrients for all age groups and are an important part of our everyday diet. However, the use of antimicrobial drugs for the treatment of food-producing animals generates the risk to human health due to the transmission of the residues and metabolites of these compounds into the food chain. Hypersensitive consumers may be subject to allergic reactions or even more severe health problems. At present, up to 10% of people have already been diagnosed to be hypersensitive or allergic to antibiotics [1, 2]. In addition to direct health problems, the presence of antibiotic residues in milk has an adverse effect on milk fermentation processes in dairy industry, as many starter cultures used for the production of fermented food products (cheese, yoghurt, etc.) may be inhibited by antimicrobial substances and the product quality will be impaired [1, 2]. Scientists and health experts also fear that wide application of antimicrobial agents is contributing to the rise and spread of antibiotic-resistant bacterial infections [3].

The overall sales of veterinary antimicrobial agents, used in food-producing animals in EU (24 countries, excluding Croatia, Greece, Malta, and Romania), were 7974.2 tons of pure ingredients in 2012. The sales of the most commonly used antibiotic classes are shown in Table 1 [4]. Based on the sales data, the classes of antibiotics shown in Table 1 are of main concern regarding the need for practical analyses of antibiotic residues in food, including milk.

Active Ingredient	Sales (Tons)	% of Total Sales
Tetracyclines	2942.8	36.9
Penicillins	1776.9	22.3
Sulphonamides	824.5	10.3
Macrolides	638.0	8.0
Polymyxins	545.0	6.8
Aminoglycosides	290.2	3.6
Lincosamides	235.0	3.0
Pleuromutilins	228.9	2.9
Fluoroquinolones	136.0	1.7
Trimethoprim	128.7	1.6
Others	228.2	2.9
Total	**7974.2**	**100**

Table 1. Sales of veterinary antibiotics used in food-producing animals in EU (24 countries, excluding Croatia, Greece, Malta, and Romania) in 2012 [4]

To protect consumers, strict legislative regulations have been imposed for the treatment of animals with antibiotics and maximum residue limits (MRLs) in foods of animal origin, which are not to be exceeded, established for residues of antibiotics or their metabolites. The residue levels in milk and other body fluids are dependent on the physicochemical properties of a particular drug, which determines the pharmacokinetics and time course of drug concentrations in the body [5]. Table 2 lists the MRL values in milk in EU for the most common antibiotics used for the treatment of dairy cows [6].

Since the introduction of EU regulations on MRLs of pharmacologically active substances in foodstuffs of animal origin (Regulation 2377/90 now replaced with the EU Regulations 470/2009 and 37/2010), it has been clear that the concept of regulating MRL values in foods can be implemented successfully if methods for quantification of these substances are available for on-site use during monitoring and testing.

Antibiotic Classes	Pharmacologically Active Substance	Maximum Residue Limit (μg/l) in Milk
Tetracyclines	Tetracycline, oxytetracycline, chlortetracycline	100*
Penicillins	Benzylpenicillin, amoxicillin, ampicillin	4
	Cloxacillin, oxacillin, dicloxacillin, nafcillin	30
Sulphonamides	Sulfadiazine, sulfamethazine, sulfadoxine, sulfamethoxazole, sulfamerazine	100**
Macrolides	Erythromycin A	40
	Spiramycin	200
	Tilmicosin, tylosin	50
Polymyxins	Colistin	50
Aminoglycosides	Dihydrostreptomycin, streptomycin	200
	Kanamycin A	150
	Gentamycin	100
	Neomycin B (incl. framycetin)	1500
Lincosamides	Lincomycin	150
	Pirlimycin	100
Pleuromutilins	Tiamulin	n.a.
Fluoroquinolones	Enrofloxacin	100
	Danofloxacin	30
	Marbofloxacin	75
	Flumequine	50
Diaminopyrimidines	Trimethoprim	50

*Sum of the parent compound and its 4-epimer

**Sum of all sulphonamides should not exceed 100 μg/l

Table 2. The MRL values for the most commonly used antibiotics in milk [6].

Routine testing of milk for the detection of residues of different antibiotics to grant the quality and safety of milk has become a major task for farmers and dairy industry. At present, two major technologies are commonly used for milk analyses: qualitative milk screening tests [7-9] and various chromatography-based techniques [7, 10]. Qualitative microbial inhibition tests have been proven to be very suitable for milk screening purposes [7-9]. These tests comprise spores of specific bacteria, sensitive to particular antibiotics on agar gel including nutrients for bacterial growth and a pH indicator. After milk is added to the test, it is incubated at the appropriate temperature to germinate and grow the spores. In the absence of antibiotic residues, the growth of bacteria can be detected visually either by the change of opacity of the agar medium or by the color change of the pH indicator, resulting from acid production and change of pH. In the presence of antibiotic residues (or any other inhibitors), the growth of bacteria is suppressed and there are no observable changes of the color. The main advantages of these tests are their low cost, simple performance, and broad selection toward different antibiotics. Although these tests are considered to be rapid, they take 3–24 hours to perform in an incubator. The bacterial strains used in tests should be constantly monitored to ensure that they have not become resistant to the antibacterials. The interpretation of test results is quite subjective and may lead to false negative or positive results. The presence of natural inhibitors in abnormal milk (e.g., milk of mastitic cows or colostrum) can be the cause of false positive results [7-9].

In addition to microbial inhibition tests; there are different rapid tests, based either on immunoassay or enzymatic operation, available for the screening of a number of antibiotic residues in milk. These tests provide results usually within 30 min [11]. As already said, milk tests are specific to the particular class of antibiotics. As a rule, the detection limits of milk tests are in the range of the established MRL values [7, 8, 11-13].

Chromatography is the most reliable technique for quantitative detection of antibiotic residues [7, 10]. Nowadays more than 80% of the analytical techniques for the determination of veterinary drugs use high-performance liquid chromatography in combination with mass spectrometry (HPLC/MS) [10]. However, chromatography-based methods require expensive equipment and trained personnel with high experience. In addition, HPLC techniques demand laborious pre-treatment of samples for the extraction of the compound analyzed, from the sample matrix.

As the milk matrix is one of the most complex ones, the application of biosensors, enabling a selective detection of particular compounds in natural or only minimally pre-treated samples is a good option for the on-site assessment of milk quality [7, 14-16]. Biosensors are compact devices transferring the selective biochemical recognition into a measurable physical signal, which can be translated into an indication of the safety or quality of milk. Physically biosensors comprise of bio-recognition and signal transduction elements. Biosensors offer an opportunity for the development of quick and portable devices for real-time analysis in complex matrixes, operating fully automatically or manually, so the user does not require special skills. At present, most of the biosensing methods focus on the detection of single antibiotic groups, but there are also studies dealing with the simultaneous determination of different groups of antibiotics.

The present review gives an overview of the developments and construction issues of antibiotic biosensors, applicable for the analysis of bovine milk quality, during the last 15 years. Due to the large number of different technologies used, these biosensors are classified into five separate groups according to the bio-recognition employed. The basic parameters of performance of different developments are collected into concise tables.

2. Receptor and enzyme-based biosensors

In these biosensors, specific receptors or enzymes are utilized to generate a bio-recognition reaction, whose signal is then detected with a suitable transducer. Receptor/enzyme-based biosensors usually employ optical or electrochemical signal detection principles [7, 15]. For optical detection, surface plasmon resonance (SPR) has been used most commonly. The application of SPR technology secures low detection limits, even below the established MRL values. The main drawbacks of SPR biosensors are their high cost; nonspecific binding of compounds of sample matrix to the sensor surface; and assay time (including chip preparation, incubation of receptors, detection, and system regeneration), which could take even a couple of days [17-23].

A receptor-based SPR biosensor, where a conjugate of cephalosporin C and a H1 monoclonal antibody (mAb) was attached to the sensor surface before injecting the milk sample mixed with DD-carboxypeptidase (EC 3.4.16.4), was developed for β-lactam antibiotics (β-Ls) by Gustavsson et al. [18]. In the presence of β-L residues in the sample, this receptor (DD-carboxypeptidase) did not bind to the complex on the sensor surface, making the sensor response inversely proportional to the β-L concentration in the sample. The limit of detection for the studied β-Ls was below or near the corresponding MRL. The same workgroup also developed an alternative SPR-based biosensor assay for β-Ls, using the catalytic properties of carboxypeptidase from *Streptomyces* R39 [19, 20, 23]. The enzyme catalyzes the hydrolysis of a tri-peptide (acetyl-L-Lys-D-Ala-D-Ala) into a di-peptide (acetyl-L-Lys-D-Ala). In the presence of β-Ls, the enzymatic activity is inhibited and less di-peptide will be formed. The assay described measured the amount of remaining enzymatic substrate (tri-peptide). The detection limits of this biosensor system for most of the studied β-Ls were below or equal to the established MRL values, except for cloxacillin (CLOX) and ceftiofur (CEFT).

A frequently used receptor for the detection of β-L residues is penicillin-binding protein (PBP), which covalently binds to penicillin G (PEN) and other β-Ls. High molecular mass PBPs exhibit DD-carboxypeptidase activity and catalyze the final steps of peptidoglycan cross-linking [17]. PBPs are used as binding reagents in receptor and enzyme-based assays, specific to β-Ls, and the biosensor system measures the inhibition of the enzymatic activity of DD-carboxypeptidase. A SPR-based biosensor to measure the inhibition of the binding of digoxigenin-labelled ampicillin (DIG-AMP) to a soluble PBP 2x* from *Streptococcus pneumonia* was proposed by Cacciatore et al. [17]. The nonspecific binding was minimized and kept on a constant level by applying heat-treatment and centrifugation steps and the addition of carboxymethylated dextran. Sample pre-preparation enabled to detect some of the studied β-Ls at MRL levels. The assay, however, does not distinguish between different β-Ls and is intended to be used as a screening method prior to identification and quantification of the individual analytes by other methods, such as by HPLC/MS [17]. To detect β-Ls, Lamar and Petz [21] immobilized PBP 2x* from *S. pneumonia* to a micro-plate. DIG-AMP complex was added to the samples. The amount of DIG-AMP bound via its AMP part to the PBP was decreasing along with the increase of β-L concentrations in the sample. In the detection step, anti-DIG *Fab* fragments marked with horseradish peroxidase (HRP) were added. The more DIG-AMP complex was bound to the receptor protein, the more antibody fragments were bound via the DIG part of the complex. A maximum color development with a chromogen agent was achieved, when no β-L residues were present and the amount of bound *Fab* fragments with peroxidase was highest. With this system, it was possible to detect the studied β-L residues at levels corresponding to 50% of their respective MRL values in milk. To eliminate the matrix interferences and increase sensitivity, it was necessary to remove fat from the milk samples. Setford et al. [22] integrated PBP with screen-printed electrodes to measure PEN. Although the system was simple to perform and easy to use, it was only possible to distinguish between PEN concentrations below or above 1.3 MRL levels.

A group of enzyme-based biosensors for the detection of β-Ls are called penicillinase (EC 3.5.2.6, PCNase) biosensors. PCNase is produced by bacteria providing resistance toward β-L antibiotics [24]. All β-L antibiotics have a common element in their molecular structure: a four-atom ring known as a β-Lactam ring. PCNase catalyzes the opening of the β-Lactam ring turning PEN into penicilloic acid and so deactivating the molecule's antibacterial properties [24]. The measurement of the hydrolysis of β-L ring by detecting the pH change has been used in biosensors for β-L antibiotics [25-28]. Chen et al. [25] proposed a PEN sensor, where hematein (pH indicator) was co-immobilized with multi-walled carbon nanotubes (MWCNTs) and PCNase onto glassy carbon electrode. MWCNTs were used to enhance electron transfer. In case PEN was present in the sample, the pH value decreased due to PEN hydrolysis into penicilloic acid catalyzed by PCNase. Once $[H]^+$ was accepted, hematein as a pH-sensitive redox probe was reduced to hematoxylin, what induced the increase of electrochemical signal. Severe interferences from the matrix of raw milk, caused by the adsorption of milk proteins and fat onto the electrode surface, probably forming a barrier between the enzyme and analyte molecules, were observed. To eliminate these effects, proteins and fat were separated from milk samples using salting-out and centrifugation processes. However, the interferences from the milk matrix were still observable and the detection limit remained quite high—9 mg/l. Hence, this sensor is not applicable for the detection of PEN in milk but only in water samples. Wu et al. [26] reported a similar PEN biosensor using single-graphene nanosheets (SGCs) instead of MWCNTs. Hematein was attached directly to graphene by adsorption, then ionic liquid was added due to its good biocompatibility, favoring further immobilization of PCNase. They found that at higher concentrations of PEN, the PCNase activity was lowered due to accumulation of the acidic products, decreasing the system sensitivity. The PEN detection limit was declared to be at 0.04 pg/l. The main drawback of pH-dependent biosensors is the fact that they exhibit only a limited range of applicability. For example, in complex fermentation media these biosensors are useless, as there can be dramatic pH changes caused by the fermentation itself.

Concalves et al. [29] proposed a PEN sensor using a cysteine-based self-assembled monolayer to immobilize PCNase onto a gold electrode. Ferrocene was added to establish the effectiveness of the biosensor development, although cysteine and PCNase inhibit the electron transfer to ferrocene. The PCNase used was a metallo-β-lactamase. The reaction could be monitored cronoamperometrically without any redox pH probe (e.g., hematein). The obtained detection limit was 1.5 μg/l.

Ismail and Adeloju [30] developed a potentiometric biosensor for PEN by exploiting the non-conducting polytyramine-PCNase film, which was placed on Pt electrode. In potentiometric setup, the main nonspecific factors influencing the detection sensitivity are buffer concentration and pH. As PCNase was used in this sensor system, pH was fixed at 7.0 to ensure the optimal enzymatic activity. The detection limit of the sensor in buffer solutions was 0.3 μM (100 μg/l). In milk samples, the PEN recovery was not reproducible, being 110±80% at 1 mg/l, 92±32% at 5 mg/l, and 78±16% at 10 mg/l PEN. The same workgroup [27] proposed also a bilayer potentiometric biosensor for PEN. The benefit of using a bilayer configuration, where the enzyme is immobilized in both layers, is a considerable enhancement of the sensitivity through the efficiency of electron transfer between the enzyme and the electrode. In the bilayer configuration, polypyrrole film was used for PCNase immobilization. The PEN detection limit was not improved in the buffer solutions, being 0.3 μM. In milk, the detection limit was found to be 5 mg/l; unfortunately, at higher PEN concentrations the recoveries in milk were very low (30–60%).

A biosensor array, based on the analyses of the patterns of change of oxidation kinetics of lactose and its metabolites, has been proposed for a rapid multiplex detection of the most common veterinary antibiotics in raw milk [31]. In this biosensor array, different oxidoreductases were used to catalyze the oxidation of lactose and its hydrolysis products galactose and glucose in separated sample flow channels. The combination of different reaction parameters of different biosensors forms a pattern of milk sample and in the presence of antibiotics this combination forms the fingerprints of particular antibiotics.

The condensed overview of receptor/enzyme-based biosensors for the determination of antibiotic residues in milk is given in Table 3

Biosensor Assay	Bio-Selective Element	Antibiotic Residues	LOD	Linear Range	Principle of Detection	Ref.
Surface plasmon resonance based inhibition sensor	DD-carboxypeptidase	PEN, AMP, AMOX, OXA, CLOX, CEFL, CEFA, CEFT	1.5 μg/l 1.2 μg/l	n.a. n.a.	The assay measures the amount of remaining enzymatic substrate (tri-peptide or di-peptide) using antibodies against the tri-peptide or di-peptide.	[19,20]
Surface plasmon resonance	DD-carboxypeptidase and monoclonal H1 antibody	PEN, OXA, CLOX, CEFL, CEFA	1-2 μg/l (PEN), 11-12 μg/l (OXA), 7-8 μg/l (CLOX),	n.a.	Microbial receptor protein with DD-carboxypeptidase activity. Using H1 antibody as substrate for the enzyme reaction.	[18]

Biosensor Assay	Bio-Selective Element	Antibiotic Residues	LOD	Linear Range	Principle of Detection	Ref.
			6-7 μg/l (CEFL), 3-4 μg/l (CEFA)			
Surface plasmon resonance bio-specific interaction assay	PBP	PEN, AMP, AMOX, CLOX, CEFL, CEFO	2 μg/l (PEN), 2 μg/l (AMP), 2 μg/l (AMOX), 15 μg/l (CLOX), 50 μg/l (CEFL), 25 μg/l (CEFO)	n.a.	Inhibition of the binding of digoxigenin-labelled AMP to soluble PBP.	[17]
Chemometric non-competitive binding assay	PBP	AMP, PEN, CEFQ, CLOX, CEFZ, CEFO	+/– tests: ≥ 1 μg/l (AMP), ≥ 1 μg/l (PEN), ≥ 1 μg/l (CEFQ), ≥ 3 μg/l (CLOX), ≥ 7 μg/l (CEFZ), ≥ 5 μg/l (CEFO)		Inhibition of digoxigenin-labelled AMP binding to soluble PBP.	[21]
Electrochemical	PBP	PEN	Half quantitative—enables to distinguish between no PEN and 1.3 MRL of PEN in samples		Assay utilizes immobilized PBP in a competitive binding assay format. No sample pre-treatment required.	[22]
Electrochemical	PBP	PEN, CEFA, CLOX, OXA	4 μg/l (PEN), 19.5 μg/l (CLOX), 6.6 μg/l (CEFA), 13.6 μg/l (OXA)	12–91 μg/l (PEN), 3–232 μg/l (CLOX), 14–123 μg/l (CEFA), 20–109 μg/l (OXA)	Quantification through competitive binding between the target and HRP-labelled specific tracer for the binding sites of the immobilized PBP.	[32]
Electrochemical	PBP	SMR, SDZ, SCP, OTC,	Half quantitative—enables to discriminate		Quantification through competitive	[33]

Biosensor Assay	Bio-Selective Element	Antibiotic Residues	LOD	Linear Range	Principle of Detection	Ref.
		TC, CEFT, CTC, SPY, CEFA	between the presence and absent of antibiotic residues at the MRL concentration level but not to identify the type of antibiotic detected		binding between the target and HRP-labelled specific tracer. Hydroquinone was used as an electron transfer mediator and peroxide as an enzyme substrate.	
Electrochemical pH sensor	PCNase	PEN	24 µM	24 µM to 0.89 mM	pH change, caused by the hydrolyses of PEN by PCNase, is measured with a pH-sensitive hematein probe.	[25]
Electrochemical pH sensor	PCNase	PEN	1 nM	1.5×10^{-13} – 7.5×10^{-3} M	pH change, caused by the hydrolyses of PEN by PCNase, is measured with a pH-sensitive hematein probe.	[26]
Fluorescence pH sensor	PCNase	PEN	n.a.	n.a.	Measures pH changes caused by hydrolyses of PEN to penicilloic acid by PCNase using photosensitive polymer matrices on optical imaging fibers.	[34]
Cronoamperometric sensor	Metallo-PCNase	PEN	1.5 ppb	3.5–16.7 µg/l	Catalytic hydrolysis of PEN was monitored using enzyme-catalyzed hydrolysis reaction.	[29]
Potentiometric sensor	PCNase	PEN	0.3 µM (≥ 5 ppm in milk)	7.5–146 µM	Change of electrode sensitivity was measured with a pyrrole-PCNase single layer and bilayer sensor.	[27]

Biosensor Assay	Bio-Selective Element	Antibiotic Residues	LOD	Linear Range	Principle of Detection	Ref.
Potentiometric sensor	PCNase	PEN	0.3 μM (≥ 20 ppm in milk)	3–283 μM	Polytyramine-PCNase film for potentiometric detection of PEN.	[30]
Biosensor array	Oxidoreductases	PEN	50 ppb		Different reaction parameters of parallel biosensors form a pattern of milk sample and in the presence of antibiotics this pattern forms the fingerprints of particular antibiotics.	[31]

PEN—penicillin G; AMP—ampicillin; AMOX—amoxicillin; OXA—oxacillin; CLOX—cloxacillin; CEFL—cephalexin; CEFA—cefapirin; CEFO—cefoperazone; CEFQ—cefquinome; CEFZ—cefazolin; SMR—sulfamerazine; SDZ—sulfadiazine; SCP—sulfachlorpyridazine; OTC—oxytetracycline; TC—tetracycline; CEFT—ceftiofur; CTC—chlortetracycline; SPY—sulfapyridine; PCNase—penicillinase; PBP—penicillin binding protein; LOD—limit of detection; n.a.—data not available.

Table 3. Receptor/enzyme-based biosensors for the detection of antibiotic residues in milk.

3. Microbial biosensors

There are a few biosensors for detecting antibiotic residues in milk based on the application of enzymatic activity of microorganisms [35-37]. Systems for the monitoring of β-Ls are based on similar principles as microbiological inhibition tests [8, 10], with the difference that the biorecognition reaction signal is detected quantitatively or semi-quantitatively. The microbial biosensors are based on the measurement of the inhibition of bacterial growth due to the presence of antibiotics [7-10].

Ferrini et al. [35] presented a hybrid biosensor combining classical microbiological screening of antibacterials with electrochemical detection and reading. In this system, *Bacillus stearothermophilus* var. *calidolactis* was used as a test microorganism and its growth was followed electrochemically measuring the quantity of CO_2 produced. The presence of microbial inhibitors (e.g., antibiotics) in the milk sample prohibits the growth of test strain and thus decreases the CO_2 production rate. This variation in CO_2 production was recorded during the initial 120 min in comparison to a control milk sample. The detection limits were at MRL levels.

A *Bacillus cereus 66* assay based on its β-lactamatic activity and using iodine as reaction indicator has been proposed by Das et al. [36]. The system was examined for different β-Ls and other antibiotics to study its selectivity. In case antibiotics were not present, cultures did not show any color change in the test ampoules meaning that the basic enzyme production by microorganisms was not sufficient to reduce the starch iodine mixture. In the presence of antibiotics, a color change was observed within 15–25 min. The growth of *B. cereus 66* was inhibited by β-Ls at ≥ 100 mg/l, which is much higher than the allowed MRL values for any of the studied β-Ls. Other antibiotics studied showed inhibition at very high concentrations ranging from 2.5 to 1000 mg/l, indicating the low sensitivity of the system.

For the detection of quinolones (Qs) and tetracyclines (TCs), an electrochemical microbial biosensor was proposed by Pellegrini et al. [37]. The detection was based on the measurement of CO_2 production rate in relation to the inhibition of microbial growth by antibiotics. The microorganism used in this study was *Escherichia coli* (ATCC 11303). *E. coli* was chosen for its good sensitivity to Qs and TCs. The inhibition degrees were evaluated after 120 min. Qs and TCs residues were detectable at 25 μg/l. The biosensor was not sensitive toward other studied antibiotics (macrolides, β-Ls, aminoglycosides, and sulfonamides).

The condensed overview of microbial-based biosensors for the determination of antibiotic residues in milk is given in Table 4.

Biosensor Assay	Bio-Selective Element	Antibiotic Residues	Assay Time	LOD	Principle of Detection	Ref.
Electrochemical assay	*B. stearothermophilus* var. *calidolactis*	PEN, AMP, OXA, CLOX, diCLOX	~ 120 min	At MRL levels	CO_2 detection (microbial growth inhibition)	[35]
Iodometric assay	*Bacillus cereus 66*	PEN, AMP, CLOX, AMOX, CEFL, CFZ	~ 10 min (+incubation 4 hours)	PEN, AMP, AMOX, CLOX ≥ 100 mg/l; CEFL, CFZ 2.5–1000 mg/l	Color change detection	[36]
Electrochemical assay	*Escherichia coli*	TC, OTC, CTC, NALA, ENRO, MAR, NOR, CIPRO, FLU, DAN	~ 120 min	≤ 25 μg/l	CO_2 detection (microbial growth inhibition)	[37]

PEN—penicillin G; AMP—ampicillin; AMOX—amoxicillin; OXA—oxacillin; CLOX—cloxacillin; diCLOX—dicloxacillin; CEFL—cephalexin; CFZ—cefazolin; TC—tetracycline; OTC—oxytetracycline; CTC—chlortetracycline; NALA—nalidixic acid; ENRO—enrofloxacin; MAR—marbofloxacin; NOR—norfloxacin; CIPRO—ciprofloxacin; FLU—flumequine; DAN—danofloxacin; LOD—limit of detection.

Table 4. Microbial biosensors for the detection of antibiotic residues in milk.

4. Immunosensors

The largest group of biosensors, used for the detection of antibiotic residues in milk, is based on the exploitation of immunochemical biorecognition reactions. The most frequently applied immunosensors are the electrochemical and optical ones, the latter most often being an SPR biosensor. Although immunosensors are very selective, the speed of analysis depends on the incubation time required to form antigen/antibody complex. In addition, the full regeneration of the sensor can also be quite time consuming.

For the detection of β-L residues in milk, a SPR biosensor based on a commercial anti-ampicillin (AMP) antibody, which had much higher affinity toward open β-lactam rings than the closed ones, was constructed by Gaudin et al. [38]. In order to open the β-lactam rings and increase the assay sensitivity, two different methods for sample pre-treatment—enzymatic (with the help of PCNase) and chemical (basic hydrolysis)—were tested. The application of pre-treatment enabled to achieve AMP detection limits at 33 μg/l and 12.5 μg/l after enzymatic and chemical pre-treatment, respectively. Another SPR immunosensor for the analysis of AMP was described by Zhang et al. [39]. This biosensor was a competitive binding assay between AMP covalently immobilized on the sensor surface and AMP-containing sample, mixed with monoclonal anti-AMP antibody. The sensor detected the amount of free antibody bound to the sensor surface after the injection of milk sample. The limit of detection of AMP with this sensor was 2.5 μg/l.

A portable SPR immunosensor for the determination of fluoroquinolone (FQ) antibiotics (enrofloxacin (ENRO), ciprofloxacin (CIP), and norfloxacin (NOR)) in milk was proposed by Fernandez et al. [40]. The assay worked in indirect inhibition format based on binding of the polyclonal anti-FQ-haptenized protein (FQ-BSA) antibody to the SPR sensing surface, activated with FQ-BSA, while the presence of FQs in sample inhibited the binding. The limit of detection was 2.0 μg/l. An earlier version of this portable SPR biosensor was constructed for the simultaneous detection of three different antibiotic classes (FQs, sulfonamides, and phenicols) [41]. This sensor was based on a similar competitive assay format. The limits of detection were 1.7 μg/l for ENRO, 2.1 μg/l for sulfapyridine (SPY), and 1.1 μg/l for chloramphenicol (CAP). Rebe Raz et al. [42] developed a microarray biosensor, based on an imaging SPR platform for the simultaneous detection of aminoglycosides in milk. The detection of antibiotics was carried out by combining seven different specific immunoassays on one sensor chip and was based on the competitive inhibition of antibody binding. The immunosensor showed ppb (μg/l)-level sensitivity toward the target compounds if 10 times diluted milk samples were used. A SPR immunosensor for CAP residues in milk, designed also as a binding inhibition assay, was proposed by Ferguson et al. [43]. The detection limit of CAP in milk with this assay was quite low—0.05 μg/l. For the screening of streptomycin (STR) residues in milk, Haasnoot et al. [44] tested both direct and competitive binding SPR immunoassays, based on monoclonal anti-dihydro STR antibodies. The limit of detection for STR was 20 μg/l, both for the direct and competitive binding assays. One more competitive STR immunosensor, exploiting commercial

Qflex™ antibodies, was reported by Ferguson et al. [45]. This assay enabled the analysis of STR in whole bovine milk (fat content 3.5%) at concentration level 30 μg/l.

A parallel affinity immunosensor array (PASA) for the analysis of 10 different antibiotics in milk (see Table 5), using multi-analyte immunoassays with an indirect competitive ELISA format, was presented by Knecht et al. [46]. Hapten conjugated with different antibiotics was attached to modified microscope glass slides to prepare disposable microarrays. Specific monoclonal antibodies against each antibiotic allowed the simultaneous detection of each individual analyte. Antibody binding was detected by a secondary antibody, labelled with horseradish peroxidase (HRP) generating enhanced chemiluminescence. The detection limits ranged from 0.12 μg/l up to 32 μg/l. Kloth et al. [47] proposed an improved PASA system, enabling the multiplexed analysis of even 13 antibiotic residues in milk (see Table 5). In these re-generable microarray chips, hapten-antibiotic conjugates were coupled onto epoxy-activated polyethylene glycol (PEG) chip surfaces. The simultaneous detection of the 13 antibiotics in raw milk samples close to the corresponding MRL values was possible within 6 min.

A disposable amperometric magneto-immunosensor, using a polyclonal sheep anti-tetracycline (TC) antibody immobilized on the surface of protein G-functionalized magnetic beads (ProtG-MBs) and screen-printed carbon electrodes (SPCEs) for the detection of TCs in milk, was described by Conzuelo et al. [48]. TC detection was performed through competitive binding between TC in the sample and a HRP-labelled specific tracer (TC-HRP) for binding sites of the capture antibodies. The detection limits were 8.9 μg/l for TC, 1.2 μg/l for oxytetracycline (OTC), 66.8 μg/l for chlortetracycline, and 0.7 μg/l for doxycycline. Conzuelo et al. [49] proposed a similar immunosensor for the specific detection and quantification of sulfonamide (SAs) residues in milk. They used polyclonal rabbit antibodies immobilized onto the electrode surface, modified with 4-aminobenzoic acid. The limit of detection was 0.15 μg/l. Another direct competitive immunoassay for the determination of SPY was based on antibody immobilized onto the surface of protein G-modified glassy carbon plates [50]. The limit of detection of this immunosensor for SPY was on a similar level—0.13 μg/l.

A wavelength interrogated optical sensor (WIOS) technology has been employed for the development of biosensors for SAs, FQs, β-Ls, and TCs. A competitive immunosensor for the simultaneous detection of three antibiotics—SPY, CIP, and OTC—in raw milk has been described by Suarez et al. [51]. These three assays were performed in indirect formats with three specific haptens. Raw milk samples were mixed with different antiserum receptors that specifically reacted with particular antibiotics and further spiked with three antibiotics at corresponding MRLs (100 μg/ml). In contact with the sensing surface, excess antibodies not bound to antibiotics were attached to the hapten-coated sensing region. The attached antibodies were revealed with a secondary antibody. Adrian et al. developed WIOS immunosensors for the detection of SAs [52] and for the simultaneous screening of SAs and other most frequently occurring antibiotic classes: FQs, β-Ls, and TCs [53]. These sensors relied on a competitive immunoassay format, where haptenized proteins for SAs, FQs, β-Ls, and TCs were immobilized on the chip surface, where they formed independent sensing zones. Milk samples were mixed with the specific antibodies and bio-receptors were added. The detection limits of these sensors were 0.5 μg/l for SPY, 1.3 μg/l for CIP, 3.1 μg/l for AMP, and 34.2 μg/l for OTC.

An impedimetric immunosensor for CIP that comprised polyclonal anti-CIP antibody immobilized on an electrogenerated N-hydroxysuccinimide-functionalized polypyrrole film was developed by Ionescu et al. [54]. The antibody-antigen affinity reaction resulted in an extremely sensitive and specific impedance response, even with CIP concentrations as low as 10 pg/ml. A flow injection impedimetric immunosensor for the direct detection of PEN in milk samples, based on immobilized monoclonal anti-PEN on self-assembled thioctic acid monolayer on gold electrode, was developed by Thavarungkul et al. [55]. Binding of PEN to anti-PEN on the electrode surface causes the impedance to increase. The limit of detection of this immunosensor was 1 pg/l, much lower than the corresponding MRL in milk, but the sensor preparation took as long as 2 days.

A competitive amperometric immunoassay for PEN in milk was developed by Merola et al. [56]. This immunosensor was based on the competitive binding of free PEN and BSA-PEN conjugate immobilized on the sensor membrane, to anti-PEN-biotin-avidin-peroxidase complex. The limit of detection of this immunosensor was as low as 5 ng/l. Another amperometric immunosensor for PEN in milk has been described by Wu et al. [57]. This biosensor was based on the interaction of PEN with the covalently bound new methylene blue (NMB) and HRP-labelled PEN polyclonal antibody (HRP-PEN-Ab) on a glassy carbon electrode. Cyclic voltammetry and impedance spectroscopy enabled to gain the detection limit 0.6 μg/l. An electrochemical magneto immunosensor for CIP has been proposed by Pinacho et al. [58]. Magnetic beads for the attachment of CIP, which are modified with antibody (Ab171) and HRP-BSA, are collected after incubation with the sample onto magnetic electrode and detected electrochemically following the H_2O_2 oxidation, catalyzed by HRP. The detection limit for CIP 9 ng/l was almost as low as gained in Ref. [56].

A nanogold resonance-scattering (RS) spectral assay for the determination of PEN was developed by Jiang et al. [59]. The binding of PEN to anti-PEN, immobilized on the surface of gold nanoparticles, leads to the cleavage of nanoparticles. Uncovered nanoparticles aggregated in the presence of PEN and generated the resonance scattering (RS) effect, measured at 560 nm. The detection limit of the assay was 0.78 μg/l. Another immunosensor, based on the application of the aggregation of nanoparticles, was developed by Chen et al. [60] for the detection of kanamycin (KAN) in milk. This sensor, using superparamagnetic iron oxide (SPIO) nanoparticles, acted as a magnetic relaxation switch. The target analyte KAN competed with KAN immobilized on the surface of the SPIO nanoparticles and hence affected the formation of SPIO aggregates. The dispersed and aggregated states of the SPIO modulate the spin–spin relaxation time (T_2) of the neighboring water molecules, which change due to the effect of the target analyte. The limit of detection with this biosensor for KAN was 0.1 μg/l.

Piezoelectric immunosensors for PEN, AMP, and the total content of penicillin antibiotics have been developed by Karaseva et al. [61]. The receptor coating of the sensors was prepared by the immobilization of PEN- or AMP-hapten-protein conjugates on the polypyrrole film via glutaraldehyde. The limits of detection obtained were 0.8 μg/l for PEN, 3.9 μg/l for AMP, and 1.7 μg/l for total penicillins.

Finally, a surface acoustic wave (SAW) biosensor has been introduced by Gruhl et al. [62] for the rapid detection of PEN in milk. This was a binding inhibition assay, using the interaction of PEN in the sample and PEN epitopes immobilized on the sensor surface with monoclonal anti-PEN. The binding of antibodies onto the sensor surface was followed by means of acoustic (gravimetric) detection. Low PEN concentrations led to a high binding load of free antibodies to the surface and hence to high signals. The detection limit for PEN in low-fat milk was 2.2 μg/l.

The condensed overview of immunosensors used to determine antibiotic residues in milk is given in Table 5.

Biosensor Assay	Bio-Selective Element	Antibiotics	LOD	Linear Range	Ref.
Piezoelectric immunosensor	Polyclonal and monoclonal antibodies	PEN, AMP, group of penicillins	0.8 μg/l (PEN), 3.9 μg/l (AMP), 1.7 μg/l (group)	2.5–250 μg/l (PEN), 2.5–500 μg/l (AMP), 1–500 μg/l (group)	[61]
Magnetic relaxation Immune-nanosensor	Monoclonal anti-KAN	KAN	0.1 μg/l	1.5–25.2 μg/l	[60]
Electrochemical immunoassay	Polyclonal rabbit antibody (As167)	SAs	0.15 μg/l	0.6–64.2 μg/l	[49]
Amperometric immunoassay	Polyclonal sheep anti-TC antibody	DXC, OTC, TC, CTC	8.9 μg/l (TC), 1.2 μg/l (OTC), 66.8 μg/l (CTC), 0.7 μg/l (DXC)	17.8–189.6 μg/l (TC), 4.0–242.3 μg/l (OTC), 144.2–2001.9 μg/l (CTC), 2.6–234.9 μg/l (DXC)	[48]
Surface plasmon resonance assay	Polyclonal anti-FQ haptenized protein (FQ-BSA) antibody	ENRO, CIP, NOR	2.0 μg/l	n.a.	[40]
Surface plasmon resonance assay	Haptenized protein antibodies	ENRO, SPY, CAP	1.7 μg/l (ENRO), 2.1 μg/l (SPY), 1.1 μg/l (CAP)	n.a.	[41]
Waveguide interrogated optical sensor	Polyclonal rabbit antibody (As155)	SAs	0.5 μg/l	1.4–26.4 μg/l	[52]
Waveguide interrogated optical sensor	Polyclonal and monoclonal antibodies	SPY, CIP, AMP, OTC	0.5 μg/l (SPY), 1.3 μg/l (CIP), 3.1 μg/l (AMP), 34.2 μg/l (OTC)	1.4–26.4 μg/l (SPY), 4.4–70.3 μg/l (CIP), 7.1–100.0 μg/l (AMP), 56.8–193.3 μg/l (OTC)	[53]
Surface plasmon resonance assay	Monoclonal and polyclonal antibodies	NEO, GEN, KAN, STR, ENRO, CAP, SMZ	ppb levels	n.a.	[42]

Biosensor Assay	Bio-Selective Element	Antibiotics	LOD	Linear Range	Ref.
Surface plasmon resonance assay	Monoclonal anti-AMP antibody	AMP	2.5 μg/l	n.a.	[39]
Waveguide interrogated optical sensor	Antiserum receptors	SPY, CIP, OTC	Positive/negative detection: 100 μg/ml		[51]
Resonance-scattering immunosensor	Goat anti-rabbit PEN	PEN	0.78 μg/l	7.5–1700 μg/l	[59]
Impedance spectroscopy immunosensor	Polyclonal anti-CIP antibody	CIP	10 pg/ml	n.a.	[54]
Impedimetric flow immunosensor	Monoclonal anti-PEN	PEN	3.0×10^{-15} M	1.0×10^{-13}–1.0×10^{-8} M	[55]
Surface plasmon resonance assay	Qflex™ CAP antibody	CAP	0.05 μg/l	n.a.	[43]
Parallel affinity immunosensor array	Monoclonal antibodies	PEN, CLOX, CEFA, SDZ, SMT, STR, GEN, NEO, ERY, TYL	0.12–32 μg/l (near to MRL for PEN), for others far below the respective MRLs	3.3–41.3 μg/l (PEN), 0.29–3.63 μg/l (CLOX), 0.12–1.45 μg/l (CEFA), 3.49–43.5 μg/l (SDZ), 4.93–63.0 μg/l (SMT), 5.06–66.0 μg/l (STR), 12.1–141 μg/l (GEN), 31.8–427 μg/l (NEO), 0.36–4.70 μg/l (ERY), 0.95–12.6 μg/l (TYL)	[46]
Surface plasmon resonance assay	Monoclonal anti-dihydro STR antibodies	STR	20 μg/l	n.a.	[44]
Surface plasmon resonance assay	Anti-AMP antibody	Penicillins	5.9 μg/l (chemical pretreatment); 14.6 μg/l (enzymatic pretreatment)	n.a.	[38]
Chemiluminescence immunosensor	Monoclonal antibodies	SMZ, SDZ, STR, CLOX, AMP, PEN, CEFA, NEO, GEN, ERY, TYL, ENRO, TC	n.a.	20–320 μg/l (SMZ), 3.1–440 μg/l (SDZ), 21–608 μg/l (STR), 0.5–320 μg/l (CLOX), 1.7–1800 μg/l (AMP), 3.0–320μg/l (PEN), 0.8–797 μg/l (CEFA), 135–2300 μg/l (NEO), 6.8–540 μg/l (GEN), 0.1–45 μg/l (ERY), 0.6–120 μg/l (TYL), 0.5–6.3 μg/l (ENRO), 0.05–43 μg/l (TC)	[47]

Biosensor Assay	Bio-Selective Element	Antibiotics	LOD	Linear Range	Ref.
Amperometric immunosensor	Methylene blue and HRP-labelled PEN polyclonal antibody	PEN	1.82 nM	5.2–41.6 nM	[57]
Amperometric immunoassay	Anti-PEN-biotin-avidin-peroxidase complex	PEN	1.5×10^{-11} M	3.0×10^{-11}–3.0×10^{-4} M	[56]
Surface plasmon resonance assay	Qflex™ STR antibody	STR, dihydro STR	30 μg/l	n.a.	[45]
Surface acoustic wave assay	Monoclonal anti-PEN (clone 8.F. 223)	PEN	2.2 μg/l	2–6 μg/l	[62]
Immunoassay	Polyclonal rabbit antibody (As167)	SPY	0.13 μg/l	n.a.	[50]
Amperometric magneto nanosensor	Ab171	CIP	0.009 μg/l	0.043–7.38 μg/l	[58]

PEN—penicillin G; AMP—ampicillin; KAN—kanamycin; DXC—doxycycline; OTC—oxytetracycline; TC—tetracycline; CTC—chlortetracycline; ENRO—enrofloxacin; CIP—ciprofloxacin; NOR—norfloxacin; SPY—sulfapyridine; SAs—sulfonamides; CAP—chloramphenicol; NEO—neomycin; GEN—gentamycin; STR—streptomycin, SMZ—sulfamethazine; AMOX—amoxicillin; CLOX—cloxacillin; CEFL—cephalexin; CEFO—cefoperazone; CEFA—cefapirin; CIP—ciprofloxacin; SDZ—sulfadiazine; SMT—sulfamethazine; ERY—erythromycin; TYL—tylosin; DTC—doxytetracycline; HRP—horseradish peroxidase; As—antiserum; Ab—antibody; LOD—limit of detection; n.a.—data not available.

Table 5. Immunosensors for the detection of antibiotic residues in milk.

5. Aptasensors

For the detection of antibiotic residues in milk, aptamer-based biosensors (aptasensors) have been developed during last 5 years [63-77]. Aptamers can be considered as chemical or "synthetic" antibodies because of their in vitro production based on the systematic evolution of ligands by exponential enrichment (SELEX) [78]. The SELEX process enables the fabrication of aptamers also for non-immunogenic and toxic targets that cannot be produced by natural immune systems [79, 80]. The selection provides specific aptamers fold into well-defined three-dimensional shapes, which can recognize their target molecules with high affinity [78, 81]. In most studied biosensing systems, the dissociation constants of the aptamer-target molecule complexes are in the nanomolar range. In addition, aptamers are quite stable and are not affected by reasonable temperature or pH shifts; at optimal conditions they can restore their original conformation. Aptamers are smaller in size compared to antibodies enabling them to reach previously blocked or intracellular targets [80].

The immobilization of aptamers is carried out with the help of two major techniques [82]: (i) direct attachment of aptamers on a bio-coated sensor surface via suitable linkers; and (ii) non-covalent conjugation of aptamers to functionally activated surfaces. To facilitate direct immobilization, the aptamer must be functionalized by adding a terminal functional group such as biotin or amine [82]. Sometimes an extra linker is used to create flexibility between the aptamer and terminal functional group. To minimize the nonspecific binding caused by the linker, the linker usually consists of a string of thymidine [82, 83]. A serious disadvantage using aptasensors for milk analyses is the presence of milk proteins and fat, and the non-transparency of the samples, which hamper the application of optical detection methods [64, 68, 71]. Commonly pretreatment of milk samples is required [64, 68-71, 73, 74, 77].

For the detection of β-L antibiotics, several aptasensors have been proposed [64, 67, 69, 71]. Dapra et al. [67] developed polymer biosensor chips comprising a micro-fluidic system and immobilized aptamers, integrated with the measurement of electrochemical impedance. Polymers exhibit excellent properties to master the task of transducing a binding event between an analyte and a biological probe into a measurable signal. Polymers have been used in biosensing as alternatives to traditional electrode materials due to their inexpensive fabrication and simple functionalization [84]. In aptasensors proposed by Dapra et al., single-stranded DNA (ssDNA) aptamers functionalized with fluorescein amidite (FAM) were used [67]. With this sensor it was possible to detect ampicillin (AMP) concentrations below the established MRL (the detection range was 35 ng/l to 350 μg/l). Song et al. [69] proposed an aptasensor for sulfadimethoxine (SDMX), which consisted of FAM-modified ssDNA aptamers attached to coordinated polymer nanobelts (CPNBs). The detection limit for this aptasensor for SDMX in milk was 10 μg/l, which is 10 times below the allowed MRL in EU. However, the aptamer–antibiotic interaction was interfered by other components of raw milk (proteins, fat, $[K]^+$, $[Na]^+$), reported also by other authors [63, 70, 71, 85]. It has been proposed that changes of $[K]^+$ and $[Na]^+$ inactivate aptamers by preventing them to fold into correct structures, and accordingly at least 10-fold diluted matrix should be used for milk analyses [63, 85].

There are numerous aptasensors for the detection of chloramphenicol (CAP) [65, 68, 70], whose use is actually banned in food-producing animals and whose presence in milk is strictly prohibited. The regulations set the MRL for CAP in milk at 0.3 μg/l, which is the LOD with HPLC-MS [6]. Wu et al. [68] proposed a fluorescence-based aptasensor, using aptamer-conjugated magnetic nanoparticles (MNPs) for both CAP recognition and concentration. In the absence of target molecules (antibiotics), MNP-aptamer complex hybridizes to its complementary DNA (cDNA) modified with upconversion nanoparticles (UCNPs) to form the duplex structure giving a maximum fluorescent signal. Upon CAP addition, the aptamer preferentially binds with CAP and causes the dissociation of some cDNA, liberating some UCNPs-cDNA complexes, and leads to a decreased fluorescence signal on the surface of MNPs. Under optimal conditions, a linear CAP detection range from 0.01 to 1 μg/l was achieved. Alibolandi et al. [70] constructed an aptasensor for the detection of CAP, where aptamers were conjugated with Cd-Te quantum dots (QDs), which exhibit high resistance to photo-bleaching, stable fluorescence, high quantum yield, narrow and symmetric emission band, and broad adsorption spectra [86]. The detection limit of CAP with this aptasensor was 0.2 μg/l [70].

Various aptasensors for the detection of tetracyclines (TCs) in milk have been proposed [73-77, 85]. Jeong and Paeng [85] used two different enzyme-linked aptasensors (ELAA) based either on a ssDNA-aptamer or on an RNA-aptamer. A competitive assay with sequential mode detection was adapted for both systems. A similar system was developed by Kim et al. [73], who used ssDNA aptamers for the detection of oxytetracycline (OTC) In terms of specificity, detection limit, and dynamic range, the results obtained with ELAA in both the above-mentioned studies were not an improvement in comparison with the commonly used immunoassays (ELISA) [73]. For the detection of TC, colorimetric aptasensors have been proposed by He et al. [74, 75]. In these systems, aptamers were adsorbed onto the gold nanoparticles (AuNPs) by electrostatic interactions. To generate and stabilize the AuNPs, poly(diallyldimethylammonium) (PDDA) [75] and hexadecyltrimethylammonium bromide (CTAB) [74] were used. The detection of TCs was based on the aggregation phenomenon of functionalized AuNPs, which can be detected by the change in color of the solution from blue to wine red. According to the information presented, if AuNPs were stabilized with PDDA, nanoparticles aggregated in the presence of TC [75]. If CTAB was used, the aggregation of nanoparticles was not favored in the presence of TC and an opposite effect was detected [74]. These assays also required pretreatment of milk samples with acetic acid in order to remove Ca^{2+}. In both systems, the average recoveries of TC were in the range 81–112%. Compared to the PDDA assay, the CTAB assay had 2.7 times higher detection limit (20 μg/l and 54 μg/l, respectively). A label-free electrochemical method based on modified glassy carbon (GC) electrode and ssDNA-aptamers has been proposed by Zhang et al., also for the detection of TCs [76]. The advantage of this electrochemical aptasensor in comparison with the optical ones is the fact that the proposed system did not need any sample pretreatment. The linear detection range for TC was 0.1–100 μg/l. An ultrasensitive resonance scattering method for the detection of TC in milk was described by Luo et al. [77]. This biosensor was based on the competition of aptamers between nanogold and TC. SsDNA-aptamer coats the surface of nanogold particles through van der Waals and intermolecular forces and prevents the aggregation of nanogold particles. But in the presence of TC, the nanogold surface becomes naked, because of the high affinity between TC and aptamer, and the nanoparticles aggregate. The constructed biosensor could be used to detect trace levels of TC in milk samples with good sensitivity (the limit of detection was 22 μg/l).

The overview of aptasensors used to determine antibiotic residues in milk is given in Table 6.

Biosensor Assay	Antibiotic Residues	K_d	Assay Time	LOD	Linear Range	Comments	Ref.
Indirect competitive enzyme-linked aptamer assay	OTC	Su3: 4.7 nM	60 min + 30 min sample pretreatment	12.3 μg/l	n.a.	The 5′-biotin ssDNA-aptamer	[73]
Electrochemical microfluidic biosensor	AMP	13.4 nM	220 min + sample pretreatment	n.a.	100 pM-1 μM (AMP)	The 5′-amino modified ssDNA-aptamer	[67]

Biosensor Assay	Antibiotic Residues	K_d	Assay Time	LOD	Linear Range	Comments	Ref.
	KAN	78.8 nM		n.a.	10 nm-1 mM (KAN)		
Fluorescence nanoparticle bioassay	CAP	n.a.	110 min + 15 min sample pretreatment	0.01 μg/l	0.01–1 μg/l	The 5′-biotin ssDNA-aptamer	[68]
Fluorescence nanobelt bioassay	SDMX	Su13: 84 nM Su11: 150 nM	95 min + 50 min sample pretreatment	10 μg/l	10–100 μg/l	The 5′-fluorescein amidite labelled ssDNA-aptamer	[69]
Fluorescence-colorimetric nanoparticle bioassay	AMP	Su4: 9.4 nM Su17: 13.4 nM Su18: 9.8 nM	200 min + 50 min sample pretreatment	5 μg/l	5–50 μg/l	The 5′-fluorescein-amidite labelled ssDNA-aptamer	[64]
Colorimetric sensor	TCs	n.a.	25 min + sample pretreatment	45.8 nM	0.01–0.4 μM	5′-amino modified ssDNA-aptamer	[75]
Quantum dots and graphene electrochemical sensor	CAP	n.a.	25 min detection +sample pretreatment	0.2 μg/l	0.1–10 nM	5′-amino modified ssDNA-aptamer	[70]
Spectrophotometric assay	KAN	n.a.	100 min + 40 min sample pretreatment	1 nM	1–8 nM and 100–500 nM	5′-thiol-stranded ssDNA-aptamer	[71]
Indirect competitive chemiluminescent enzyme immunoassay	ENRO	Su17: 188 nM	65 min + 30 min sample pretreatment	2.26 μg/l	6.4–90.0 μg/l	ssDNA-aptamer	[63]
Ultrasensitive resonance scattering assay	TCs	n.a.	40 min + 40 min sample pretreatment	11.6 nM	Up to 250 nM	ssDNA-aptamer	[77]
Electrochemical assay	TCs	n.a.	n.a.	1 μg/l	0.1–100 μg/l	ssDNA-aptamer	[76]
Colorimetric nanoparticle sensor	TCs	n.a.	60 min + sample pretreatment	122 nM	0.01–0.5 μM	ssDNA-aptamer	[74]
Competitive enzyme-linked aptamer assay	TCs	Su76: 63 nM Su57: 770 pM	16 h + 160 min + sample pretreatment	45.7 μg/ml 16.8 μg/ml	3.2×10^{-4}-3.2×10^{-7} M 1.0×10^{-4}-1.0×10^{-7} M	3`-end biotin ssDNA-aptamer 5` dimethoxytrityl-	[85]

Biosensor Assay	Antibiotic Residues	K_d	Assay Time	LOD	Linear Range	Comments	Ref.
						biotin amidite RNA-aptamer	

CAP—chloramphenicol; AMP—ampicillin; TCs—tetracyclines; OTC—oxytetracycline; ENRO—enrofloxacin; KAN—kanamycin; SDMX—sulfadimethoxine; ssDNA—single-stranded DNA; n.a.—data not available; K_d—dissociation constant for the aptamer; LOD—limit of detection.

Assay time does not include aptamer and electrode preparation.

Table 6. Aptasensors for the detection of antibiotic residues in milk.

6. Molecularly Imprinted Polymer (MIP) sensors

A recent development in the biosensing of antibiotic residues in milk is the application of molecularly imprinted polymer sensors [87-93]. Molecular imprinting is a technique for the creation of synthetic materials containing specific receptor sites having high affinity toward the target molecule. Molecularly imprinted polymers (MIPs) are cross-linked organic structures containing pre-designed molecular recognition sites complementary in shape, size, and functional groups to the template molecule [94]. MIPs are effective alternatives for the natural receptors in biosensor assays.

For the detection of TCs, a photonic MIP sensor was developed by combining colloid crystal templating and molecular imprinting techniques [93]. Three structurally similar TCs (tetracycline, oxytetracycline, and chlortetracycline) were used as template molecules for MIPs synthesis. Target antibiotic molecules generated optical changes in the Bragg diffraction peak, detected directly using fiber optic spectroscopy. No labelling or sample pretreatment was necessary. The MIPs prepared enabled to discriminate between the very similar TCs, indicating that the cooperative effect of shape, size, and interaction sites of the formed binding areas plays a critical role in the selective molecular recognition process of MIPs. Hydrogen bonding is suggested to be the main interaction responsible for the retention of TCs in MIPs. Because hydroxyl groups form hydrogen bonds with carboxylic groups of the polymer, the difference in the amount of OH^- groups along with the molecular structure of particular TCs result detectable variations in retention [93].

A voltammetric MIP sensor for the detection of sulfadiazine (SDZ) based on carbon paste electrode modified with SDZ-MIP and measuring the potential change due to the binding of antibiotics was proposed by Sadeghi and Motaharian [92]. However, the selectivity of the MIP sensor was rather poor, as sulfonamides including two aromatic cycles generated a difference in the measured currents less than 20% in the case of SDZ and sulfapyridine, and around 30% in the case of benzenesulfonamide and 4-methyl-benzenesulfonamide. The milk matrix did not interfere with the measurements.

An electrochemical MIP sensor was fabricated for the detection of erythromycin (ERY) by decorating a gold electrode with chitosan-platinum nanoparticles/graphene-gold nanoparticles double-layer nanocomposite [91]. The MIPs were prepared using $HAuCl_4$ and 2-mercaptonicotinic acid, and ERY as a template. The selectivity of the sensor was evaluated toward four different antibiotics (ERY, kanamycin, neomycin, and spiramycin). The response of the MIP electrode to ERY was approximately 80% higher than toward the other studied compounds, revealing the good selectivity of the MIP sensor. The limit of ERY detection in milk was around 75 μg/l [91].

MIP sensors have also been developed for the detection of different aminoglycosides. For tobramycin (TOB), an MIP-based pyrrole glassy carbon electrode [89] and a quartz crystal microbalance nanosensor [88] have been proposed. These MIP sensors were very sensitive, with the detection limits in milk under 1 μg/l. The selectivity and stability of these sensors were also excellent. Liu et al. [87] proposed a method for the detection of streptomycin (STR) residues in milk, based on nanogold-encapsulated poly(*o*-phenylenediamine) shell on magnetic iron oxide core. The assay was carried out in a competitive-type mode between the target molecule and glucose oxidase-labelled STR. The sensor response was based on the catalytic oxidation of glucose by glucose oxidase, amplifying the signal and improving the sensitivity. The application of magnetic beads facilitates the construction of the sensors and enables the concentration of samples. In milk, the detection limit for STR of this system was 1 μg/l [87]. For the detection of neomycin (NEO), a multilayer sensor system of gold electrode, composite material (chitosan-silver nanoparticles/graphene-multiwalled carbon nanotubes), and poly(pyrrole)-based MIP were used [90]. The selectivity of this sensor was good toward NEO, and in milk the detection limit was 6 μg/l.

The MIP sensors, which have been developed, are very sensitive with low detection limits. However, the linear ranges of detection of these sensors tend to be below the allowed MRL values, so complicating their practical applications. The condensed overview of MIP sensors used to determine antibiotic residues in milk is given in Table 7.

Sensor Assay	Antibiotic Residues	Stability of the MIP	Monomer	LOD	Linear Range	Assay/Recovery	Ref.
Quartz crystal microbalance nanosensor	TOB	After 45 days, the value of mass shift was approx. 98% of the initial	2-hydroxyethyl methacrylate and N-methacryloyl-L-glutamic acid (ethylenglycol dimethacrylate as a linker)	5.7×10^{-12} M	1.7×10^{-11}–1.5×10^{-10} M	Gold electrode, recovery in milk 95–101%	[88]
Electrochemica l assay	NEO	After 2 weeks at 4°C the current response decreased by about 7.3%	Pyrrole	7.6×10^{-9} M	9×10^{-9}–7×10^{-6} M	Gold electrodes decorated with chitosan-silver nano-particles/graphene-multiwalled carbon nanotubes	[90]

Sensor Assay	Antibiotic Residues	Stability of the MIP	Monomer	LOD	Linear Range	Assay/Recovery	Ref.
						composites, recovery in milk 98–103%	
Electrochemical assay	TOB	After 45 days, the peak current was approx. 96%	Pyrrole	1.4×10^{-10} M	5×10^{-10}–1×10^{-8} M n.a.	Glassy carbon electrode, recovery in milk 88–100%	[89]
Voltammetric assay	SDZ	At ambient conditions for more than 18 weeks without significant change in the response	Ethylene glycol dimethacrylate and 2-azobisiso-butyronitrile	1.4×10^{-7} M	2×10^{-7}–1×10^{-4} M	Carbon paste electrode, recovery in milk 97–104%	[92]
Electrochemical assay	STR	After 25 days at 4°C the response decreased by 10%	*o*-phenylene-diamine	10 pg/ml	0.05–20 μg/l	Gold-promoted magnetic nanospheres, recovery in milk 94–113%	[87]
Electrochemical assay	ERY	After stored at 4°C for 2 weeks the current response decreased by 9%	2-mercapton-cotinicacid and $HAuCl_4$	2×10^{-8} M	7×10^{-8}–9×10^{-5} M	Chitosan-platinum nanoparticles/ graphene-gold nanoparticles composites, recovery in milk 96–101%	[91]
Optical spectrometric assay	TC, OTC, CTC	n.a.	Acrylamide (N,N ′-methylene bisacrylamide as linker)	n.a.	0.05–20 μg/l		[93]

TOB—tobramycin; NEO—neomycin; SDZ—sulfadiazine; STR—streptomycin; ERY—erythromycin; TC—tetracycline; OTC—oxytetracycline; CTC—chlortetracycline; LOD—limit of detection; n.a.—data not available; MIP—molecularly imprinted polymers

Table 7. Molecularly imprinted polymer sensors for the detection of antibiotic residues in milk.

7. Conclusions

Due to the urgent need for on-site analysis of milk quality and safety, the research activity for the development of biosensors for the detection of antibiotic residues in milk has been very high during the last years. Most studies are focusing on the detection of antibiotics, which are most commonly used for the treatment of food-producing animals: β-lactams, tetracyclines, sulphonamides, and aminoglycosides. Biosensors proposed for these antibiotics often exhibit detection limits below or equal to the allowed maximum residue levels in milk. There are almost a hundred studies within the last decade reporting about biosensors for the detection of different antibiotic residues in milk; unfortunately, we could not find any studies dealing with biosensing of colistin or trimethoprim.

In most studies, milk samples spiked with a selected antibiotic or antibiotics have been used for the "proof of concept" and validation of the proposed technology. Analyses of the milk of animals, who are undergoing antibiotic treatment, are scarce, although these real samples can be a key factor to indicate the applicability of biosensor technology for practical analyses, as these "natural" samples can contain in addition different metabolites of antibiotic compounds. Some complex biosensing systems require additional pretreatment of milk samples to remove fat and proteins—the implication of these technologies for on-site analyses also seems problematic. The average detection time (excluding pretreatment) is 30–40 min or even up to 2 hours in case a longer incubation period for bio-recognition is required.

The biggest problem in biosensing is the stability of compounds, used for bio-recognition and the life-time of sensors. In terms of stability, the most prospective are MIP biosensors, where the biological material is used only as a template, and aptasensors. Concerning the regeneration of biosensor systems, there is very little data available. Usually, the full dissociation of complexes, formed during the bio-recognition, takes 30–40 min. In terms of rapid analyses, single use biosensors have a great advantage.

In analytical developments, reliability and robustness are the keywords of future trends. We hope that along with automation and independent action, these features enable biosensors to become an essential tool for on-site analysis of milk and other substances.

Acknowledgements

This work was supported by the Estonian Science Foundation grant N°9061, Estonian Research Council grant IUT 20-17, and Estonian Centre of Excellence "Mesosystems—Theory and Applications."

Author details

Kairi Kivirand*, Margarita Kagan and Toonika Rinken

*Address all correspondence to: kairi.kivirand@ut.ee

Institute of Chemistry, University of Tartu, Tartu, Estonia

References

[1] E. L. Miller. The penicillins: A review and update. Journal of Midwifery & Women's Health 2002; 47, 426–434.

[2] A. D. Dayan. Allergy to antimicrobial residues in food: Assessment of the risk to man. Veterinary Microbiology 1993; 35, 213–226.

[3] B. Mole. MRSA: Farming up trouble. Nature 2013; 499, 398–400.

[4] Sales of veterinary antimicrobial agents in 26 EU/EEA countries in 2012. 4th ESVAC Report, European Medicines Agency 2014.

[5] Chemical Analysis of Antibiotic Residues in Food, Edited by Jian Wang, James D. MacNeil, and Jack F. Kay (Wiley, Hoboken, 2012).

[6] Commission Regulation (EU) No. 37/2010, ec.europa.eu/health/files/mrl/mrl_20101212_consol.pdf 2010

[7] R. Babington, S. Matas, M. P. Marco, and R. Galve. Current bioanalytical methods for detection of penicillins. Analytical and Bioanalytical Chemistry 2012; 403, 1549–1556.

[8] C. Chafer-Pericas, A. Maquieira, and R. Puchades. Fast screening methods to detect antibiotic residues in food samples. Trends in Analytical Chemistry 2010; 29, 1038–1049.

[9] M. C. Beltran, M. I. Berruga, A. Molina, R. L. Althaus, and M. P. Molina. Performance of current microbial tests for screening antibiotics in sheep and goat milk. International Dairy Journal 2015; 41, 13–15.

[10] L. Kantiani, M. Farre, and B. Damia. Analytical methodologies for the detection of [beta]-lactam antibiotics in milk and feed samples. Trends in Analytical Chemistry 2009; 28, 729–744.

[11] Chemical Analysis of Antibiotic Residues in Food, Edited by J. Wang, J. D. MacNeil, and J. F. Kay (Wiley, Hoboken, 2012).

[12] ThermoFisher; Rapid Antibiotic Residue Testing in Milk, www.thermofisher.com.au/show.aspx?page=/ContentAUS/Scientific/Microbiology-Products/Microbial-Safety-Testing/Rapid-Antibiotic-Residue-Testing-Milk/Rapid-Antibiotic-Residue-Testing-Milk.html 2015

[13] DSM: Reliable Antibiotic Residue Tests for Milk, www.dsm.com/content/markets/foodandbeverages/en_US/solutions/responsible-food-delivery/reliable-antibiotic-residue-tests.html?wt.srch=1&wt.mc_id=GA-Branding 2015

[14] T. F. McGrath, C. T. Elliott, and T. L. Fodey. Biosensors for the analysis of microbiological and chemical contaminants in food. Analytical and Bioanalytical Chemistry 2012; 403, 75–92.

[15] K. Narsaiah, S. N. Jha, R. Bhardwaj, R. Sharma, and R. Kumar. Optical biosensors for food quality and safety assurance – A review. Journal of Food Science and Technology 2012; 49, 383–406.

[16] K. Reder-Christ and G. Bendas. Biosensor applications in the field of antibiotic research – A review of recent developments. Sensors 2011; 11, 9450–9466.

[17] G. Cacciatore, M. Petz, S. Rachid, R. Hakenbeck, and A. A. Bergwerff. Development of an optical biosensor assay for detection of beta-lactam antibiotics in milk using the penicillin-binding protein 2x*. Analytica Chimica Acta 2004; 520, 105–115.

[18] E. Gustavsson, P. Bjurling, J. Degelaen, and A. Sternesjö. Analysis of beta-lactam antibiotics using a microbial receptor protein-based biosensor assay. Food and Agricultural Immunology 2002; 14, 121–131.

[19] A. Sternesjö and E. Gustavsson. Biosensor analysis of beta-lactams in milk using the carboxypeptidase activity of a bacterial penicillin binding protein. Journal of AOAC International 2006; 89, 832–837.

[20] E. Gustavsson, J. Degelaen, P. Bjurling, and A. Sternesjö. Determination of beta-lactams in milk using a surface plasmon resonance-based biosensor. Journal of Agricultural and Food Chemistry 2004; 52, 2791–2796.

[21] J. Lamar and M. Petz. Development of a receptor-based microplate assay for the detection of beta-lactam antibiotics in different food matrices. Analytica Chimica Acta 2007; 586, 296–303.

[22] S. J. Setford, R. M. Van Es, Y. J. Blankwater, and S. Kröger. Receptor binding protein amperometric affinity sensor for rapid beta-lactam quantification in milk. Analytica Chimica Acta 1999; 398, 13–22.

[23] E. Gustavsson, P. Bjurling, and A. Sternesjö. Biosensor analysis of penicillin G in milk based on the inhibition of carboxypeptidase activity. Analytica Chimica Acta 2002; 468, 153–159.

[24] S. M. Drawz and R. A. Bonomo. Three decades of beta-lactamase inhibitors. Clinical Microbiology Reviews 2010; 23, 160–201.

[25] B. Chen, M. Ma, and X. Su. An amperometric penicillin biosensor with enhanced sensitivity based on co-immobilization of carbon nanotubes, hematein, and beta-lactamase on glassy carbon electrode. Analytica Chimica Acta 2010; 674, 89–95.

[26] Y. Wu, L. Tang, L. Huang, Z. Han, J. Wang, and H. Pan. A low detection limit penicillin biosensor based on single graphene nanosheets preadsorbed with hematein/ionic liquids/penicillinase. Materials Science and Engineering: C 2014; 39, 92–99.

[27] F. Ismail, S. B. Adeloju, and A. N. Moline. Fabrication of a single layer and bilayer potentiometric biosensors for penicillin by galvanostatic entrapment of penicillinase into polypyrrole films. Electroanalysis 2014; 26, 2607–2618.

[28] S. Zhou, Y. Zhao, M. Mecklenburg, D. Yang, and B. Xie. A novel thermometric biosensor for fast surveillance of beta-lactamase activity in milk. Biosensors and Bioelectronics 2013; 49, 99–104.

[29] L. M. Goncalves, W. F. A. Callera, M. D. P. T. Sotomayor, and P. R. Bueno. Penicillinase-based amperometric biosensor for penicillin G. Electrochemistry Communications 2014; 38, 131–133.

[30] F. Ismail and S. B. Adeloju. Galvanostatic entrapment of penicillinase into polytyramine films and its utilization for the potentiometric determination of penicillin. Sensors 2010; 10, 2851–2868.

[31] T. Rinken and R. Jaaniso. On-line system, method of its calibration and simultaneous detection of antibiotic residues and their concentration in milk, 23-10-2013, EP2529217 B1.

[32] M. Gamella, S. Campuzano, F. Conzuelo, M. Esteban-Torres, B. Las Rivas, A. J. Reviejo, R. Muņoz, and J. M. Pingarron. An amperometric affinity penicillin-binding protein magnetosensor for the detection of beta-lactam antibiotics in milk. Analyst 2013; 138, 2013–2022.

[33] F. Conzuelo, V. Ruiz-Valdepenas Montiel, S. Campuzano, M. Gamella, R. M. Torrente-Rodriguez, A. J. Reviejo, and J. M. Pingarron. Rapid screening of multiple antibiotic residues in milk using disposable amperometric magnetosensors. Analytica Chimica Acta 2014; 820, 32–38.

[34] B. G. Healey and D. R. Walt. Improved fiber-optic chemical sensor for penicillin. Analytical Chemistry 1995; 67, 4471–4476.

[35] A. M. Ferrini, V. Mannoni, G. Carpico, and G. E. Pellegrini. Detection and identification of beta-lactam residues in milk using a hybrid biosensor. Journal of Agricultural and Food Chemistry 2008; 56, 784–788.

[36] S. Das, N. Kumar, R. H. Vishweswaraiah, L. Haldar, M. Gaare, V. K. Singh, and A. K. Puniya. Microbial based assay for specific detection of beta-lactam group of antibiotics in milk. Journal of Food Science and Technology 2014; 51, 1161–1166.

[37] G. E. Pellegrini, G. Carpico, and E. Coni. Electrochemical sensor for the detection and presumptive identification of quinolone and tetracycline residues in milk. Analytica Chimica Acta 2004; 520, 13–18.

[38] V. Gaudin, J. Fontaine, and P. Maris. Screening of penicillin residues in milk by a surface plasmon resonance-based biosensor assay: Comparison of chemical and enzymatic sample pre-treatment. Analytica Chimica Acta 2001; 436, 191–198.

[39] Z. F. Zhang, J. Liu, K. X. Xu, and T. Shi. Development of a miniature optical biosensor-based inhibition immunoassay for detection of ampicillin residues in milk. Nanotechnology and Precision Engineering 2009; 7, 483–489.

[40] F. Fernandez, D. G. Pinacho, F. Sanchez-Baeza, and M. P. Marco. Portable surface plasmon resonance immunosensor for the detection of fluoroquinolone antibiotic residues in milk. Journal of Agricultural and Food Chemistry 2011; 59, 5036–5043.

[41] F. Fernandez, K. Hegnerova, M. Piliarik, F. Sanchez-Baeza, J. Homola, and M. P. Marco. A label-free and portable multichannel surface plasmon resonance immunosensor for on site analysis of antibiotics in milk samples. Biosensors and Bioelectronics 2010; 26, 1231–1238.

[42] S. R. Raz, M. G. E. G. Bremer, W. Haasnoot, and W. Norde. Label-free and multiplex detection of antibiotic residues in milk using imaging surface plasmon resonance-based immunosensor. Analytical Chemistry 2009; 81, 7743–7749.

[43] J. Ferguson, A. Baxter, P. Young, G. Kennedy, C. Elliott, S. Weigel, R. Gatermann, H. Ashwin, S. Stead, and M. Sharman. Detection of chloramphenicol and chloramphenicol glucuronide residues in poultry muscle, honey, prawn and milk using a surface plasmon resonance biosensor and Qflex kit chloramphenicol. Analytica Chimica Acta 2005; 529, 109–113.

[44] W. Haasnoot, E. E. M. G. Loomans, G. Cazemier, R. Dietrich, R. Verheijen, A. A. Bergwerff, and R. W. Stephany. Direct versus competitive biosensor immunoassays for the detection of (dihydro)streptomycin residues in milk. Food and Agricultural Immunology 2002; 14, 15–27.

[45] J. P. Ferguson, G. A. Baxter, J. D. G. McEvoy, S. Stead, E. Rawlings, and M. Sharman. Detection of streptomycin and dihydrostreptomycin residues in milk, honey and meat samples using an optical biosensor. Analyst 2002; 127, 951–956.

[46] B. G. Knecht, A. Strasser, R. Dietrich, E. Märtlbauer, R. Niessner, and M. G. Weller. Automated microarray system for the simultaneous detection of antibiotics in milk. Analytical Chemistry 2004; 76, 646–654.

[47] K. Kloth, M. Rye-Johnsen, A. Didier, R. Dietrich, E. Märtlbauer, R. Niessner, and M. Seidel. A regenerable immunochip for the rapid determination of 13 different antibiotics in raw milk. Analyst 2009; 134, 1433–1439.

[48] F. Conzuelo, M. Gamella, S. Campuzano, A. J. Reviejo, and J. M. Pingarron. Disposable amperometric magneto-immunosensor for direct detection of tetracyclines antibiotics residues in milk. Analytica Chimica Acta 2012; 737, 29–36.

[49] F. Conzuelo, M. Gamella, S. Campuzano, G. Pinacho, A. J. Reviejo, M. P. Marco, and J. M. Pingarron. Disposable and integrated amperometric immunosensor for direct determination of sulfonamide antibiotics in milk. Biosensors and Bioelectronics 2012; 36, 81–88.

[50] F. Conzuelo, L. Stratmann, S. Grützke, J. M. Pingarron, and W. Schuhmann. Detection and quantification of sulfonamide antibiotic residues in milk using scanning electrochemical microscopy. Electroanalysis 2014; 26, 481–487.

[51] G. Suarez, Y. H. Jin, J. Auerswald, S. Berchtold, H. F. Knapp, J. M. Diserens, Y. Leterrier, J. A. E. Manson, and G. Voirin. Lab-on-a-chip for multiplexed biosensing of residual antibiotics in milk. Lab on a Chip – Miniaturisation for Chemistry and Biology 2009; 9, 1625–1630.

[52] J. Adrian, S. Pasche, J. M. Diserens, F. J. Sanchez-Baeza, H. Gao, M. P. Marco, and G. Voirin. Waveguide interrogated optical immunosensor (WIOS) for detection of sulfonamide antibiotics in milk. Biosensors and Bioelectronics 2009; 24, 3340–3346.

[53] J. Adrian, S. Pasche, G. Voirin, J. Adrian, D. G. Pinacho, H. Font, F. J. Sanchez-Baeza, M. P. Marco, J. M. Diserens, and B. Granier. Wavelength-interrogated optical biosensor for multi-analyte screening of sulfonamide, fluoroquinolone, beta-lactam and tetracycline antibiotics in milk. Trends in Analytical Chemistry 2009; 28, 769–777.

[54] R. E. Ionescu, N. Jaffrezic-Renault, L. Bouffier, C. Gondran, S. Cosnier, D. G. Pinacho, M. P. Marco, F. J. Sanchez-Baeza, T. Healy, and C. Martelet. Impedimetric immunosensor for the specific label free detection of ciprofloxacin antibiotic. Biosensors and Bioelectronics 2007; 23, 549–555.

[55] P. Thavarungkul, S. Dawan, P. Kanatharana, and P. Asawatreratanakul. Detecting penicillin G in milk with impedimetric label-free immunosensor. Biosensors and Bioelectronics 2007; 23, 688–694.

[56] G. Merola, E. Martini, M. Tomassetti, and L. Campanella. Simple and suitable immunosensor for beta-lactam antibiotics analysis in real matrixes: Milk, serum, urine. Journal of Pharmaceutical and Biomedical Analysis 2014; 106, 186-196.

[57] H. Wu, S. Fan, W. Zhang, H. Chen, L. Peng, X. Jin, J. Ma, and H. Zhang. Amperometric immunosensor based on covalent immobilization of new methylene blue and penicillin polyclonal antibody for determination of penicillin G in milk. Analytical Methods 2014; 6, 497–502.

[58] D. G. Pinacho, F. Sanchez-Baeza, M. I. Pividori, and M. P. Marco. Electrochemical detection of fluoroquinolone antibiotics in milk using a magneto immunosensor. Sensors (Switzerland) 2014; 14, 15965–15980.

[59] Z. Jiang, Y. Li, A. Liang, and A. Qin. A sensitive and selective immuno-nanogold resonance-scattering spectral method for the determination of trace penicillin G. Luminescence 2008; 23, 157–162.

[60] Y. P. Chen, M. q. Zou, C. Qi, M. X. Xie, D. N. Wang, Y. F. Wang, Q. Xue, J. F. Li, and Y. Chen. Immunosensor based on magnetic relaxation switch and biotin-streptavidin system for the detection of Kanamycin in milk. Biosensors and Bioelectronics 2013; 39, 112–117.

[61] N. A. Karaseva and T. N. Ermolaeva. Piezoelectric immunosensors for the detection of individual antibiotics and the total content of penicillin antibiotics in foodstuffs. Talanta 2014; 120, 312–317.

[62] F. J. Gruhl and K. Länge. Surface acoustic wave (SAW) biosensor for rapid and label-free detection of penicillin G in milk. Food Analytical Methods 2014; 7, 430–437.

[63] H. Ni, S. Zhang, X. Ding, T. Mi, Z. Wang, and M. Liu. Determination of enrofloxacin in bovine milk by a novel single-stranded DNA aptamer chemiluminescent enzyme immunoassay. Analytical Letters 2014; 47, 2844–2856.

[64] K. M. Song, E. Jeong, W. Jeon, M. Cho, and C. Ban. Aptasensor for ampicillin using gold nanoparticle based dual fluorescence-colorimetric methods. Analytical and Bio-analytical Chemistry 2012; 402, 2153–2161.

[65] S. Pilehvar, J. Mehta, F. Dardenne, J. Robbens, R. Blust, and K. De Wael. Aptasensing of chloramphenicol in the presence of its analogues: Reaching the maximum residue limit. Analytical Chemistry 2012; 84, 6753–6758.

[66] A. N. Berlina, N. A. Taranova, A. V. Zherdev, Y. Y. Vengerov, and B. B. Dzantiev. Quantum dot-based lateral flow immunoassay for detection of chloramphenicol in milk. Analytical and Bioanalytical Chemistry 2013; 405, 4997–5000.

[67] J. Dapra, L. H. Lauridsen, A. T. Nielsen, and N. Rozlosnik. Comparative study on aptamers as recognition elements for antibiotics in a label-free all-polymer biosensor. Biosensors and Bioelectronics 2013; 43, 315–320.

[68] S. Wu, H. Zhang, Z. Shi, N. Duan, C. Fang, S. Dai, and Z. Wang. Aptamer-based fluorescence biosensor for chloramphenicol determination using upconversion nanoparticles. Food Control 2015; 50, 597–604.

[69] K. M. Song, E. Jeong, W. Jeon, H. Jo, and C. Ban. A coordination polymer nanobelt (CPNB)-based aptasensor for sulfadimethoxine. Biosensors and Bioelectronics 2012; 33, 113–119.

[70] M. Alibolandi, F. Hadizadeh, F. Vajhedin, K. Abnous, and M. Ramezani. Design and fabrication of an aptasensor for chloramphenicol based on energy transfer of CdTe quantum dots to graphene oxide sheet. Materials Science and Engineering C 2015; 48, 611–619.

[71] N. Zhou, J. Zhang, and Y. Tian. Aptamer-based spectrophotometric detection of kanamycin in milk. Analytical Methods 2014; 6, 1569–1574.

[72] S. Jeong and I. Rhee Paeng. Sensitivity and selectivity on aptamer-based assay: The determination of tetracycline residue in bovine milk. The Scientific World Journal 2012; 2012, 159456.

[73] C. H. Kim, L. P. Lee, J. R. Min, M. W. Lim, and S. H. Jeong. An indirect competitive assay-based aptasensor for detection of oxytetracycline in milk. Biosensors and Bioelectronics 2014; 51, 426–430.

[74] L. He, Y. Luo, W. Zhi, Y. Wu, and P. Zhou. A colorimetric aptamer biosensor based on gold nanoparticles for the ultrasensitive and specific detection of tetracycline in milk. Australian Journal of Chemistry 2013; 66, 485–490.

[75] L. He, Y. Luo, W. Zhi, and P. Zhou. Colorimetric sensing of tetracyclines in milk based on the assembly of cationic conjugated polymer-aggregated gold nanoparticles. Food Analytical Methods 2013; 6, 1704–1711.

[76] J. Zhang, B. Zhang, Y. Wu, S. Jia, T. Fan, Z. Zhang, and C. Zhang. Fast determination of the tetracyclines in milk samples by the aptamer biosensor. Analyst 2010; 135, 2706–2710.

[77] Y. Luo, L. He, S. Zhan, Y. Wu, L. Liu, W. Zhi, and P. Zhou. Ultrasensitive resonance scattering (RS) spectral detection for trace tetracycline in milk using aptamer-coated nanogold (ACNG) as a catalyst. Journal of Agricultural and Food Chemistry 2014; 62, 1032–1037.

[78] M. Famulok and G. Mayer. Aptamer modules as sensors and detectors. Accounts of Chemical Research 2011; 44, 1349–1358.

[79] W. Zhou, P. J. Jimmy Huang, J. Ding, and J. Liu. Aptamer-based biosensors for biomedical diagnostics. Analyst 2014; 139, 2627–2640.

[80] S. Y. Toh, M. Citartan, S. C. B. Gopinath, and T. H. Tang. Aptamers as a replacement for antibodies in enzyme-linked immunosorbent assay. Biosensors and Bioelectronics 2015; 64, 392–403.

[81] D. Shangguan, Z. Tang, P. Mallikaratchy, Z. Xiao, and W. Tan. Optimization and modifications of aptamers selected from live cancer cell lines. ChemBioChem 2007; 8, 603–606.

[82] S. Balamurugan, A. Obubuafo, S. A. Soper, and D. A. Spivak. Surface immobilization methods for aptamer diagnostic applications. Analytical and Bioanalytical Chemistry 2008; 390, 1009–1021.

[83] S. Centi, S. Tombelli, M. Minunni, and M. Mascini. Aptamer-based detection of plasma proteins by an electrochemical assay coupled to magnetic beads. Analytical Chemistry 2007; 79, 1466–1473.

[84] W. Xiuyun and U. Shunichi. State of the Art in Biosensors – General Aspects, Edited by Toonika Rinken (InTech, Rijeka, 2013), Chap. 3, p. 67.

[85] S. Jeong and I. Rhee Paeng. Sensitivity and selectivity on aptamer-based assay: The determination of tetracycline residue in bovine milk. The Scientific World Journal 2012; 2012, 159456.

[86] M. F. Frasco and N. Chaniotakis. Semiconductor quantum dots in chemical sensors and biosensors. Sensors 2009; 9, 7266–7286.

[87] B. Liu, D. Tang, B. Zhang, X. Que, H. Yang, and G. Chen. Au(III)-promoted magnetic molecularly imprinted polymer nanospheres for electrochemical determination of streptomycin residues in food. Biosensors and Bioelectronics 2013; 41, 551–556.

[88] M. L. Yola, L. Uzun, N. Özaltin, and A. Denizli. Development of molecular imprinted nanosensor for determination of tobramycin in pharmaceuticals and foods. Talanta 2014; 120, 318–324.

[89] V. K. Gupta, M. L. Yola, N. Özaltin, N. tar, Z. Üstündag, and L. Uzun. Molecular imprinted polypyrrole modified glassy carbon electrode for the determination of tobramycin. Electrochimica Acta 2013; 112, 37–43.

[90] W. Lian, S. Liu, J. Yu, J. Li, M. Cui, W. Xu, and J. Huang. Electrochemical sensor using neomycin-imprinted film as recognition element based on chitosan-silver nanoparticles/graphene-multiwalled carbon nanotubes composites modified electrode. Biosensors and Bioelectronics 2013; 44, 70–76.

[91] W. Lian, S. Liu, J. Yu, X. Xing, J. Li, M. Cui, and J. Huang. Electrochemical sensor based on gold nanoparticles fabricated molecularly imprinted polymer film at chitosan-platinum nanoparticles/graphene-gold nanoparticles double nanocomposites modified electrode for detection of erythromycin. Biosensors and Bioelectronics 2012; 38, 163–169.

[92] S. Sadeghi and A. Motaharian. Voltammetric sensor based on carbon paste electrode modified with molecular imprinted polymer for determination of sulfadiazine in milk and human serum. Materials Science and Engineering: C 2013; 33, 4884–4891.

[93] L. Q. Wang, F. Y. Lin, and L. P. Yu. A molecularly imprinted photonic polymer sensor with high selectivity for tetracyclines analysis in food. Analyst 2012; 137, 3502–3509.

[94] E. Turiel and A. Martin-Esteban. Molecularly imprinted polymers for sample preparation: A review. Analytica Chimica Acta 2010; 668, 87–99.